LEGAL CONSIDERATIONS FOR FIRE AND EMERGENCY SERVICES

LEGAL CONSIDERATIONS FOR FIRE AND EMERGENCY SERVICES

J. Curtis Varone

THOMSON

DELMAR LEARNING   Australia Brazil Canada Mexico Singapore Spain United Kingdom United States

THOMSON

DELMAR LEARNING

Legal Considerations for Fire and Emergency Services
J. Curtis Varone

Vice President, Technology and Trades ABU:
David Garza

Director of Learning Solutions:
Sandy Clark

Managing Editor:
Larry Main

Acquisitions Editor:
Alison Pase

Product Manager:
Jennifer A. Starr

Marketing Director:
Deborah S. Yarnell

Marketing Manager:
Erin Coffin

Marketing Coordinator:
Patti Garrison

Director of Production:
Patty Stephan

Production Manager:
Stacy Masucci

Content Project Manager:
Jennifer Hanley

Technology Project Manager:
Kevin Smith

Editorial Assistant:
Maria Conto

Library of Congress Cataloging-in-Publication Data:

Varone, J. Curtis
   Legal considerations for fire and emergency services / J. Curtis Varone. — 1st ed.
      p. cm.
   Includes index.
   ISBN 1-4018-6571-2
   1. Emergency medical services—Law and legislation—United States. 2. Labor laws and legislation—United States. 3. Fire departments—Law and legislation—United States. I. Title.
   KF3826.E5V37 2006
   344.7303'218—dc22
                                          2006028927

NOTICE TO THE READER

CONTENTS

PREFACE

INTENT OF THIS BOOK

Legal Considerations for Fire and Emergency Services is intended as a textbook for college-level classes in fire service law, and meets the Fire and Emergency Services Higher Education (FESHE) requirements issued by the United States Fire Administration. Yet this book can travel much farther. Firefighters of all ranks, and from all types of fire departments—large and small, urban and rural, union and nonunion, career and volunteer—can use it as a guide to the legal issues associated with the fire service. Some may also use it as a legal reference book. It addresses issues of great concern to all firefighters.

Written for firefighters who want to understand the legal issues that are intertwined with fire service, the book covers many of the pivotal questions confronting today's fire service, including legal liability, sovereign immunity, overtime laws, collective bargaining, OSHA compliance, workers' compensation, physical abilities testing, medical examinations, drug testing, discrimination, and sexual harassment.

WHY I WROTE THIS BOOK

While I have represented firefighters in my law practice for more than 21 years, I have also spent my entire adult life in the company of firefighters. *Legal Considerations for Fire and Emergency Services* is a product of having lived, worked, socialized, celebrated, and mourned with firefighters.

When I first began teaching fire law in the fire science program at Providence College, I struggled to find the right textbook. Some very fine attorneys have written a number of books on legal issues in the fire service, but I felt that those books lacked a perspective that connected with firefighters. When I finished reading those books, I was left with more questions than I had when I started. Firefighters are curious by nature. *Legal Considerations for Fire and Emergency Services* caters to the curiosity of firefighters. It answers the legal questions that are asked in firehouses from coast to coast, with down-to-earth explanations. It also explains where grey areas exist, and why. *Legal Considerations for Fire and Emergency Services* makes the connection between law as an intellectual study, and law as a reality that impacts firefighters each and every day on so many different levels.

HOW TO USE THIS BOOK

The book is organized in a logical sequence, starting with an overview of laws, our court systems, fire service organizations, and administrative agencies—with a particular focus on the Occupational Safety and Health Administration (OSHA)—followed by an overview of criminal law and criminal procedure. The book next addresses civil liability, with separate chapters on intentional torts, negligence, and immunity. A chapter on contract law addresses mutual aid agreements, fire insurance law, municipal purchasing, employment agreements, employment at will, and due process. The final chapters address labor law, workers' compensation, employment discrimination, sexual harassment, the Fair Labor Standards Act, and public accountability laws.

While the book is well laid out for use in a one-semester course in fire law, it can also be used as an adjunct for modular courses on specific fire service issues, such as arson, search and seizure, fire service negligence, risk management, the Fair Labor Standards Act, collective bargaining, discrimination, or sexual harassment. It includes an extensive glossary, as well as an appendix containing legal documents, laws, and reference information. Many firefighters may find *Legal Considerations for Fire and Emergency Services* useful as a desk reference.

FEATURES OF THIS BOOK

- *Actual case examples* are used throughout the book. The history of the fire service in the United States can be told through the lawsuits and cases that begin almost as soon as the last section of hose is packed on the first-in engine company. The book features actual cases presented in a concise, reader-friendly manner from some of the most significant fires and disasters that this country has ever seen: the Cocoanut Grove fire in Boston, in which 492 people died; the Beverly Hills Supper Club fire; the Catlett, Virginia train derailment, and the Worcester Cold Storage fire. These cases have been included in *Legal Considerations for Fire and Emergency Services* not only to help students learn important legal concepts, but to also to help them appreciate how law impacts the fire service.

- *Supporting photos* depicting law situations and drawn from the actual case incidents serve to engage the student in the readings and further emphasize the legal consequences of decisions made during incident response, and the daily operation of the fire department and other emergency service institutions.

- *Case highlights* at the end of each case provide a quick summary of the main points—the case title, the court in which the case was held, and the final decision of the court—to provide review for the student, and a quick reference for the active firefighter.

- *Key terms* are boldfaced throughout the text as well as noted in the margins, complete with definitions, to ensure that students learn the legal terminology essential to effective communication.

- *Sidebars* highlight the critical points that an informed emergency responder should know in order to address legal matters effectively while on the job.

- *Examples* integrated throughout the text provide students with insight into various legal concepts as they apply to response situations.

- *Review questions* allow students to evaluate their knowledge of the concepts learned in the chapter.

SUPPLEMENT TO THIS BOOK

For those interested in teaching fire service law, there is a comprehensive **e-resource** on CD-ROM available to instructors. The CD-ROM includes:

- *Instructor's Guide* containing lesson plans, points of discussion, and answers to the chapter review questions contained in the book.

- *PowerPoint presentations* for each chapter, including graphics, to highlight important legal concepts.

- *Test bank* organized by chapter and containing approximately 300 questions in multiple-choice, true/false, and scenario-based formats. The test bank is in ExamView format, allowing instructors to edit, randomize, and add questions with ease to meet their specific training needs.

- *FESHE correlation grid* that aligns the chapters is this book with the course objectives for Legal Aspects of the Fire Services.

The instructor materials provide additional insight into many of the incidents depicted in the book and offer numerous tips and exercises to engage students in the study of fire service law.

(Order #: 1-4018-6571-2)

ABOUT THE AUTHOR

I am a third-generation firefighter and a second-generation attorney. Like many firefighters, I became a volunteer as soon as I turned 16. At the time, my grandfather was the fire chief, and I was just another young gun hanging on the back step of a 1941 Ward LaFrance pumper. Over the next few years, the town in which I lived consolidated the five volunteer fire companies into one fire department and established a part-time call department. By the

Author Curt Varone on the scene of an incident. (Photo by Ray Taylor.)

time I started law school, I had already been a firefighter for 10 years, and had become a career firefighter in Providence, Rhode Island. I went to law school armed with the questions that many firefighters have about the legal system, and made it my mission to find the answers. With each class I took, I became more aware of the interrelationship between the law and the fire service.

When I began my law practice in 1985, most of my clients were firefighters. I also began representing fire departments, volunteer fire companies, firefighter labor unions, and fire chiefs. This experience has given me great insight into the legal issues confronting the fire service, and has sharpened my ability to make these issues understandable to firefighters.

I am a Deputy Assistant Chief in the Providence (RI) Fire Department, and currently serve as A-group shift commander. Besides teaching in the Executive Fire Officer Program at the National Fire Academy, I have been teaching fire law at Providence College for more than 10 years, and continue to lecture on a variety of legal issues. However, I also teach fire tactics, incident command, and firefighter safety. I like to think that teaching these other fire courses—together with continuing to work a shift in a busy urban fire department—has helped to keep me balanced, and that balance is reflected in this text.

ACKNOWLEDGMENTS

The assistance of the following individuals with the preparation of this book is gratefully acknowledged:

Gerard P. Cobleigh, Esq.,
Cobleigh & Giacobbe, Warwick, RI

Gerald. J. Coyne, Esq.,
Deputy Attorney General,
State of Rhode Island

Steven H. Crawford, Esq.,
Murphy & Fay, Providence, RI,
and Providence Fire Department

Steven T. Cross,
Chief of Investigations,
Rhode Island Ethics Commission

Paul A. Doughty, Esq.,
Providence, RI, and President of
Providence Firefighters Local 799, IAFF

John J. Enright,
United States Attorney's Office,
Providence, RI

Mary E. Hoye,
Area Director,
Occupational Safety and Health
Administration, USDOL

Charles S. Kirwan, Esq.,
Charles S. Kirwan & Associates,
Providence, RI

Paul J. Narducci, Esq.,
Assistant State's Attorney,
State of Connecticut

Charles M. Nystedt, Esq.,
Providence, RI

John A. Varone, Esq.,
Bluffton, SC

Captain J. Jeffrey Varone,
Providence Fire Department

Detective Sergeant J. William Varone,
Groton, CT Police Department

Richard T. Varone,
Director, Department of Workers'
Compensation, State of Rhode Island

Vanessa J. Varone, Esq.,
Cranston, RI

The author and Thomson Delmar Learning wish to thank the committee that participated in reviewing the manuscript throughout the development process:

Steve Hirsch,
Training Officer,
Sheridan County Fire Department, Hoxie, KS

John Mack,
Emergency Services Instructor,
Chemeketa Community College, Salem, OR

Robert McGraw,
Program Manager,
Fire Science/EMS, Seminole Community
College, Sanford, FL

Gail Ownby-Hughes,
Assistant Professor/Fire
Science Coordinator,
University of Alaska-Anchorage,
Anchorage, AK

Clint Smoke,
Program Chair (Emeritus),
Asheville–Buncombe Technical
Community College, Asheville, NC

Thanks also to the following for assistance with the photographs used in the book:

Lt. Rick Blais,
Providence Fire Department

Bonnie Benson,
Firefighter,
Providence Fire Department

Inspector Paul Calardo,
Providence Fire Department

Tom Carmody,
Firefighter,
Cranston Fire Department

Detective James E. Clift,
Providence Police Department

Fire Chief Bob Donahue and Capt. Paul
Moore, MASSPORT Fire-Rescue

Lou Emerson and Paul Smith,
Editors, *Fauquier Citizen*

Fire Marshal George Farrell,
Providence Fire Department

Joe Molis, Firefighter,
Providence Fire Department and NFPA

Geoffrey Read,
Providence Citywide Fire Network

Dan Rinaldi,
Firefighter,
Providence Fire Department

Glenn Scheiwe,
Scheiwe's Print Shop

Ray Taylor,
Providence Citywide Fire Network

Tim Whalen, Firefighter,
Providence Fire Department

To my wife Julie, my sons Andrew and Matt, and to my extended family: the members of A-Group of the Providence Fire Department: thank you all for your patience with me and your support through the endeavor of writing this book. Thanks also to Chiefs David Costa, Mark Pare, and Mike Dillon for your assistance.

NOTE TO READERS

If there are topics or questions that you feel should have been addressed in *Legal Considerations for Fire and Emergency Services*, please let me know. In addition, if you are aware of cases that would be well suited for inclusion in the second edition, by all means please contact me. My e-mail address is jcatlaw@aol.com.

INTRODUCTION

Legal Considerations for Fire and Emergency Services is the culmination of two careers: fire and law. The two topics have always had an inescapable connection to me, and I believe that after studying this book the connection will be apparent to you as well.

UNDERSTANDING THE CASES IN THIS BOOK

Reading published court cases can seem intimidating to non-attorneys. I remember well the trials and tribulations of reading cases for the first time in college-level fire science courses. I have taught a college-level course in fire law for the past 10 years, so I know firsthand the capabilities of firefighters to grasp legal concepts when they relate to important fire service issues. Cases are nothing more than stories, and the human mind craves learning through storytelling. It is unfortunate that all learning cannot take place through such storytelling, because our minds soak up the facts of a story the way a dry sponge soaks up water.

Most of the cases chosen for this book are fire cases, and tell stories of interest to firefighters. Some arose out of the most significant—and tragic—fires in our nation's history: the Cocoanut Grove fire, the Beverly Hills Supper Club fire, and the Worcester Cold Storage fire, to name a few. Other cases involve issues that are of major concern to firefighters, such as residency, grooming, physical abilities testing, overtime, discrimination, and fire service liability. Throughout these cases, the connection between the law and the fire service is evident.

To the greatest extent possible, I have removed extraneous material from the cases, including citations, references, and footnotes, to make the cases more reader-friendly. Three periods in a row (. . .), also called ellipsis points, indicate that material has been omitted from a sentence. Four periods in a row (. . . .) indicate that more than a single sentence has been omitted. Any words or letters encased within brackets [like this] are mine, and not part of the actual published court decision.

My advice to those concerned about reading cases is to read them the way you would read a novel, and not the way you would read a textbook. When you come across a word you do not recognize in a novel, you simply read past it. Do not focus on the minutiae of the cases, as you might normally do when reading a textbook for a class. Focus on the storyline. Never lose sight of where the story is going!

HOW TO READ AND BRIEF CASES

When reading cases, read past footnotes, citations, and references. Allow your eyes to skim right over them. They are merely there for documentation purposes, in the event you wish to follow up on a certain case or law. Do not let them distract you from the storyline. Focus on the big picture, and follow the story. Search for four important pieces of information, namely: the *facts,* the *issues,* the *holding,* and the *rationale.* Identifying these four factors is commonly referred to as *briefing* a case. You may choose to write down the four pieces of information for future reference as you read each case.

The *facts* are the court's explanation about what happened and the history of the case. The facts give us a sense of what occurred, and what caused the case to get to court. Try to figure out who are the parties involved and what took place.

The second important piece of information to look for is the court's identification of the *issue* or *issues* in the case. The issue is the legal question, which the judge or judges have concluded must be addressed in order to decide the case. Some cases have one issue; others have multiple issues. Do not be surprised if the parties disagree about the "real" issues in a case. It is the court's job to clarify the issues. Spotting the issues may be a challenge initially, but your ability to issue-spot will improve rapidly with practice.

The third item to look for in a case in the *holding,* or ruling of the court. The holding is essentially the court's answer to the issue or issues that have been identified. Each issue in a case will have a holding. Since many cases in this book have multiple issues, they will have multiple holdings.

The final component of a case is the *rationale,* or the reasoning stated by the court in reaching its conclusion. The rationale provides an explanation for why the court ruled the way it did, and adds a very important perspective to the decision.

MAKING THE MOST OF THE LESSONS IN THIS BOOK

Legal Considerations for Fire and Emergency Services is organized to be read and taught by starting with Chapters 1 through 4. These chapters review the basics of the study of law as it relates to the fire service, and provide a foundation for the remaining 11 chapters. The remaining chapters build upon the basic concepts explained in Chapters 1 through 4, but may be addressed in whatever order your instructor considers prudent.

Most chapters include one or more cases. The cases are intended to reinforce and complement the text material, helping to anchor a legal concept in one's mind in the way that a picture or diagram can provide perspective about how to operate a piece of equipment. It is possible to use this textbook successfully without reading the cases. However, the richness of the experience will be lost, as will the opportunity to connect a legal concept to a real-life story. It is not uncommon for lawyers and judges in their sixties and seventies to recall cases they studied in their early twenties, having read them but once in law school. Such is the power of storytelling through studying cases.

Following each case is a Case Highlight, which is included to help provide perspective about the case. The Case Highlight will explain the relevance of the case in order to help you understand how it relates to the material in the chapter. The Case Highlight is different from the holding. My advice is to refer to the Case Highlight only if you are having difficulty understanding the point of the case relative to the chapter, or to confirm that you have correctly understood the relevance of the case.

Remember that no book on the law can substitute for competent legal advice. Seemingly insignificant facts can make a tremendous difference in your legal rights and responsibilities. If you or your organization have legal questions, seek the advice of an attorney.

Good luck! Be safe!

Curt

FIRE AND EMERGENCY SERVICES HIGHER EDUCATION (FESHE)

In June 2001, the U.S. Fire Administration hosted the third annual *Fire and Emergency Services Higher Education Conference* at the National Fire Academy campus in Emmitsburg, Maryland. Attendees from state and local fire service training agencies, as well as colleges and universities with fire-related degree programs, attended the conference and participated in work groups. Among the significant outcomes of the working groups was the development of standard titles, outcomes, and descriptions for six core associate-level courses for the *model fire science* curriculum that had been developed by the group the previous year. The six core courses are: *Fundamentals of Fire Protection, Fire Protection Systems, Fire Behavior and Combustion, Fire Protection Hydraulics and Water Supply, Building Construction for Fire Protection,* and *Fire Prevention.*[1] The committee also developed similar outlines for other recommended courses offered in fire science programs. These courses included: *Fire Administration I, Occupational Health and Safety, Legal Aspects of the Emergency Services, Hazardous Materials Chemistry, Strategy and Tactics, Fire Investigation I,* and *Fire Investigation II.*

FESHE Content Area Comparison The following table provides a comparison of the Legal Aspects FESHE course with this text, *Legal Considerations for Fire and Emergency Services.*

---

[1] *2001 Fire & Emergency Services Higher Education Conference Final Report* (Emmitsburg, Maryland: U.S. Fire Administration, 2001), page 12.

FIRE AND EMERGENCY SERVICES HIGHER EDUCATION (FESHE) COURSE CORRELATION GRID

Name:	Legal Aspects of the Fire Service	Legal Considerations for Fire and Emergency Services Chapter Reference
Course Description:	This course introduces the Federal, State, and local laws that regulate emergency services, national standards influencing emergency services, standard of care, tort, liability, and a review of relevant court cases.	
Prerequisite:	None.	
Outcomes:	1. Define the different types of laws, explain their basic differences, and how the law functions in society.	1
	2. Become familiar with federal, state, and local laws, which regulate or influence emergency services.	1, 3, 4, 11, 12, 13, 14, 15
	3. Explain the role and purpose of national codes and standards concerning their legal influence.	1
	4. Become familiar with legal decisions that have or will affect the fire service.	1–15
	5. Discuss the organization and legal structure of the fire department.	3
	6. Define the liabilities of firefighters.	7, 8, 9
	7. Recognize legal duties of emergency service members.	5, 7, 8
	8. Discuss negligence in an emergency setting.	7, 8, 9
	9. Define discrimination and identify areas of potential discrimination in the emergency service.	12, 13
	10. Identify, explain, and discuss the legalities of entrance requirements, residency, grooming, and drug testing.	12, 13, 14
	11. Discuss the scope of the Civil Rights Act.	12, 13
	12. Discuss the parameters and explain the basic intent of the American Disabilities Act, Fair Labor Standards Act, and Family Medical Leave Act.	12, 13, 14
	13. Explain the at-will doctrine.	10
	14. Explain the purpose of labor and employment laws.	11
	15. Identify and analyze the major causes involved in the line-of-duty firefighter deaths related to health, wellness, fitness, and vehicle operations.	Firefighter Safety Section

FIREFIGHTER SAFETY SECTION

FESHE OBJECTIVE 15: Identify and analyze the major causes involved in line-of-duty firefighter deaths related to health, wellness, fitness, and vehicle operations.

Each year, approximately 100 firefighters are killed and 100,000 are injured. The United State Fire Administration (USFA), the National Fire Protection Association (NFPA), and the International Association of Firefighters (IAFF) maintain statistics on these deaths and injuries. Collectively, these studies indicate the following:

Sudden heart attack is the most common cause of death of firefighters killed on the job. Heart attacks claim anywhere from 40 to 50 percent of the firefighters who die in the line of duty each year. A 10-year study by the NFPA, covering 1995 to 2004, found that 440 of the 1,006 line-of-duty deaths (43.7 percent) were related to sudden cardiac events, while many other deaths were the result of stress-related conditions such as strokes and aneurysms. Most of these victims had known heart diseases, or had conditions that could have been diagnosed had the members undergone mandatory fitness-for-duty medical examinations, as recommended by the NFPA.

Not surprisingly, the most likely place for a firefighter to be killed or injured is at an emergency scene. Fireground deaths account for between 28 and 40 percent of firefighter deaths each year, with nonfire emergencies accounting for an additional 8 to 10 percent.

Approximately 25 to 35 percent of the firefighters who die each year succumb while responding to an alarm, or returning from an alarm. Vehicle accidents cause many of these deaths. Research has identified tanker rollover accidents as accounting for a statistically significant number of firefighter fatalities over the years. Training deaths commonly account for between 10 and 20 percent of firefighter fatalities annually.

Age has been shown to be a risk factor for firefighters. NFPA studies have concluded that a firefighter over the age of 60 is three times more likely to die than a firefighter who is between 40 and 49. Older firefighters are more likely to die from heart-related causes, while younger firefighters are more likely to die from trauma.

While firefighter line-of-duty deaths and injuries can never be eliminated completely, fire departments can achieve positive results by adopting comprehensive occupational safety and health programs. These programs address a wide range of firefighter activities, including:

- periodic medical examinations
- physical fitness programs
- health and wellness programs
- accident prevention programs
- effective training programs
- acquisition of proper equipment

- proper equipment maintenance programs
- enforcement of operational procedures
- risk management programs
- proper staffing of apparatus

The single most important measure that fire departments and firefighters can take to reduce the death and injury rate is to implement the NFPA 1500 Standard on Occupational Safety and Health Programs.

SOURCES

LeBlanc, Paul and Rita Fahy. U.S. Firefighter Fatalities for 2004, *NFPA Journal*, July/August 2005.

LeBlanc, Paul and Rita Fahy. U.S. Firefighter Fatalities for 2003, *NFPA Journal*, July/August 2004.

LeBlanc, Paul and Rita Fahy. U.S. Firefighter Fatalities for 2002, *NFPA Journal*, July/August 2003.

National Institute for Occupational Safety and Health. *Fire Fighter Fatality Investigation and Prevention Program—Annual Report: 2003.*

United States Fire Administration. *Firefighter Fatalities in the United States in 2003,* FA-283, Item # 9-0981.

United States Fire Administration. *Firefighter Fatalities: A Retrospective Study 1990–2000,* FA-220, Item # 9-0437

TYPES AND SOURCES OF LAWS

Learning Objectives

Upon completion of this chapter, you should be able to:

- Identify the primary sources of law in the United States.
- Identify the three levels of government in the United States.
- Identify the three branches of government and their roles.
- Identify the difference between civil and criminal laws.
- Distinguish between standards and codes.
- Identify the differences in jurisdiction between Federal, state, and local government.

INTRODUCTION

It is a quiet Tuesday evening down at the local firehouse. Several younger fire-fighters are sitting at the kitchen table drinking coffee and discussing local politics. "They can't do that. That's against the law," says one firefighter in an authoritative tone. Everyone nods in agreement.

From off in the corner, a cagey veteran firefighter chimes in, "Do you know what you are talking about? 'The law'—what do you mean by 'the law'? What is 'the law'? What law are you talking about?"

WHAT IS "THE LAW"?

What does it mean when someone says "That's against the law"? Is the law a tangible object like the Bible that someone can pick up and read? Is there one book or one place that we can go to in order to find out about the law? And perhaps more important, why does a firefighter need to know about the law?

Firefighters need to have a good understanding of law and legal issues, because fire departments in large measure are creatures of the law. In other words, fire departments are created by laws and governed by laws. The right of firefighters to perform their duties is based upon the law **(Figure 1-1)**. The

Figure 1-1

The right of firefighters to perform their duties is based upon the law. (Photo by Rick Blais.)

Figure 1-2
Laws give fire apparatus certain privileges on the road.

authority for firefighters to drive their fire apparatus on the road with red lights and sirens **(Figure 1-2)**, to enter into people's homes and businesses, and to deliver emergency medical care is based upon the law. Many fire departments are responsible for enforcing laws, such as fire codes. Fire departments are also subject to a variety of laws, such as Occupational Safety and Health Administration (OSHA) regulations, the Fair Labor Standards Act, and right-to-know laws.

In order to understand the law, we first have to understand something about basic civics. We also need a working definition of a "law." There are a variety of definitions of law, but for our purposes we will consider a **law** to be that which must be obeyed subject to sanction or consequences by the government. There are a wide variety of things that meet such a definition. For example, there are laws that authorize firefighters to order buildings and areas evacuated **(Figure 1-3)**.

law
that which must be obeyed subject to sanction or consequences by the government

WHERE DO LAWS COME FROM?

Who or What Is the Government?

Just as an oversimplified understanding of laws may lead some people to refer to "the law" as if it were a single book, some people oversimplify the organization of our government by referring to it as "the government." In the United

Figure 1-3
Laws authorize firefighters to order buildings and areas evacuated.

States, there are actually three distinct levels of government: Federal, state, and local. In some jurisdictions, there are several layers of local government, including county government, cities, towns, townships, fire districts, and regional authorities. Laws can be made at all three levels of government.

Supreme Law: The United States Constitution

In 1776, the 13 colonies declared their independence from England. However, it was not until 1791, after years of debate, that we adopted our present Constitution **(Figure 1-4)**. The United States Constitution serves two primary functions. First, it serves as the supreme law of the land. In other words, in order for any other law in the United States to be valid, it must not violate any provision in the United States Constitution. In the event of a conflict between any law and the United States Constitution, the other law would be declared invalid and unenforceable (Figure 1-4).

Second, the Constitution establishes our democratic form of government. It creates three branches of the Federal government: the legislative branch, the executive branch, and the judicial branch, and defines their roles. The legislative branch is responsible for creating laws and imposing taxes. The executive branch is responsible for enforcing the laws and running

PREAMBLE – An Introduction explaining that the purpose of the Constitution is to establish a more perfect union complete with justice, tranquility, and liberty.

ARTICLE I – Establishes the Legislative Branch.

ARTICLE II – Establishes the Executive Branch.

ARTICLE III – Establishes the Judicial Branch.

ARTICLE IV – Establishes the relationship between the Federal government and the states, and establishes the procedure for admitting new states to the Union.

ARTICLE V – Establishes how the Constitution may be amended.

ARTICLE VI – Establishes the Constitution as the supreme law of the United States, authorizes Congress to borrow money, and requires that public officials take an oath to support the Constitution.

ARTICLE VII – Establishes requirements for ratification of the Constitution.

AMENDMENTS 1–10 (sometimes referred to as the Bill of Rights (added in 1791) – Establishes the following rights:
Amendment 1 – Freedom of religion, press, speech
Amendment 2 – Right to bear arms
Amendment 3 – Limits the quartering of soldiers
Amendment 4 – Search warrants upon probable cause
Amendment 5 – Right against self-incrimination and double-jeopardy, and right of due process.
Amendment 6 – Right to counsel, a speedy trial by jury, and confrontation of witnesses
Amendment 7 – Trial by jury in civil cases
Amendment 8 – Prohibits excessive bail, and cruel and unusual punishment
Amendment 9 – Establishes that the rights of the people are not limited to the rights listed here
Amendment 10 – The powers of the Federal Government are limited to those listed in the Constitution, and all other powers belong to the states or the people themselves

AMENDMENTS 11–27
Amendment 11 (1798) – Establishes limits on the Jurisdiction of Federal Courts to hear cases involving suits against states
Amendment 12 (1804) – Establishes the system for electing the president, vice president, commonly known as the Electoral College
Amendment 13 (1865) – Abolished slavery

Figure 1-4
An overview of the U.S. Constitution.

(continued)

Amendment 14 (1868) – Established the right of citizenship to all people born in the United States or naturalized, and prohibited states from denying any person due process

Amendment 15 (1870) – All men may vote without regard to race

Amendment 16 (1913) – Established Income taxation

Amendment 17 (1913) – Established method for electing Senators

Amendment 18 (1919) – Prohibited liquor (repealed by 21st Amendment)

Amendment 19 (1920) – Established right of women to vote

Amendment 20 (1933) – Established new terms of office for the president and Congress

Amendment 21 (1933) – Repealed the 18th Amendment

Amendment 22 (1951) – Limited the term of office of the President

Amendment 23 (1961) – Citizens of Washington, D.C., given the right to vote in Presidential elections

Amendment 24 (1964) – Elimination of Poll taxes (tax assessed by some communities when people vote)

Amendment 25 (1967) – Addressed issues of presidential disability and succession

Amendment 26 (1971) – Lowered voting age to 18 years old

Amendment 27 (1992) – Any Congressional pay increases cannot go into effect until the next Congressional session.

Figure 1-4
(*continued*)

the day-to-day operations of government. The judicial branch is responsible for conducting trials, handling appeals, and interpreting laws within the context of cases **(Figure 1-5)**.

Balance of Power

balance of power
a series of checks and balances built into our Constitution that are designed to keep any one branch of government from becoming too powerful

In theory, each branch of government is considered to be co-equal, with a series of checks and balances that are designed to keep any one branch from becoming too powerful. This arrangement is often referred to as the **balance of power.** The founders of our nation feared that if any one branch or person became too powerful, there could be a return to a monarchy or even a dictatorship. The balance of power is effective because no one branch can function without reliance upon the other branches.

Congress has the power to pass laws and impose taxes. The president has the power to sign acts of Congress into law, or he may veto them. A veto effectively prevents an act of Congress from becoming a law, unless the Congress overrides the president's veto by a two-thirds majority vote.

The president and the executive branch are responsible for running the day-to-day operations of government, enforcing the laws that Congress passes, and collecting the taxes that Congress imposes. Congress itself has no

Legislative Branch	Executive Branch	Judicial Branch
• Makes laws • Imposes taxes	• Enforces the Law • Collects the taxes • Runs day-to-day operations of government	• Interprets and applies the law
Balance of Power:	*Balance of Power:*	*Balance of Power:*
• Holds hearings to consider new laws • Provides funding for other branches • Can override presidential veto	• Power to veto legislation • Discretion in enforcing laws	• Judicial review of legislative and executive actions

Figure 1-5
The three branches of government and the balance of power.

authority to enforce the laws it passes, or to collect taxes. The judicial branch applies the laws of Congress in the context of a case, and interprets the meaning of a law where it is questioned. The judicial branch has no authority to apply or interpret the laws in the absence of someone bringing a case to the court. This concept effectively limits the power of the judicial branch to cases that are brought before it.

While only the executive branch can enforce the laws, the judicial branch serves as a check on the power of the executive branch to enforce the laws, since the courts stand in judgment of someone's innocence or guilt. In other words, it is the executive branch that charges someone with committing a crime, but only the courts can convict and sentence that person for having committed the crime.

📖   **SIDEBAR**

Let's suppose that Congress passes a law mandating that a halon fire extinguishing system be installed in every airport control tower. Judge Snipe, an overzealous environmentalist who believes halon should be banned due to concerns over depletion of the ozone layer, disagrees with such a law, and decides he will strike it down. Can Judge Snipe simply convene a hearing and issue a ruling striking down the law?

ANSWER: No. In the absence of someone bringing a case to his court challenging the validity of such a law, Judge Snipe cannot strike down this act of Congress. A court has jurisdiction to act only when a case or controversy has been brought before it. This is an important limitation on the authority of our courts.

State Constitutions

In addition to the United States Constitution, all states have adopted constitutions as well. Just as the United States Constitution is the supreme law of the United States, a state constitution serves as the highest law within the state, subject only to the United States Constitution. Any laws in a state that violate the state constitution will be invalid.

State constitutions also function to organize state government into three branches, just as we saw at the Federal level: executive, legislative, and judicial. The organization and roles of each branch may differ somewhat from the Federal Constitution, and states may differ from each other in terms of their internal organization. State constitutions provide the legal authority and jurisdiction for the various parts of state government, including political subdivisions such as cities and towns.

Local Charters

The term **constitution** refers to an organizing law that is established by the people themselves. On the local level, the most commonly used organizing law is not a constitution, but rather is a **charter**. A charter differs from a constitution in that a charter is an organizing law that is issued by a sovereign, such as the state.

Where utilized, local charters are issued by the state legislature pursuant to state constitutional authority. The charter specifies the legal authority of the municipality and designates its boundaries. In states such as Kansas, where formal charters are not issued, the legal authority and boundaries of a municipal entity are controlled by the state legislature through statutes. In either case, municipalities are considered to be subdivisions of the state because they are created by the state, for the purpose of helping the state address the safety and well-being of the residents.

Amendments to local charters can only be made by the state legislature, unless state law allows a municipality to adopt a **home rule charter**, or otherwise amend its charter without legislative approval. Home rule charter laws allow the citizens of a municipality to adopt or amend their own charter, without having to go back to the state legislature. The creation of a home rule charter, or the passage of charter amendments, typically requires voter approval in the form of a **referendum**. A referendum is a process for the eligible voters in a community to vote on a given law.

Municipal charters serve as the supreme law of the local jurisdiction and set up the form of local government. Charters must comply with the United States Constitution and state constitutions. While some local jurisdictions have three branches of government, many do not require such a complex organization to manage the day-to-day operations of government. At a minimum, local governments must have one branch, namely a legislative branch. In such cases, the

constitution
the organizing document for the Federal government, or for a state government, that serves as the supreme law and establishes the organization of government

charter
the organizing document for a political sub-division of a state, serving the equivalent function of a constitution

home rule charter
a charter that allows the citizens of a municipality to adopt or amend their own charter, without having to go back to the state legislature

referendum
a process for the eligible voters in a community to vote on a given law

statutes

laws passed by Congress or state legislatures

title number

the laws of Congress are commonly cited as 29 USC 654, with the first number identifying the title number; 29 USC 654 is therefore Title 29, Section 654 of the United States Code

section number

the laws of Congress are commonly cited as 29 USC 654, the second number indicating the section number; 29 USC 654 is therefore Title 29, Section 654 of the United States Code; the symbol § is commonly used to denote a section, such as §654

general law

a term commonly used for state statutes

ordinances

laws passed by local legislative bodies

legislative branch, such as a town council or board of selectmen, exercises both legislative and executive powers. Other local jurisdictions may function with two branches, legislative and executive. Concerns over the balance of power are not as significant at the local level as they are at the state and Federal levels.

Beyond the Constitution: Statutory Law

As we have seen, the United States Congress is responsible for passing laws. The laws that Congress passes are called **statutes**. Federal statutes are organized into a large body of laws called the United States Code. The United States Code is abbreviated as U.S.C. When referring to laws that are passed by Congress, we would commonly cite them as 29 USC 654, with the first number identifying the **title number**, and the second number indicating the **section number**. Therefore, 29 USC 654 is Title 29, Section 654 of the United States Code. The symbol § is commonly used to denote a section, such as §654.

At the state level, states have legislatures that serve an analogous function to Congress. State legislatures enact laws which are also called statutes, or **general laws**. State statutes are organized into codes, such as Rhode Island General Laws (RIGL), Illinois Compiled Statutes (ILCS), Florida Statutes (FS), Connecticut General Statutes (CGS), and the Official Code of Georgia Annotated (OCGA). State statutes must comply with the United States Constitution as well as the state constitution.

On the local level, the legislative body of a municipality, such as a city or town council, has the authority to pass laws called **ordinances**. Ordinances must comply with both the state constitution and the U.S. Constitution. In addition, local ordinances cannot violate any state statutes.

📖   **SIDEBAR**

Why Do Some Federal Statutes Have Two Sets of Numbers?

When Congress enacts a law, the law is assigned a public law number based upon the chronological order in which the law is passed. In addition, the numbering system used within each act is unique to the act. For example, Title VII of the Civil Rights Act of 1964 addresses employment discrimination, and provides

SEC. 703. (a) It shall be an unlawful employment practice for an employer

(1) to fail or refuse to hire or to discharge any individual, or otherwise to discriminate against any individual with respect to his compensation, terms, conditions, or privileges of employment, because of such individual's race, color, religion, sex, or national origin; or

(2) to limit, segregate, or classify his employees in any way which would deprive or tend to deprive any individual of employment opportunities or otherwise adversely affect his status as an employee, because of such individual's race, color, religion, sex, or national origin.

The public law designation for this law is Public Law 88-352, Section 703.

In addition to the public law format, Federal statutes are entered into the United States Code, and assigned a number that is organized according to topic. Section 703 of the Civil Rights Act of 1964 is codified in the United States Code as 42 USC 2002e-2. States follow a comparable numbering system and codification process for state statutes.

Beyond Statutes and Ordinances

administrative agencies
agencies that exist within the executive branch of government to assist the executive in carrying out responsibilities imposed by law; agencies are created by the legislative branch, through laws called enabling acts

Are all laws constitutions, charters, statutes, or ordinances? If only it were that easy. Life today is quite complex. In order for our society and our economy to function properly, legislative authority could not rest solely with Congress, state legislatures, and local city or town councils. Think of the complexity needed to administer commercial airline traffic, or manage the magnitude of radio traffic, in the United States. No legislative body could possibly be expert enough in all subject areas to enact all the laws needed. The solution: Congress created **administrative agencies**, such as the Federal Aviation Administration and the Federal Communications Commission. We will discuss these administrative agencies in considerable detail in Chapter 4. However, what is important for our discussion of laws is to realize that Federal administrative agencies can be authorized by Congress to pass laws of their own, called **regulations**.

regulations
laws that are created by administrative agencies pursuant to authority delegated by the legislature

Federal regulations are laws that are created by Federal administrative agencies. Federal regulations are compiled and placed into a code called the **Code of Federal Regulations**, abbreviated CFR. Perhaps you are familiar with the following: 29 CFR 1910.120. That is the Federal regulation issued by the Occupational Safety and Health Administration (OSHA) that addresses hazardous materials in the workplace. State administrative agencies are similarly authorized to pass state regulations, and local agencies may be authorized to pass local regulations as well.

Code of Federal Regulations
the codified compilation of regulations created by Federal administrative agencies, abbreviated CFR

SIDEBAR

You are asked by your fire chief to serve on a regional committee that is researching a new radio system for fire departments in your area. At the first meeting of the committee, you are handed copies of some "laws" that the chairman of the committee says must be followed. Someone in the room yells out "Who says we have to follow this law? We don't even know who adopted it. Maybe it's not a law at all, but a guideline." As you glance at front page you see the following: 47 USC 154. What does this tell you? What if it said 47 CFR 300?

ANSWER: If the document says "USC" it means it is from the United States Code, and is a Federal statute. It is a law passed by Congress. If it says "CFR" it is a Federal regulation passed by a Federal administrative agency, in this case probably by the Federal Communications Commission, or FCC.

Case Law

case law
the written decisions of judges issued in the process of deciding actual cases

Another important source of law is case law. **Case law** refers to the written decisions of judges issued in the process of deciding actual cases. These written decisions can set binding precedent for future cases decided in the same jurisdiction, thereby creating another source of law. We will discuss the concept of precedence and jurisdiction further in Chapter 2. Courts exist at the Federal, state, and local levels. Courts at each of these levels are capable of producing a significant body of case law, which serves as a source of law for future cases.

case books
books that contain the actual text of the written decisions by judges in deciding questions of law

Written decisions are commonly organized into series of books called **case books** or **reporters**. Cases are identified and referred to by the names of the parties involved in the suit. Generally, the name of a case is either underlined or placed in italics to indicate that a case is being referenced, and a "v." is placed between parties on each side of the suit to indicate "versus." Following the name of the case, a series of numbers and abbreviations may follow. These numbers and abbreviations identify where the case has been published.

reporters
these books may also be called case books

For example, the case of *Grogan v. Commonwealth of Kentucky,* 577 S.W.2d 4, (KY, 1979) was a lawsuit brought by Robert Grogan against the Commonwealth of Kentucky, arising out of the Beverly Hills Supper Club fire that killed 165 people on May 28, 1977 **(Figure 1-6)**. We will be studying

Figure 1-6
The Beverly Hills Supper Club fire killed 165 people on May 28, 1977. (Photo by Corbis.)

the case in detail in Chapter 9. The suit alleged that the Commonwealth was at fault for the deaths and injuries due to the failure to properly enforce the fire code. The abbreviation S.W.2d refers to a series of books known as the Southwest Reporter Series, Second Edition. The number 577 refers to the volume number of the Southwest Reporter Series, Second Edition, while the 4 refers to the page number upon which the case begins. The information in parentheses (KY, 1979) indicates that the case was decided by the Kentucky Supreme Court in 1979.

For additional information about the abbreviations used in citations, see Appendix B, Chapter 1.

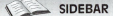 **SIDEBAR**

Is Case Law Really That Important a Source of Law Compared with Statutes?

ANSWER: Case law is an absolutely critical source of law. Some of the most important and controversial laws that exist are the result of case law, not acts of Congress. Examples include *Miranda v. Arizona,* which established that a person who is arrested has to be informed of their rights before being questioned; *Brown v. Board of Education,* which held that separate but equal education for whites and blacks is unconstitutional; and *Roe v. Wade,* which affirmed a woman's right to choose to have an abortion.

Common Law

common law
judge-made law that developed in England, and was in existence prior to our Declaration of Independence on July 4, 1776

arson
(at common law) The willful and malicious burning of the dwelling of another

A source of law that is similar to case law is what is called the **common law.** Common law refers to judge-made law that developed in England, and was in use in the American colonies prior to the Revolution. On July 4, 1776, when we declared our independence from England, we had our own country, but at the time we had absolutely no laws. Was that a problem? It's July 5, 1776, and someone lights another person's house on fire. Could the arsonist be arrested and charged? We had no written laws. The common law declared that the willful and malicious burning of the dwelling of another constituted the common-law crime of **arson**. Arson was a capital offense for which a person could be executed **(Figure 1-7)**.

Out of necessity, our legal system adopted the English common law that was in effect on July 4, 1776. It made sense, because all the lawyers and judges in the colonies were trained in the English system. It also meant that the new legislatures in the colonies could focus on the more important tasks that confronted them, rather than developing an entirely new legal system.

Figure 1-7
In 1776, arson was a capital offense in common law, for which a perpetrator could be executed. (Photo by Corbis.)

Except where specifically abolished or amended, the common law remains a valid source of law today. The following law is from Section 2.01 of the Florida Statutes, addressing common law.

2.01 Common law and certain statutes declared in force.—The common and statute laws of England which are of a general and not a local nature, with the exception hereinafter mentioned, down to the 4th day of July, 1776, are declared to be of force in this state; provided, the said statutes and common law be not inconsistent with the Constitution and laws of the United States and the acts of the Legislature of this state.

Less Common Sources of Law

executive order
a law issued by an executive, such as the president, governor, or mayor

Legislatures at the Federal, state, and local levels may authorize the executive, be it the president, governor, or mayor, to issue **executive orders**, which have the effect of law. In the absence of such explicit authorization by the legislature, the constitutionality of the executive issuing edicts remains an unsettled area. Nonetheless, presidents have routinely issued executive orders since 1789. An example of an executive order is Homeland Security Presidential Directive 5, issued by President George W. Bush on February 28, 2003, mandating the implementation of the National Incident Management System (NIMS) by all Federal agencies.

injunction
a court order directing a party to a lawsuit to do or refrain from doing some act

preliminary injunction
injunction issued (usually at the beginning of a lawsuit) to maintain the status quo while the case is heard

temporary restraining order
a form of injunction that is issued by a court on an emergency basis for a very brief period of time (7 to 14 days), until the court can hold a full hearing on the matter

warrant
a court order that commands law enforcement personnel to carry out a specific task, such as arrest a person (arrest warrant), search a certain area and seize certain evidence (search warrant), or execute a criminal defendant (death warrant)

subpoena
a court order commanding a person to appear at a certain time and place to give testimony or produce certain evidence

International treaties that are signed by the president and ratified by two-thirds of the Senate are another source of law. Although treaties rarely impact state and local governments directly, they are considered to be law and must be honored by all states.

The lawful orders of a court are also a source of law for the persons involved in the controversy. If a court orders a person to do something, the order must be followed subject to sanction or legal consequences. Examples of court orders include injunctions, temporary restraining orders, warrants, and subpoenas. An **injunction** is a court order directing a party to a lawsuit to do or refrain from doing some act. Injunctions may be temporary (also called preliminary injunctions), or they may be permanent. **Preliminary injunctions** are usually issued at the beginning of a lawsuit to maintain the status quo while the case is heard. A **temporary restraining order** is a form of injunction that is issued by a court on an emergency basis for a very brief period of time (7 to 14 days), until the court can hold a full hearing on the matter.

A **warrant** is a court order that commands law enforcement personnel to carry out a specific task, such as arrest a person (arrest warrant), search a certain area and seize certain evidence (search warrant), or execute a convicted criminal (death warrant). A **subpoena** is a court order commanding a person to appear at a certain time and place to give testimony or produce certain evidence.

Unlike the executive and legislative branches of government, the judicial branch has the inherent power to enforce its own orders. Courts may cite someone who fails to comply with their order by finding them to be in contempt of court. Citations for contempt can include fines or incarceration, or both.

Table 1-1 summarizes the various sources of law.

Table 1-1   *Sources of law.*

Type of Law	Federal	State	Local
Supreme Law	Constitution	Constitution	Charter
Legislature	Statute	Statute	Ordinance
Administrative Agencies	Regulations	Regulations	Regulations
Judicial branch	Case Law	Case law	Case law
	Court Order	Court Order	Court Order
Executive	Executive Order	Executive Order	Executive Order
Others	Common Law	Common Law	Common Law
	Foreign Treaties		

STANDARDS AND CODES

standards
voluntary guidelines and recommendations that do not carry the force and effect of law

codes
a compilation of standards, regulations, or statutes

Closely related to our discussion of laws is the subject of standards and codes. While laws are defined as that which must be obeyed subject to sanction or consequences by the government, the term **standards** refers to voluntary guidelines and recommendations that do not carry the force and effect of law. Standards may be created by governmental organizations such as the National Institute of Occupational Safety and Health (NIOSH), by private organizations such as the National Fire Protection Association (NFPA) and Building Officials and Code Administrators International (BOCA), or by partnerships between government and private organizations. Standards and compilations of standards may also be referred to as **codes**, such as the National Electrical Code, which is a standard produced by the NFPA.

The term "standard" is confusing, for several reasons. First of all, some standards are adopted into law and thus have the full effect and weight of law. For example, some states have adopted NFPA Standard 1, National Fire Code, and NFPA Standard 101, Life Safety Code, to serve as the state fire code. In such a state, the standard carries the weight of law. See Appendix B, Chapter 1 for a sample of a statute that adopts a standard into law. On a local level, most communities have adopted a building code standard such as the BOCA code. In communities where NFPA Standard 101 and the BOCA code have been adopted into law, both are standards and laws at the same time.

Second, the term "standard" is not consistently applied, resulting in ambiguity. For example, OSHA regulations are commonly referred to as "OSHA standards." As we discussed above, OSHA regulations are laws. They are not voluntary, nor are they guidelines. Use of the term "codes" is even more confusing, because some codes are voluntary standards while other codes are actual laws, such as the United States Code, the Code of Federal Regulations, and state fire codes. As a result, there is a great deal of confusion and ambiguity concerning the terms "standard" and "code."

Where Do Standards Come From?

Consensus standards
standards that are developed through a formal process, such as required by American National Standards Institute (ANSI), and represent generally accepted industry-wide practices and recommendations

Organizations such as the NFPA create consensus standards. **Consensus standards** represent generally accepted, industry-wide practices and recommendations. The process of developing consensus standards involves a formalized procedure that includes the opportunity for industry, users, government, and the public to have input and comment. The American National Standards Institute (ANSI) is an organization of standards-making organizations that creates procedures and standards for organizations to follow when developing consensus standards. The consensus standards-making process must include procedures that ensure openness, a balance of interests on the standards-making committee, due process, and an appeals process. The term "consensus" refers to the need for a general agreement, but not a unanimous vote of all members involved in the development of the standard.

Besides consensus standards, there are standards that are developed through nonconsensus procedures. Standards developed through nonconsensus processes include industry-developed standards, standards developed by committees of technical experts, government standards, and company standards. In each of these cases, the standards that are developed are the result of a process that does not include a consensus-based process as required by ANSI.

When Standards Become Laws

Governments may adopt standards into law in one of two ways. First, the legislative branch may adopt a standard through legislation. Second, an administrative agency may adopt a standard through a rulemaking process (described in Chapter 4). In either event, an important constitutional issue is often raised: Has the legislature or agency unconstitutionally delegated lawmaking power to an unelected entity, namely the standards-making organization?

In order for a standard to be adopted without violating the constitution, the legislature or agency must be very specific in citing the standard and the edition being adopted. Once a standard is adopted into law, any subsequent revision of that standard adopted by the standards-making organization is not legally enforceable unless and until the legislature or agency formally convenes and adopts the newer revision.

EXAMPLE

In 1988, the Rhode Island legislature adopted NFPA 1500, 1987 Edition, into law for all fire departments, excluding volunteer fire departments. RIGL 23-28.4. NFPA 1500 was rewritten in 1992, 1997, and 2002. However, since the legislature never reenacted RIGL 23-28.4 to include the newer editions, the 1987 Edition of NFPA 1500 is still the edition that must be complied with.

The theory behind the constitutional limit on delegation of lawmaking powers is that if such a delegation were allowed, lawmaking power would essentially be delegated to an unelected, nongovernmental organization. The organization, such as the NFPA, would then be free to pass whatever laws that it deemed appropriate, thereby usurping the constitutional authority granted to the legislature, or in the case of the agency, usurping the authority delegated by the legislature to the agency. By requiring the adoption of a specific edition of a standard, the legislature or the agency has the opportunity to examine and review the proposed standard, and choose to adopt or not adopt the standard as written.

Figure 1-8
NFPA headquarters in Quincy, Massachusetts. (Photo by Joseph L. Molis.)

📖   **SIDEBAR**

What Is the National Fire Protection Association (NFPA)?

The National Fire Protection Association (NFPA) is a private, nonprofit organization whose goal is to reduce the impact of fire and other disasters upon people and their property **(Figure 1-8)**. The NFPA is not part of government. The NFPA is made up of voluntary, dues-paying members, many of whom are members of the fire service, insurance industry, building trades, and others concerned about the problem of fire in our society. The NFPA fulfills its mission by researching and developing consensus standards that address all aspects of fire and life safety, as well as developing educational information and materials.

CIVIL VERSUS CRIMINAL

criminal law
law designed to prevent harm to society in general and which declares certain actions to be criminal

Our legal system can be divided into two broad categories, civil law and criminal law. **Criminal law** refers to laws that are designed to prevent harm to society in general, and which declare certain actions to be criminal. **Civil law** refers to laws that are designed to address the relative rights between parties where one party has been wronged by another, as well as matters related to probate, domestic relations, workers' compensation, and other noncriminal matters. Civil law includes a system designed to compensate someone who

Table 1-2 *Criminal versus civil law.*

Type of Law	Brought By	Burden of Proof	Outcome
Criminal	Government	Beyond a reasonable doubt	Jail, probation, fine or restitution
Civil	Party who has been wronged	More likely than not	Monetary damages or comply with court order

civil law

law that seeks to enforce private rights and remedies for some wrong done to an individual or organization; depending upon the context of how the term is used, civil law can also be viewed as all law that is not criminal law

defendant

the party who is required to answer a complaint or case in court

plaintiff

the party who files a civil lawsuit

jurisdiction

the legal authority of a person, official, court or agency to take an official action; the term may have different meaning depending upon the context of its usage; the jurisdiction of a court refers to the legal authority of a court to hear a matter; the jurisdiction of an administrative agency refers to its legal authority to take action on a certain matter

has been injured or wronged, at the expense of the wrongdoer. The same act by a wrongdoer can give rise to both a civil and a criminal action.

The differences between a criminal case and a civil case are quite dramatic (see **Table 1-2**). A criminal case must be brought by the government, and must be based upon the violation of a specific criminal law. Once charged, the person who is alleged to have committed the criminal act, called the **defendant**, is entitled to a jury trial. At trial, the government must prove the defendant committed the crime beyond a reasonable doubt. If convicted, the guilty party may be sentenced to prison, placed on probation, have to pay a fine, and/or pay restitution.

In a civil case, a person or organization, called the **plaintiff**, files suit against another party named the defendant, alleging that the defendant has somehow wronged the plaintiff. The plaintiff may be a private party or it may be a part of government. The case may be tried before a jury or a judge, depending upon the issues involved, and the plaintiff need only prove that it was more likely than not that the defendant was wrong. This civil burden of proof is often described as "by fair preponderance of the evidence." If the plaintiff wins, the defendant may have to pay damages to the plaintiff, or be required to comply with a court order.

JURISDICTION

In the United States, each of our three levels of government—Federal, state, and local—has a specific area of authority, or **jurisdiction.** Jurisdiction refers to the legal authority of a person, official, court, or agency to take an official action. The jurisdiction of each level of government is initially defined by the United States Constitution, state constitution, and local charter. Jurisdiction is further clarified by court decisions that have defined and refined what jurisdiction each governmental level has. Each level of government can only create laws and exercise the authority for which it has the legal jurisdiction.

Federal Jurisdiction

The Federal government is said to be a government of limited powers, with most jurisdictional power left to the states. Federal jurisdiction is limited by the language of the United States Constitution. According to the Constitution, the jurisdiction of Congress to pass laws is limited to matters involving interstate commerce, bankruptcy, currency, banking, immigration, patent and copyright laws, the military, and admiralty, to name a few. Federal jurisdiction also extends to the enforcement of Federal constitutional and civil rights.

While in theory the Federal government is one of limited jurisdiction, in reality the Federal government wields considerable jurisdictional powers that extend well beyond the original intent of the framers of the Constitution. One of the principal reasons for the extension of Federal jurisdiction is the expansive interpretation of the interstate commerce clause in the Constitution. The interstate commerce clause authorizes Congress to regulate interstate commerce. Back in the 1700s, interstate commerce was a relatively simple concept, since most commerce was of a local nature.

Today, virtually every business or activity in our society has some relationship to interstate commerce. We depend upon interstate commerce for the goods we buy. Our communications carriers and our banks are all engaged in interstate commerce. It is hard to conceive of any activity that at some level does not involve interstate commerce. As a result, Federal jurisdiction has the potential to be quite broad.

State Jurisdiction

It was a quiet Sunday in the firehouse. The housework was completed, and everyone was sitting reading the Sunday paper. Bill looked at an ad for radar detectors that read: "Warning: not legal in all states." Bill said, to no one in particular, "You know, I just don't understand how something can be legal in one state, and illegal in another. There should be one set of laws. It would make things a lot easier."

Ron looked up from his paper and said, "Really? I kind of like being able to ride my motorcycle without a helmet on, and in some states they require helmets."

Mary added, "Yeah—and some states are real strict on gun control and hunting, while others aren't. How would you keep everyone happy?"

"Hey Cap, why is it that different states can have different laws on the same subject?"

The Tenth Amendment states that ". . . the powers not delegated to the United States by the Constitution, nor prohibited by it to the States, are reserved to the States respectively, or to the people." While the Federal government is one of limited jurisdiction, the states have inherent authority to exercise all jurisdiction that is not specifically given to the Federal government by the Constitution. One of the most important jurisdictional powers left to the states is the exercise of what is called **police powers**. The subject of police powers will be discussed further in Chapter 5. However, when it comes to matters of public safety, and the health and welfare of the people, states have broad powers to pass laws as they see fit to protect their citizens. This police powers jurisdiction includes matters pertaining to fire protection, emergency medical response, fire codes, and criminal offenses.

police powers
the authority of each state to govern matters related to the welfare and safety of residents; states are empowered with broad police powers to pass laws as they see fit to protect their citizens; police powers jurisdiction includes matters pertaining to fire protection, emergency medical response, fire codes, and criminal offenses

Most criminal offenses are made and enforced at the state level. These include the principal crimes of murder, robbery, burglary, larceny, rape, assault, and arson, but also minor offenses, such as disorderly conduct, traffic offenses, and public drunkenness. The Federal government is limited in what criminal laws it can pass, because it does not have jurisdiction to exercise broad police powers as do states. There must be some Federal reason for there to be Federal jurisdiction. As a result, the Federal government does not have the authority to make arson or assault a Federal crime, in the absence of some Federal connection. Arson of a Federal property, or assault of a Federal official, would have the necessary Federal connection. Absent such a connection, basic criminal jurisdiction is reserved to each state.

Local Jurisdiction

Like the Federal government, local jurisdictions have limited authority. Local jurisdictions have only what authority and responsibility that state law grants to them. Generally, local jurisdiction is limited to matters affecting zoning, planning, economic development, and relatively minor matters. It is the individual states that have the broadest discretion to pass laws governing activities that occur within their borders.

State Freedom to Pass Laws

States may differ considerably in what laws its residents may desire. States are not bound by the desires and wishes of those in other states in passing laws, so long as the laws they choose to pass do not violate the United States Constitution. Some states have strict laws on gun control, while other states believe that citizens should have a right to carry firearms. The ability of a state to pass laws independently of the Federal government, as well as independently from what other states have done, is considered by many to be one of the strengths of our Federal system of democracy.

In a similar way, the supreme court of one state is not bound by the decisions of the supreme court of another state, even where the supreme court of the other state has already interpreted the exact same language in an identical statute passed by its state legislature. States may have identical statutes on a given subject that have been interpreted differently by their supreme courts, leading to completely different results.

SUMMARY

There is no single source of law in the United States. On the contrary, our laws come from a variety of different sources. Laws are created at the Federal, state, and local levels. Laws are created by all three branches of government: legislative, executive and judicial. It is through a complex interplay of these various sources of law that we gain an understanding of what the law is. It is important for firefighters to recognize that the law is dynamic and continuously evolving, because as the law evolves and changes, so do their rights and responsibilities.

REVIEW QUESTIONS

1. What are the three levels of government in the United States?

2. What are the three branches of government in the United States?

3. Identify 10 sources of law in the United States.

4. Explain how the concept of balance of power has been incorporated into the United States Constitution.

5. What are the two functions that are served by the United States Constitution?

6. Explain why the terms "standard" and "code" are ambiguous.

7. Does Congress have the authority to make arson a Federal offense?

8. Identify three differences between civil law and criminal law.

9. Where does a law that is cited as 47 CFR 300 come from?

10. What level of government has inherent police powers?

DISCUSSION QUESTIONS

1. What would be the harm if a legislature adopted into law an NFPA standard that automatically included the most up-to-date version of the standard that the NFPA might issue in the future?

2. What are the advantages to having three levels of government, as opposed to simply a national government? What are the disadvantages?

Chapter 2

COURTS AND COURT SYSTEMS

Learning Objectives

Upon completion of this chapter, you should be able to:

- Describe courts in general.
- Explain why the jurisdiction of Federal courts is limited, but the jurisdiction of state courts is general.
- Explain the difference between trial courts and appellate courts.
- Explain the three types of evidence.
- Identify the four phases of a civil lawsuit.
- Describe the Federal and state court systems.
- Explain stare decisis and precedence.
- Explain how stare decisis does not bind one trial court to follow the precedent set by another.
- Explain the difference between questions of law and questions of fact.
- Explain the difference between the role of a judge and jury at a trial.
- Explain how our knowledge of lawsuits can be distorted by the media.
- Describe the sources available for conducting legal research.

INTRODUCTION

"Company Officer Held Liable for Accident Victim's Injuries." The headlines were read aloud again for everyone in the fire station kitchen to hear. A group of firefighters stood staring at the newspaper clipping pinned on the station bulletin board.

"How can that be?" asked the lieutenant. "This article doesn't say anything about the driver of the apparatus or the fire department being liable, it just says the officer was liable when the apparatus he was in command of hit a car."

"See, there's another reason why I won't take a promotional exam," says Pete.

"What is going on with the court system?" said Ron.

"Runaway juries is the problem," replied Pete.

As we learned in Chapter 1, there are three levels of government in the United States: Federal, state, and local. Federal and state governments are created by constitutions, and local governments are created by acts of the state legislature, most commonly through the issuance of charters.

Constitutions and charters generally establish three branches of government: the executive branch, the legislative branch, and the judicial branch. Each branch is designed to be coequal with systems in place to ensure a balance of power. The function of the judicial branch of government is to apply the law to actual cases and controversies. In addition, the judicial branch serves as a balance of power against the legislative branch's creation of laws and the executive branch's enforcement of the law.

COURTS IN GENERAL

The judicial branch of government is made up of courts **(Figure 2-1)**. Not all courts are the same, nor do they operate in the same way. Some courts have very broad subject matter jurisdiction, and are able to hear a variety of different types of cases. Other courts have very limited jurisdiction and can only hear particular types of cases. Some courts have judges and juries, and some courts just have judges. Some courts have one judge, some have three judges, and some have up to nine judges. The legal power of a court is based upon the court's jurisdiction.

Figure 2-1
Courthouse.

JURISDICTION

The jurisdiction of a court refers to the legal authority to hear a case and render a decision. The jurisdiction of a court is based initially upon a constitutional delegation of judicial power. In the case of the Federal courts, the Constitution establishes the United States Supreme Court, and grants to Congress the authority to establish "such inferior courts" as Congress believes are needed. Therefore, Congress has a large say in how many Federal courts there will be, and what their jurisdiction is.

Each state's constitution establishes the authority and framework for the state court system. State legislatures typically have the constitutional authority to create state courts, and assign their jurisdictional authority. Generally, it is state law, and to a lesser extent local charters, that establish the framework for local courts.

The subject matter jurisdiction of courts can be grouped into two broad categories: courts with general jurisdiction and courts with specific jurisdiction. Most state trial courts are said to be courts of **general jurisdiction**, meaning that they have jurisdiction over all cases and controversies unless specifically exempted. Courts with **specific jurisdiction** have only such jurisdiction as the legislature or enabling authority provides. Examples of courts with specific jurisdiction would be probate courts, family courts, and workers' compensation courts **(Figure 2-2)**.

general jurisdiction
jurisdiction of a court that has the legal authority to hear and decide all cases and controversies unless specifically exempted

specific jurisdiction
jurisdiction of a court that is limited by legislation or enabling authority to hear only certain types of cases and controversies

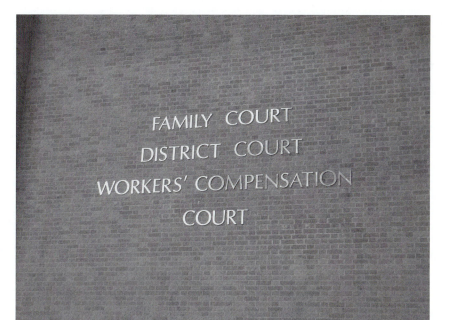

Figure 2-2 *There are a variety of courts, with different types of subject matter jurisdiction.*

TRIAL COURTS AND APPELLATE COURTS

Trial courts are courts where lawsuits are initially filed, and where most of the activity that we commonly associate with our court system takes place. It is where the judge and jury will apply the law to the facts of the case and render a decision. In the Federal court system, the trial court is the Federal district court. On the state level, trial courts are commonly named superior court or district court, but in some states there are unique names, such as the Court of Common Pleas (Pennsylvania), and the Supreme Court (New York).

Appellate courts are where appeals from trial courts are heard. All court systems allow for at least one level of appeals. There are two levels of appellate courts in the Federal system: the Circuit Courts of Appeal, and the United States Supreme Court. Appellate courts at the state level are commonly called the court of appeals, the supreme court, or the supreme judicial court.

EVIDENCE

evidence
any type of proof that can legally be presented at a trial or hearing

Lawsuits are driven by **evidence**. In the absence of evidence, even the most meritorious of cases cannot be pursued. The availability and admissibility of evidence ultimately determine what cases will be brought to court and

direct evidence

evidence that proves a fact, without the necessity for an inference or presumption

real evidence

tangible evidence that can be brought into court and examined; real evidence may also be referred to as demonstrative evidence

demonstrative evidence

another name for real evidence

circumstantial evidence

evidence that requires someone to infer, deduce, or presume something from the direct evidence that is presented

which side will prevail. Conclusions about the probative value and admissibility of evidence also cause cases to be settled.

Evidence can be grouped into three categories: direct, real, and circumstantial. **Direct evidence** is evidence that proves a fact, without the necessity for an inference or presumption. Generally, direct evidence comes in the form of testimony from a witness who has firsthand knowledge of an event or fact, through that person's senses of sight, hearing, smell, taste, or touch. An example of direct evidence is the testimony of a witness who personally observed a person lighting a fire.

Real evidence is tangible evidence that can be brought into court and examined. Real evidence is also referred to as **demonstrative evidence**. Examples of real evidence in an arson case would be torn-up sections of flooring showing a pour pattern, or a timing device used to set a fire **(Figure 2-3)**.

Circumstantial evidence is evidence that requires someone to infer, deduce, or presume something from the direct evidence that is presented. An example

Figure 2-3 *Torn-up sections of flooring showing a pour pattern would constitute real evidence. (Photo by Tim Whalen.)*

of circumstantial evidence would be testimony that the defendant entered a vacant house at 10:00 p.m. with a fluid container in his hand, left the house at 10:05 p.m. without the container, and at 10:10 p.m. the house was observed to be on fire. This testimony does not present direct evidence that the defendant set the fire. Rather, it requires that an inference be drawn from the testimony.

For evidence to be admissible in a court, it must be relevant, probative, and competent. In other words, the evidence must tend to prove or disprove the existence of a certain fact that is in controversy in the case.

Evidence may be inadmissible in court for a variety of reasons. Typically the reasons that evidence can be excluded pertain to the validity of the evidence, the manner in which the evidence was obtained (particularly in criminal cases), public policy considerations about the use of certain types of evidence, and procedural rules aimed at providing a fair trial. The rules of evidence differ in civil and criminal cases.

ANATOMY OF A CIVIL LAWSUIT

There are typically four parts to a civil lawsuit: the pleadings phase, the discovery phase, the trial, and the appeal.

Pleadings

PLEADINGS ⟶ DISCOVERY ⟶ TRIAL ⟶ APPEAL

complaint
a formal allegation or series of allegations by a plaintiff that the defendant did certain things that somehow injured or damaged the plaintiff, and constitute a cause of action upon which the court should take action; the complaint is what formally initiates a lawsuit

Lawsuits are initiated by a plaintiff filing a **complaint** or petition with the trial court. The complaint is nothing more than a series of allegations by the plaintiff that the defendant did certain things which somehow injured or damaged the plaintiff, and that constitute a claim upon which the court should take action. The complaint must also contain a specific request that the plaintiff is seeking the court to order, such as the payment of money damages or the issuance of an injunction.

serve
to deliver a summons, complaint, subpoena, notice, or other official document to a person

Once the plaintiff files the complaint with the court, the defendant is **served** with a copy of the complaint, together with a court-issued document called a **summons**. The summons serves as official notice to the defendant of the existence of the lawsuit. The summons is also a court order that requires the defendant to answer the complaint within a designated time period or lose the case on a default. The summons and complaint are served upon the defendant in a method prescribed by law. The action of serving a party to a lawsuit with the summons and complaint is called **service of process**. Different jurisdictions have different requirements for service of process. Some courts require that a court official personally hand the documents to the defendant. Other courts may simply require service by certified mail.

summons
a court order that requires the defendant to answer the complaint within a designated time period or lose the case on a default

service of process
giving a person legal notice of a court or administrative action

answer
a legal document in which a defendant in a civil case either admits or denies the allegations that the plaintiff made in the complaint

default judgment
a judgment in a lawsuit against a party who fails to answer or defend against allegations made by the other party, resulting in the other party winning by default

pleadings
consists of the complaint and the answer, along with any counter-claims, third-party complaints, or motions to dismiss in a lawsuit

discovery
the phase in a lawsuit in which each side has the opportunity to learn about the evidence and witnesses that the other side has

deposition
a procedure by which a party to a lawsuit may compel another party or a witness to appear to answer questions under oath as part of a process known as discovery; most depositions are taken before a court reporter (stenographer), but in some cases videotaped depositions are also allowed

Once served, a defendant commonly has from fourteen to thirty days to respond to, or **answer**, the complaint. The defendant's answer is a legal document in which he or she either admits or denies the allegations that the plaintiff made in the complaint. The failure of the defendant to file an answer within the time allowed will result in a **default judgment**, which essentially means the plaintiff automatically wins the case. A plaintiff who wins by default must simply prove the amount of damages he or she sustained.

In addition to filing an answer, a defendant may file a counterclaim against the plaintiff, alleging that the plaintiff has somehow wronged the defendant. The defendant may also file additional claims against third parties who may bear some responsibility to the matter.

The complaint and the answer make up what is known as the **pleadings** of the case. Samples of a complaint, answer, and summons are included in Appendix B, Chapter 2. During the initial pleadings phase, either party may file a motion to dismiss the case or to decide the case in their favor based upon the facts alleged in the pleadings. Once the pleadings phase is complete, the case enters the discovery phase.

Discovery

PLEADINGS $\longrightarrow$ *DISCOVERY* $\longrightarrow$ TRIAL $\longrightarrow$ APPEAL

The **discovery** phase of a lawsuit is most commonly the longest phase, lasting from several months to several years. Discovery is the opportunity for each side to learn about the evidence and witnesses that the other side has. The underlying philosophy behind discovery is that full disclosure encourages the parties to reach an equitable settlement. If each side knows in advance what the other side has for evidence, the parties should be more inclined to resolve the case without having to go to trial.

Discovery includes several specific activities, including depositions, interrogatories, requests for production, and requests for admissions.

Depositions A **deposition** is a procedure by which a party to a lawsuit may compel another party or a witness to appear to answer questions under oath. Generally, depositions take place at an attorney's office, or in a neutral location, but not in court. A judge is not present, but a court-approved stenographer capable of administering oaths is required. The attorneys will ask the person who is being deposed questions, and the entire process is recorded by the stenographer. Recognizing technological advances that have occurred over the years, some jurisdictions now allow depositions to be videotaped. In limited situations, the videotapes may actually be played in court in lieu of a live witness appearing in person. The use of video testimony is an evolving area of the law.

interrogatories
written questions that each party to a lawsuit is entitled to ask any other party in the suit to answer in writing under oath

request for production
a method of discovery that allows a party to obtain evidence from any other party including documents, photographs, physical evidence, or other items relevant to the case, so that they may be examined, reviewed and in some cases analyzed

request for admissions
a method of discovery that allows a party to a lawsuit to formally ask an opposing party to admit to certain statements of facts in order to narrow the issues in dispute in trial

motion for summary judgment
a request by a party to a lawsuit for the court to rule that as a matter of law, he or she should prevail; motions for summary judgment may be granted when there is no genuine issue as to any material fact and the moving party is entitled to a judgment as a matter of law

Interrogatories Each party to a lawsuit is entitled to ask any other party in the suit to answer questions in writing under oath. These written questions, called **interrogatories**, must be answered in writing. The party answering the interrogatories has a designated period of time to answer the questions. Generally the maximum number of questions that can be asked is limited to 30, unless the court grants permission to ask more. Interrogatories are used as an inexpensive discovery tool to gather basic information relevant to the case. Sample interrogatories are included in Appendix B, Chapter 2.

Requests for Production A party in a lawsuit may demand that the other side provide documents, photographs, physical evidence, or other items relevant to the case, so that they may be examined, reviewed and in some cases analyzed. This demand is made in the form of a **request for production**. The failure to produce evidence when properly requested may result in the party withholding the evidence being sanctioned by the court, including the exclusion of that evidence at trial.

Requests for Admissions A party to a lawsuit may formally ask an opposing party to admit to certain statements of facts, called a **request for admissions**, in order to narrow the issues in dispute in trial. Once admitted, these facts are considered by the court to be true, and testimony concerning them at trial is not necessary.

Pretrial Motions During the discovery phase, a variety of motions are commonly made, some of which concern discovery, and some of which involve the procedure of the case. For example, the judge may be asked to exclude or limit certain evidence or testimony. In addition, one party may ask that the case be decided in their favor based upon a clear legal principle. Another party may ask that they be dropped from the suit because the evidence produced during discovery shows they were not liable as a matter of law. Such a request to decide the case prior to trial is termed a **motion for summary judgment**, and may be granted when there is no genuine issue as to any material fact and when the moving party is entitled to a judgment as a matter of law. (See Appendix B, Chapter 2.)

📖 **SIDEBAR**

Motion for Summary Judgment

Mrs. Smith lives in the Town of Smithville. The Town of Smithville does not provide fire or emergency services. Fire services are provided by several independent volunteer companies who are financially supported by the county. Mrs. Smith's house caught on fire, and the Rough & Ready Volunteer Fire Company responded. Mrs. Smith believes the firefighters did unnecessary damage in chopping a hole in the roof and breaking windows. She sues the Town of Smithville, believing that the firefighters who responded to her house were full-time employees of the town.

When the Town of Smithville answers Mrs. Smith's complaint, they state that the town does not provide fire protection, that fire protection is actually provided by the Rough & Ready Volunteer Fire Company, and that the town was in no way involved in the incident in question. The town then files a motion for summary judgment. The court considers motion for summary judgment, and finding no genuine issue as to any material fact, renders a decision in favor of the town.

Once discovery is complete, the case moves on to the trial phase.

Trial

PLEADINGS ⟶ DISCOVERY ⟶ *TRIAL* ⟶ APPEAL

When people think of our judicial system, they immediately think of the jury trial. While jury trials play a vital role in our judicial system, there is much more going on in our courts than most people are aware of. At trial, the key players are the plaintiff, the defendant, the judge, and the jury. In our judicial system, juries exist only at the trial court level. However, not all trial courts allow for juries. Some courts, such as bankruptcy courts, probate courts, divorce courts, and workers' compensation courts, do not provide for juries. The size of a jury varies with the jurisdiction, commonly ranging from six to twelve members.

In a trial, the roles of the judge and jury are very clear and very distinct, but often misunderstood. The role of the judge is to oversee the progress of the case through the pleadings, discovery and trial phases, and in the process make a ruling on questions of law. Ruling on **questions of law** refers to the interpretation of laws and the application of legal principles to the issues in the case.

The role of the jury is to decide questions of fact during the trial phase. A **question of fact** is a determination about some factual event, as opposed to the interpretation of a law. While a question of law is a legal interpretation, a question of fact is a determination of whether or how something happened. Some examples should help explain the difference.

question of law
the interpretation of laws and the application of legal principles to the issues in the case; the judge decides questions of law

question of fact
a factual determination that must be made in a case; in a jury trial, the jury decides questions of fact

---

📖   **SIDEBAR**

Questions of Fact versus Questions of Law

Questions of fact: A fire truck is returning to quarters after a fire, and collides with an automobile at an intersection that is controlled by a traffic light **(Figure 2-4)**. The firefighters claim they had a green light, and thus the right of way. The driver of the other vehicle claims he had a green light, and never saw the fire truck until it was too late. There are other witnesses whose stories conflict. In such a case, who had the red light and who had the green light is a question of fact that the jury must decide based upon the evidence presented.

Questions of law: A fire truck is returning to quarters after a fire, and collides with an automobile at an intersection that is controlled by a traffic light. The firefighters and the other driver admit that the fire truck had the red light, and caused the accident in question. The only question is whether or not the immunity law makes the firefighters exempt from being sued. This is a question of law. It involves the interpretation and application of legal concepts, not the determination of facts. In this case, everyone agrees upon the facts.

burden of proof
the obligation of a party in a lawsuit to affirmatively prove the facts of the case to a required degree in order to prevail

by fair preponderance of the evidence
the burden of proof in a civil case, also referred to as the "more likely than not" standard

Burden of Proof In a trial, the party seeking to prove their case must affirmatively establish the facts to a certain degree in order to prevail. This is called the **burden of proof (Figure 2-5)**. In cases where there is conflicting factual testimony, responsibility for the burden of proof can be a critical factor in determining who wins the case.

The plaintiff in the case generally has the burden of proof. In civil cases, the burden of proof is **by fair preponderance of the evidence**. The civil standard is also referred to as the "more likely than not" standard. In other words, the party with the burden of proof in a civil case need only prove the facts in their case are "more likely than not." All that is necessary is to tip the proverbial scales of justice just slightly in order to prevail using this standard.

Figure 2-4 *Who had the red light would be a question of fact.*

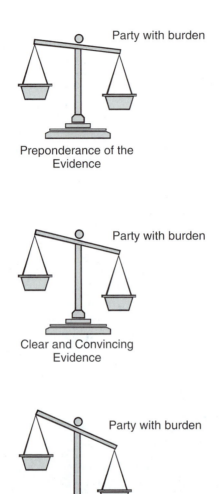

Party with burden

Preponderance of the
Evidence

Party with burden

Clear and Convincing
Evidence

Party with burden

Beyond a Reasonable
Doubt

Figure 2-5 *Burden of proof.*

Consider the example of the apparatus accident at the traffic light where the parties cannot agree on who had the red light. Assume the driver of the other vehicle sues the fire department, claiming the fire apparatus was at fault for the accident and the department should pay for the damages. If the evidence is conflicting as to who had the red light, and the jury simply cannot determine that it was more likely than not that the firefighters were at fault, then the party with the burden of proof (the driver of the other car) has not met his burden, and will lose the case against the firefighters.

In a criminal case, the burden of proof is said to be **beyond a reasonable doubt**. This heightened burden of proof is considered an important safeguard

beyond a reasonable doubt
the burden of proof in a criminal case

against wrongful criminal convictions. The burden is always on the prosecution in a criminal case. This means that the prosecution must prove each and every element of its case beyond a reasonable doubt. The exact definition of the phrase "beyond a reasonable doubt" is left up to the fact finder, although some courts will attempt to help clarify its meaning with generalized statements to assist the fact finder. However, courts frown upon attempts to quantify the appropriate burden of proof by use of percentages or other mathematical formulas.

A third burden of proof is recognized in some jurisdictions in certain types of cases. This third burden of proof is called the **clear and convincing evidence** standard. Clear and convincing is considered to be an intermediate standard between the normal civil burden and the criminal burden, requiring the finder of fact to conclude that the party with the burden of proof has proven its case substantially more likely than not. The clear and convincing evidence standard is the burden of proof used in special situations, such as that necessary to establish that a person is incompetent, needs a guardian, or should be committed involuntarily. Some states require clear and convincing evidence to establish civil fraud or to award punitive damages.

clear and convincing evidence

the burden of proof used in certain types of civil cases that requires the finder of fact to conclude that the party with the burden of proof has proven its case substantially more likely than not; it is an intermediate burden of proof that is greater than the normal civil burden but less than the criminal burden

Logan v. Peterson Construction
604 P.2d 488 (1979)
Supreme Court of Utah

WILKINS, Justice:

On July 7, 1976, a fire broke out at the site of plaintiffs' house under construction on Dimple Dell Road in Salt Lake County. At the time of the fire, the construction of the house was approximately 95 percent completed; the fire completely destroyed the structure. Three bales of straw used in the construction were stacked on the outside of the house, placed there by defendant, the general contractor and builder. Defendant testified that on the morning of July 7 he had moved these bales away from the house for the purpose of clearing the area for the pouring of cement; that at that time there was no loose straw, and that the area had been raked before the workmen left the site on July 2, the Friday before the Bicentennial holiday. Other witnesses testified that they had observed loose straw scattered at the southeast corner of the building, partly under the eaves, on July 5, and on July 7, prior to the fire.

Alfred Levin, one of the subcontractors testified that he and his son were at the site installing a shower door on the evening of July 7. When he arrived, he observed the three stacked bales of straw, and some loose straw in the vicinity of the bales. He heard what he thought was firecrackers a short time after his arrival, and a popping noise later, as he was preparing to leave, when the southeast corner, of the house burst into flame. He did not know whether the fire started in the straw. Neither Levin nor his son are smokers.

(continued)

(*continued*)

Batallion Chief Herbert Nichols of the Salt Lake County Fire Department, who investigated the fire, testified that the fire started at the southeast corner of the structure. He said that he could not determine the cause of the fire, but that it probably started from a cigarette in combustible material, such as loose straw. He saw no loose straw and found no evidence of negligence on the part of the workmen, or the contractor. He stated that the fire could have started from a number of causes including matches or firecrackers, and he could not be positive that the straw was the first to ignite. The bales of straw were blackened on two sides, but were not entirely burned as they had been moved away from the fire by the time he arrived.

The Fire Chief and all other witnesses testified that the construction site was clean and reasonably well maintained for a construction site.

Plaintiffs brought action alleging negligence on the part of defendant, contending that the bales and loose straw constituted a nuisance. . .

The District Court entered judgment on the basis of its findings of fact and conclusions of law, which included the following: that defendant was not in exclusive control of the premises, as workmen in control of plaintiff were on the site, and it was open to the public; that . . . that there was no showing of nuisance; that there was no showing that defendant was negligent in his maintenance and care of the construction site, which was clean and orderly; that plaintiffs offered no evidence of the specific causation of the fire; that there had been no showing that the actions and/or omissions of defendant were the proximate cause of the fire . . . .e . . . .

We do not perceive that the District Court's findings are incomplete in any material respect. Why? Because the record supports a basis for the Court's finding that plaintiffs failed in sustaining their burden concerning causation.

CROCKETT, C. J., and MAUGHAN, HALL
and STEWART, JJ., concur.

Case: Logan v. Peterson Construction

Court: Supreme Court of Utah

Summary of Main Points: The plaintiff has the burden of proof to establish that the defendant caused or was legally responsible for the damages. If the plaintiff fails to meet this burden, he/she will lose the case. In this case the defendant (a general contractor who was building plaintiff's house) was not in exclusive control of plaintiff's property at the time a fire broke out. The defendant did not create a nuisance, and plaintiff has offered no proof that the defendant was negligent or otherwise responsible for the fire. Therefore, court ruled in favor of defendant.

The Merits of a Case When a jury has decided a case based upon the facts, we say the case has been "decided upon its merits." In other words, the jury has looked at the evidence and made a determination as to whose version of the events it believes. Should a judge dismiss a case, or grant summary judgement prior to the case reaching the jury, we say the case "never reached its merits."

SIDEBAR

Case Not Reaching Its Merits

Firefighter Jones was killed when the aerial ladder he was climbing collapsed at a training exercise. The aerial was made by the Big Stick Ladder Company. Firefighter Jones's widow sues the fire department and the Big Stick Ladder Company. She alleges the Big Stick Ladder Company manufactured a defective ladder, and the fire department was negligent in maintaining it. The fire department files a motion for summary judgment, claiming it has immunity from liability. During the discovery phase of the case, the trial court rules in favor of the fire department on the motion for summary judgment, essentially dismissing the case against the fire department. In this case, we would say that the suit against the fire department was dismissed without having reached the merits of the case. The jury never reached the question of whether the fire department was negligent.

SIDEBAR

Jury Trials

The jury system in the United States has been called the greatest vehicle ever invented for getting to the truth. No doubt our jury system is a tremendously valuable tool in many respects. However, as most attorneys and judges will agree, putting any case to a jury is a risky proposition. Is that a bad thing? Should a system be devised that will make a jury's verdict more predictable? If, through discovery, both sides know what the evidence will show, but they also know that letting a case go to the jury is unpredictable, are the parties more likely or less likely to settle? The unpredictability of a jury can exert a certain amount of pressure on both parties to compromise their positions and encourage them to reach a settlement. Many legal scholars believe the unpredictability of a jury is not necessarily a bad thing, since it can serve to encourage parties who may otherwise be set in their positions to reach a compromise to settle their case.

Juries Juries have traditionally consisted of 12 jurors who must reach a unanimous verdict in a case. Some jurisdictions now allow for juries of less than 12 in both civil and criminal cases, with six or eight persons being the most common. The rules and the constitutional considerations for the size and the unanimity of verdicts are different for civil and criminal cases. While most jurisdictions still require unanimous verdicts in criminal cases, some states allow civil juries to decide cases on less than unanimous verdicts.

To Jury or Not to Jury State constitutions generally provide that all persons have a right to a jury trial in most civil matters and in serious criminal matters. There is no Federal constitutional right to a jury trial for minor criminal matters that are considered to be petty offenses. In Federal court, the right to a jury trial in a civil case attaches if the amount in controversy exceeds $20. In cases where there are factual issues to be decided, and a jury is not requested or allowed, the judge will fill the role of the jury and will decide questions of fact.

Are there reasons why parties to a lawsuit may not want a jury trial, even though they have a right to one? There are several reasons that parties may choose to forgo their right to a jury trial and leave the case in the hands of a judge. One common reason occurs in a criminal case where the defendant has a serious criminal record and has a concern that the jurors may be so outraged by his prior criminal record that they won't be objective about the facts in the case before them. Another situation where the right to a jury trial is typically waived is a case that is very complex, such as in cases involving patents, chemical formulas, and complex business transactions.

Appeals

PLEADINGS ⟶ DISCOVERY ⟶ TRIAL ⟶ *APPEAL*

Critical to the judicial system is the appellate process. The purpose of the appellate process is to ensure that the procedure followed at trial, and the rulings of law made by the trial judge, follow established and fair legal procedures. The appellate process is not an opportunity to have a jury's decision overruled. In fact, there is no appeal of a decision of fact made by a jury. Only questions of law can be appealed. For example, if a jury decides in a given case that the light was red when the defendant drove through it, then the light was red, and that decision is not subject to appeal. The only exception to this is where there was no evidence presented at trial from which a reasonable jury could conclude that the light was red. This is a very, very limited exception.

Because the focus of appeals is on questions of law, there is no need for a jury on appeals. Typical appellate-level courts consist of between three and nine judges. The appellate court will accept written briefs from both sides that state their relative positions and cite supporting case law. Where appropriate, appellate courts will allow each side to make oral arguments before the judges. The court will then issue a written ruling that becomes binding precedent in that jurisdiction for future cases. Appellate decisions are published in books, and most are now available online for review. Collectively, these written decisions constitute case law, which, as we learned in Chapter 1, is a very important source of laws in the United States.

FEDERAL COURT SYSTEM

As we learned in Chapter 1, the Federal government is one of limited jurisdiction. The jurisdiction of Federal courts is similarly limited to "Federal" matters. In other words, there must be a Federal issue involved for the case to be brought in Federal court. One exception to this rule is that lawsuits between individuals who are residents of different states can be brought in Federal court. However, barring such an exception, cases that lack a Federal issue cannot be brought in Federal court, but may be brought in state court. Typical Federal jurisdiction cases involve discrimination; civil rights; fair labor standards; patent, trademark and copyright infringement; immigration; Social Security; and Federal criminal cases.

The Federal court system includes 94 Federal district courts. Decisions of the Federal District Court can be appealed to one of the 12 Circuit Courts of Appeal. Final appeal in the Federal system is to the United States Supreme Court **(Figures 2-6 and 2-7)**.

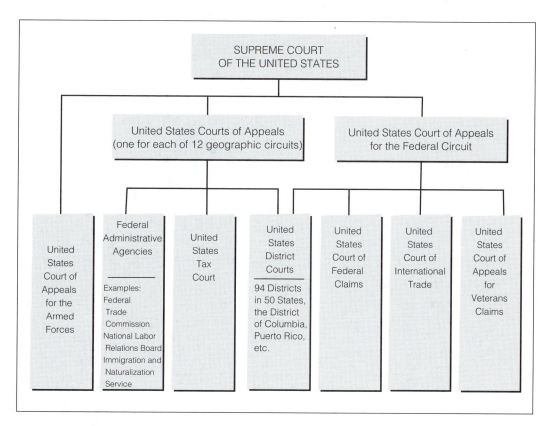

Figure 2-6   *Organization of the Federal court system.*

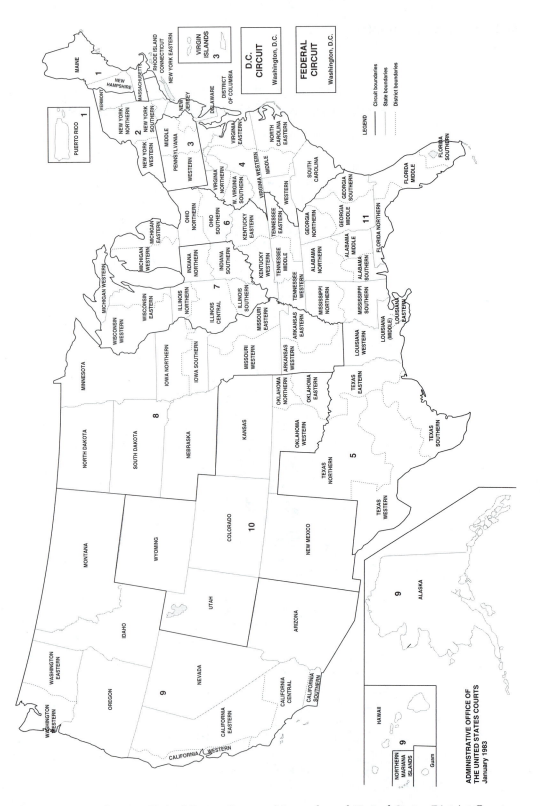

Figure 2-7 *Map showing United States Courts of Appeals and United States District Courts.*

There are also various other Federal courts, such as Federal Bankruptcy Court, Federal Tax Court, the U.S. Court of Claims, and the Court of International Trade, each of which serves specific, limited judicial functions.

STATE COURTS

All states have trial courts, but they may have different names. Some states call their trial courts superior courts, while some call them district courts, or courts of common pleas. To add to the confusion, in New York State the trial court is called the Supreme Court. Compounding that confusion is that some states have two levels of trial courts, one for minor matters (district court) and another for more serious matters (superior court).

Unlike the Federal District Courts, state trial courts are typically considered to be courts of general jurisdiction. All matters that are not otherwise limited to specific courts can be brought in state trial courts. Most states have created a series of specialized courts to handle specific types of matters, such as divorce, workers' compensation, and probate. Such specialized courts allow the judges to become expert in these subject areas, and in turn take a burden off the main trial courts.

Each state also has appellate courts, which again vary in name. Appellate courts are known by a variety of names in different jurisdictions, including courts of appeal, supreme court, or supreme judicial court. Some states have one level of appeals, and some, like the Federal court system, allow for two levels of appeal **(Figure 2-8)**.

Regardless of the name, a state's highest appellate court is the supreme authority on matters of state law. Even the United States Supreme Court cannot overturn a state supreme court decision that is based on state law, unless it violates the United States Constitution.

stare decisis
the principle that once a court establishes a legal principle or interpretation, other courts at the same or lower levels in the same jurisdiction must apply that decision in future cases on the same facts; stare decisis is Latin for "let the decision stand"

STARE DECISIS AND PRECEDENCE

Stare decisis is a Latin term for "let the decision stand." It refers to the principle that once a court establishes a legal principle or interpretation, other courts at the same or lower level in the same jurisdiction must apply that principle or interpretation in future cases on substantially the same facts. In other words, when deciding a case, a judge must follow and apply the prior decisions and legal interpretations of courts in that jurisdiction, rather than establish new legal principles or new interpretations.

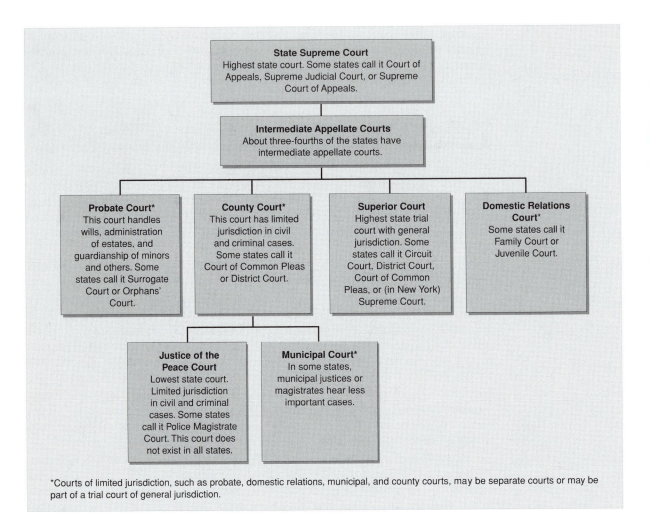

Figure 2-8   *State court system. (Reproduced with permission of West Group.)*

precedent
an important legal principle that should be followed; precedent is established by court decisions that have been decided in the past in similar cases in the same jurisdiction

The policy reason behind the principle of stare decisis is that the average person needs to be able to rely upon the settled law on a given topic with some degree of certainty, without having to guess what a court might rule the next time the same issue comes up. Adhering to stare decisis allows court decisions to set **precedent** on a given matter.

⚖ EXAMPLE

In the case of *Michigan v. Tyler,* decided in 1978, the United States Supreme Court ruled that under the United States Constitution, firefighters and investigators who enter a building that is on fire can lawfully seize evidence that is in plain view, despite not having obtained a search warrant.

In a subsequent case from a different state, a defendant is charged with arson. He seeks to have evidence obtained by firefighters at the fire scene excluded from trial because the firefighters did not obtain a search warrant. The defendant alleges that the United States Constitution requires that a search warrant be obtained prior to firefighters being able to seize any evidence at a fire scene.

According to stare decisis, the judge in the subsequent case should apply the precedent established in *Michigan v. Tyler,* and rule that the evidence is admissible. We will discuss *Michigan v. Tyler* extensively in Chapter 6.

Stare decisis is applicable only for courts in the same jurisdiction that are at the same or lower rank as the court rendering the original decision. In other words, if a state supreme court renders an interpretation of a state law, all inferior state courts must also follow that decision, as must the supreme court in the future. However, the state supreme court would not be bound by stare decisis to follow a decision of an inferior state court, such as a superior or district court.

In addition, the United States Supreme Court is the final interpreter of the United States Constitution. Thus all courts must apply stare decisis to decisions of the Supreme Court on matters related to the United States Constitution.

HOW "DECISIS" IS "DECISIS"?

In many jurisdictions, the principle of stare decisis does not bind one trial court to follow the precedent set by another trial court handing a similar case. In such jurisdictions, trial courts are only bound by the precedent set by an appellate court.

In addition, while an appellate court is theoretically bound by its own prior decisions, courts do from time to time revisit their own precedents. Courts will usually try to distinguish the case they are deciding from the case precedent, based upon the facts, before they deviate from the precedent. However, the reality is that courts do not always adhere to precedent and will establish new precedent when circumstances warrant.

Figure 2-9 *The media have a difficult job providing a complete and balanced perspective while ensuring that the stories are interesting to the public and meet deadlines.*

Courts in one state are not bound to comply with decisions from courts in other states. However, attorneys advocating a case, and courts deciding a case, will frequently cite out-of-state cases, finding that the reasoning used in such cases is persuasive, but not binding.

LAWSUITS AND THE MEDIA

Have you ever been to a fire, then read about it in the newspaper the next day, and asked yourself "Are they talking about the same fire that I went to?" The news media has a tough job. Journalists have to condense a lot of facts into a short, readable article. Often, the writers and editors are operating under time constraints that prioritize a story that is 90 percent correct but meets the deadline over a story that is 100 percent correct but misses the deadline. We see this time and time again when the media covers the emergencies that we respond to.

When covering lawsuits, the media may also provide less than a complete and balanced perspective **(Figure 2-9)**. When covering legal cases, there are factors beyond deadlines to contend with. Perhaps it is the media's propensity to sensationalize stories, perhaps it is an attempt to exploit a sociological bias against lawyers and lawsuits, or perhaps it's the technical nature of our legal system, but far too often the outrageous headlines we see in the media in regard to lawsuits simply do not do justice to the truth about what a court's or jury's decision was.

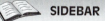 **SIDEBAR**

A woman is injured when a fire truck responding to a fire collides with her vehicle. Let's assume it is unclear who was at fault for the accident. The woman sues the city, the driver of the fire truck, and the company officer.

The city is sued because it owns the truck, it employs the driver, and therefore would be liable under the doctrine of respondeat superior (to be discussed in Chapter 8), if the driver was liable. The driver is sued because the injured person claims he was negligent in driving the apparatus. The officer is sued for negligently failing to properly supervise the driver.

During the pleadings phase, the officer's attorney files a motion for summary judgment that requests that the case against the officer be dismissed, alleging that the accident happened so fast that he could not possibly have done anything different to have prevented the accident. The trial judge considers the motion, and grants the motion for summary judgment in the officer's favor. If this case was reported in the media, what would the headlines say? The headlines would probably something like "Officer Not Liable for Accident Victim's Injuries."

Now let's say the plaintiff's attorneys appeal the trial judge's ruling, arguing that a legitimate question of fact exists for the jury to decide, namely whether the officer had time to tell the driver to slow down or do something different. The issue could also be raised that this driver has a history of driving recklessly, and the officer was aware of this history and should have taken steps to prevent another accident. After considering the issues, the appeals court rules that the case should go back to the trial court and that the jury should decide if the officer was negligent or not, based upon the facts. Now what would the headlines say? "Appellate Court Rules Company Officer Liable for Accident Victim's Injuries."

Of course, that is not what the appellate court really said. The court said the officer "could" be liable, if the facts at trial indicate he had the opportunity to do something and he negligently failed to do so. On the other hand, the headline is not entirely false, either. The newspaper editor's job is to create headlines that will attract readers, hence the dilemma.

It is important for firefighters to understand this dilemma, and how it affects the accuracy of articles and media stories about lawsuits. The reality is that after our formal education is over, a great deal of what we learn comes from the media. A healthy dose of skepticism, an appreciation of the journalistic process, and an understanding of the court system can go a long way toward ensuring that media stories about lawsuits are kept in perspective. A wise old attorney once said that the study of law is just an advanced course in common sense. If you hear about something in the law that doesn't make sense to you, then you don't have all the information. Everything in the law does make sense once you have all the information. This wise old attorney, John A. Varone, was a second-generation firefighter himself.

LEGAL RESEARCH

The ability to conduct legal research is an easily learned skill that can come in handy for any fire service professional. The sources of material from which one can conduct legal research are changing dramatically with the availability of online research tools. Prior to the Internet, lawyers had to be able to navigate through large volumes of books in order to conduct research. While the ability to find cases and laws in books remains an important skill, it is quickly becoming supplanted by online and CD- or DVD-based legal research tools **(Figure 2-10)**.

There are a variety of sources of materials from which to conduct legal research.

- *Case books* **(Figure 2-11)** are books that contain the actual text of the written decisions by judges in deciding questions of law. Most case books contain decisions by appellate courts, but some contain written

Figure 2-10
Traditional legal research requires a law library.

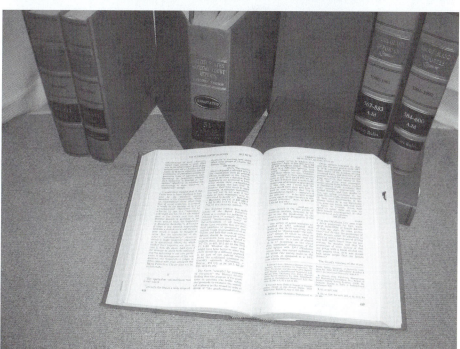

Figure 2-11 *Case books contain the actual text of decisions written by judges.*

bill
a proposed law submitted before a legislative body, such as Congress or a state legislature

codified
said of laws that have been passed by the legislative branch into comprehensive codes

search and seizure
the area of the law that addresses the rights and responsibilities of law enforcement authorities and governmental actors to conduct searches for evidence, and the seizure of any evidence that is found

decisions of trial court judges ruling on matters of law. Case books are often referred to as reporters. Some are published by the jurisdictions themselves, and some are published by major legal publishing companies, such as Thomson West.

- *Statute books* **(Figure 2-12)** are books that contain the text of statutes that are passed by legislative bodies. When a law is passed by the legislature, it has an individual **bill** number, as well as public law number. (See Chapter 1.) These laws then must be **codified** and published in statute books. In other words, they are assigned a title and chapter number, and formally published. Like case books, some statute books are published by the jurisdictions themselves, and some are published by major legal publishers.

- *Digests* **(Figure 2-13)** are books that contain summaries of cases. They are organized by topic, so that you can look under a subject such as **search and seizure** and find cases that address the issues you are researching.

- *Legal encyclopedias* **(Figure 2-14)** are generalized books that explain the law on various subjects. Like general encyclopedias, they are commonly organized by topic in alphabetical order.

Figure 2-12 *Statute books contain codified versions of the laws passed by the legislature.*

Figure 2-13 *Digests contain summaries of cases, organized by topic.*

Figure 2-14 *Legal encyclopedia. (Reproduced with permission of West Group.)*

- *Practice books* **(Figure 2-15)** are books that are intended to help a legal practitioner advise and represent a client. Practice books focus on certain areas of the law. They often include a synopsis of the law, legal checklists, forms, and other documents that would assist an attorney in handling a case.

- *Online resources* **(Figure 2-16)** and electronic databases utilizing CDs and DVDs **(Figure 2-17)** electronically duplicate the various text-based research tools discussed previously, plus add the search power associated with computers. Online legal resources include paid Web sites and free Web sites.

  - Paid Web sites contain powerful search engines capable of rapidly searching cases, statutes, constitutions, regulations, and articles in professional journals. These sites allow researchers great flexibility to limit their searches to certain courts and jurisdictions, or to expand their searches to the entire nation—and perhaps beyond—for cases, articles, and materials on the topic they are searching for. However, some online legal resources are quite expensive.

  - Most states and many universities have free Web sites that contain statutes, regulations, and some case law. However, the free

Figure 2-15 *Practice books are intended to help an attorney advise and represent a client.*

Figure 2-16 *Westlaw, a paid Web site for legal research. (Reproduced with permission of West Group.)*

sites do not have the quality search capability that the paid sites have. Nevertheless, free Web sites can be an important tool for researching law economically.

- CD- and DVD-based programs are sold by some legal publishing companies and may contain cases, statutes, and other legal resources.

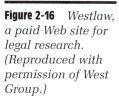

Figure 2-17 *West CD-ROM libraries.*

LEGAL ARGUMENTS

When reading case law, one of the first things that stands out to nonlawyers is the way that lawyers make legal arguments. Lawyers will rarely argue with each other over facts or over the same legal principle. On the contrary, lawyers try to characterize an issue in a case in such a way that a common-sense answer will favor their case. This is often called **framing the issue**. Lawyers try to frame the issue in such a way that their clients will prevail. As a result, the battle in a lawsuit is frequently over how to frame the issue, more so than both sides arguing different conclusions from the same facts.

framing the issue
characterizing an issue in a case in such a way that a commonsense answer will favor a particular perspective

---

SIDEBAR

How to Frame the Issue

A firefighters' union sues the city over what it perceives as unsafe staffing levels that have resulted in repeated OSHA violations. The union sees the issue as one of safety for the personnel and the citizens. The city sees the issue as one of who should make policy decisions about how taxpayers' dollars should be allocated and spent, and whether a court is the proper forum to make these kinds of decisions.

Which issue will the court see as the "real" issue? Can the city prevail on the firefighter safety issue? Should the city's attorneys even argue the point, or should they instead focus their attention on the issue they have framed? Can the union prevail on the public policy issue, or should it focus on the issue of firefighter and citizen safety? It becomes the court's role to decide whose issue trumps whose. Both issues are equally meritorious, but only one side can prevail.

SUMMARY

The judicial system in the United States is made up of courts. Courts exist at the Federal, state, and local levels. Some courts have general jurisdiction and some have specific jurisdiction. Some courts are trial courts while others are appellate courts. There are four distinct phases of a civil lawsuit: pleadings, discovery, trial, and appeal. At trial, judges make decisions on questions of law, and juries make decisions on questions of fact. Only questions of law can be appealed. Appellate courts establish precedent when deciding cases that come before them by virtue of stare decisis.

REVIEW QUESTIONS

1. Explain the basis for the jurisdiction of a court.

2. Explain the difference between direct evidence, real evidence, and circumstantial evidence.

3. Describe the four phases of civil lawsuit in the order in which they will occur.

4. Identify four common methods used to conduct discovery.

5. Explain the difference between the roles of a judge and a jury during a trial.

6. Explain why someone might not want a jury trial even though they are entitled to one.

7. What are the three burdens of proof and when is each used?

8. Appeals are limited to what types of questions?

9. Explain the meaning of stare decisis and how it affects other courts in the same jurisdiction, as well as outside of that jurisdiction.

10. Explain the difference between a case book and a statute book.

DISCUSSION QUESTIONS

1. Why is framing the issue so important to attorneys? In what other areas beside law do you see people framing issues?

2. Should a person who is dissatisfied with a jury verdict on a question of fact be allowed to appeal that decision? What are the consequences to our legal system if appeals over questions of fact were allowed?

Chapter

3

TYPES OF FIRE DEPARTMENTS

Learning Objectives

Upon completion of this chapter, you should be able to:

- Distinguish between fire departments based upon the type of entity, type of fire department organization, the funding source, and the employment status of firefighters.
- Distinguish between public sector and private sector entities.
- Identify the four types of corporations, and how each is created.
- Identify the reason that the use of an association creates unnecessary risks for firefighters.
- Distinguish between municipal fire departments, county fire departments, regional fire departments, fire districts, volunteer fire companies, industrial fire departments, and fire brigades.
- Explain the importance of agreements between a volunteer fire company and the jurisdictions they protect.

INTRODUCTION

The mood in the meeting room was tense. An older firefighter stood up and asked the question everyone else was thinking: "How can the town council take away the money for our new truck? They promised us that money last year."

"Yeah, and they are even forcing Meadowlands Fire Department to lay off two of their full-time personnel," added another firefighter.

"Why are we being cut and yet the Highlands Fire District isn't having any problems with funding? They are in the town. They're talking about hiring more personnel. Why doesn't the council cut their budget?"

The chief answers, "Well, the Highlands Fire District is their own fire district."

"Fire company, fire department, fire district, why does what you call yourself have anything to do with it?"

WHAT IS A FIRE DEPARTMENT?

What is a fire department? The question seems so simple that one expects there would be a short, simple answer. Unfortunately there is no short, simple answer **(Figure 3-1)**.

Figure 3-1 *What is a fire department? Is a fire station, or even a fire truck, required?*

Figure 3-2 *Some fire organizations are relatively new, the result of the consolidation of fire protection services.*

There are a variety of organizations and entities that provide fire protection services in the United States. Some of these entities are governmental, and some are privately owned. Some are over 200 years old, and actually predate the existence of the local municipal government in their jurisdiction. Other fire organizations are relatively new, the result of the consolidation of fire protection services **(Figure 3-2)**.

Compounding our difficulties in defining a fire department is the fact that previous attempts to categorize the various types of fire departments have assigned conflicting and at times contradictory names and definitions to many of these entities. Authors have sometimes merely described local fire service entities they were familiar with, and assumed that the rest of the country operated the same way. As a result, it is hard to find two or more books that list the same types of fire departments. Some authors have confused the

- type of entity
- type of fire department organization
- source of funding from which the department sustains itself
- employment status of personnel

Table 3-1 outlines the ways to distinguish fire service entities.

Table 3-1   *Ways of distinguishing fire service entities.*

Types of Entities	Types of Fire Department Organizations	Sources of Revenue	Types of Personnel
State or Federal Government	Agency of municipality or county fire district regional fire department	Tax from Federal or state government	Career
Municipal corporation		Tax from municipal government	Part-time (call)
Quasi-municipal corporation		Tax from county	Volunteer
For-profit corporation	State or Federal fire department	Ability to tax (fire tax)	Combination
Nonprofit corporation	Quasi-governmental authority	Agreement with local community	
Association	Volunteer fire company	Fundraising	
	Industrial fire department	Subscription	
	Fire brigade	Business enterprise	

THE ROLE OF A FIRE DEPARTMENT

Fire departments have taken on an expanded role in society compared with 50 years ago, including providing hazardous materials response, emergency medical service, and numerous technical rescue disciplines **(Figure 3-3)**. While clearly a fire department that also provides technical rescue services remains a fire department, does a volunteer organization with no engine

Figure 3-3   *The services provided by fire departments include numerous technical rescue disciplines.*

Figure 3-4 *While a fire department that provides hazardous-materials response is still a fire department, what about an organization that responds only to hazmat incidents?*

company, no truck or ladder company, no hose, no water, no pumps, just a service truck that carries technical rescue equipment qualify as a volunteer fire company under state law because it responds to fires? What about an organization that responds only to hazardous materials incidents **(Figure 3-4)**, dive rescue incidents, or emergency medical incidents? Does the fact that a volunteer fire company runs a bingo fundraiser on a monthly basis to subsidize its operations change its status? What if the volunteer fire company holds a nightly bingo game generating $1.5 million a year in income, but has stopped responding to alarms because the local municipality has hired full-time firefighters?

Understanding the type of entity that delivers fire protection services is important, because it impacts the legal authority of the department, the laws that apply to the department, the department's ability to provide services, the liability of the department, and the very important issue of immunity from lawsuits.

PUBLIC SECTOR VERSUS PRIVATE SECTOR

Initially, it is critical for us to understand the difference between public sector and private sector entities. The distinction between public sector entities and private sector entities triggers different treatment under a variety of laws, including workers' compensation, wage and hour, civil liability, sovereign immunity, Occupational Safety and Health Administration (OSHA), collective bargaining, and the Internal Revenue Code.

Public sector entities are agencies and instrumentalities of the Federal, state, or local government. Public sector entities include the Federal government, Federal agencies, states, political subdivisions of states, cities and towns, special districts, and public authorities.

Private sector entities, on the other hand, are privately owned, operated, and managed. "Privately owned" in this case does not mean that just one or two people own the company. Rather, privately owned refers to the fact that the owners, shareholders, or—in the case of a nonprofit—their members, are individuals or other private companies, not a part of government.

The distinction between public and private sector entities becomes blurred, particularly when it comes to volunteer fire companies. However, understanding the distinction between the public sector and the private sector is absolutely essential to understanding the types of fire departments and the laws that impact fire departments.

TYPES OF ENTITIES

When a group of people come together for a common purpose—whether to manufacture a product for profit, to organize a fundraising effort for a charity, or to create a fire department—two primary entities can be created: a corporation or an association. A variety of other possible business entities exist, such as partnerships, limited partnerships, and limited-liability companies, but these entities are not used to provide fire protection services.

Corporations

A **corporation** is a legally created entity that is accepted, approved, and recognized by a state through a formalized process. Once created, corporations have the legal capacity to enter into contracts, and can sue or be sued. Corporations can be one of four types: municipal, quasi-municipal, for-profit, and not-for-profit.

1. **Municipal corporations** are corporations created by the state legislature, usually at the request of the inhabitants, or at least with the approval of the inhabitants through a referendum vote. Municipal corporations are created in an effort to govern or address local needs. Municipal corporations are typically issued a charter by the state legislature, and have authority for self-government subordinate to the state. Typical examples of municipal corporations are cities and towns. Municipal corporations are always considered to be part of the public sector.

---

public sector
entities that are agencies and instrumentalities of the Federal, state, or local government

private sector
entities that are privately owned, operated, and managed

corporation
a legally created entity that is accepted, approved, and recognized by a state through a formalized process

municipal corporation
a corporation that is created by the state legislature, usually at the request of the inhabitants, or at least with the approval of the inhabitants through a referendum vote; municipal corporations are usually characterized by having inherent powers such as self-government, law making, and the ability to tax

quasi-municipal corporation

a corporation that is created by the state legislature, but not always at the request of the local inhabitants to aid the state in its administrative functions, as opposed to serving the needs of the inhabitants for governance; quasi-municipal corporations have some but not all of the powers and authority of municipal corporations

2. **Quasi-municipal corporations** are, similarly, corporations created by the state legislature, but not always at the request of the local inhabitants. Not all states recognize quasi-municipal corporations, and those that do differ on the exact definition. Quasi-municipal corporations may also be referred to as quasi-public corporations. In some states, the purpose of quasi-municipal corporations is to aid the state in its administrative functions, as opposed to serving the needs of the inhabitants for governance. In other states, quasi-municipal corporations have some of the powers and authority of municipal corporations, but are not municipal corporations per se.

Many books on fire service law apply different definitions to quasi-municipal and municipal corporations. Authoritative sources also differ in their listing of common examples of quasi-municipal corporations, probably as a result of local differences under state law.

Municipal corporations are usually characterized by having inherent powers such as self-government, law making, and the ability to tax. A good rule of thumb to remember is that typical municipalities—cities and towns, whose purpose is to provide self-governance for their populations—are usually municipal corporations; and special entities, such as transit authorities and school districts, are often classified as quasi-municipalities.

Quasi-municipal corporations are often treated differently under state laws than municipal corporations, particularly in areas related to inherent authority, liability, and sovereign immunity. Municipal corporations are generally less subject to state control and regulation than quasi-municipal corporations. Both municipal and quasi-municipal corporations are considered part of the public sector.

The distinction between municipal and quasi-municipal corporations is important when issues of legal authority and liability are raised under statutes that give authority or immunity to one but not the other. As if all this were not confusing enough, some states classify fire districts (discussed below) as quasi-municipal corporations, while other states consider fire districts to be municipal corporations, because they have the power to tax and their officials are elected.

for-profit corporation

a corporation that is owned by stockholders, whose goal is to make money for the stockholders

3. **For-profit corporations** are corporations owned by stockholders. The goal of a for-profit corporation is to make money for the stockholders. The stockholders elect a board of directors to oversee management of the corporation. Corporate officers (president, vice-president, secretary, and treasurer) are employed to run the day-to-day operations. In smaller corporations, the stockholders themselves generally serve as the board of directors and officers. In larger corporations, business professionals are generally hired to fill these positions. For-profit corporations, also known as business corporations, are part of the private sector.

nonprofit corporation
a corporation formed for some charitable, benevolent, or other purpose not designed to make profits that will benefit its owners, directors, or officers; also called a not-for-profit corporation

articles of incorporation
the organizing documents for a corporation, which when duly filed with the secretary of state's office, create the corporation; once accepted by the state, the articles of incorporation become the equivalent of a charter or constitution for the corporation

4. **Nonprofit corporations** are corporations formed for a charitable, benevolent, or some other purpose that will not make profits that benefit its organizers or members. Nonprofit corporations have no stockholders, but may have "members." However, members are prohibited from sharing in any of the proceeds or profits of the corporation. Nonprofit corporations are not prohibited from paying their employees, directors, and officers. Rather, the corporation is required to utilize its net profits to benefit the charitable or nonprofit cause, rather than give dividends, bonuses, or property to its members. The rights and duties of a member, as well as limitations upon who may be a member, will vary from corporation to corporation. As a general rule, the board of directors of a nonprofit manages the corporation, and in many cases the members elect the board of directors. Nonprofit corporations, even large benevolent corporations such as the American Red Cross or the Salvation Army, are generally considered to be part of the private sector.

While municipal and quasi-municipal corporations are created by an act of the state legislature, for-profit and not-for-profit corporations are created by individuals pursuant to procedures that are set by state law and are overseen by a state agency, commonly the secretary of state **(Figure 3-5)**. In order to create a for-profit or not-for-profit corporation, the individuals who desire to create the corporation must file a document called the **articles of incorporation** with the state, along with appropriate fees and supporting documentation.

Figure 3-5 *Both for-profit and nonprofit corporations are overseen by a state agency, commonly the secretary of state's office.*

Table 3-2   *Types of corporations.*

Entity	Created by	Purpose
Municipal corporation	State legislature	Self-governance by local inhabitants
Quasi-municipal corporation	State legislature	Assist state with some administrative functions
For-profit corporation	Individuals, approved by state	Make money for stockholders
Not-for-profit corporation	Individuals, approved by state	Charitable, benevolent, or nonprofit purposes

Once accepted by the state, the articles of incorporation become the equivalent of a charter or constitution for the corporation, and the corporation comes into legal existence. The articles of incorporation can be amended only by filing an amendment to the articles of incorporation with the state.

Corporations also have the authority to create bylaws, which are basically the internal rules by which the corporation is run. Bylaws must be consistent with the articles of incorporation, but can be amended more easily.

Table 3-2 provides a quick reference for the four types of corporations.

Corporate Existence   The formation of a corporation creates a new, legally recognized entity separate and apart from those who created the corporation. A corporation is the functional equivalent of a person: it can open a bank account, hold title to land, sue and be sued, hire and fire employees, and file and pay income taxes, all independent of the persons who organize, own, or manage the corporation. In addition, the debts and obligations of the corporation belong to the corporation, not to the directors, officers, shareholders, or individual members. At most, a shareholder can lose his or her investment in the corporation, but cannot be held liable for the debts of the corporation.

Provided a corporation adheres to state laws, it can continue to exist perpetually. Most states require all corporations to file an annual report. The annual report provides updated information about the corporation, and may require the payment of an annual fee. Failure to file an annual report can result in a corporation's charter being revoked. A corporation's charter can also be revoked for misconduct or illegal activities.

If the state revokes the corporation's charter, the corporation's existence ceases. Any further activities or operations by owners, officers, directors, members, or employees would be considered to be in the nature of an unincorporated association, discussed below.

association
a group of individuals who act in furtherance of a specific purpose without incorporating; associations may also be referred to as unincorporated associations

Associations

Whenever a group of individuals choose to act in furtherance of a specific purpose without incorporating, they have formed an **association**. Such associations are also referred to as unincorporated associations. Associations can be informal or formal.

Associations become formalized when their members adopt formal rules, regulations, or bylaws. Formal associations may also adopt articles of association that serve as the equivalent of a charter. The association may elect or appoint officers, hold formal meetings, and even transact business according to Atwood's Rules for Meetings or Robert's Rules of Order. Informal associations have no such regulations, and exist merely in the minds and actions of their members.

SIDEBAR

The Problem with Unincorporated Associations

You are driving down the street past a playground, and you see a group of youngsters playing baseball. One of the players hits a ball out of the playground and it strikes your windshield, breaking it. The youngsters scatter. Who is responsible for the damage to your windshield? The youngsters are analogous to an informal association. It is not clear who is part of the association, and who is not. They are engaged in a joint enterprise, in this case to play baseball. There is no legally recognized entity that organized the youngsters, and from whom damages could be sought. The exact scope of the joint enterprise in not clear. Was this a one-time ball game, or was it part of an ongoing competition among the youngsters? Does the same group play baseball in the summer and football in the winter?

---

joint enterprise
individuals acting under an express or implied agreement with a common purpose

When individuals engage in a **joint enterprise**, all members may be held liable for the acts of any one member of the enterprise, providing the responsible member was acting within the scope of the enterprise. In addition, any member may bind the joint enterprise to a contract. In other words, when one of the members in a joint enterprise creates a debt for the enterprise, it is possible that all of the members may be held liable.

Unincorporated associations can be construed as joint enterprises. The legal status of unincorporated associations varies from state to state. Some states have specific laws that provide a degree of liability protection upon the officers and members of associations for the debts of the association. All states have mechanisms whereby an unincorporated association can be sued without having to sue all of the members individually. However, the risk of doing business as an unincorporated association is that a court will deem the association to be in the nature of joint enterprise among the individuals who make up the association. If this is the case, the individual associates can be jointly liable for the debts of the association.

SIDEBAR

Twenty firefighters decide to form a motorcycle club called the Hot Dawgs. They each come up with $500 to start the club. They sign a five-year lease agreement on a hall to use as their club headquarters. They begin renovating the inside of the hall. One of the club members signs an agreement on behalf of the club with a liquor distributor to provide a beer tap. Another member signs an agreement on behalf of the club to lease a commercial refrigerator. Yet another member signs a lease for some furniture, a photocopy machine, and some office equipment. The club is in business for a month when one member gets involved in a vehicle accident while running an errand for the club. The accident was clearly the member's fault. A person injured in the accident decides to sue to club. Soon the club runs out of money, and the owner of the hall, as well as the vendors who have contracts and leases with the club, demand to be paid.

If the club was organized as an unincorporated association, the members' activities could be viewed as a joint enterprise. The exact result would vary somewhat from state to state, based on statutory differences. Generally, in the first instance, any debts that the club owed would have to be satisfied out of the association's funds. This could be accomplished by suing one or more of the association officers.

However, if the association's debts exceeded its assets, it would be possible for a vendor to go after the individual association members to recoup the balance owed. Some states would allow the vendors to go after the individual association members in the first instance.

The major point to take away from the preceding sidebar is that if the club was set up as a corporation, and the bills were lawfully authorized by corporate officers, then the individual club members would be liable only to the extent of their $500 investment in the club. Probably the most important benefit of doing business as a corporation is the liability protection afforded to the shareholders, members, and officers of a corporation.

Firefighters need to consider the risk of liability exposure to themselves personally if they are involved with:

- volunteer fire companies operating as unincorporated associations
- volunteer fire companies who were incorporated, but whose corporate existence has lapsed or been revoked
- organizing seminars or training programs independently from their department
- organizing fundraisers or community events independently from their department

Corporate Name

A corporation must adopt a name that includes one of several specific terms designed to distinguish itself from other types of legal entities. The terms include: "Incorporated" or "Inc."; "Company" or "Co."; and "Limited" or "LTD." Use of these terms is necessary in order to "put the world on notice" that the entity is a corporation, not an association or sole proprietorship.

Many fire service organizations choose to incorporate themselves, but use a name that includes the term "association" in their name, such as the "Smithville Volunteer Firefighters Association, Inc." Use of the term "association" in the name does not change the basic organizational structure of an entity as a corporation. Many volunteer fire companies, firefighter unions, and fire chiefs' organizations refer to their corporate organization as an association.

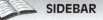 **SIDEBAR**

Why Incorporate?

What advantage does someone gain by incorporating? The Hot Dawgs case is one example. In a business sense, incorporating is a smart move because it creates a new entity separate and apart from the people who invest in or own the business. It creates a layer of protection for those involved in the corporation. Following is another example:

Pete Smith has been a fire department paramedic for 20 years and wants to open his own ambulance company. He owns a home, and after he invests $75,000 in the ambulance business, he has $100,000 in the bank. If he operates the business as an unincorporated sole proprietorship, and one of the ambulances gets in an accident causing damages that are in excess of his insurance coverage, the injured parties can sue Pete and potentially take his house and $100,000, as well as any money that is left in the business.

On the other hand, if Pete organizes the business as a corporation, and one of his ambulances gets in that same accident, the injured parties can only get what money is left in the business. In other words, Pete could lose his $75,000 investment, but the injured parties could not take his house or bank accounts.

PUBLIC SECTOR FIRE ORGANIZATIONS

Municipal Fire Departments

A municipal fire department is an agency of a municipality **(Figure 3-6)**. The municipality may be a city, town, township, or some similar political subdivision of the state. The fire department could be fully career, fully volunteer,

Figure 3-6
Municipal fire departments are agencies of local government.

part-time paid, or a combination department. The name of the department is not a major factor. The critical determining factor is whether the fire department is an agency of the municipality or not. Usually the charter or municipal ordinances will establish the fire department as an agency of the municipality.

A municipal fire department is responsible to the municipal executive. Its funding comes directly from the municipal budget, without passing through the bank accounts of another entity. The fire chief is appointed by, and answers to, the municipal executive, a fire or public safety commissioner appointed by the municipal executive, or a public safety board appointed by local government. Individual firefighters are employed by, or volunteer directly for, the municipality.

County Fire Departments

In some parts of the United States, responsibility to provide fire protection services falls on county government. In such jurisdictions, the county fire department may be an agency of the county, in the same way a municipal fire department is an agency of the municipality. County fire departments may employ career, part-time, or volunteer firefighters, or a combination of the three.

County fire departments may also be organized as fire districts, discussed below. In addition, there are a variety of jurisdictions in the United States where fire protection is organized on a county level, but the service is actually provided by a network of individual volunteer fire companies operating within the county. These fire companies may be funded by, and be under varying degrees of control of, the county fire authority. The degree of control and the amount of funding that the county provides may vary from jurisdiction

Figure 3-7 *In Catlett, Virginia on September 28, 1989, a fire truck responding to a fire was struck by an Amtrak train, resulting in numerous deaths and injuries. (Photo courtesy of* The Fauquier Citizen.*)*

to jurisdiction, and from time to time, as various laws and ordinances are passed that impact the county's role as fire service organizer. However, when the actual provider of the emergency service is a private volunteer fire company, it is not the county that is providing the service. We will discuss this issue further in Chapter 9 with regard to the Catlett, Virginia train accident case **(Figure 3-7)**. There a fire company responding to a fire was struck by an Amtrak train, resulting in a derailment and numerous casualties. Amtrak sued the fire company, claiming it was separate from the county and liable for damages.

Transitional Situations

There are many locations around the country where volunteer fire companies have operated for years, but now find themselves in the process of being incorporated into a county-wide fire department. In such cases, it is the county that is, or will soon be, providing the fire protection service, at which time the volunteer fire companies become relegated to a supporting role. Complex legal issues can arise during such transitions, such as when the county has taken over fire department operations but the apparatus and stations are owned by the fire company, or vice versa.

A variety of legal issues also arise when municipal fire departments are consolidated into county fire departments, and where municipalities decide to form fire departments that will split off from county fire departments.

Fire Districts or Fire Protection Districts

fire district

a political subdivision of the state that has the authority to impose taxes and organize fire and emergency services

The term fire district has different definitions in different states. As used here, the term **fire district** refers to a political subdivision that has the authority to impose taxes and organize fire and emergency services **(Figure 3-8)**. Fire districts are created by the state legislature, and have limited powers of self-government. Fire districts are managed by officials who are elected by the voters in the district. The elected officials may be known by various names, such as the board of directors, board of fire wardens, or board of engineers.

Fire districts are granted the power to tax, similar to other municipalities. Fire districts may have some limited authority to pass ordinances, depending upon their charter and state law. They must comply with state open meeting laws, open records laws, election laws, and conflict-of-interest laws, because they are political subdivisions of the state.

Fire districts are municipal corporations themselves, but limited municipal corporations, insofar as their purpose is not to provide general self-governance to the local population. Some states consider fire districts to be quasi-municipal corporations. Fire districts have the limited responsibility of providing fire protection, emergency medical, and rescue services. Some fire districts may also be chartered to provide other services such as public water, street lights, and related municipal services.

Fire districts may be subject to different laws than county or municipal fire departments, particularly in states where fire districts are considered to

Figure 3-8 *Fire districts are political subdivisions of the state.*

be quasi-municipal corporations. (In West Virginia, for example, this issue was decided in *State v. Board of Parks,* 47 S.E. 2d 689 WV, 1948.) The difference between a municipal corporation and a quasi-municipal corporation can be particularly important when it comes to liability and immunity.

fire protection district
an entity established in some states to ensure fire protection in rural or suburban areas that do not have a fire department; a fire protection district has the authority to collect and utilize taxes to contract with a nearby fire department to provide fire protection for the district

Another entity with a confusingly similar name is a **fire protection district**. In some states, a fire protection district may be established to ensure fire protection in rural or suburban areas that do not have a fire department. A fire protection district has the authority to collect and utilize taxes to contract with a nearby fire department to provide fire protection for the district. However, the fire protection district does not directly provide the service itself. The similarity between the names *fire district* and *fire protection district* has led to some confusion, and the distinction between the two names is not universally recognized.

Complicating this matter even further is the existence of fire districts that function to organize, oversee, and fund one or more volunteer fire companies that maintain their autonomy from the fire district **(Figure 3-9)**. In attempting to determine if the actual fire protection services are provided by the fire district (public sector) or the volunteer fire companies (private sector), the following issues are relevant:

- Does the fire district actually provide the service, or does it simply serve as a funding source for the volunteer fire company?
- Who owns the fire apparatus? Who owns the stations?
- Who appoints or elects the fire chief? To whom does the fire chief report?
- How are new members appointed or admitted to the department? Are they voted on by the membership of the fire company, or is the decision

Figure 3-9   *Some fire districts exist to organize, oversee, and fund volunteer fire companies that operate within their jurisdiction.*

made by the fire chief, the fire district board, or a membership commit-tee? If there is a membership committee, is it a committee appointed by the fire company or the fire district board?

- Who establishes and oversees operational policy?
- Who purchases the equipment, and who pays the bills?

Regional Fire Departments

In some jurisdictions, neither a municipal nor a county fire department suits the needs of the community. Where several communities can agree upon the need to consolidate their fire departments, a regional fire department can be created. Several options are available for the organization of such a depart-ment. The first is to create a special fire district capable of imposing taxes and financing itself. Fire districts are not limited by city, town, or county bound-aries, but must be created by the state legislature.

A second option is to create a nontaxing, quasi-municipal district similar to those commonly used to create a school district. The quasi-municipal dis-trict would ordinarily need to be established by the legislature, and would depend upon the annual allocation of funds from each municipality for its operation. A third option is for one fire department (perhaps a municipal or county fire department) to provide fire protection to an adjoining community or communities in exchange for compensation.

Regional fire departments often run into legal problems comparable to those faced by county fire departments. At various times, existing fire depart-ments need to be incorporated into the regional organization, and at various times municipalities may choose to withdraw from the regional organization. A myriad of legal issues must be addressed, including ownership of apparatus and stations, radio channel licensing, and collective bargaining agreements.

Public Safety Departments

Some communities choose to combine the operations of their police depart-ment, fire department, and in some cases emergency medical services into one entity, utilizing cross-trained personnel. While not officially a fire de-partment, such an organization provides fire protection services and as such may qualify as a fire department under state law for certain purposes, such as immunity protection and workers' compensation.

State and Federal Fire Departments and Related Entities

Both state and Federal governments may provide fire protection to certain types of locations, such as airports, institutions, special economic devel-opment zones, and military installations, through the use of state or Federal employees. In addition, many wildland firefighters are state and Federal em-ployees **(Figure 3-10)**.

Figure 3-10 *Many wildland firefighters are state or Federal employees.*

There are also special entities created by state and Federal governments that oversee or manage a particular site or location, such as airports, seaports **(Figure 3-11)**, economic development zones, or important facilities. The organizations that manage these sites are quasi-municipal or quasi-governmental corporations, as opposed to governmental agencies. The responsibilities

Figure 3-11 *The Massachusetts Port Authority is an independent public authority that manages airports and seaports in and around Boston. (Photo courtesy of Massport Fire and Rescue.)*

and funding sources for these corporations will vary from organization to organization. In some jurisdictions, these corporations are referred to as "public authorities."

The firefighters who are employed by these public authorities may be state or Federal employees, but commonly they are considered to be a separate category of public employee. As such, they are not classified as typical state or Federal employees, but they are not private sector employees either. Rather, they are considered to be public employees of the particular quasi-governmental corporation. Employees of these public authorities are sometimes referred to as quasi-public employees or quasi-governmental employees.

PRIVATE SECTOR FIRE ORGANIZATIONS

Volunteer Fire Companies

Volunteer fire companies may be either corporations or associations. The key distinction between a volunteer fire company and a volunteer municipal fire department is that the volunteer fire company is not an agency or part of local government. As such, a volunteer fire company is a private sector entity **(Figure 3-12)**. Even though the fire company may be funded and in many ways controlled by the municipality, if it retains its own autonomy it remains a part of the private sector.

Figure 3-12
Volunteer fire companies are private sector entities that deliver a public service.

The following case provides an interesting historical perspective on private volunteer fire companies in Pennsylvania, while addressing one of the most important reasons for understanding the distinction between public and private sector entities: governmental immunity.

Zern v. Muldoon
516 A. 2d 799, 101 Pa. Commw. 258 (1986)

The Citizens Fire Company # 1 of Palmyra was summoned on [September 8, 1978] to fight a fire in a building owned by Edward Zern. The fire had started in the kitchen of a restaurant located therein, which was operated by Charles Muldoon. When the firefighters arrived on the scene, they observed flames in the exhaust system of the kitchen area. The fire chief went to the basement and turned off the electricity, disengaging the exhaust system. Other firefighters climbed to the roof and attempted to fight the fire through a vent. These and other efforts failed. The building suffered extensive damage and was subsequently razed.

Eight separate lawsuits were subsequently filed . . . in which the Plaintiffs, the majority of whom were other tenants in the building, sought to recover damages from Muldoon and/or Zern. Muldoon in turn [sued] the fire company as an additional defendant in each suit. The fire company filed motions for summary judgment in which it sought that the actions be dismissed . . . because . . . the suits against it were barred by governmental immunity.

[T]he trial court granted summary judgment in favor of the fire company, concluding as a matter of law that volunteer fire companies were entitled to the common law defense of governmental immunity. . . .

To decide these appeals, we must define the legal relationship between volunteer fire companies and the local municipalities they serve. In pursuit thereof, we will first examine the origin and development of volunteer fire companies in this Commonwealth.

History of Firefighting in this Commonwealth

An early, exhaustive recitation of the history of firefighting in the City of Philadelphia is found in *Harmony Fire Co. v. Trustees of the Fire Association*, 35 Pa. 496 (1860). Justice Read there referred to the first laws governing fire protection which required homeowners to store two leather buckets in each dwelling so that an ample supply would be available when the primitive human "hand to hand" water chain was needed to fight a community fire. Between 1700 and 1736, the city, by then incorporated, provided the means to fight fires. It purchased its first engine in 1718. As the population surged and technology advanced, the need for a stronger firefighting force grew. An infamous fire at "Budd's long row" in 1736 was the disaster which focused the imperative demand for a more practical, better organized system for the control of fire in this densely populated residential,

(continued)

(*continued*)

commercial and industrial area. The local citizenry banded together and formed independent organizations whose unpaid members formed the first "Volunteer Fire Companies," e.g., Benjamin Franklin's Union Fire Company, the Hand in Hand, the Heart in Hand and the Friendship.

These volunteer companies effectively replaced the City's early efforts to combat fires and, in 1811, the City, recognizing their contribution, began appropriating monies to the companies. The divergent companies then combined to form the "Fire Association of Philadelphia," which administratively governed disputes among the member companies and regulated their financial and equipment requirements.

The City of Philadelphia, pursuant to Section 42 of the Act Incorporating the City of Philadelphia, in 1855 assumed total responsibility for fire prevention and control of volunteer fire companies by establishing and funding its Fire Department. Its officers consisted of a chief engineer, his seven assistants, and a secretary, all of whom were salaried. These officers were responsible to the City Administration for the organization and control of the volunteer fire companies which, having met certain equipment standards, were assigned fire duty in specific geographic areas. The volunteer companies, which still performed the actual firefighting duties, were funded by a conglomerate of interests viz. the City, insurance companies, businessmen, property owners, and the actual firefighters.

On March 15, 1871, following numerous hostile clashes among volunteer fire companies over territorial jurisdiction, equipment need and even political differences, the City enacted a resolution which established a permanent fully funded fire department. Although this marked the demise of the volunteer force in Philadelphia, volunteer fire companies still maintain a strong presence to this day throughout this Commonwealth. Many of these volunteer fire companies are supported by the statutorily created relief associations, which control the economic and social agenda of the companies.

The unique structural development of these volunteer fire companies presented difficult questions of law respecting funding and liability for damages caused by the discharge of their duties. . . .

Early Immunity of Fire Companies

The earliest decisions treating the issue of the liability of a volunteer fire company for damages caused by its negligence are found in *Boyd v. Insurance Patrol of Philadelphia,* 113 Pa. 269, 6 A. 536 (1886) (Boyd I), and *Fire Insurance Patrol v. Boyd,* 120 Pa. 624, 15 A. 553 (1888) (Boyd II). Boyd was fatally injured when he was struck on the head by a tarpaulin dropped from a building by an employee of the Fire Insurance Patrol. In Boyd I, the Supreme Court discussed the evolution of the doctrine of charitable immunity as it applied to volunteer organizations, but concluded that the trial court record was insufficient for it to apply the doctrine to the Fire Insurance Patrol. In Boyd II, the Supreme Court, after reviewing the funding appropriation and purpose of the Fire Insurance Patrol, held that it

(*continued*)

(*continued*)

was a public charitable institution and immune from the negligence of its employees under the doctrine of charitable immunity.

However, Justice Paxson, writing for the Court, in addressing the status of volunteer fire companies, also indicated that they would be immune under the doctrine of governmental immunity:

> *Our conclusion is that the Fire Insurance Patrol of Philadelphia is a public charitable institution; that in the performance of its duties it is acting in aid and in ease of the municipal government in the preservation of life and property at fires. . . . Upon this point we are free from doubt. It has been held in this state that the duty of extinguishing fires and saving property therefrom is a public duty, and the agent to whom such authority is delegated is a public agent and not liable for the negligence of its employees. This doctrine was affirmed by this court in* Knight v. City of Philadelphia, *15 W.N. 307, where it was said: 'We think the court did not commit any error in entering judgment for the defendant. . . . The members of the fire department are not such servants of the municipal corporation as to make it liable for their acts or negligence. Their duties are of a public character, and for a high order of public benefit. The fact that this act of assembly did not make it obligatory on the city to organize a fire department, does not change the legal liability of the municipality for the conduct of the members of the organization. The same reason which exempts the city from liability for the acts of its policemen, applies with equal force to the acts of the firemen.' And it would seem from this and other cases to make no difference as respects the legal liability, whether the organization performing such public service is a volunteer or not. . . . But I will not pursue this subject further, as there is another and higher ground upon which our decision may be placed.*

Boyd II at 646-47, 15 A. at 556-57.

Present Legislative Aid to Volunteer Fire Companies

The legislature has also continued to aid volunteer fire companies financially and functionally. A pertinent recitation of statutory assistance to volunteer fire companies is found in *Harmony Volunteer Fire Co. v. Pennsylvania Human Relations Commission,* 73 Pa. Commonwealth Ct. 596, 602-04, 459 A.2d 439, 443 (1983), where this Court stated:

> *Numerous legislative enactments further interweave the functioning of the government and the fire company. Several statutes provide the fire company with particular benefits and powers: volunteer firefighters may become special fire police with full power to regulate traffic, control crowds and exercise all other police powers necessary to facilitate the fire company's work at a fire or any other emergency; volunteer fire associations are exempt from vehicle title and registration fees; and fire companies are eligible for low interest state loans in order to purchase equipment. Other statutes also recognize the*

(*continued*)

(*continued*)

> *intimate relationship between a volunteer fire company and governmental entities; the borough is liable for the negligent operation of equipment by a volunteer firefighter responding to an emergency; an employer may not terminate a volunteer firefighter for missing work while responding to a fire call; firefighters are government employees under the workmen's compensation act; firefighter relief associations are entitled to receive a two percent tax on all foreign fire insurance premiums; the borough may make regulations for fire safety and may make appropriations to volunteer fire companies; the state may regulate relief companies; and the fire station is exempt from property taxes.* (Footnotes omitted.)

Moreover, the legislature has solidified the existence of volunteer fire companies by decreeing that local volunteer fire companies may not be replaced by a paid firefighting force unless a majority of the voters in the local municipality vote by referendum in favor of the change. . . .

We draw from this plethora of statutory and judicial pronouncements an important conclusion—the history, structure, organization and public duty of volunteer fire companies distinguish them as an entity from any other organization in existence in this Commonwealth today.

Issue

From this focus point, we now turn to the question of whether volunteer fire departments are entitled to immunity for damages arising from its actions.

Discussion

In [cases decided in 1951 and 1971 the Pennsylvania Supreme Court] restated the doctrine that volunteer fire companies are immune from liability for torts committed by their servants in furtherance of their corporate purpose. . . .

However, since [that time] . . . our Supreme Court, following the national trend, abolished the immunity doctrine as it applied to charitable organizations [in 1965] . . . local municipalities [1973] . . . and the Commonwealth itself [in 1978]. . . .

Although the legislature responded quickly to restore immunity, with certain specific exceptions to local municipalities and to the sovereign, the fire that destroyed Mr. Zern's building took place during the period between the judicial and legislative action, i.e., September 8, 1978.

Therefore, we are constrained to hold that the common pleas court erred in granting summary judgment in favor of the fire departments . . . because at the time the fire occurred, the doctrine of governmental immunity was not in existence due to the judicial decision . . . [on] May 23, 1973, and had not yet been revived through the November 26, 1978 Political Subdivision Tort Claims Act.

We recognize the common pleas court's able attempt to . . . create an exception to protect volunteer fire companies during the void in the governmental immunity protection. We can, however, find no appellate decisions to support the exception. . . .

Reversed and remanded.

Case Name: Zern v. Muldoon

Court: Commonwealth Court of Pennsylvania

Summary of Main Points: The legal relationship between volunteer fire companies in Pennsylvania, and the communities they represent, is the result of hundreds of years of history, a history that can be examined and understood by looking at a series of cases extending back into the 1800s. The relationship is important to this case because it impacts the issue of immunity protection for the volunteer fire companies who may be sued for their actions at the scene of a fire. The trial court found that immunity protection existed for volunteer fire companies based upon that rich history, despite the fact that at the time of the fire, the old immunity law had been declared invalid, and a new immunity law had not yet been passed. The appellate court, after examining the history of volunteer firefighting in Pennsylvania, reversed concluding that no such immunity protection existed absent explicit legislative authorization.

We will discuss the issue of immunity protection in the fire service at length in Chapter 9. Our focus in reading the *Zern* case is to gain perspective of the rich legal history of the fire service, and to see an example of how volunteer fire companies can be recognized as independent private sector entities for some purposes, but considered to be engaged in a public sector function for others.

The case of *Steffy v. City of Reading*, 46 A2d 182 (PA, 1946), presents another good example of different treatment of volunteer fire companies under different laws. In *Steffy*, a paid driver employed by a volunteer fire company sought to contest his termination. Previous case law determined that paid employees of volunteer fire companies were to be treated as municipal employees of the city in which they volunteer, for purposes of workers' compensation benefits. As such, the municipality had to provide workers' compensation coverage for such employees. In *Steffy*, the court ruled that despite this law, the paid employee of a volunteer fire company was not an employee for civil service purposes. As such, the member was not entitled to a civil service hearing when he was terminated by the volunteer fire company.

Some of the other areas where the status of volunteer fire companies as public or private sector entities is important are:

- Do Federal OSHA regulations apply to a volunteer fire company?
- Do the fire service exemptions under the Fair Labor Standards Act apply to paid employees of a volunteer fire company?
- Does a volunteer fire company have to comply with open meetings, open records, and state ethics laws?
- Do state collective bargaining laws apply to paid employees of volunteer fire companies, or does the National Labor Relations Act apply?

The above questions can only be answered on a state-by-state, case-by-case basis. The analysis must look at the specific laws in question and the relationship between the volunteer fire company and the municipality, county, or fire district. In the chapters that follow, we will discuss the factors that can cause a case to go one way or the other for a volunteer fire company.

Agreements between Volunteer Fire Companies and Jurisdictions Protected   Many volunteer fire companies operate under formal agreements with the jurisdiction or jurisdictions they serve. These agreements provide the opportunity to address many of the issues and problems that can arise between the fire company and the local authorities, and clearly define the rights, roles, and relationships between the parties.

Unfortunately, in many jurisdictions the parties have been operating for years on the basis of an informal understanding. The lack of a formal agreement can become a problem for both parties when a dispute arises. In Texas in 1999, the state attorney general issued an advisory opinion stating that, in the absence of a formal agreement between a county and a volunteer fire company, taxpayer funds could not be given to a volunteer fire company (see Appendix B, Chapter 3). Some states have laws that specifically authorize municipalities to make payments to volunteer fire companies even without formal agreements (see Rhode Island General Laws, RIGL 45-18-1 in Appendix B, Chapter 3).

Volunteer Fire Company Oversight Boards   In some jurisdictions, municipal and county officials have sought to gain control over volunteer fire companies through oversight boards. These boards may be called a board of engineers, board of wardens, or some similar term, but in essence they serve as public bodies, elected by the public or appointed by elected officials, whose purpose is to coordinate and consolidate control over volunteer fire companies. Oversight boards may be created in the municipal charter or by ordinance.

Volunteer Firefighter Rights   The authority, liability, and rights of volunteer firefighters differ greatly from state to state. Consider the following case from New York State. New York is unusual in that there is a state statute that says that volunteer firefighters cannot be removed from office except for cause, and after a hearing.

Crawford v. Jonesville Board of Fire Commissioners
645 N.Y.S.2d 586, 229 A.D.2d 773, 1996.NY.33214 (1996)

After a hearing, petitioner, a volunteer firefighter in the Jonesville Fire District, was found guilty of misconduct based upon his refusal to participate in sexual harassment training provided by the District. Petitioner concedes that he refused to

(continued)

(*continued*)

participate in the training, but contends that the notice of charges served on him was inadequate and that respondent Jonesville Board of Fire Commissioners (hereinafter respondent) never adopted a rule which required that all firefighters participate in the training. We reject both arguments.

It is undisputed that volunteer firefighters are entitled to due process in disciplinary proceedings and, therefore, the notice of charges must reasonably apprise the accused of the alleged misconduct so that an adequate defense can be prepared and presented. . . . The notice to petitioner in this case stated, inter alia:

> *You are charged with misconduct by reason of your refusal to abide by rules and regulations of the * * * District requiring all members * * * to complete sexual harassment training in accordance with the * * * District Sexual Harassment Policy adopted by the * * * District by resolution dated April 12, 1994.*

In contrast to the case of *Matter of Bigando* . . . upon which petitioner relies, the notice in this case contains a single unambiguous charge which cites to a specific requirement of a District rule or regulation. We conclude that the notice was reasonably calculated to apprise petitioner of the charges against him so as to enable him to adequately prepare and present a defense. . . .

Unable to deny that he had refused to participate in the sexual harassment training provided by the District, petitioner mounted the only real defense available to him, which is that the District never adopted a rule or regulation mandating the training for all District members. As petitioner points out, the sexual harassment policy formally adopted by the District in April 1994 contains no express training requirement. The minutes of respondent's April 1994 meeting at which the policy was adopted indicate that an inquiry was to be made regarding the use of a local consultant to provide sexual harassment training for the District. In July 1994, respondent adopted a resolution which authorized the District to contract with the consultant. Although the resolution itself contained no statement regarding the mandatory nature of the training, the minutes of respondent's meeting at which the resolution was adopted states:

> *All firefighters/employees will be required to have this training on sexual harassment. Any firefighter having received such training at his/her work place may be excused from the drill upon submission of documentation from his/her place of employment.*

Subsequent editions of the District's newsletter sent to each member included a notice concerning mandatory sexual harassment training. A November 1994 letter from the District concerning the training and its mandatory nature was sent to each member. In February 1995, respondent formally adopted a resolution which "establish[ed] July 1, 1995 as the deadline to complete the required sexual harassment training". Subsequent editions of the District's newsletter included a notice concerning the July 1, 1995 deadline to complete "the required training on sexual harassment".

At the hearing, petitioner conceded that he was aware of both the sexual harassment training provided by the District and the District's position that the

(*continued*)

(*continued*)

training was mandatory for all firefighters. Petitioner also conceded that he was aware that he could satisfy the requirement by submitting documentation of the sexual harassment training he had received at his work place, but that he elected not to do so. He maintained that he was never made aware of respondents' [sic] formal adoption of a specific rule or regulation which mandated the training.

We are of the view that petitioner's argument is one of form over substance. The record provides a rational basis for respondent's determination that mandatory sexual harassment training had been established by District rule or regulation and that petitioner knowingly violated the requirement. . . . As to petitioner's final argument, the penalty of a one-year suspension was not so disproportionate to the offense as to be shocking to one's sense of fairness. . . .

Case Name: Crawford v. Jonesville Board of Fire Commissioners

Court: State of New York, Appellate Division

Summary of Main Points: Some states, such as New York, provide members of private volunteer fire companies with rights, such as a right to a hearing prior to being disciplined or terminated from a volunteer fire company. However, even with those rights a firefighter who refuses to comply with the lawful requirements of the fire company may be disciplined.

In most states, volunteer firefighters have no statutory or constitutional right to a hearing before being disciplined or terminated, unless they directly serve a municipality or fire district. New York State is an unusual exception. Some volunteer fire companies may provide for a right to a disciplinary hearing in their bylaws or charter. As has been said, issues involving volunteer fire companies can only be addressed on a state-by-state, case-by-case basis.

Subscription Fire Departments

Subscription fire departments are private corporations or associations that charge people a fee for responding to their emergencies. Subscription fire departments do not receive funding from the local municipality, nor do they have the authority to impose taxes themselves. Rather than pay a per-incident fee, residents are offered an annual subscription fee, typically at a greatly reduced price. Those who choose not to subscribe generally receive the same service as those who do subscribe, but receive a bill for any services provided. The bill may be on a flat fee per call basis, or it may be based upon the services actually rendered.

A subscription fire department must be distinguished from other fire department organizations that offer subscription fees for certain types of services, particularly subscription fees for emergency medical treatment and transportation. A subscription fire department is a fire department funded through this method of billing and maintaining subscription fees. Subscription fire departments are ordinarily organized as private nonprofit corporations. There must be a legal basis for a subscription department to impose a fee for service, such as a state statute that expressly authorizes such a practice.

For-Profit Corporations

There are some for-profit businesses that provide fire protection in the United States. Generally, these entities provide fire protection service under a service contract with a local community, or for a governmental agency. Some of these businesses are large multinational corporations that provide a variety of security and protective services, with fire protection being just one of them. These businesses may provide fire protection services to military bases and other installations under a contract, or may provide the services to a municipality. Typically these for-profit companies must competitively bid for their contracts with terms ranging from 2 years to 10 years.

Industrial Fire Departments

Industrial fire departments are fire departments that are owned and operated by the owner of an industrial site that lacks adequate municipal fire protection. The business establishes its own full-time fire department to address its need, as a cost of doing business. Some industrial fire departments are large operations employing hundreds of firefighters, operating multiple stations, and protecting hundreds of square miles of industrial, commercial, or special facility property. Other industrial fire departments are relatively small operations that are reliant upon assistance from the local fire department.

The relationship between an industrial fire department and the local fire department can vary dramatically, depending upon the circumstances. In some cases, there is no local fire department due to the size, layout, or jurisdiction of the facility falling under a Federal agency, such as the Department of Energy, NASA, or the military. Where the local fire department is a municipal agency or fire district with a legally mandated responsibility that includes the facility, the industrial fire department will be subservient to the local fire department. However, where the local fire department is a volunteer fire company with no statutory authority, and no formal agreement with the local municipality to serve as exclusive fire service provider, the case is less clear.

Industrial fire departments, and fire departments that respond to facilities that have industrial fire departments, are well advised to have agreements in place that address issues such as response plans, chain of command, operational procedures, reimbursement, and liability.

Fire Brigades

Fire brigades are not true fire departments. They are emergency response teams set up at manufacturing or industrial facilities, and are designed to provide an initial response to fires at the facility. Members of fire brigades are employees of the facility that they protect. Fire brigade members have specific full-time job assignments, but in the event of a fire or disaster, they are pulled away from their main jobs to help mitigate the emergency.

CONCLUSIONS ON FIRE DEPARTMENT AUTHORITY

Under our Federal system of government, all powers not specifically delegated by the Constitution to the Federal government are reserved to the states. One of the areas where the states have traditionally exercised exclusive power is what is called the state police powers. The term does not refer simply to law enforcement, or the power to arrest, but rather refers to the authority of each state to govern matters related to the welfare and safety of residents.

How does each state choose to protect its citizens from fire? What fire and building codes are to be followed, and who will ensure compliance? The authority for a fire department to carry out its responsibilities comes primarily from the state, and to a lesser extent from local law. States differ in how they delegate responsibility for fire protection. State law may assign certain fire-related responsibilities to a state agency, to local municipalities, or to county government—or state law may be silent about the subject. Ultimately, it is the state that has the inherent authority to address fire protection by virtue of its police powers.

VOLUNTEER FIRE COMPANY AUTHORITY

In many states, there has been no formal delegation of responsibility to provide fire protection services to counties, cities, or towns. As we saw in the *Zern* case, responsibility for firefighting in Pennsylvania evolved informally over the years. In the absence of formal authority, private sector volunteer fire companies were formed to fulfill a critical role in our society, with little or no governmental oversight.

The relationship between volunteer fire companies and the communities they protect varies from locale to locale **(Figure 3-13)**. In some jurisdictions, authority over volunteer fire companies impliedly exists in the delegation of responsibility to a municipality to care for the public safety and welfare. In other jurisdictions, state statutes and municipal charters are completely

Figure 3-13 *The relationships between volunteer fire companies and the communities they protect vary from one locale to another.*

silent about providing fire protection, creating ambiguity about who has what authority and responsibility. In some instances, municipal charters specifically mandate that firefighting will be performed by volunteer fire companies, and that the town's control will be limited to funding the fire companies.

Why is this important? States possess inherent police powers, but if the responsibilities for fire protection have not been delegated to municipalities, and are being exercised on a daily basis by volunteer fire companies, what happens in the event of a conflict between the municipality and the local volunteer fire company? Consider the following questions:

- Does the municipality in which the fire company operates have any inherent authority over the volunteers, beyond withholding funding? Does the municipality have a legal obligation to fund the volunteer fire company? Can the fire company enforce this obligation in court?

- Can the municipality disband a local volunteer fire company, or prevent the company or its members from responding to alarms? Can the municipality unilaterally change the company's response district?

- How much control can the town exercise over the volunteer fire companies in matters beyond fire suppression, such as fire code enforcement, plan review, hazardous materials permitting and response, and emergency medical services? Who is authorized to provide plan review and inspection services, the town or the volunteer fire company? If the volunteer fire company wants to perform plan review and code enforcement, can the town take that responsibility away?

- How much control does the municipality have over who is a volunteer?

- Does the volunteer fire company have to respond to every address in the community? Can the volunteer fire company refuse to respond to certain addresses, for example a high-hazard occupancy for which they are ill-prepared or unsuited?

- Can an established volunteer fire company prevent a new volunteer fire company from forming within its own response district?

Fortunately or unfortunately, there is virtually no case law that addresses any of these questions. Even if there were a large body of case law, the cases would be of little value outside the state in which they were decided, because each case will turn upon the interplay of the specific facts with the specific state and local laws. Certainly, agreements between volunteer fire companies and the local jurisdiction may be able to resolve most of these issues. However, in the absence of authority delegated by statute or charter to a municipality for fire protection, some ambiguity remains regarding the ability of a municipality to enter into an agreement concerning matters over which they have no inherent authority.

SUMMARY

There are a variety of organizations and entities that provide fire protection and emergency services in the United States. It is impossible to accurately and completely categorize all of these entities, but the primary organizations providing fire protection services are municipal fire departments, county fire departments, fire districts, and volunteer fire companies. The legal issues surrounding such fire organizations must be viewed on a state-by-state, case-by-case basis.

REVIEW QUESTIONS

1. Describe the differences between the four types of corporations.

2. What is the distinction between public sector and private sector entities, and why is it so important?

3. What is the difference between a municipal volunteer fire department and a volunteer fire company?

4. What are the consequences of a volunteer fire company having its corporate charter revoked by the state?

5. What is the most important benefit that a volunteer fire company would get by being a corporation as opposed to an association?

6. Explain the similarities and differences between industrial fire departments and fire brigades.

7. Explain the differences between a fire district, a municipal fire department, and a volunteer fire company.

8. What level of government has the inherent authority to address public safety, including matters relating to fire safety?

9. Is an organization named the Smithville Volunteer Fireman's Association, Inc. a corporation or an association?

10. Explain why firefighters who are employed by public authorities are sometimes referred to as quasi-public employees.

DISCUSSION QUESTIONS

1. What are the legal challenges that confront regional fire departments and county fire departments when coverage must be provided to a new response area? What are the challenges that arise when a municipality for whom coverage is presently provided suddenly wants to withdraw and start its own fire department?

2. Why are agreements between volunteer fire companies and the jurisdictions they protect beneficial to both parties?

Chapter

4

ADMINISTRATIVE AGENCIES

Learning Objectives

Upon completion of this chapter, you should be able to:

- Identify administrative agencies as part of the executive branch.
- Explain that administrative agencies exist at the Federal, state, and local levels.
- Explain how agencies are created, and the purpose of enabling acts.
- Describe the separation of powers concerns created by administrative agencies, and the methods used to address these concerns.
- Explain the jurisdiction of OSHA and OSHA's three primary activities.
- Define approved-plan state and non-approved plan state.
- Explain why the term "OSHA state" is ambiguous.
- Explain the function of OSHRC.
- Explain the application of OSHA to volunteer and part-time firefighters.

INTRODUCTION

Firefighters arrive on the scene of a large construction site for a worker who has been critically injured. The worker fell approximately 25 feet from some scaffolding while painting a ceiling. As several members attend to the injured man, a foreman comes over to the company officer and asks, "Do you know if OSHA has been notified yet?"

The officer responds, "Our dispatch center notifies OSHA immediately whenever we are sent out to a workplace accident like this."

"Oh," replies the foreman, "I don't know how this could have happened. By the way, this is the first job our company has had in this state. Is this an OSHA state?"

The lieutenant responds, "No, we're a non-OSHA state."

"Great, so who is coming, state inspectors?"

With a perplexed look on his face, the lieutenant says, "No, the Feds will come on this one."

"Well, I thought you said you were a non-OSHA state."

ADMINISTRATIVE AGENCIES

The role of the executive branch of government is to enforce the laws and manage the day-to-day operations of government. Obviously, the president of the United States cannot manage the entire Federal government alone, nor can governors or mayors manage their states or cities alone. The solution to managing the day-to-day operations of our government rests in the creation of administrative agencies.

Administrative agencies exist within the executive branch to help run the day-to-day operations of government **(Figure 4-1)**. Agencies are created by the legislative branch, through laws called **enabling acts**. Besides creating administrative agencies, enabling acts provide the legal basis for an agency to operate. Enabling acts provide a statutory framework for the organization of the agency, and delegate the authority that the agency can exercise.

Not all administrative agencies are organized the same way, nor do all agencies exercise the same powers. Different agencies are given widely different powers, and function in vastly different ways. Some agencies are authorized to pass laws, called *regulations*. Some agencies are authorized to enforce certain laws, and have the power to investigate and prosecute violations of these laws. Some agencies are given the power to hold hearings that are comparable in many ways to trials, and in which legal determinations are made. These administrative tribunals may also assess fines and penalties.

enabling acts
statutes enacted by Congress or state legislatures that authorize the creation of administrative agencies and provide the legal authority for the agency to operate

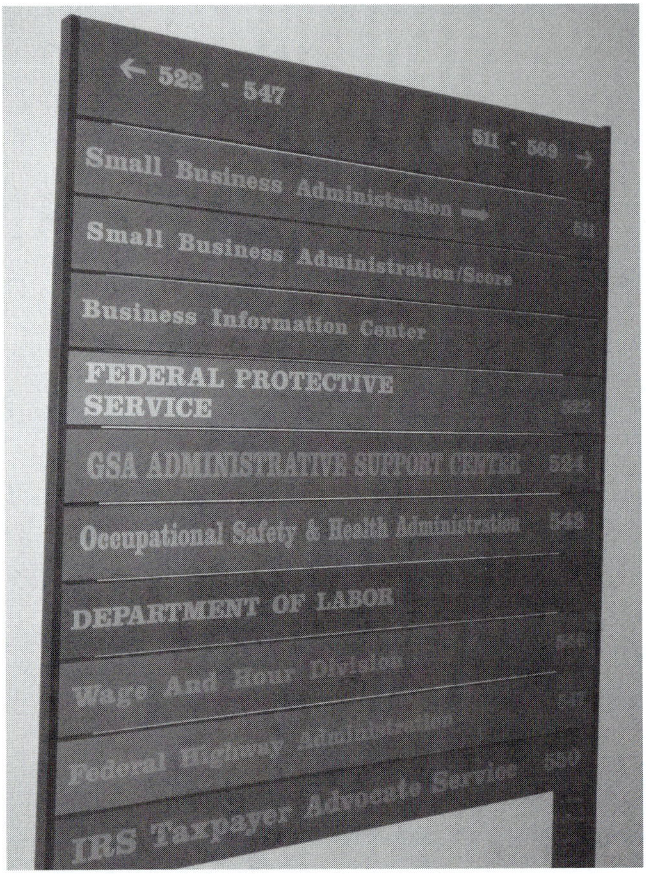

Figure 4-1
Administrative agencies exist within the executive branch to help run the day-to-day operations of government.

The delegation of legislative powers from Congress or a state legislature to an administrative agency makes a lot of sense. Congress cannot be expected to possess the expertise needed in every area over which laws may be needed. Think about the variety of things that we expect our government to regulate, from overseeing nuclear power plants and prescribing flight patterns over metropolitan areas to ensuring automobile safety and even regulating commercial fishing.

Unlike Congress, administrative agencies can be staffed with workers who are experts in the appropriate fields. Furthermore, when creating regulations or making other types of decisions, agencies are less likely to be as rushed or be as subject to political influence as Congress may be.

Throughout this chapter, we will look in detail at one specific Federal agency, the Occupational Safety and Health Administration (OSHA), to help us understand how all agencies function **(Figure 4-2)**.

Figure 4-2 *OSHA is an example of a Federal agency.*

OCCUPATIONAL SAFETY AND HEALTH ADMINISTRATION

Congress created the Occupational Safety and Health Administration in 1970 by passing the Williams-Steiger Occupational Safety and Health Act (OSH Act). In the Act, Congress authorized the secretary of labor to adopt occupational safety and health standards, to investigate alleged violations of occupational safety and health regulations, and to enforce compliance through the issuance of citations. The secretary of labor is appointed by the president, subject to Senate confirmation. The secretary of labor reports directly to the president and is a member of the president's cabinet.

The Department of Labor is the principal agency under the control of the secretary of labor. Pursuant to the OSH Act, the Occupational Safety and Health Administration was created within the Department of Labor. OSHA was given the authority to exercise the duties assigned by the OSH Act to the secretary of labor.

In addition to the secretary of labor, the president also appoints nine assistant secretaries of labor, one of whom is designated as the assistant secretary of labor for occupational safety and health. This person has the responsibility to directly oversee OSHA.

The focus of OSHA is to ensure and enhance workplace health and safety by

- Adopting appropriate safety regulations
- Enforcing those regulations through inspections, and investigating complaints and accidents
- Issuing citations for noncompliance

GETTING THE JOB DONE

The OSH Act imposes two important duties on employers: (1) to comply with all duly promulgated OSHA regulations and (2) to maintain a workplace that is free from recognized hazards. While the first duty is quite clear and well understood, the second duty is less well known, yet in many ways places a more significant burden on employers. This second duty, to maintain a workplace that is free from recognized hazards, is referred to as the "general duty clause."

The general duty clause places an affirmative duty on employers to provide a safe workplace beyond merely complying with OSHA regulations. The failure to correct a problem that causes a worker to be injured, even if the correction is not required by an existing OSHA regulation, could be a violation of the general duty clause and thereby trigger an OSHA citation. For example, failure to comply with a manufacturer's safety instruction for the use of dangerous equipment could give rise to a general duty clause violation.

Furthermore, the failure to comply with a nationally recognized safety standard applicable to a certain industry could be considered a violation of the general duty clause. The rationale for OSHA to cite employers who fail to comply with generally accepted industry standards is that the existence of an industry-wide safety standard is evidence that experts in that industry have recognized that a hazard exists, and that precautions are warranted. As applied to the fire service, the failure to comply with a National Fire Protection Association (NFPA) standard, such as NFPA 1500, Standard on Occupational Safety and Health Programs for Fire Departments, could give rise to a general duty clause violation. The fact that there is no specific OSHA regulation that addresses a particular issue is not a defense.

OSHA INSPECTIONS

OSHA inspectors have the authority to inspect any place of employment at any reasonable time. However, they need some basis on which to initiate an inspection. If an employer objects to an inspection, inspectors must obtain a

Figure 4-3 *OSHA inspections may be triggered by a specific complaint or an accident, or may be part of a program to address a specific hazard in a targeted industry.*

warrant to carry out the inspection. Any number of events may initiate inspections, including a specific complaint (often filed by an employee or union), an accident, or if the OSHA inspection is part of a program aimed at a specific hazard in a targeted industry **(Figure 4-3)**. In addition, in the event of a serious injury to three or more employees, or of a fatality to a worker, OSHA inspectors will conduct an investigation.

OCCUPATIONAL SAFETY AND HEALTH REVIEW COMMISSION

In addition to OSHA, the OSH Act created a separate Federal agency, the Occupational Safety and Health Review Commission (OSHRC), consisting of three members appointed by the president. The function of OSHRC is to oversee a tribunal of administrative law judges who handle the adjudication of OSHA citations. In other words, although it is an administrative agency, OSHRC fulfills a role analogous to that of a court. OSHRC administrative law judges will hear cases involving OSHA citations, make factual determinations, and render legally binding decisions.

OSHRC AND OSHA

The secretary of labor, through OSHA, issues citations to employers who violate an OSHA regulation or the general duty clause. An employer who is cited by OSHA may contest the citation, in which case the matter is brought before an OSHRC administrative law judge. The decision of an administrative law judge is considered to be a final decision by the OSHRC. Thereafter, any decision rendered by the OSHRC can be appealed to a court of law.

OSHA VIOLATIONS AND SANCTIONS

OSHA violations carry the possibility of fines or even imprisonment. OSHA can issue citations of up to $70,000 per violation. Violations of OSHA regulations that are deemed to be willful can be prosecuted by the United States Attorney's Office as criminal offenses. If convicted, an employer can be fined up to $10,000 and be sentenced up to six months in prison, and for repeated offenses, up to $20,000 and up to one year in prison.

In the aftermath of the Station Nightclub fire **(Figure 4-4)**, which killed 100 people—including seven employees—on February 20, 2003 in West Warwick, Rhode Island, OSHA cited the club owners for one willful and six serious violations, with fines totaling $85,200 (see Appendix B, Chapter 4).

Figure 4-4   *The Station Nightclub fire in West Warwick, Rhode Island on February 20, 2003 killed 100 people and injured 200 more. (Photo by Geoffrey P. Read.)*

ADMINISTRATIVE AGENCIES AND SEPARATION OF POWERS

separation of powers
a fundamental principle underlying the Constitution, designed to prevent any of the three branches of government from becoming too powerful; to accomplish this, the Constitution specifies each branch's separate powers and responsibilities

A key principle upon which our government was founded is that of separation of powers. The Constitution prescribes the powers and responsibilities of each branch of government. **Separation of powers** is a fundamental tenet embodied in the Constitution that prevents any one branch of government from becoming too powerful. Actions by one branch of government that fall outside of that branch's constitutional authority, or that intrude on the authority of another branch, are considered to be invalid and unconstitutional exercises of powers. For example, the president is not allowed to preside over any type of court hearing, nor can the judiciary pass any type of statute.

It should be apparent that administrative agencies risk violating the principle of separation of powers. Remember, administrative agencies are part of the executive branch. The congressional grant of power in creating the agency may authorize the agency to exercise legislative power, executive/enforcement powers, and at times judicial/adjudicatory powers. These types of delegations have been challenged from time to time as being a violation of the separation of powers. In other words, someone may challenge the legality of an administrative agency (as part of the executive branch) exercising quasi-legislative or quasi-judicial power.

Generally, for a legislature to constitutionally delegate legislative power to an administrative agency, the legislature must set forth the legislative policy in the enabling act, and establish the primary standards that the agency must follow in carrying out the policy. In this way it is the elected officials of the legislature who are setting the policy, and the agency officials are merely carrying out the policy within prescribed guidelines. The failure of the legislature to set the policy, or the failure of agency officials to follow the policy, can result in an agency's actions being nullified by a court.

Grants of legislative and judicial authority to administrative agencies must be narrowly written and contain adequate safeguards and limitations in order to be constitutional. For example, Congress could not authorize an administrative agency such as OSHA to develop regulations regulating whatever topics the secretary of labor deemed appropriate, nor could Congress authorize an agency such as the OSHRC to hear general civil litigation cases. However, Congress can lawfully authorize OSHA to pass and enforce workplace safety regulations, and authorize the OSHRC to hear cases involving contested OSHA citations.

Additional oversight of administrative agencies comes through the political process. The political process helps to govern administrative agencies, because agencies exist within and under the control of the executive branch. Be it the president, governor, or mayor, ultimate authority for the actions of the agency rest with the executive.

Agencies are also subject to considerable ongoing legislative oversight. Congress and state legislatures routinely hold hearings into how agencies are functioning, what their funding needs are, and whether they have too little or too much in the way of authority. Legislatures can and do curb agency power, and sometimes overrule agency decisions by enacting new legislation. Legislatures also control agencies through funding.

Finally, just about every decision made by an administrative agency is subject to some type of judicial review by a court. Anyone aggrieved by an agency decision has the option to challenge that decision, either through an administrative review process culminating in some type of judicial review, or through immediate judicial review. This judicial review option is critical in ensuring that separation of powers concerns are addressed.

EXHAUSTION OF REMEDIES

The exhaustion of remedies doctrine is a legal principle that requires a party who has an administrative remedy available to seek relief through the administrative remedy first, before proceeding to court. It requires a court to dismiss a lawsuit if the party bringing the action failed to utilize an available administrative tribunal to hear the case first.

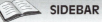 **SIDEBAR**

ABC Construction is cited by OSHA for violating 29 CFR 1910.146, the Confined Space Standard, and 29 CFR 1910.134, the Respiratory Protection Standard, after workers who entered an underground electrical vault at a construction site were overcome due to low oxygen levels in the vault.

ABC Construction claims that neither standard was violated, and files suit in Federal District Court to challenge the OSHA citations. The court will dismiss the action, because ABC Construction has an available administrative option, namely requesting a hearing before the OSHRC.

The rationale underlying the exhaustion of remedies doctrine is to encourage the use of administrative options, and to prevent litigants from taking detours around the administrative steps by going directly to court. Administrative tribunals are created to handle specific types of cases. As a result, their administrative law judges and hearing officers develop expertise in the given subject area, while at the same time taking a burden off the civil court system. The policy behind the exhaustion of remedies doctrine is that judicial review should be the last resort of an aggrieved party, not the first.

The exhaustion of remedies doctrine is a common theme throughout the law, particularly in labor relations and contract law. A party to an agreement may be barred from filing suit in court where the agreement requires the matter be submitted to an arbitrator or grievance procedure.

JURSIDICTION AND THE TERM "OSHA STATE"

Congress expressly limited OSHA's jurisdiction to enforce workplace safety under the OSH Act to private sector employers **(Figure 4-5)**. In other words, the OSH Act does not apply to state or local employers or entities. OSHA has no jurisdiction to investigate or cite state or local agencies. Federal OSHA enforces OSHA regulations only against private sector employers and non-exempt Federal agencies.

Most states have chosen to adopt Federal OSHA standards for state and municipal employers, and have assigned a state agency—such as the state's department of labor—the responsibility for enforcing the Federal OSHA regulations over state and municipal employers. Enforcement mechanisms vary

Figure 4-5 *The jurisdiction of Federal OSHA inspectors is limited to private sector employers.*

greatly from state to state. Some states take a hard-line approach, similar to the Federal OSHA practice, which involves conducting formal investigations into alleged public sector violations, issuing citations when violations are found, and leveling fines against cities, towns, and state agencies. Other states take a less adversarial approach, choosing instead to rely upon political and bureaucratic mechanisms to ensure compliance.

The OSH Act encourages individual states to take over enforcement of Federal OSHA regulations, and offers states a financial incentive to do so. States may opt to take over enforcement of Federal OSHA regulations completely, in which case they must agree to enforce OSHA regulations over public employers to the same extent that they do against private sector employers. In exchange, the Federal government agrees to pay up to 50 percent of salary costs for state inspectors, plus the state is allowed to keep fine money that it assesses. In order to qualify for such a Federal subsidy, a state must submit a plan to OSHA explaining how it will enforce Federal OSHA regulations against both private and public sector employers in a manner consistent with Federal OSHA requirements.

Twenty-one states and Puerto Rico have taken advantage of the Federal government's offer. These states are referred to as "approved-plan OSHA states" or "state-plan states." States that enforce Federal OSHA regulations against public sector agencies but are not approved-plan OSHA states are considered to be "non-approved plan OSHA states" (see **Table 4-1**).

Federal OSHA has also offered states some financial incentive to adopt a "public sector only" enforcement model. Under this arrangement, the state agrees to enforce Federal OSHA regulations against state and municipal employers, and the Federal government will help subsidize some of the costs associated with this enforcement. States opting to become public sector state-plan states must agree to enforce Federal OSHA regulations against state and local employers in a manner at least as effective as Federal OSHA requirements.

Table 4-1 *Who enforces federal OSHA regulations in various states.*

Employer	State-Plan or Approved-Plan State[1]	Public-Sector-Only Approved-Plan States[2]	Non-Approved Plan States
Public sector	State enforces[3]	State enforces[3]	State enforces
Private sector	State enforces[3]	Federal enforces	Federal enforces

[1] AK, AZ, CA, HI, IN, IA, KY, MD, MI, MN, NV, NM, NC, OR, SC, TN, UT, VT, VA, WA, WY; Puerto Rico
[2] CT, NJ, NY; Virgin Islands
[3] State agrees to enforce Federal OSHA regulations in a manner that is at least as effective as Federal OSHA.

> ### 📖 SIDEBAR
>
> The terms "OSHA state" and "non-OSHA state" have been criticized for being confusing and ambiguous. Unfortunately, use of these terms is widespread. Many private sector employers refer to an "OSHA state" as a state where Federal OSHA enforces OSHA regulations, and a "non-OSHA state" as one where a state agency has enforcement jurisdiction over private sector employers. To the typical firefighter, an OSHA state is a state where the fire department may be cited for failure to comply with an OSHA regulation, which in essence is an approved-plan state. Thus, the term "OSHA state" can mean different things to different people.
>
> A better distinction, which avoids the confusion associated with the term "OSHA state" is to use approved-plan OSHA state and non-approved plan OSHA state. It is recommended that fire service and safety professionals avoid using the term "OSHA state" due to its ambiguity.

STATE AND LOCAL AGENCIES

Administrative agencies also exist at the state and local levels **(Figure 4-6)**. State agencies are created by acts of the state legislature, in a manner that is analogous to how Congress creates Federal agencies. Examples of state administrative agencies are the state department of labor, state department of transportation,

Figure 4-6
Administrative agencies exist at the state and local levels as well as at the Federal level.

Figure 4-7 *The fire service has close ties to a number of administrative agencies.*

and state department of health. Like their Federal counterparts, state agencies are authorized to act by enabling legislation, which sets forth the specific duties and authority of the agency. Some state agencies are authorized to create regulations, some to conduct investigations, and some to hold hearings.

Local agencies are usually created by a municipality's charter, but may also be created by municipal ordinance. Municipal fire departments are administrative agencies, as are the building department, planning department, school department, and other such departments **(Figure 4-7)**.

ADMINISTRATIVE LAW MAKING

When Congress considers passing a law, there is a formalized procedure that must be followed in order for the proposed legislation to become law. The legislative procedure is open to public scrutiny and comment. Public debate over Congressional action is allowed, and hearings are held at which people can testify and have input prior to a law being passed. Most important, if people do not approve of a law, they can vote in the next election against those responsible for the law.

What happens when an administrative agency decides to pass a law? How does OSHA go about passing a new regulation governing, say, respiratory protection practices or infection control? Can a group of unelected OSHA experts go into a closed-door meeting, and come out with a new regulation that the entire nation must comply with?

Rulemaking

Administrative agencies may be authorized to pass laws through a procedure called rulemaking. The rulemaking power of agencies refers to the development of three basic types of rules:

1. *Regulations* are rules that carry the weight of law and are designed to carry out the basic quasi-legislative authority of the agency.

2. *Procedural rules* govern the agency's internal operations and procedures.

3. *Interpretative rules* are basically statements of the agency's perspective or interpretation of how the agency would like to see the law applied to certain fact patterns. Interpretive rules are not laws per se. Rather, they embody how the agency believes the law should be interpreted.

In 1946, Congress passed the Administrative Procedures Act, or APA. This law created a framework for all Federal administrative agencies to operate, spelling out how agencies are to pass rules and regulations and hold adjudicatory hearings. All states have adopted comparable administrative procedures acts, which govern how state agencies are to operate. In general, the APA requires administrative agencies to conduct rulemaking in a publicly open method.

The APA provides for two types of rulemaking procedures, *formal* and *informal*. The *informal rulemaking* procedure is the more expedited and commonly used procedure. Proposed Federal rules and regulations, as well as proposed amendments to rules and regulations, are published in the *Federal Register,* a document published daily by the Federal Government. Public comment is allowed for a designated period of time (commonly thirty days) through written comments to the proposed rules and regulations. The final rules and regulations are then published in the *Federal Register.* Once adopted, the regulations are added to the Code of Federal Regulations.

Formal rulemaking is reserved for situations where Congress expressly orders it. The biggest difference between formal and informal rulemaking is that under formal rulemaking, public hearings must be held on proposed rules, and the agency must state in writing all findings of fact, conclusions of law, and decisions rendered on the proposed rules, based upon evidence introduced into the record at the hearing.

Judicial Review

Most decisions made by an administrative agency are subject to review by a court, including the validity of regulations. However, in order to challenge the validity of a regulation or other agency decision in court, a party must have standing, meaning that the party seeking to challenge an agency action is somehow adversely affected or aggrieved by that action.

The APA provides for the judicial review of an agency action in the event that a party with standing challenges an agency's action. Congress, through the APA, has established a "substantial evidence standard" of judicial review for Federal agencies. Simply stated, the substantial evidence standard requires that enough evidence be in the record such that a reasonable person reviewing this evidence could reach the same conclusion as the agency. Under this standard, the agency decision is entitled to great deference by the reviewing courts, particularly where the enabling act authorizes the agency to make the types of decisions that are under review. Where there is substantial evidence to support an administrative agency's determination, the agency decision will not be overturned by a court.

OSHA AND THE NFPA STANDARDS

One of the primary missions of the National Fire Protection Association (NFPA) is to develop standards to help government and industry protect people and property from fire. NFPA standards include recommended fire codes, building codes, electrical codes, sprinkler requirements, and storage practices for industry. The NFPA also has a variety of fire service related standards, including standards for fire apparatus, protective clothing, equipment, hose, ladders, medical requirements, and training and safety programs for firefighters.

Questions commonly arise over the relationship of OSHA regulations to NFPA standards. The NFPA is not a governmental agency, and NFPA standards are not law, unless a particular jurisdiction has adopted them. However, as discussed previously, the failure of a fire department to comply with NFPA standards pertaining to firefighter safety could serve as the basis for an OSHA general duty clause violation.

NFPA standards are developed through a comprehensive process that permits input from a variety of parties who are interested in and knowledgeable on a given subject. Part of the standards-making process includes looking at applicable laws, including OSHA regulations, and ensuring that NFPA standards meet or exceed OSHA requirements. On occasion, there may be a conflict between an OSHA regulation and an NFPA standard. In such cases, the violation of an OSHA regulation may not result in a violation citation if the applicable NFPA standard was complied with. OSHA will look at the facts on a case-by-case basis. Where an employer has complied with an industry-wide standard that is aimed at protecting employees, OSHA may defer to the industry-wide standard.

APPLICATION OF OSHA TO VOLUNTEERS AND PART-TIME PERSONNEL

The treatment of volunteer and part-time firefighters under OSHA is a difficult and challenging area **(Figure 4-8)**. Does OSHA apply to volunteer firefighters at all? Does it apply to part-time employees? Does it matter if the fire department is 100 percent volunteer, or if it has some part-time or paid firefighters? Does it matter if the volunteers receive some nominal compensation for their time? Does it matter whether the firefighters are volunteering for an independent volunteer fire company, a municipality, or a fire district?

There are many issues that must be considered in determining if OSHA regulations apply to a fire department, let alone a volunteer fire department. The final conclusion will be based upon the totality of the circumstances, including the facts and laws that are found to apply. Even on identical facts, the results will vary from state to state, depending upon state law. Those interested in getting a specific answer to a specific situation must consult competent local counsel. That said, certain general principles are applicable.

Figure 4-8   *The application of Federal OSHA regulations to volunteers raises several complex issues.*

Volunteers in General

Federal OSHA has jurisdiction over volunteer employees who are working for an employer in the private sector. The best non-fire-service analogy to this is people who volunteer to work at hospitals, so-called "candy stripers." Hospitals, as employers, must comply with OSHA regulations for such volunteer employees.

OSHA's focus in such cases is on whether or not there is an employer-employee relationship between the parties. The existence of an employer-employee relationship between a volunteer and an organization is based upon a number of factors, which have been identified by the United States Supreme Court in *Nationwide Mutual Insurance Company v. Darden,* 503 US 318, 112 S.Ct. 134, 117 L.Ed.2d 581 (1992) and *Community for Creative Non-Violence v. Reid,* 490 US 730, 109 S.Ct. 2166 (1989). These factors are:

- the right to control the manner and means by which the work is completed
- the level of skill required to perform the job effectively
- the source of the tools and instruments required to perform the work
- the location at which the work is performed
- the duration of the relationship between the employer and the workers
- the right of the employer to assign new projects to the workers
- the extent of the workers' control over when and how long to work
- the method of payment or compensation
- the workers' role in hiring and paying assistants
- whether the work performed is the regular work of the employer
- whether the employer is in business
- whether benefits are provided to employees
- the tax treatment of the worker.

All factors of the relationship must be considered, and no one factor will be considered to be determinative. Thus it is possible, based upon the facts of a given case, for a volunteer or part-time firefighter who serves in a private nonprofit volunteer fire company that is not sufficiently affiliated with a state or municipality to fall under Federal OSHA.

State and Municipal Employee Exception

As we have discussed above, Federal OSHA does not have jurisdiction over fire departments in the public sector. For any fire department that is an agency of a state, county, or municipal government, or for a fire district, the determination of whether volunteer and part-time firefighters fall under

OSHA is a matter for state OSHA officials to decide. Any OSHA requirements that exist for public sector firefighters come from state OSHA.

Private Volunteer Fire Companies

In order to determine if a private volunteer fire company is subject to Federal OSHA, a closer consideration of the nexus between the fire company and the community it serves is necessary. Among the factors that must be examined is the agreement between the municipality and the fire company **(Figure 4-9)**; whether taxpayers' dollars are used to fund the operations of the department; and whether, under state law, the fire company is considered to be a **quasi-public entity**. To OSHA, the term quasi-public entity means that the entity has sufficient connection to the public sector to be treated as part of the public sector. Federal OSHA will not exercise jurisdiction if the fire company is determined to be a quasi-public entity.

In some states, volunteer firefighters who volunteer for a private volunteer fire company are considered to be employees of the state or municipality by virtue of their volunteer status with the volunteer fire company. Recall

quasi-public entity
an entity that has sufficient connection to the public sector to warrant it being treated as being part of the public sector for a specific purpose, such as for OSHA coverage

Figure 4-9 *Even volunteer fire companies may be subject to OSHA regulations. There are many factors in the relationship between the municipality and the fire company that must be examined.*

the *Steffy* case from Chapter 3, in which members of private volunteer fire companies were found to be employees of the municipality in which they volunteered for purposes of workers' compensation coverage. In such a case, Federal OSHA will consider them to be quasi-public employees, and therefore exempt from Federal OSHA coverage.

However, if a private volunteer fire company is not considered a quasi-public entity, and the volunteers are considered to be "employees" of the fire company pursuant to the *Nationwide Mutual Insurance Company v. Darden* test, then Federal OSHA would have jurisdiction over the fire company.

Of course, this discussion of volunteer fire companies is irrelevant in approved-plan states. In approved-plan states, state OSHA exercises the same jurisdiction over both public and private sector entities, and thus OSHA would apply, provided the volunteers are considered to be "employees" of the fire company pursuant to the *Nationwide Mutual Insurance Company v. Darden* test.

SUMMARY

Administrative agencies exist at the Federal, state, and local levels of government. Agencies are created by the legislative branch of government through enabling acts. Agencies may be given a number of powers, including the power to create laws called regulations, the power to enforce laws, and the power to adjudicate contested matters.

OSHA is a Federal agency that has a great impact on the fire service. Federal OSHA jurisdiction is limited to the private sector, but OSHA regulations may apply to public sector fire departments because some states have chosen to adopt and enforce OSHA regulations. The application of Federal OSHA to volunteer firefighters is a complex topic, and while it is usually a question of state law, it is possible that Federal OSHA will apply to some volunteer fire companies.

REVIEW QUESTIONS

1. What branch of government are administrative agencies considered to be part of?
2. At what levels of government do we find administrative agencies?
3. Describe the role of enabling acts in creating administrative agencies.
4. Why do administrative agencies risk violating the principle of separation of powers?
5. What are the three primary activities in which Federal OSHA engages to address workplace safety and health?
6. Explain why the term "OSHA state" is ambiguous.
7. What is OSHRC, when was it created, and what does OSHRC do?
8. Does Federal OSHA apply to volunteer and part-time firefighters?
9. How does the political process serve as a check on the power of administrative agencies?
10. What is the exhaustion of remedies doctrine?

DISCUSSION QUESTIONS

1. What is the difference between an approved-plan OSHA state and a non-approved plan OSHA state, and how does the difference between the two impact fire departments?

2. What incentives does a state have to become an approved-plan OSHA state? Are there any disincentives?

3. Explain how OSHA's general duty clause can be use to cite an employer who fails to comply with industry-wide standards. In particular, explain the relationship between NFPA standards and the general duty clause.

Chapter 5

CRIMINAL LAW

Learning Objectives

Upon completion of this chapter, you should be able to:

- Distinguish between violations of civil and criminal law.
- Distinguish between felonies and misdemeanors.
- Identify the three types of elements that make up a crime.
- Explain when an omission can give rise to criminal liability.
- Identify the four criminal mental states.
- Identify the elements for the following crimes:
    - first-degree murder, second-degree murder, voluntary manslaughter, involuntary manslaughter, battery, assault, sexual assault, rape, and child molestation
    - larceny, robbery, obtaining money under false pretenses, extortion, embezzlement, burglary, false imprisonment, kidnapping, RICO
    - arson

INTRODUCTION

Jones and Jake were ecstatic when they got off the plane from New Orleans. The last-minute arrangements to go to the Super Bowl had worked out wonderfully. As they walked down the corridor in the terminal, two plainclothes police officers stopped them, and informed Jones that he was under arrest. Jones protested, "There must be some mistake."

One of the officers said, "There's no mistake. Next time you call in sick to work, make sure you are not seen on national television cheering at the Super Bowl."

Mystified, Jones asked, "What's the charge?"

"Obtaining money under false pretenses," replied the officer.

Jones looked at Jake, who shrugged at Jones and said, "Don't look at me—I told you that you should have used vacation time."

Our legal system can be broken down in a number of ways, but perhaps the simplest is to divide it into two major areas: civil law and criminal law. *Criminal law* relates to laws designed to prevent harm to society in general, and which declare certain actions to be crimes. **Crimes** are activities that violate a criminal law, for which society may penalize an offender. The goal of criminal law is the vindication of a wrong done to society, and the prevention of future bad acts by both this offender and others.

Civil law pertains to laws that seek to enforce private rights and remedies for some wrong done to an individual or organization. The same act can constitute both a wrong to society and a wrong to an individual and thus may give rise to both a criminal action and a separate civil action. Our focus on the criminal side is to ensure that the wrongdoer is punished, that the wrongdoer and others are deterred from committing the same wrong, and that self-help by the victim and others who may want to act on behalf of the victim against the perpetrator is unnecessary. Compensation to the victim of a crime has also become a consideration on the criminal side.

crimes
activities that violate a criminal law for which society may penalize an offender

STATUTORY AND COMMON-LAW CRIMES

Most crimes are defined by statutes. These statutes, which are acts of Federal and state legislatures, define what actions are criminal and specify the penalties. However, there are certain common-law crimes that still exist and are not the result of a legislative act. Rather, common-law crimes are defined historically by cases and the common law. The crime of arson was one of those common-law crimes. Common-law arson was defined as the willful and malicious burning of the dwelling of another. Under common law, if the fire was set in a barn or store, it was not arson. If a person set his own house on fire, it was not arson. If the fire burned only contents but did not cause charring of structural members, it was not arson. This definition illustrates the problem of applying common-law crimes in the modern context, as well as the reason why all states have replaced or supplemented common-law arson with statutory arson offenses.

FEDERAL VERSUS STATE JURISDICTION

As we have discussed throughout the book, the Federal government is one of limited jurisdiction. This limitation on jurisdiction is a very important consideration in the realm of criminal law. Federal criminal jurisdiction is limited to Federal issues, namely those items provided for in the United States Constitution. These include matters affecting interstate commerce, drugs, immigration, counterfeiting, and the military, to name a few.

The remainder of criminal jurisdiction is left to the states. States are generally viewed as having broad police powers in order to address matters pertaining to the health, safety, and welfare of their inhabitants. Most criminal laws exist and are enforced at the state level. These include the crimes of murder, manslaughter, assault, robbery, larceny, burglary, rape, and arson. States may provide local municipalities with the authority to create some criminal laws, but local criminal offenses are generally limited to relatively minor matters.

CRIMES MUST BE SPECIFICALLY DECLARED TO BE CRIMINAL

Criminal laws must specifically describe the conduct that is prohibited, and declare its commission to be a crime. The mere violation of a law does not necessarily constitute a criminal act. The conduct must be declared by the statute to be a criminal act. For example, a car dealer who fails to comply with a so-called "lemon law" may violate state law by not honoring his obligation to provide a new car to someone whose car is a "lemon," but it does not rise to the level of a criminal violation. The "lemon law" is a civil law, and its violation is not declared by the law to constitute a crime.

FELONIES AND MISDEMEANORS

Crimes can be divided into two major categories, felonies and misdemeanors. Felonies are considered to be the more serious crimes, while misdemeanors are the less serious. The general rule is that crimes punishable by more than one year in jail are felonies, while crimes punishable by sentences of one year or less are misdemeanors. Some states also look at the dollar value of the fine, using either $500 or $1,000 as the threshold to distinguish a felony from a misdemeanor.

ELEMENTS

element of a crime

a fact that must be proven in order for a person to be convicted of a crime; crimes are made up of elements; there are three categories of elements: acts, mental states, and attendant circumstances; all crimes have act requirements and mental state requirements; certain crimes have attendant circumstance requirements

presumption of innocence

the principle by which a person is presumed to be innocent, until his or her guilt is established by proof beyond a reasonable doubt as to each and every element of the crime

Whether a crime is defined by a statute or the common law, crimes are made up of elements. An **element of a crime** is a fact that must be present in order for a crime to be committed. In order for someone to be convicted of a crime, each element of the crime must be proven beyond a reasonable doubt. If any element of the crime is not proven beyond a reasonable doubt, the accused cannot be convicted. This concept is referred to as the **presumption of innocence**. A person is presumed to be innocent until his or her guilt is established by proof beyond a reasonable doubt as to each and every element of the crime. There is no responsibility on the part of an accused to prove his or her innocence. The burden of proving a criminal case rests solely with the prosecution.

Elements fall into three categories: (1) act; (2) mental state; and (3) attendant circumstances.

Act

One essential element of every crime is an act. An act can be any affirmative action taken, or the failure to act when one is legally required to act. An act must be voluntary in nature. The crime of arson requires that the perpetrator start a fire. Starting the fire is the act. The act could be striking a match, setting a timer, or pouring the accelerant.

The act requirement for murder is an action that results in the death of the victim. It could be shooting the victim. It could be striking the victim with a baseball bat. It could be placing poison in the victim's food. Any voluntary act that causes the victim's death can satisfy the "act" requirement.

Omissions and the Duty to Act Can the failure of someone to act satisfy the "act" requirement for a criminal offense? If a child is trapped inside a burning building, can a passerby who fails to rescue the victim be charged criminally in connection with the death? As a general rule, no one has an affirmative responsibility to come to the aid of anyone who needs help. A person could be face down in a puddle of water on a busy street, about to drown, and no one has a legal responsibility to come to that person's aid.

However, a duty to act may arise from the relationship between the parties. The relationship may be a family relationship, a contractual relationship, or a voluntary assumption of responsibility. For example, parents have a legal duty to come to the aid of their child. Lifeguards have a duty to come to the aid of someone in distress in the pool they are responsible for watching. Similarly, an on-duty firefighter has a duty to respond to alarms when dispatched. The failure to act, when one is under a legal duty to act, can satisfy the act requirement of a crime.

SIDEBAR

QUESTION: Able is a lifeguard at the state beach. Baker dates Able's former girlfriend, Charleen. As a result, Able hates Baker. While swimming at the state beach, Baker starts to drown. Able sees him in distress, but laughs and waves to Baker as he drowns. Can Able's failure to help Baker satisfy the act requirement if Able is charged with Baker's death?

ANSWER: Yes. Under the circumstances, Able had a legal duty to come to the aid of Baker, and his failure to do so would satisfy the act requirement.

A duty to act may also arise where a statute imposes a duty to act. Many states impose a duty to render assistance based upon specific circumstances. For example, many states require that the operator of a car or boat that is involved in an accident render aid to anyone who is injured. Other states, such as Rhode Island, have adopted a more generalized "duty to act" or "duty to render assistance" law, which requires that any person come to the aid of another person who is in danger, provided such assistance can be rendered in relative safety to the rescuer (see RIGL 11-56-1 in Appendix B, Chapter 5). Such "duty to render assistance" laws create the potential to expose those who fail to act to criminal liability, as well as civil liability, for their failure to render aid.

Commonwealth v. Levesque
436 Mass. 443 (2002)
Massachusetts Supreme Judicial Court

COWIN, J. A grand jury in Worcester County returned six indictments against each defendant [Thomas Levesque and Julie Barnes] for involuntary manslaughter. . . . The indictments were based on grand jury testimony concerning the defendants' conduct in starting by accident and then failing to report a fire in the Worcester Cold Storage factory building (warehouse), which took the lives of six Worcester fire fighters. [sic] The defendants moved to dismiss the manslaughter indictments. . . . [T]he Superior Court allowed the motions to dismiss . . . because the defendants had no legal duty to report the fire and their failure to act did not

(*continued*)

(continued)

Figure 5-1 *The fire at the Worcester Cold Storage building on December 3, 1999 claimed the lives of six Worcester firefighters. (Photo by Tom Carmody.)*

satisfy the standard of wanton and reckless conduct required for manslaughter charges. The Commonwealth appealed. . . .

The evidence presented to the grand jury . . . indicated the following. . . . For several months prior to December 3, 1999, the defendants lived in a room on the second floor of the vacant five-story warehouse. The warehouse was a cold storage building and, as such, had brick walls, wood framing, and a compartmentalized floor plan with many small windowless rooms insulated with cork and styrofoam. [sic] The second floor where the defendants stayed had some windows, but those windows were boarded up. The room occupied by the defendants contained a bed, closet, and personal effects, including clothing, blankets, a radio, a wooden end table, and a kerosene heater. The defendants had an operable cellular

(continued)

(*continued*)

telephone, food, and pets. Because there was no electricity, a flashlight, candles, and a heater were used for light. On more than one occasion, the defendants had an overnight guest in these quarters. . . .

On the afternoon of December 3, 1999, between 4:15 P.M. and 4:30 P.M., the defendants had a physical altercation in their bedroom at the warehouse that resulted in the knocking over of a lit candle. A fire started and the defendants tried unsuccessfully to put the fire out with their feet and a pillow. The fire spread rapidly until everything in the room began to burn. The defendants searched for the cat and dog that lived in the warehouse with them but the search was futile. The defendants left the warehouse and did not report the fire to the authorities.

After leaving the warehouse, the defendants passed several open businesses and shopping mall stores where public telephones were available. Between 4 and 5 P.M., the general manager of Media Play store saw the defendants in his store and heard Julie Barnes say, "I can't believe I lost all my stuff. . . . I lost everything. I don't have anything. I lost all my stuff. I can't believe I lost everything." Thomas Levesque replied, "Don't worry about it. Let's go." After leaving Media Play, the defendants walked around the mall until they left to get dinner. They returned to the mall where they first went back to Media Play to listen to more music, and then went to a Sports Authority store to get a job application. The defendants subsequently went to Regina Guthro's house where Levesque remained until the next morning. Barnes spent the night with Bruce Canty at a hotel where both Barnes and Canty viewed the ongoing warehouse fire from their hotel room window.

Levesque made three telephone calls from his cellular telephone the day of the fire. One call was made at approximately 6 A.M. The record is unclear whether the other calls, at 11:20 and 11:28 (made to the hotel where Barnes was staying), were made in the morning or the evening. The next telephone call was placed from Levesque's cellular telephone four days after the fire.

The fire was not reported until 6:13 P.M. that evening when an emergency caller reported the fire. Sergeant O'Keefe, an expert in arson and fire investigations, stated that "the significance of the delay in reporting [the fire] ha[d] a great deal to do with what kind of fire the Worcester Fire Department got to that day." After arriving on the scene, fire fighters were informed that there might be homeless persons inside the warehouse. The fire fighters entered the warehouse in an effort to locate any persons that might have been inside, and to evaluate their tactics to combat the fire. It was during these efforts that six fire fighters went into the building and never returned. Rescuers recovered their remains during the eight days that followed.

A joint investigation by the Worcester fire department, [sic] the Massachusetts State police fire and explosion investigation section, and the United States Bureau of Alcohol, Tobacco and Firearms revealed that the warehouse fire, which originated in the defendant's [sic] second-floor "makeshift" bedroom, was most likely accidental and the result of an open candle flame in contact with combustible material. Remnants of the defendants' belongings, including plastic milk crates, an outline of a bed-type structure, the remains of a dog and cat, a candle, and a telephone calling card, were found among the debris.

(*continued*)

(*continued*)

Figure 5-2   *A tower ladder operates at the Worcester Cold Storage building fire on December 3, 1999. (Photo by Tom Carmody.)*

Standard of review. Our inquiry here is limited to whether the evidence presented to the grand jury was sufficient to support the defendants' indictments for involuntary manslaughter. . . .

Sufficiency of the evidence. Because Massachusetts has not defined manslaughter by statute, its elements are derived from the common law. . . . Involuntary manslaughter is "an unlawful homicide, unintentionally caused . . . by an act which constitutes such a disregard of probable harmful consequences to another as to constitute wanton or reckless conduct." . . . [T]he defendants contend that they did not have a duty to report the fire, thus rendering the evidence insufficient to demonstrate wanton and reckless conduct, and that the evidence does not support a finding that the fire fighters' deaths were caused by their failure to report the fire.

Duty to report the fire. Wanton or reckless conduct usually consists of an affirmative act "like driving an automobile or discharging a firearm," . . . An omission, however, may form the basis of a manslaughter conviction where the defendant has a duty to act. . . . It is true that, in general, one does not have a duty to take affirmative action, however, a duty to prevent harm to others arises when one creates a dangerous situation, whether that situation was created intentionally or negligently.

(*continued*)

(*continued*)

When formulating duties in the criminal context, we have, in the past, drawn on the duties imposed by civil law. For example . . . we [have] held that parents can be convicted of the involuntary manslaughter of their child if the failure to provide medical care for the child results in death. . . . The court found a duty by drawing on a parent's established civil duty to support sufficiently his or her child. . . .

Our law, both civil and criminal, imposes on people a duty to act reasonably. . . . The civil law creates a specific duty that we may apply to the situation in this case. The Restatement (Second) of Torts § 321 (1) (1965) reads: "If the actor does an act, and subsequently realizes or should realize that it has created an unreasonable risk of causing physical harm to another, he is under a duty to exercise reasonable care to prevent the risk from taking effect." Other jurisdictions have used this principle as a basis for criminal liability. . . .

Although we have yet to recognize explicitly § 321 as a basis for civil negligence . . . we have expressed agreement with its underlying principle. It is consistent with society's general understanding that certain acts need to be accompanied by some kind of warning by the actor. . . . [W]here one's actions create a life-threatening risk to another, there is a duty to take reasonable steps to alleviate the risk. The reckless failure to fulfil this duty can result in a charge of manslaughter.

Where a defendant's failure to exercise reasonable care to prevent the risk he created is reckless and results in death, the defendant can be convicted of involuntary manslaughter. Public policy requires that "one who creates, by his own conduct . . . a grave risk of death or injury to others has a duty and obligation to alleviate the danger." . . . We are not faced with the situation of a mere passerby who observes a fire and fails to alert authorities; the defendants started the fire and then increased the risk of harm from that fire by allowing it to burn without taking adequate steps either to control it or to report it to the proper authorities.

Whether a defendant has satisfied this duty will depend on the circumstances of the particular case and the steps that the defendant can reasonably be expected to take to minimize the risk. Although, in this case, the defendants apparently could not have successfully put out the fire, they could have given reasonable notice of the danger they created. It was for the grand jury (and later, the petit jury) to decide whether the defendants' failure to take additional steps was reasonable, and if not, whether the defendants' omission constituted wanton or reckless conduct.

Wanton or reckless conduct. A defendant's omission when there is a duty to act can constitute manslaughter if the omission is wanton or reckless. . . . To constitute wanton or reckless conduct, "the risk of death or grave bodily injury must be known or reasonably apparent, and the harm must be a probable consequence of the defendant's election to run that risk or of his failure reasonably to recognize it." . . . Under Massachusetts law, recklessness has an objective component as well as a subjective component. A defendant can be convicted of manslaughter even if he was "so stupid [or] so heedless . . . that in fact he did not realize the grave danger . . . if an ordinary normal man under the same circumstances would have realized the gravity of the danger." . . .

Although it is true that recklessness must involve an intentional act or omission, a finding of recklessness is grounded in intent to engage in the reckless

(*continued*)

(*continued*)

conduct, and not intent to bring about the harmful result. . . . Thus, the grand jury needed to determine only that the defendants' choice not to report the fire was intentional, not that the fire was intentionally set. . . .

The Commonwealth has presented sufficient evidence to allow a grand jury to conclude that the defendants' choice not to report the fire was intentional and reckless. The following testimony was sufficient in this regard: the defendants attempted to put out the fire and were unsuccessful, thus demonstrating they were cognizant of the fire's rapid spread; they observed the fire consume their possessions over a short period of time; they were forced to abandon their attempts to rescue their pets, again evidencing their awareness of the peril posed by the fire's rapid spread; they possessed a cellular telephone and passed several open stores after their exit from the warehouse, thus allowing the grand jury to infer that the defendants had multiple opportunities and the means to call for help if they chose to do so. Further, the testimony that the defendants went shopping and calmly ate a meal after leaving the building refutes any suggestion that panic might explain a failure to report the fire. Finally, the fact that the defendants may have faced criminal liability for trespass had they informed authorities that they had been living in the warehouse provided a motive for their failure to report the fire.

The defendants also assert that their conduct could not have been reckless because it was unforeseeable that such grievous harm would result to the fire fighters who responded. The Superior Court judge agreed, noting that fire fighters ordinarily do not lose their lives in the course of fighting a fire, and that even the fire fighters themselves failed to appreciate the gravity of the danger. However, an uncontrolled fire is inherently deadly to all who may come into contact with it, whether fire fighters or ordinary citizens. The defendants are charged with this knowledge. . . .

Figure 5-3 *An impromptu shrine dedicated to the six Worcester firefighters who died in the December 3, 1999 fire at the Worcester Cold Storage building. (Photo by Tom Carmody.)*

(*continued*)

(*continued*)

The defendants argue that certain actions by the fire department contributed to the fire fighters' deaths, such as the fire fighters' inability to navigate the maze-like building, and the fire fighters' ignorance of the true extent of the danger posed by the fire. However, "the intervening conduct of a third party will relieve a defendant of culpability only if such an intervening response was not reasonably foreseeable." . . . The inability of the fire fighters to navigate the building or estimate the true caliber of the danger was foreseeable to those in the defendants' position as a part of the normal risks of combating a fire. . . .

Conclusion. The order allowing the defendants' motions to dismiss is reversed. We remand the case to the Superior Court for further proceedings consistent with this opinion.

Case Name: Commonwealth v. Levesque

Court: Massachusetts Supreme Judicial Court

Summary of Main Points: The act requirement for the crime of involuntary manslaughter can be satisfied by the failure to report a fire that one is responsible for causing, provided that the failure to report the fire was wanton or reckless with regard to the risk posed to the victims.

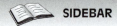 **SIDEBAR**

On May 30, 2002, in the aftermath of the Massachusetts Supreme Judicial Court's decision, Superior Court Judge Daniel F. Toomey decided to postpone any plea agreements or trial, and continued the involuntary manslaughter cases against Levesque and Barnes for five years. Provided Levesque and Barnes comply with the terms of their pretrial release, and have no further problems with the law during that period, the charges will be dismissed at the end of five years.

Mental State

When we think about a crime, we often think that committing the act constitutes committing the crime. However, in addition to an act, every crime also has a mental state requirement, called a **mens rea**. The specific mental state needed for a crime varies from crime to crime. According to the Model Penal Code, there are four basic criminal mental states, although various jurisdictions and authorities may cite slightly different terms and definitions. The four criminal mental states are purposeful, knowing, reckless, and negligent.

Purposeful means that the person intended or desired to engage in certain conduct or cause a certain outcome.

mens rea
the criminal mental state necessary to satisfy the mental state element for a crime

> ### 📖 SIDEBAR
>
> Dave hates Ed and shoots him with a gun on purpose, desiring to kill or injure him. Dave's mental state toward killing or injuring Ed was purposeful.

Knowingly means that the person knew to a substantial certainty that a certain outcome would follow from his or her act, even if the person did not intend or desire it.

> ### 📖 SIDEBAR
>
> Fred lit his house on fire. After the fire, George, the investigator, seized evidence that needed to be sent to the crime lab for analysis. Fred does not want the evidence to reach its destination. He plants a bomb in George's car and blows it up as George is driving to the crime lab. Fred does not desire to harm George; he only wants to destroy the evidence. However, he knows to a substantial certainty that blowing up George's car while it is en route to the crime lab will result in George's death. Fred has acted knowingly with regard to George's death. Incidentally, with regard to setting the fire and blowing up the car, Fred acted purposefully.

recklessness
an aggravated form of gross negligence; recklessness requires that the actor have knowledge that harm was likely to result from his behavior, and a conscious choice to act (or refusal to act when under a duty to act) despite the risk; it is the knowledge that harm is likely to result that separates recklessness from gross negligence (note: recklessness can also be viewed as a criminal mental state)

Recklessness is the mental state in which a person consciously disregards a known and substantial risk of harm involving a gross deviation from the standard of care of the law-abiding citizen. While somewhat less clear-cut than purposeful or knowingly, recklessness is an important criminal mental state for many crimes.

> ### 📖 SIDEBAR
>
> Holly drives her sports car through a crowded neighborhood at 95 miles per hour, and accidentally strikes Inez, killing her. The speed limit in the neighborhood is 25 miles per hour. Holly's mental state in relation to the death of Inez is an example of recklessness.

Some jurisdictions draw a distinction between ordinary "willful and wanton recklessness," and an extreme type of recklessness known as depraved heart recklessness. The classic example of depraved heart recklessness involves someone playing a game of Russian roulette. With one bullet in a revolver that holds six bullets, it cannot be said that the shooter purposefully intends to shoot the victim. It can also not be said that the shooter knew to a substantial certainty that the victim would be shot, since at best there was a one-in-six chance that the gun would go off. Nevertheless, the mental state of the person playing Russian roulette is so outrageous and callous with regard to the consequences as to warrant the same punishment as someone who acts purposefully or knowingly.

As we will discuss below, depraved heart recklessness will usually be enough to be convict a perpetrator of a serious crime requiring intentional conduct, such as murder. Ordinary recklessness will usually not be enough to convict for murder, but is usually adequate for involuntary manslaughter.

Negligence is a criminal mental state in which the perpetrator should have been aware that a substantial and unjustifiable risk of harm would result from his or her conduct. The criminal mental state of negligence involves a gross deviation from what the reasonably prudent person would have done under the circumstances. The definition of negligence for purposes of criminal mental state is similar—but not identical to—the definition of negligence as a civil action or tort, as discussed in Chapter 7. Some states equate criminal negligence with the civil standard for gross negligence.

In the classic *Dante's Inferno*, hell is depicted as concentric circles going deeper and deeper into the underground. The worse one's sins, the deeper one's place in hell, and the worse the punishment that one has to endure. In an analogous way, criminal law looks at a perpetrator who acts purposefully as being more culpable than one who acts knowingly, and one who acts knowingly as being more culpable than one who acts recklessly. Negligence is the least culpable of the four criminal mental states. **See Figure 5-4.**

Which Mental State Is Required? Each crime carries a different mental state requirement. Which mental state is required for a given criminal offense? Terminology

negligence (criminal mental state)
a criminal mental state in which the perpetrator should have been aware that a substantial and unjustifiable risk of harm would result from his or her conduct

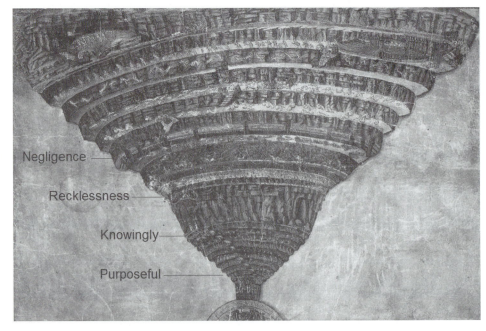

Figure 5-4 *Dante's Inferno, as depicted in this Botticelli drawing, envisions hell as concentric circles going deeper and deeper into the underground. The worse one's sins, the deeper one's place in hell.*

Negligence

Recklessness

Knowingly

Purposeful

> ### SIDEBAR
>
> Why isn't the act requirement alone enough to convict someone? Why do we need to show an actor's mental state? Isn't it what a person does, not why he did it, that should be punished? Consider someone whose act is flipping a switch that turns on the lights in an adjacent room. The room is full of gasoline vapors within the flammable range. If person does not know that the room is full of gasoline vapors, and that turning on the switch will ignite the vapors, he does not have a criminal mental state. If the person knows that turning on the switch will ignite the gas, and desires to burn the building down, he is acting purposefully. Thus, in some ways, we are more concerned with a person's mental state than we are with the act.

varies greatly from state to state, but frequently the term "intentional" is used to describe the mental state necessary for someone to be convicted of most serious crimes. How does the term "intentional" relate to purposeful, knowing, reckless, or negligent? Generally, purposeful and knowing are always enough to convict someone of a crime. Recklessness, if significant enough, can be enough to convict someone of murder. Typically, if a person's recklessness is so outrageous as to indicate the act was committed with a "depraved heart," recklessness will be deemed the equivalent of purposeful or knowing, and will be deemed intentional. Negligence is generally not enough to convict someone of more serious crimes. However, for certain crimes—such as negligent homicide or vehicular homicide—negligence may be enough, depending upon the law of the specific jurisdiction.

Concurrence of the Act and the Mental State   For a crime to be committed, the act must be committed with the requisite mental state. The fact that the perpetrator commits the act and has the criminal mental state at about the same time is not enough if the act does not "concur" with the mental state.

> ### SIDEBAR
>
> Assume that Jack desires to kill Kevin. Jack drives over to Kevin's house with a gun to shoot him. While Jack is driving to Kevin's house, a figure runs out in front of Jack's car, and Jack's car strikes the person, killing him instantly. Upon stopping his vehicle, Jack approaches the victim, and only then does he realize that it is Kevin. While Jack possessed the desire to kill Kevin, and while Jack actually did kill Kevin, Jack did not act with the required mental state in order to be convicted in Kevin's death. The mental state must concur with the act giving rise to the death.

Strict Liability Crimes   There is one category of crime for which a criminal mental state is not required to be proven. Such crimes are referred to as strict liability crimes, because proof of the defendant's mental state is not required. Generally, strict liability can only be imposed on relatively minor crimes, traditionally of a regulatory nature. Traffic offenses are often considered to be strict liability offenses, as are violations of many environmental, food, and health regulations. The focus of a strict liability crime is solely on the act.

Attendant Circumstances

Some crimes may also require that certain facts be present in addition to an act and a mental state. For example, under common-law arson, the building that is set on fire must be a dwelling. Thus, in addition to proving that the fire occurred, and that it was intentionally set by the defendant, the prosecution must prove that the building was a dwelling. For the crime of assaulting a firefighter, one of the attendant circumstances is that the person assaulted must be a firefighter.

CRIMINAL OFFENSES

Homicide

murder
the intentional killing of another without justification

first-degree murder
murder that is committed with malice aforethought

malice aforethought
the mental state requirement for murder, and refers to the fact that the crime was premeditated

second-degree murder
any murder committed without malice aforethought; it can also be defined as any murder which is not first-degree murder

Homicide is broadly defined as the killing of another person. The term homicide includes two primary groups of offenses, murder and manslaughter. Each state differs in its specific definitions of homicide crimes. Some states have created additional categories, such a vehicular homicide and negligent homicide. However, there are some general principles that transcend state-to-state differences that are worth discussing.

Murder is the intentional killing of another without justification. Most states recognize two degrees of murder, first-degree and second-degree. **First-degree murder** is murder that is committed with **malice aforethought**. Malice aforethought refers to the fact the crime was premeditated. Any murder committed without malice aforethought is considered to be **second-degree murder.**

Malice aforethought, or premeditation, can be inferred from the perpetrator's actions. Premeditation need only occur for a moment before the act is committed. There is no specific time requirement placed on premeditation, so long as the actor had time to think about the consequences of his or her actions before committing the act. Using poison, or lying in wait to kill someone, are examples of how someone's actions demonstrate premeditation. More difficult are the cases where the perpetrator acted without much time to reflect upon the consequences of his or her actions. Each case must be evaluated by the jury to determine if the perpetrator acted with premeditation.

felony murder rule
a rule adopted in many states that any killing, whether accidental or intentional, that occurs during the commission of a felony shall be considered to be first-degree murder

Another type of murder that most states consider to be first-degree murder is any killing, whether accidental or intentional, that occurs during the commission of a felony. This is known as the **felony murder rule**. For example: A person robs a bank using a gun. During the robbery, a bank teller has a heart attack and dies. The robber could be charged with first-degree murder via the felony murder rule. Any death that occurs during the commission of a felony can result in a conviction of first-degree murder, even when the death is accidental or completely beyond the control of the perpetrator. Some states hold that if the underlying felony is not inherently dangerous, the perpetrator can be charged only with second-degree murder.

A more appropriate example for us would be the accidental death of a firefighter at a building fire that is deliberately set. Provided the perpetrator is charged with a felony arson charge, the felony murder rule could be used to charge the arsonist with first-degree murder.

Second-degree murder is simply defined as any murder that is not first-degree. It can also be defined as murder without premeditation. A few states limit first-degree murder to special circumstances, such as the murder of a police officer in the line of duty. In such states, premeditated murder would be chargeable as second-degree murder.

manslaughter
homicide crime that does not constitute murder, but may nevertheless be chargeable as a crime; there are two categories of manslaughter, voluntary and involuntary

Homicides that do not constitute murder may nevertheless be chargeable as **manslaughter**. There are two categories of manslaughter, voluntary and involuntary.

Voluntary manslaughter is the intentional killing of another, committed in the heat of passion as a result of a severe provocation. The classic example of voluntary manslaughter occurs when a husband comes home to find another person in bed with his wife, and shoots one or both of them. It is for the jury to decide whether the perpetrator acted in the heat of passion as a result of severe provocation.

voluntary manslaughter
the intentional killing of another, committed in the heat of passion as a result of a severe provocation

Involuntary manslaughter is an unintentional killing that results from the reckless conduct of the defendant. It is an unintentional killing committed under circumstances in which the recklessness of the actor make him legally responsible.

involuntary manslaughter
an unintentional killing that results from the reckless conduct of the defendant

On November 28, 1942, 492 people died in a fire at the Cocoanut Grove nightclub in Boston, Massachusetts **(Figure 5-5)**. The club had a legal capacity of 600, but there were more than a thousand people in the club at the time of the fire. Some exit doors had been locked to keep nonpayers out. Other exit doors that did operate opened inward, in clear violation of the fire code. The building also had highly combustible furnishings that contributed to the fire's rapid spread.

The cause of the fire has never been conclusively proven, but most authorities believe that a busboy started the fire accidentally. The Commonwealth of Massachusetts indicted a number of individuals for involuntary manslaughter, including the owner of the club.

Figure 5-5 *On November 28, 1942, 492 people died in the Cocoanut Grove nightclub fire in Boston, Massachusetts. (Photo by Corbis.)*

Commonwealth v. Welansky
316 Mass. 383 (1944)
Massachusetts Supreme Judicial Court

LUMMUS, J. On November 28, 1942 . . . a corporation named New Cocoanut Grove, Inc., maintained and operated a "night club" [sic] in Boston, having an entrance at 17 Piedmont Street. . . . It employed about eighty persons. The corporation [was] completely dominated by the defendant Barnett Welansky. . . . He owned, and held in his own name or in the names of others, all the capital stock. He leased some of the land on which the corporate business was carried on, and owned the rest, although title was held for him by his sister. He was entitled to, and took, all the profits. Internally, the corporation was operated without regard to corporate forms, as though the business were that of the defendant as an individual. It was not shown that responsibility for the number or condition of safety exits had been delegated by the defendant to any employee or other person. . . .

(continued)

(*continued*)

The physical arrangement of the night club on November 28, 1942. . . . The total area of the first or street floor was nine thousand seven hundred sixty-three square feet. Entering the night club through a single revolving door at 17 Piedmont Street, one found himself in a foyer or hall having an area of six hundred six square feet. From the foyer, there was access to small rooms used as toilets, to a powder room and a telephone room, to a small room for the checking of clothing, and to another room with a vestibule about five feet by six feet in size adjoining it, both of which were used as an office in the daytime and for the checking of clothing in the evening. In the front corner of the foyer, to the left, beyond the office, was a passageway leading to a stairway about four feet wide, with fifteen risers. That stairway led down to the Melody Lounge in the basement, which was the only room in the basement open to the public. . . .

We now come to the story of the fire. A little after ten o'clock on the evening of Saturday, November 28, 1942, the night club was well filled with a crowd of patrons. It was during the busiest season of the year. An important football game in the afternoon had attracted many visitors to Boston. Witnesses were rightly permitted to testify that the dance floor had from eighty to one hundred persons on it, and that it was "very crowded." . . . Upon the evidence it could have been found that at that time there were from two hundred fifty to four hundred persons in the Melody Lounge, from four hundred to five hundred in the main dining room and the Caricature Bar, and two hundred fifty in the Cocktail Lounge. . . . There were about seventy tables in the dining room, each seating from two to eight persons. There was testimony that all but two were taken. Many persons were standing in various rooms. The defendant testified that the reasonable capacity of the night club, exclusive of the new Cocktail Lounge, was six hundred fifty patrons. . . .

A bartender in the Melody Lounge noticed that an electric light bulb which was in or near the cocoanut husks of an artificial palm tree in the corner had been turned off and that the corner was dark. He directed a sixteen year old bar boy who was waiting on customers at the tables to cause the bulb to be lighted. A soldier sitting with other persons near the light told the bar boy to leave it unlighted. But the bar boy got a stool, lighted a match in order to see the bulb, turned the bulb in its socket, and thus lighted it. The bar boy blew the match out, and started to walk away. Apparently the flame of the match had ignited the palm tree and that had speedily ignited the low cloth ceiling near it, for both flamed up almost instantly. The fire spread with great rapidity across the upper part of the room, causing much heat. The crowd in the Melody Lounge rushed up the stairs, but the fire preceded them. People got on fire while on the stairway. The fire spread with great speed across the foyer and into the Caricature Bar and the main dining room, and thence into the Cocktail Lounge. Soon after the fire started the lights in the night club went out. The smoke had a peculiar odor. The crowd were panic stricken, and rushed and pushed in every direction through the night club, screaming, and overturning tables and chairs in their attempts to escape.

(*continued*)

(*continued*)

The door at the head of the Melody Lounge stairway was not opened until firemen broke it down from outside with an axe and found it locked by a key lock, so that the panic bar could not operate. Two dead bodies were found close to it, and a pile of bodies about seven feet from it. The door in the vestibule of the office did not become open, and was barred by the clothing rack. The revolving door soon jammed, but was burst out by the pressure of the crowd. The head waiter and another waiter tried to get open the panic doors from the main dining room to Shawmut street [sic], and succeeded after some difficulty. The other two doors to Shawmut Street were locked, and were opened by force from outside by firemen and others. Some patrons escaped through them, but many dead bodies were piled up inside them. A considerable number of patrons escaped through the Broadway door, but many died just inside that door. Some employees, and a great number of patrons, died in the fire. Others were taken out of the building with fatal burns and injuries from smoke, and died within a few days.

I. The pleadings, verdicts, and judgments.

The defendant, his brother James Welansky, and Jacob Goldfine, were indicted for manslaughter in sixteen counts of an indictment. . . . [I]t specified among other things that the alleged misconduct of the defendant consisted in causing or permitting or failing reasonably to prevent defective wiring, the installation of inflammable decorations, the absence of fire doors, the absence of "proper means of egress properly maintained" and "sufficient proper" exits, and overcrowding. . . .

II. The principles governing liability.

The Commonwealth disclaimed any contention that the defendant intentionally killed or injured the persons named in the indictments as victims. It based its case on involuntary manslaughter through wanton or reckless conduct. The judge instructed the jury correctly with respect to the nature of such conduct.

Usually wanton or reckless conduct consists of an affirmative act, like driving an automobile or discharging a firearm, in disregard of probable harmful consequences to another. But where, as in the present case, there is a duty of care for the safety of business visitors invited to premises which the defendant controls, wanton or reckless conduct may consist of intentional failure to take such care in disregard of the probable harmful consequences to them or of their right to care. . . .

To define wanton or reckless conduct so as to distinguish it clearly from negligence and gross negligence is not easy. . . . What must be intended is the conduct, not the resulting harm. . . . The words "wanton" and "reckless" are practically synonymous in this connection, although the word "wanton" may contain a suggestion of arrogance or insolence or heartlessness that is lacking in the word "reckless." . . .

The standard of wanton or reckless conduct is at once subjective and objective. . . . Knowing facts that would cause a reasonable man to know the danger is equivalent to knowing the danger. . . . The judge charged the jury correctly when he said, "To constitute wanton or reckless conduct, as distinguished from mere negligence, grave danger to others must have been apparent, and the defendant must have chosen to run the risk rather than alter his conduct so as to avoid the

(*continued*)

(*continued*)
act or omission which caused the harm. If the grave danger was in fact realized by the defendant, his subsequent voluntary act or omission which caused the harm amounts to wanton or reckless conduct, no matter whether the ordinary man would have realized the gravity of the danger or not. But even if a particular defendant is so stupid [or] so heedless . . . that in fact he did not realize the grave danger, he cannot escape the imputation of wanton or reckless conduct in his dangerous act or omission, if an ordinary normal man under the same circumstances would have realized the gravity of the danger. A man may be reckless within the meaning of the law although he himself thought he was careful."

The essence of wanton or reckless conduct is intentional conduct, by way either of commission or of omission where there is a duty to act, which conduct involves a high degree of likelihood that substantial harm will result to another. . . . Wanton or reckless conduct amounts to what has been variously described as indifference to or disregard of probable consequences to that other. . . .

The words "wanton" and "reckless" are thus not merely rhetorical or vituperative expressions used instead of negligent or grossly negligent. They express a difference in the degree of risk and in the voluntary taking of risk so marked, as compared with negligence, as to amount substantially and in the eyes of the law to a difference in kind. . . . For many years this court has been careful to preserve the distinction between negligence and gross negligence, on the one hand, and wanton or reckless conduct on the other. . . . . In pleadings as well as in statutes the rule is that "negligence and wilful and wanton conduct are so different in kind that words properly descriptive of the one commonly exclude the other." . . .

[I]t is now clear in this Commonwealth that at common law conduct does not become criminal until it passes the borders of negligence and gross negligence and enters into the domain of wanton or reckless conduct. There is in Massachusetts at common law no such thing as "criminal negligence." . . .

Wanton or reckless conduct is the legal equivalent of intentional conduct. . . . If by wanton or reckless conduct bodily injury is caused to another, the person guilty of such conduct is guilty of assault and battery. . . .

To convict the defendant of manslaughter, the Commonwealth was not required to prove that he caused the fire by some wanton or reckless conduct. Fire in a place of public resort is an ever present danger. It was enough to prove that death resulted from his wanton or reckless disregard of the safety of patrons in the event of fire from any cause. . . .

There is nothing in the point that because the corporation might have been indicted and convicted, the defendant could not be. The defendant was in full control of the corporation, its officers and employees, its business and its premises. He could not escape criminal responsibility by using a corporate form. . . .

Other assignments of error, relied on by the defendant but not discussed in this opinion, have not been overlooked. We find nothing in them that requires discussion.

Judgments affirmed.

Case Name: Commonwealth v. Welansky

Court: Massachusetts Supreme Judicial Court

Summary of Main Points: To convict the defendant of manslaughter in this case, the prosecution was not required to prove that the defendant caused the fire. Rather, it was enough to prove that the victims' deaths resulted from the defendant's wanton or reckless disregard for the safety of persons (in this case patrons of a night club) whom the defendant was under a legal duty to protect. As the court noted: "It was enough to prove that death resulted from his wanton or reckless disregard of the safety of patrons in the event of fire from any cause . . . ."

Battery

Battery is the unpermitted, offensive touching of another. When we think of a battery, we normally think of a punch or a slap, but any unpermitted, offensive touching can constitute a battery, including a push, a shove, poking someone with a finger, or even spitting. Two important issues that arise in the context of battery are consent and implied consent. Consent is a defense to battery. A person may consent to being touched, or even hit. Consent may be given in writing, in situations ranging from a doctor treating a patient to participants in a boxing match.

Consent may also be implied through social custom or engaging in an activity. A person who willingly gets on a crowded subway train impliedly consents to the normal jostling and contact that occurs on crowded subway trains. In many sports and in our daily lives, people come into physical contact with one another. Contact that is expected to occur in, say, a football game or a hockey game, is not considered a battery even if injury results, because the parties, by virtue of participating in the event, are deemed to have consented to such contact. On the other hand, where the parties do not consent to such contact, the act of touching, even in the absence of physical injury, is a battery.

Consent can also be implied when a person is unconscious and in need of medical attention. If a person is unconscious or otherwise incapable of consenting to, or declining, medical treatment, there is implied consent to treat the person. This implied consent is limited to the treatment necessary to save life and limb. Thus, if an unconscious person were brought to a hospital, a plastic surgeon who performs a cosmetic facelift has grossly exceeded the implied consent and could be guilty of battery.

For consent to be valid, it must be given knowingly and voluntarily, and be free from fraud, mistake, or deceit. The classic cases in this area involve people who pretend to be physicians, and revolve around the fact that the perpetrators obtain the victims' consent to touch them through the use of fraud or deceit. In such cases, the consent is considered to have been obtained unlawfully, and any contact that occurs is chargeable as battery, even though the patient consented at the time.

SIDEBAR

QUESTION: Jones is a new firefighter with no first aid training. While off duty in another community, he comes upon the scene of a car accident involving several female college cheerleaders returning from a football game. He stops, identifies himself as a nurse, and begins examining each cheerleader. The cheerleaders do not object. Could Jones be guilty of battery?

ANSWER: Yes, Jones runs the risk that he could be charged with battery. The cheerleaders' consent was obtained through Jones's misrepresentation about his medical skill and training. Jones could be charged with battery even though all contact was appropriate and in accordance with how a nurse or paramedic would have conducted the examinations.

Contact that goes beyond the scope of the consent granted constitutes battery. For example, in a hockey game with full-contact checking permitted, one player strikes another player over the head with his stick during a timeout. This act could be chargeable as a battery, even if the injury is no worse than the type of injuries that could have happened from contact that was consented to. Consent to the normal contact associated with a sport is not consent to all contact. The exact scope of contact that was consented to is a question of fact for the jury to decide.

One important limitation upon consent is that a person cannot consent to an illegal act. Any such consent to an illegal act is invalid and may result in criminal liability for the perpetrator. For example, someone cannot validly consent to being killed. In addition, someone cannot consent to engage in an illegal, unsanctioned boxing match. The perpetrator of such an act can be charged despite the victim's consent.

Assault

Assault is an attempted battery or other action that places the victim in fear of imminent bodily contact. Physical contact is not necessary for assault. In fact, if physical contact occurs, the assault becomes a battery. Some jurisdictions do not draw a clear distinction between assault and battery, and categorize either action as an assault. Simple assault, or simple battery, is usually chargeable as a misdemeanor. All states have additional felony assault and battery offenses where certain attendant circumstances are present, such as assault of a uniformed police officer, assault of a firefighter in the line of duty, or assault with a deadly weapon. Domestic assaults are an additional category of assaults.

False Imprisonment

False imprisonment was a common-law offense consisting of the unlawful restraint upon a person's freedom and ability to come and go as they please. False imprisonment has traditionally been a misdemeanor, and loosely associated with the more serious offense of kidnapping. Historically, a distinction has been drawn between false imprisonment and kidnapping, such that with kidnapping, the victim had to be transported some distance, whereas with false imprisonment the victim was held in place or transported only a short distance.

Kidnapping

asportation
the requirement that the victim of a kidnapping be transported from the site of the kidnapping to another location; at common law, the asportation had to be across some geographical boundary, such as out of a country, or across state or county lines

At common law, kidnapping was the forcible taking of another person, and transporting that person against his or her will across a boundary. Statutes have eliminated the boundary requirement, and typically treat kidnapping as an aggravated form of false imprisonment, usually requiring some sort of **asportation** or transportation of the victim. Kidnapping also requires the use of force or threatened use of force. In lieu of asportation, some states consider it to be kidnapping when a person is secretly isolated and confined against their will, without being transported. Kidnapping is a felony.

Consent and Patient Treatment

The crimes of battery, assault, false imprisonment, and kidnapping all come into play for firefighters and emergency medical personnel when it comes to treating, restraining, and transporting combative patients **(Figure 5-6)**. A key focus in such cases involves the capacity of the victim to decline medical treatment. Competent adult patients have the absolute right to decline medical treatment, and forcing treatment and transportation upon a patient who does not desire it could constitute criminal wrongdoing. On the other hand, identifying patients who lack the capacity to consent or decline treatment—due to age, intoxication, or mental impairment—can be a real challenge. This subject will be discussed in considerable detail in Chapter 7.

Rape

Common-law rape was defined as sexual intercourse without the other party's consent. While rape was a capital offense (a serious felony for which the perpetrator could be executed or sentenced to life in prison), it had a very narrow definition. Common-law rape was limited to sexual intercourse perpetrated by a male upon a female. At common law, many acts that were equally lewd and deserving of punishment went unpunished.

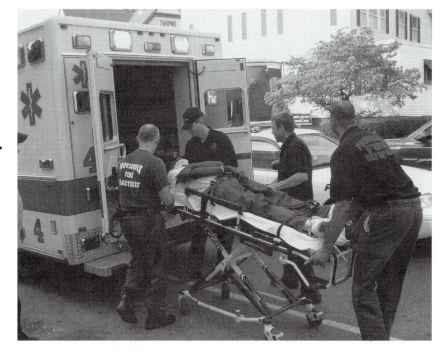

Figure 5-6 *The crimes of battery, assault, false imprisonment, and kidnapping all come into play for firefighters and emergency medical personnel when it comes to treating, restraining, and transporting combative patients.*

The modern trend is toward an expanded definition of rape, and reclassification of what we commonly referred to as "rape" as sexual assault. Statutory sexual assault laws commonly assign degrees to the different types of sexual conduct. First-degree sexual assault is typically the most serious offense, involving sexual penetration through the use of force. Sexual penetration is now interpreted to mean the penetration of any body opening by a sexual organ, or penetration of a sexual organ by any object. There is no requirement that the perpetrator be a man and that the victim be a woman.

In addition to the various degrees of sexual assaults, many states now recognize sexual contact with minors as child molestation, a separate charge. Child molestation similarly has various degrees assigned, based upon the severity of the conduct.

Larceny

At common law, larceny was the taking and transporting of property belonging to another with intent to permanently deprive the owner. An essential element of larceny was that the perpetrator had to have the specific intent to permanently deprive the owner of the item. For example, "borrowing" a horse was not considered larceny if the person intended to return it to the rightful owner.

While the common-law definition of larceny may have made sense in the 17th, 18th, and 19th centuries, modern times called for refinement. From the common-law crime of larceny, we now have a broad range of statutory larceny and theft crimes. The exact names and offenses of these modern theft crimes differ from state to state, but typically include stealing/larceny, embezzlement, obtaining money under false pretenses, receiving stolen goods, and extortion.

Larceny crimes are generally chargeable either as felonies or misdemeanors, depending upon the amount in question. Some state statutes hold that if the amount stolen exceeds $500, it is considered grand larceny, which is a felony. If the amount involved in the theft is less than $500, it is considered petty larceny, which is a misdemeanor.

Larceny or stealing remains the most basic of the theft crimes. True to its common-law definition, larceny is still the taking of the property of another with intent to permanently deprive the owner.

Embezzlement

The crime of embezzlement is larceny that occurs when someone who lawfully has possession of another's property by virtue of a trust relationship fraudulently appropriates that property for his or her own use. Embezzlement requires that the perpetrator legally have possession of the property to begin with, and then at some point the perpetrator wrongfully appropriates the property for him- or herself.

Obtaining Money Under False Pretenses

Obtaining money under false pretenses is a form of larceny that involves obtaining money, goods, wares, or other property through any false pretense or misrepresentation. The key element is a "false pretense," which is a false statement of material fact intended to deceive someone into giving the perpetrator the money or property. The false statement must be intended to induce the owner to give the perpetrator the money or property. Obtaining money under false pretenses frequently is an issue in arson for profit and insurance fraud cases, in which a fire is set purposefully to obtain money from the insurer. It can also be applicable in cases of welfare fraud or for personnel who pretend to be sick or injured in order to receive compensation from their employer.

Robbery

Robbery can be viewed quite simply as larceny through the use or threatened use of violence. It is the taking of property from another through the use, or threat, of immediate force. Robbery can be distinguished from the related

---

📖 **SIDEBAR**

Which of the following would be considered embezzlement?

- A firefighter at the scene of a car fire removes the battery from the car, and hides it on the fire truck. Upon return to the station he installs it in his personal vehicle.

  ANSWER: Not embezzlement, but larceny. The firefighter never had lawful possession of the battery.

- A firefighter finds cash at a fire scene, places it in his turnout gear, and doesn't turn it in. The firefighter later uses the money to buy presents for his children.

  ANSWER: Not embezzlement, but larceny. However, a case for embezzlement could be made if it could be proven that the firefighter initially intended to turn the money in, and later changed his mind.

- What if, in the above example, the firefighter turned the cash over to his officer, who turned it over to a police officer, who later decides to keep half of it for himself?

  ANSWER: Embezzlement on the part of the police officer.

- A union treasurer writes himself a check from the union's bank account to pay for his wife's credit card bill.

  ANSWER: Embezzlement.

---

theft crime of extortion through the immediacy of the threat. A robber threatens immediate physical harm to a person, while someone who commits extortion may threaten harm in the future. Robbery can be distinguished from larceny by virtue of the fact that in a robbery, the property is taken from the immediate physical presence of the victim, whereas in a larceny the property may be taken even though the victim is not present.

Extortion

Extortion is the obtaining of money or property—or otherwise requiring someone to do something they are not legally required to do—by means of a threat. The types of threats sufficient to establish extortion include threats to:

- inflict future bodily injury, or damage to property
- accuse another of a crime
- reveal confidential or embarrassing information about the victim
- take or withhold any action as an official, or cause an official to take or withhold any action

A common example of extortion occurs when a perpetrator informs the victim that he will expose damaging information about the victim if the victim does not pay money. It does not matter if the threat pertains to the release of truthful or false information.

Extortion can be a tricky subject. One point needs reinforcing: It is extortion to threaten to report someone to the police or authorities if they refuse to do something they are not legally required to do. If someone has violated a law, the proper course of action is to either report them to the proper authorities, or not report them. However, threatening to report someone in order to get them to comply with your demands could be extortion.

While extortion often involves a perpetrator seeking money or property, a profit motive is not required. Threats aimed at compelling any person to do something they are not required by law to do, or that prohibit a person from carrying out a duty imposed by law, is enough to constitute extortion. One type of threat that is *not* considered to be grounds for extortion is a threat to file a civil lawsuit.

📖 SIDEBAR

Which of the following would be considered extortion?

- A fire inspector orders a building owner to remove combustible materials stored in stairways, or else he will issue a citation for violating the fire code.

  ANSWER: Not extortion, because the fire inspector is ordering the person to comply with the law.

- A firefighter informs a homeowner that if he does not buy tickets to the firemens' ball, he will report him to the police for having too many dogs.

  ANSWER: Extortion.

- A firefighter informs her officer that if she is not granted a transfer, she will publicly accuse him of having had an extramarital relationship with her.

  ANSWER: Extortion (whether the accusation is true or false).

- A town official informs a fire inspector that if he cites a particular building for fire code violations, he will not receive a promotion.

  ANSWER: Extortion.

Here is one of the most difficult examples of extortion that fire officials have to deal with: A firefighter contends that she has been sexually assaulted and harassed by a superior officer. She demands that the fire department terminate the officer, and that she be compensated for her damages. She threatens that if her request is not honored, she will file suit against the officer and the fire department. This is *not* extortion. However, if instead of simply threatening to file a civil suit, she threatens to inform the officer's wife, report the matter to the local newspaper, or bring criminal charges against the officer if her demands are not met, she has committed extortion.

Burglary

Burglary was a common-law crime that was limited to breaking and entering the dwelling of another in the nighttime with intent to commit a felony therein. In much the same way as common-law rape limited the scope of who could be charged with rape, the definition of common-law burglary limited the scope of who could be charged with burglary. As a result, all states have adopted statutory burglary and "breaking and entering" statutes, which expand the definition to include breaking into commercial buildings, daytime break-ins, and break-ins where it is unclear whether or not additional felony was intended. Many states continue to recognize common-law burglary in addition to the newer statutory crimes.

Arson

As discussed previously, arson was traditionally a common-law crime, and a capital offense. The common-law crime of arson consisted of the willful and malicious burning of the dwelling of another. This definition severely restricted the ability of society to punish arsonists. It was not arson at common law to purposefully set fire to a commercial building, nor even to set fire to one's own house. If the fire did not char structural wood in the dwelling, it was not arson.

All states have supplemented or replaced their common-law arson crimes with comprehensive statutory arson charges **(Figure 5-7)**. While each state differs in the organization and wording of its statutory arson crimes, in practice each state's arson laws criminalize essentially the same offenses, in most respects. Statutory arson charges typically include separate offenses and penalties for anyone who intentionally burns an occupied building, unoccupied building, vehicle **(Figure 5-8)**, or wildlands. Arson that is intended to defraud an insurer may be punishable as a separate and additional offense. These laws may organize arson in various degrees, according to the relative threat to life and property. Some states also provide for an aggravated form of arson when an occupied building is involved, or when personal injury or death occurs to occupants or firefighters. (See Appendix B, Chapter 5 for examples of statutory arson charges.)

Act Requirement for Arson What acts will satisfy the act requirement for arson? Consider the following scenarios:

- A person places dynamite in a building and sets it off, resulting in an explosion but no fire.

Figure 5-7 *The common-law definition of arson severely limited who could be charged, leading all states to adopt various statutory arson charges. (Photo by Tom Carmody.)*

Figure 5-8 *Modern arson laws commonly include degrees of arson for fires involving vehicles and wildlands. (Photo by Ray Taylor.)*

- A person pours gasoline in a building, and ignites it. The gasoline explodes instantly, destroying the building without any lingering fire.
- A couple decides to spend a romantic evening in front of their fireplace. About an hour after lighting the fire, the couple engage in an argument, and one of them decides to pull the burning logs out of the fireplace into the living room, thereby igniting the house.

- A person observes his cat accidentally knock over a candle, igniting some combustibles in the house. Realizing what will happen if he does nothing, the person decides to leave the house, hoping that the ensuing fire will burn the house down in order to collect the insurance money.

- What if the person whose cat knocked over a candle left for fear of his safety rather than with the intent to collect on the insurance?

Common-law arson required the charring of structural wood in order for arson to have occurred. If there was no charring of a structural member, there was no arson. Thus at common law, an explosion that did not cause the charring of wood did not constitute arson. Modern arson statutes are usually directed at anyone who causes, procures, aids, counsels, or creates by means of fire or explosion, a substantial risk of serious physical harm to any person or damage to any building **(Figure 5-9)**. Any act that causes a fire to transmit to

Figure 5-9 *Modern arson laws include any act that risks injury to any person or damage to any building by means of fire or explosion. (Photo by Ray Taylor.)*

a building, as well as any act that causes an explosion that damages a building, satisfies the act requirement for arson.

Thus, whether an arsonist uses gasoline or dynamite to cause damage to a building, the resulting explosion falls within the statutory definition of arson, even if there is no ensuing fire. The act of taking burning logs out of a fireplace and placing them onto other combustibles would be enough to satisfy the act requirement for arson. The failure to take steps to extinguish the fire accidentally caused by the cat knocking over a candle could satisfy the act requirement, if the person was under a legal duty to act. Consider the following case:

COMMONWEALTH v. CALI.
247 Mass. 20, 141 N.E. 510 (1923)
Supreme Judicial Court of Massachusetts, Worcester

BRALEY, J. The defendant having been indicted, tried and convicted . . . of burning a building in Leominster belonging to Maria Cali, which at the time was insured against loss or damage by fire, with intent to injure the insurer. . . .

The only evidence as to the origin, extent and progress of the fire were the statements of the defendant. . . . The jury who were to determine his credibility and the weight to be given his testimony could find notwithstanding his explanations of its origin as being purely accidental, that when all the circumstances were reviewed he either set it, or after the fire was under way purposely refrained from any attempt to extinguish it in order to obtain the benefit of the proceeds of the policy, which when recovered, would be applied by the mortgagee on his indebtedness. . . .

The intention . . . could be formed after as well as before the fire started. On his own admissions the jury were to say whether when considered in connection with all the circumstances, his immediate departure from the premises for his home in Fitchburg, without giving any alarm, warranted the inference of a criminal intent or state of mind, that the building should be consumed. . . .

Exceptions overruled.

Case Name: Commonwealth v. Cali

Court: Massachusetts Supreme Judicial Court

Summary of Main Points: A jury is entitled to infer from the fact that a person leaves an insured house (to which he has a financial interest) that is on fire, without attempting to extinguish or report the fire, that the person intended to burn the property to defraud an insurer. The prosecution need not prove that the fire was set with the required intent. The intent may have developed after the fire began, and the defendant chose to refrain from extinguishing or reporting the fire.

The rulings in the *Cali* case, and in the Worcester Cold Storage fire case, demonstrate that when a perpetrator causes or somehow bears responsibility for an accidentally caused fire, he or she has a duty to act reasonably. In such cases, the failure to take reasonable steps to extinguish the fire or alert the fire department may satisfy the act requirement for a charge of arson. There is very little case law on the subject, and it should not be assumed that all jurisdictions would follow such reasoning.

In the previous example, where the cat knocks over the candle and starts a fire, the homeowner who leaves the house with the intent to collect the insurance proceeds could be guilty of arson in a jurisdiction that recognizes a duty to act on the part of someone who accidentally or negligently creates a fire. The act requirement is satisfied by the failure to take steps to extinguish or report the accidental fire, which the perpetrator bears some responsibility for causing.

Mental State Requirements for Arson Most arson crimes require that someone act intentionally in setting a fire. As a general rule, the criminal mental states of purposefully or knowingly setting or causing a fire constitute "intent." In some jurisdictions, the mental state of recklessness—particularly "depraved heart" recklessness—will be adequate to establish intent. As was seen in the *Cali* case, it is possible for a fire to be started accidentally, yet still give rise to a charge of arson when the intent relates to the failure to extinguish or report the fire.

As a general rule, negligence as a criminal mental state is not adequate to convict someone of arson. However, some states, such as Minnesota, have added statues that make it a crime to start or cause a fire through gross negligence. (See Minnesota Statutes, Ch. 609.576, located in Appendix B, Chapter 5.)

Attendant Circumstances for Arson In addition to the act and mental state requirements, many statutory arson crimes have additional elements that must be proven in order establish a particular category or degree of arson. These additional elements may include matters such as the fact that the building was occupied, that the fire was set with intent to defraud an insurer, that lives were endangered, or that someone was killed.

STATE V. ANTONIO CAPRIO, JR.
477 A.2d 67 (1984)
Supreme Court of Rhode Island

KELLEHER, Justice. This appeal follows a verdict by a Superior Court jury which found the defendant, Antonio Caprio (Caprio), guilty of having committed arson in the first degree. Such a crime is defined [under RIGL§ 11-4-2] . . . as follows:

"Arson—-First degree—-Any person who knowingly causes, procures, aids, counsels or creates by means of fire or explosion, a substantial risk of serious

(continued)

(continued)

physical harm to any person or damage to any building the property of himself or another, whether or not used for residential purposes, which is occupied or in use or which has been occupied or in use during the six (6) months preceding the offense or to any other residential structure, shall, upon conviction, be sentenced to imprisonment for not less than five (5) years and may be imprisoned for life and shall be fined not more than five thousand dollars ($5,000) or both; provided, further, that whenever a death occurs to a person as a direct result of said fire or explosion or to a person who is directly involved in fighting said fire or explosion, imprisonment shall be for not less than twenty (20) years."

Caprio's appeal arises from . . . a challenge to . . . the proper meaning . . . of the terms "substantial risk of serious physical harm to any person" and "occupied or in use * * * during the six months preceding the offense."

The fire that is the subject of these proceedings occurred on August 11, 1981, at 43 Quonset Avenue in Warwick, Rhode Island. The structure, a single-family home, was located in a residential area of Warwick's Oakland Beach section. . . .

At trial the state's key witness was Jeanne Burns (Burns). . . . She was with Caprio on the night of the fire. . . . Burns arrived late in the afternoon and . . . stayed for dinner; but before dessert was served, Caprio asked her to drive him to the Quonset Avenue property. She agreed to act as his chauffeur.

Burns told the jury that she watched Caprio enter the Quonset Avenue premises through a side door and that although the house was dark, she could see the gleam of Caprio's flashlight through a window. He soon left the house, returned to the car, and asked Burns for a screwdriver. She allowed him to forage through the car's trunk, where he found a fishing knife with which he returned to the house.

After a lapse of a few minutes, Caprio returned to the car and said to Burns, "Let's go." Apparently Burns did not follow this command quickly enough, for he anxiously told Burns, "We have to get the hell out of here. I just set the place on fire."

Burns was, at first, incredulous, but Caprio insisted that the deed had been done. She suggested that they return to put it out, but Caprio said, "No, it's too late." They returned to Burns's home, where the police, alerted by the description of Burns's vehicle, were waiting for the couple.

The indictment charged that Caprio "did knowingly cause, procure, aid, counsel and create by means of fire substantial risk of serious physical harm to firefighters responding to fight said fire at and damage to the building located at 43 Quonset Ave. which was occupied and in use during the six (6) months preceding August 11, 1981." Caprio asserts that firefighters are not within the targeted class of protected individuals under § 11-4-2 and that the mere fact that the house was occupied at some point in time in the preceding six months is not sufficient occupation under the statute. . . .

Rhode Island's statutory-arson scheme seeks to rectify a variety of problems caused by the common-law definition of the crime. . . . [T]he current Rhode Island scheme punishes for the commission of the crime according to the risk involved

(continued)

(*continued*)

in a variety of factual situations. In addition to first-degree arson defined above, second-degree arson punishes for the destruction by fire or explosion of an unoccupied building, and fourth-degree arson punishes for the destruction of another's personal property valued in excess of $100.

Caprio argues that it was not the intent of the Legislature to punish for any act of arson as first-degree arson merely because of the risk to a firefighter who responds to the conflagration. The defendant insists that if this were so, any degree of the crime could be raised to the first merely by the arrival of members of the fire department. Caprio suggests that if the Legislature had intended to include firefighters, it would have done so explicitly.

Various states responded in a variety of ways to the problem with the crime of arson as defined in the common law. . . . Washington specifically includes "firemen" and also has several degrees of arson, including reckless burning. . . . "[E]xperience teaches that one of the certainties attendant upon a hostile fire is that firemen will be called and will come. Danger inheres in fire fighting." *State v. Levage,* 23 Wash.App. 33, 35, 594 P.2d 949, 950 (1979).

Our statute, like the Washington statute, places potential arsonists on notice that the foreseeable consequences of their actions will give them access to greater crimes, with greater penalties. This, of course, does not mandate the conclusion at which Caprio arrives. Not all fires that have a fire-personnel response will be susceptible of the first-degree charge. The statute requires a *substantial* risk of *serious* physical harm. Each of these elements must be proved by the state.

[W]e . . . conclude that the very breadth of the term "any person" defies the exclusion of any class of persons. That term is so broad as to require exclusion, not specific inclusion. The last line of the statute indicates precise knowledge of the probability that firefighters will most likely come to the scene of a fire. The Legislature made a conscious choice that if a firefighter were, in fact, placed in substantial risk of serious harm as the result of an intentionally set fire, first-degree arson is the appropriate crime. For this court to limit such a choice would be a serious encroachment on the legislative prerogative.

In construing the appropriate definition of the term "during the six (6) months preceding the offense," we shall apply the same set of standards as those applied above. The requirement of occupancy during those six months was also a statutory response to common-law shortcomings. As mentioned previously, arson was an offense against habitation. Therefore, if no one was living at the structure, or if the structure was abandoned, arson would not be the appropriate crime. . . . Since this arguably left a wide variety of structures outside the common-law definition, including office buildings and vacation homes, legislatures sought to statutorily fill the gap. . . .

Section 11-4-2 changes those results. Arson in the first degree now encompasses more than merely those fires that destroy only those dwellings used for habitation. Caprio claims that the state's interpretation of the phrase, finding the first-degree statute applicable in those situations in which the structure is in use

(*continued*)

(*continued*)
only once in the previous six months, would unduly expand the legislation. He asserts that this interpretation will cause the first-degree charge to be appropriate even though the burned structure had been abandoned for five months.

This complaint is a complaint with the time line drawn by the Legislature. The Legislature clearly felt that the public policy of this state would be to encompass as much activity within the first-degree arson section as was reasonably related to the protection of human life. The experience of the instant case demonstrates the sensibility of this action. Forty-three Quonset Avenue was used as rental property, and it was located in a residential neighborhood. It is conceivable that such a dwelling would be unoccupied in the interim during which one tenant was moving in and another moving out. Yet, these temporary situations do not diminish the danger to the surrounding homes.

Again, the Legislature has made its choice, and that choice is clear. The word "during" does not mean a continuous occupancy throughout the entire six-month period; rather, any point during the six-month period suffices. It is not the function of this court to quarrel with the Legislature when it has discharged its function with clarity. . . .

The defendant's appeal is denied and dismissed, the judgment entered in the Superior Court is affirmed, and the case is remanded to Superior Court for further proceedings.

Case Name: State v. Caprio

Court: Rhode Island Supreme Court

Summary of Main Points: First-degree arson includes the attendant circumstance that the building must have been "occupied or in use" during the six months preceding the fire. The court interpreted this attendant circumstance to mean that the building need not have been occupied continuously for the six months prior to the fire, but rather that it was occupied at any point during the six months prior to the fire. In addition, the court determined that firefighters are among the class of people to whom the language "a substantial risk of serious physical harm to any person" in the new arson law applies.

Racketeer Influenced and Corrupt Organizations Act

In 1970, Congress passed a historic law aimed at combating organized crime, called the Racketeer Influenced and Corrupt Organizations Act, or RICO Act. The RICO Act provides both criminal and civil tools to attack and dismantle criminal enterprises.

The statute defines certain types of crimes as "racketeering activities," and makes it a separate criminal offense for anyone to be employed by or be associated with an enterprise that engages in a pattern of racketeering activities. The term "enterprise" includes any organization, corporation, association, partnership, individual, sole proprietorship, union, or group. The term "pattern of racketeering activity" requires that the enterprise have engaged in at least two acts of racketeering activity, occurring within 10 years of each other.

Racketeering activity is defined as (A) any act or threat involving murder, kidnapping, gambling, arson, robbery, bribery, extortion, dealing in obscene matter, or dealing in a controlled substance or listed chemical, which is chargeable under state law and punishable by imprisonment for more than one year; or (B) any act that is indictable under any of the designated provisions of Title 18 of the United States Code, (including bribery, counterfeiting, embezzlement, fraud, sexual exploitation of children, white slave traffic, and aiding or assisting certain aliens to enter the United States, to name a few).

The penalty for RICO violations may include up to 20 years in prison. The sentence may be extended to life in prison if any of the underlying racketeering activities the enterprise engaged in carries a life sentence. The RICO Act authorizes the forfeiture of any proceeds obtained from the racketeering, and authorizes the seizure of any business property and assets of the enterprise. The RICO Act also provides for civil penalties and liability for the enterprise and those persons associated with the enterprise.

Federal RICO jurisdiction is limited to enterprises that are engaged in, or effect, interstate or foreign commerce. Many states have enacted comparable RICO statutes that allow state and local law enforcement personnel to pursue criminal enterprises. Each state's act, while modeled on the Federal statute, may have different statutory requirements. Let's take a look at how the typical state RICO act can be used against an arson-for-profit ring **(Figure 5-10)**.

📖 **SIDEBAR**

ABC Management Company is owned by A, B, and C. They own about 50 pieces of residential property, which they rent out to low-income tenants. Most of the properties are old and very poorly maintained. ABC Management Co. employs D and E, who collect rent from the tenants. D and E regularly use force and intimidation in collecting the rent from late payers. When a particular property is no longer profitable to ABC Management, D and E are instructed to light it on fire. ABC Management then collects on the fire insurance.

The state fire marshal can make a case against D for one arson fire, and against E for a separate arson fire. He can also prove that ABC Management Co. properties have had 15 arson fires over the past five years, resulting in more than $3 million in insurance claims. The attorney general has several witnesses who were assaulted and threatened by D and E in the course of collecting late rent. D and E refuse to testify against A, B, and C. Without RICO, it would be very difficult to implicate A, B, and C. Each act of arson would be viewed as a separate act, and would have to be prosecuted separately. Attorneys for D and E would argue that allowing any evidence not pertaining to the arson charges against their clients would prejudice their right to a fair trial.

However, utilizing the RICO act, the entire pattern of criminal activity becomes relevant, including the multiple arson cases and the strong-arming of tenants. In this way, A, B, and C can be convicted, as well as D and E. The property owned by ABC Management Co.—as well as any proceeds from the criminal enterprise in the possession of A, B, C, D, or E—can be forfeited to the government. The very threat of this tool may be enough to induce D and E to become more cooperative with the prosecution of A, B, and C.

Figure 5-10 *RICO acts can be used effectively to help combat arson-for-profit rings. (Photo by Ray Taylor.)*

SUMMARY

Criminal offenses arise either from the common law or from statutes passed by Federal, state, or local legislatures. Statutory crimes can be distinguished from violations of civil law in that the statute must specifically declare that any violation is a crime. Criminal offenses include both felonies and misdemeanors. Crimes are made up of elements: an act, a mental state, and attendant circumstances. Often when we think of a crime, we think solely of the act, but in many ways the mental state of the perpetrator is even more important. Arson was one of the common-law crimes, but has been replaced or supplemented with a variety of statutory arson charges. These statutory charges provide arson investigators and prosecutors with a variety of options for prosecuting those who set or cause fires.

REVIEW QUESTIONS

1. When does the violation of a law constitute a criminal offense?

2. What is the difference between the jurisdiction of Federal, state, and local government in regard to criminal law?

3. Identify and define the three elements of a crime.

4. Identify five examples of persons or positions who have a "duty to act."

5. Define the four criminal mental states and give examples of each.

6. What factors in the Worcester Cold Storage case did the court rely on to find that the defendants had a duty to report the fire?

7. Would the Worcester Cold Storage case have come out differently if the defendants had been charged with arson?

8. Why wasn't the defendant in the Cocoanut Grove case charged with murder?

9. How does implied consent serve as a possible defense for assault, battery, false imprisonment, and kidnapping charges in the context of an ambulance crew transporting a violent patient?

10. What were the limitations with the common-law crime of arson?

DISCUSSION QUESTIONS

1. Why should a person's mental state be a factor in whether or not they committed a crime? Why doesn't society simply punish people for committing the act?

2. The judge in the Worcester Cold Storage fire case said that there is an intentional act involved in every act of recklessness. Can you explain this comment?

3. Research your own state's arson statutes. How many separate arson crimes are there in your state, and what are they? What acts will satisfy the act requirement for each of the arson charges? Are there any attendant circumstances for the most serious of the statutory arson charges?

Chapter
6

CRIMINAL PROCEDURE

Learning Objectives

Upon completion of this chapter, you should be able to:

- Define arrest, and explain the authority of a firefighter to make an arrest.
- Explain the difference between criminal and administrative search warrants.
- Identify at least six exceptions to the search warrant requirement.
- Explain the constitutional limitations upon a firefighter conducting a cause and origin determination as part of an investigation after a fire.
- Explain what is required to constitute an attempted crime.
- Define accessory before the fact, accessory after the fact, and an aider and abettor.
- Define a criminal conspiracy and explain the liability of each co-conspirator.

INTRODUCTION

The engine company was returning from a run when they saw the smoke coming from the third floor of a five-story apartment building. The light-colored smoke indicated that it was probably still a contents fire that had not yet extended to the structure. The driver brought the truck to a stop, the officer reported the situation over the radio to the dispatch office, and the firefighters went to work.

On the third-floor landing, the smell of the smoke was not what any of them had expected. Instead of burnt food or wood smoke, it smelled more like a brush fire. The smoke was puffing from under the door as the officer gave the order to force entry.

Once the door was opened, the source of the smoke became immediately apparent. Six youths, smoking from large pipes, were lying in various locations around the apartment, oblivious to the presence of the firefighters. A large quantity of marijuana and crack cocaine was present on a table.

In Chapter 5 we discussed basic criminal law principles and the most common criminal offenses. In this chapter we will discuss criminal procedure, including matters pertaining to arrests, how formal criminal charges are brought, search and seizure, accomplice liability, and criminal defenses.

WHAT IS AN ARREST?

An **arrest** is the lawful seizure of one person by another, thereby depriving the person who is seized of his or her liberty. An arrest involves having the authority to make an arrest, and then asserting that authority to effectuate the restraint of the person.

AUTHORITY TO MAKE AN ARREST

By law, all citizens have the inherent authority to make a lawful arrest. States differ somewhat regarding the basis for which a citizen can make an arrest, but generally anyone may make a **citizen's arrest** when the person being arrested has committed a misdemeanor in their presence, or when probable cause exists to believe that the person being arrested has committed a felony.

Law enforcement officers, frequently termed **peace officers** in state statutes, have broader powers to detain and arrest those who have committed a crime. Generally, these officers have the authority to detain anyone they reasonably believe is committing, has committed, or who is about to commit a

arrest
the lawful seizure of one person by another, thereby depriving the person who is seized of his or her liberty

citizen's arrest
an arrest made by someone other than a duly authorized peace officer; citizens may make a lawful arrest when the person being arrested has committed a misdemeanor in their presence, or when probable cause exists to believe that the person being arrested has committed a felony

peace officer
another term for law enforcement officer in many states

crime, for a reasonable period to investigate the circumstances, without charging them with a crime. This period of detention must be reasonable, and in some states is limited to two hours.

Peace officers may also arrest a person if they have reasonable cause to believe the person is committing or has committed a felony, or for a misdemeanor committed in their presence. Some states grant peace officers the authority to arrest someone who has committed a misdemeanor, regardless of whether it was in their presence or not. Furthermore, peace officers have the authority to arrest persons for whom an arrest warrant has been issued.

In some jurisdictions, firefighters who are assigned to fire investigation or arson squad positions, as well as members of the state fire marshal's office, are given the authority to make arrests. In such cases, these firefighters and investigators become "peace officers" **(Figure 6-1)**.

The authority of law enforcement officials to make an arrest is generally concurrent with state and Federal laws. State law enforcement officials have the authority to make arrests for criminal violations of state law, and Federal law enforcement officials have the authority to arrest for Federal crimes. The jurisdiction of local law enforcement officials is commonly limited to arrests for violations of state or local law, and may be limited by jurisdiction or county within the state, depending upon the state's laws.

Many crimes are violations of both state and Federal law, including drug, kidnapping, bank robbery, and weapons charges. In such a case, either state or Federal officers may arrest a perpetrator based upon a violation for which they have authority to arrest. Some states explicitly grant to Federal

Figure 6-1 *Fire investigators in some departments are sworn peace officers.*

law enforcement officials full peace officer arrest powers within their states for state law crimes, while some states do not grant this power. However, there are procedures whereby state law enforcement officials can be deputized to make arrests for Federal offenses, and Federal officers can be deputized to make arrests on state charges. This practice is routinely followed when Federal and state law enforcement officials work together on joint task forces.

ASSERTING THE AUTHORITY

Arrests may be effectuated through the use of force, threatened used of force, intimidation, or the exercise of legal authority, such as a police officer saying "You are under arrest" **(Figure 6-2)**. Anyone making a lawful arrest is entitled to use reasonable force to detain the suspect. The reasonableness of the force used can be judged only by looking at the circumstances of a particular case. Using excessive force can result in criminal charges being brought against the person making the arrest. The use of excessive force can also give rise to civil liability, and when committed by Federal, state, or local officials in the performance of their duties, may result in a Federal constitutional rights violation lawsuit.

Figure 6-2

An arrest may be effected through the use or threatened use of force.

FALSE ARREST

When a lawfully appointed peace officer has made an arrest that later turns out to be improper, the officer is immune from civil or criminal liability for false arrest, as well as for any assault and battery that may otherwise have occurred for using reasonable force, provided that the officer acted in good faith with probable cause. The immunity from liability for false arrest is the result of a privilege granted to peace officers when making arrests in good faith.

However, when ordinary persons make a citizen's arrest, they are entitled to no such immunity protection. Therefore, even where the citizen acted with the utmost in good faith and with probable cause, if the person arrested did not commit a crime, the citizen who made the arrest could be liable for false arrest. In addition, the citizen making the arrest could be liable for assault and battery if force was used in effecting the arrest. Thus, in a citizen's arrest, the risk of mistake is squarely on the person making the arrest.

FIREFIGHTERS AND ARREST POWERS

When it comes to the power to arrest, firefighters and emergency medical personnel are entitled to make citizen's arrests, like any other citizen. However, since firefighters and emergency medical personnel are not peace officers, they can expose themselves to liability for false arrest in the event that a mistake is made.

ARREST WARRANTS

arrest warrant
an order issued by a judge or magistrate, authorizing any peace officer to take a suspect into custody; law enforcement officers must establish that probable cause exists to believe a crime has been committed, and that the person named in the warrant committed it, in order to obtain an arrest warrant

When police have probable cause to believe that a crime has been committed and that a certain person committed it, but the perpetrator is not in their presence, they need to obtain an arrest warrant. To obtain an **arrest warrant**, police must apply to a judge or magistrate. The judge or magistrate will review the information submitted by the police and make a determination whether or not the police have probable cause to believe that a crime has been committed and that the accused person committed it. If the judge or magistrate finds probable cause, they can issue the warrant, which will authorize any peace officer to take the suspect into custody.

CRIMINAL PROCEDURE

Upon making an arrest, law enforcement officers will take the suspect to the police station, where he is formally processed, or booked. Forms are prepared to be forwarded to the court, and the defendant is photographed and fingerprinted, and may be subject to further interrogation.

criminal complaint

a formal, court-approved document commonly prepared by the police that includes a statement describing what the defendant allegedly did, and the specific laws the defendant is charged with violating; it is the most common method of charging a defendant with a misdemeanor or minor offense

grand jury

a type of jury used in certain types of criminal cases to determine if there is probable cause to believe that a crime has been committed and that the defendant committed it, and if so, to indict the defendant

indictment

the formal allegation of criminal wrongdoing issued by a grand jury

information charging

a procedure in which the prosecutor can issue formal charges against a defendant; a probable cause hearing is then scheduled before a magistrate or judge, who determines the sufficiency of the evidence based upon the evidence that prosecutors have presented in the "information"

The defendant is entitled to an initial court appearance before a judge or magistrate to formally inform the defendant of the charges, determine if the defendant has counsel, and determine if the defendant is eligible for bail. This hearing must occur within a reasonable time, typically no later than 48 hours after the arrest. Some states have a rule that a defendant who does not receive a hearing within the prescribed time frame must be released. Unless the defendant is a flight risk or a threat to society, he is entitled to bail at an amount to be set by the magistrate or judge. If the defendant is wanted on other charges, or is already on bail or on probation for other charges, he may be held pending a determination whether the new infraction has violated the conditions of his previous release.

Formal Charges

There are three mechanisms by which a criminal defendant can be formally charged with a crime. They are:

1. Criminal complaint
2. Grand jury indictment
3. Information charging

The procedure used in a given case depends upon the seriousness of the charges. A **criminal complaint** is the most common method of charging a defendant with a misdemeanor or minor offense. It is a formal, court-approved document prepared by the police that includes a statement describing what the defendant allegedly did, and the specific laws that the defendant is charged with violating.

The **grand jury** is a type of jury used in certain types of criminal cases. The purpose of a grand jury is to determine if there is probable cause to believe that a crime has been committed and that the defendant committed it, and if so, to indict the defendant. An **indictment** is the formal allegation of criminal wrongdoing issued by a grand jury.

The grand jury process serves two functions. It is, first and foremost, aimed at protecting citizens from criminal charges being brought by overzealous prosecutors and police officers with insufficient evidence. Second, it is an investigative process whereby witnesses can be subpoenaed to testify, evidence can be gathered, and a determination can be made about whether or not sufficient evidence exists to charge someone with the commission of a crime.

According to the Fifth Amendment of the United States Constitution, grand juries are required for Federal offenses that constitute capital or infamous crimes. Many states have done away with grand juries, and the remaining states generally limit their use to the most serious of state crimes. In place of grand juries, states have developed a procedure, often called **information charging**, by which the prosecutor can issue formal charges. A probable cause hearing is then scheduled before a magistrate or judge, who determines the sufficiency of the evidence that prosecutors have presented in the "information." The standard at the probable cause hearing is the same as that required by the grand jury: whether or not there is probable cause to believe that a crime has been committed, and probable cause to believe that the accused committed it.

Grand Jury Proceedings

A grand jury proceeding is unlike a trial, on several counts. First, there is no judge. Second, the target of the grand jury is not permitted to be in the room, and may not even be aware that the proceedings are going on. While the target may be given the opportunity to rebut or explain what happened, that is up to the discretion of the grand jury. Third, it is the prosecutor who convenes and directs the grand jury, although once empanelled a grand jury is allowed to "go where the evidence takes them." Finally, the standard used by the grand jury is whether or not there is probable cause to believe that a crime has been committed and probable cause to believe that the accused committed it.

Historically, grand juries are larger than the juries used at trial, hence the name "grand." In fact, the jury used at trial is sometimes referred to as a **petit jury**. Grand juries commonly range in size from 16 to 23 members. A concurrence of 12 or more jurors is required for a finding of a "true bill," which means that the evidence is sufficient for an indictment to be issued. If 12 jurors cannot agree, then a "no bill" or "no true bill" is issued, and the target will not be indicted.

Arraignments and Bail

Once the indictment is issued, or the formal charges via an "information" or criminal complaint have been brought, the defendant will be arraigned before a judge or magistrate. In an **arraignment**, the accused is formally advised of the charges, and may have the opportunity to enter a plea to the court. If the defendant pleads **guilty**, which is admitting to commission of the crime, or **nolo contendere**, which is neither admitting nor denying the commission of the crime but agreeing that the evidence is sufficient to establish guilt, the case will enter the sentencing phase. If the defendant enters a not guilty plea, the judge will make a bail determination.

Criminal defendants are entitled to be released on bail pending trial, provided they do not present a flight risk, are not considered to be a danger to society, or are not otherwise subject to incarceration for other criminal matters. Defendants have a right to be free from excessive bail. The amount of bail is based upon several factors, including the nature and circumstances of the offense, the strength of the prosecution's case, the financial ability of the defendant to pay, the risk of flight, the character of the defendant, and the Eighth Amendment's constitutional prohibition against excessive bail.

Once the arraignment is completed, the case will then enter the discovery phase, analogous to the discovery phase in a civil trial discussed in Chapter 2.

petit jury
a jury whose purpose is to serve as the fact finder at a trial

arraignment
the initial procedure in a criminal case held before a judge or magistrate, in which the accused is formally advised of the charges, and may have the opportunity to enter a plea to the court

guilty
a plea entered in court serving as an admission to the commission of the crime

nolo contendere
a plea entered in court, which neither admits nor denies the commission of the crime but agrees that the evidence is sufficient to establish guilt, and that the defendant does not wish to contest the matter further

The exact procedure that occurs between the initial arraignment and the trial varies somewhat from state to state, and case to case, but generally it will involve a series of motions and meetings between the prosecutor and the defense attorney to effect discovery, attempt to reach a plea bargain, and ultimately prepare for trial.

One of the biggest issues that will be addressed during this period involves the admissibility of evidence. The rules of evidence are much more strict in criminal cases than in civil cases. In particular, the **exclusionary rule** prohibits the use of evidence in a criminal case that was obtained in violation of a defendant's constitutional rights. This brings us to the important subject of search and seizure.

exclusionary rule
a rule adopted by the United States Supreme Court, which prohibits the use of evidence in a criminal case that was obtained in violation of a defendant's constitutional rights

SEARCH AND SEIZURE

The Fourth Amendment to the United States Constitution reads:

> The right of the people to be secure in their persons, houses, papers, and effects, against unreasonable searches and seizures, shall not be violated, and no warrants shall issue, but upon probable cause, supported by oath or affirmation, and particularly describing the place to be searched, and the persons or things to be seized.

As written, the Fourth Amendment clearly prohibits any unreasonable search and seizure conducted by the government without a warrant. However, the Constitution does not explain what should happen if the government violates the Fourth Amendment's warrant requirement. In addition, it does not define what constitutes a "reasonable search," which arguably does not require a warrant.

Through case law, the United States Supreme Court has clarified what the Fourth Amendment requires, and established the exclusionary rule. Under the exclusionary rule, evidence seized in violation of the defendant's constitutional rights cannot be used against her in a criminal case. Also, if the illegally seized evidence leads to any additional evidence, that additional evidence is also subject to the exclusionary rule. Evidence subject to exclusion under the latter rule is commonly referred to as "fruit from the poisoned tree."

The United States Constitution, including the Fourth Amendment's warrant requirement, is applicable to actions by the Federal government. However, the Fourteenth Amendment states: "No State shall . . . deprive any person of life, liberty, or property, without due process of law." Through the Supreme Court's interpretation of the Fourteenth Amendment's due process clause, the right to be free from unreasonable searches and seizures has been determined to be an essential part of due process that state and local law enforcement officials must honor. As a result, the search warrant requirement and the exclusionary rule apply to Federal, state, and local law enforcement officials.

Search Warrant Requirement

As a general rule, the Fourth and Fourteenth Amendments require that before police may search a suspect or the property of a suspect, they must obtain a **search warrant**. A search warrant is a legal order issued by a judge or magistrate that authorizes law enforcement officials to search a certain location.

Criminal search warrants may only be issued under circumstances in which the judge or magistrate agrees that police have probable cause to believe a crime has been committed, and evidence of that crime will be located in the area to be searched. Police officers who seek to have a search warrant issued must submit an affidavit under oath in support of their request, detailing the evidence they have to support their request.

The purpose of the search warrant requirement is to prevent unreasonable searches and seizures and to protect individual liberties. While the search warrant requirement seems to be a reasonable limitation upon police and government officials, consider how it might impact a post-fire cause and origin determination. Is a search warrant required in order to conduct a cause and origin determination after every fire?

Exceptions to the Warrant Requirement

There are several exceptions to the Fourth Amendment's search warrant requirement. These exceptions are based upon commonsense considerations, by which a strict application of the warrant requirement and/or the exclusionary rule would not further secure individual protections from illegal searches and seizures. Most of the exceptions can best be understood by taking the perspective that the Supreme Court has repeatedly used: that of analyzing whether someone has a reasonable expectation of privacy in a given situation.

The first exception to the search warrant requirement is the **plain view doctrine**. The plain view doctrine authorizes police to use evidence seized without a warrant when it is in plain view of the officers. Obviously, someone who leaves something out in plain view does not have a reasonable expectation of privacy. Examples would be a gasoline can left on a front porch, or a handgun left on the dashboard of a car.

Assuming that the police officer is lawfully in a location from which he can plainly see the evidence, the seizure of that evidence is lawful. However, if the officer needed a warrant to be in the location from which he observed the evidence, the plain view doctrine would not apply. The plain view doctrine has been expanded to include the use of aerial surveillance, drug- or explosive-sniffing canines, electronic tracking devices, and even searching through trash bags that have been left for garbage collectors or thrown in dumpsters.

The next exception to the warrant requirement is consensual searches. A **consensual search** occurs when the defendant, or someone who has the right to control access to a particular area, consents to a search being conducted. The burden is on the police officers to prove that the consent to search was voluntarily obtained.

search warrant
a legal order issued by a judge or magistrate that authorizes law enforcement officials to search a certain location

plain view doctrine
an exception to the search warrant requirement that allows police to use evidence seized without a warrant when it is located in plain view of the officers

consensual search
a warrantless search that occurs when the defendant, or someone who has the right to control access to a particular area, consents to a search being conducted

Issues often arise in consensual search cases when someone other than the defendant consented to the search. The consent of a third party will be considered valid, provided that the officer reasonably believes the person has common authority over the area. An example would be a wife consenting to a search of a house that she shares with her husband, and where evidence against the husband is found.

EXAMPLE

LOCKER SEARCHES: Do firefighters have a reasonable expectation of privacy in their lockers in the fire station?

In the case of *Chicago Firefighters IAFF Local 2 v. City of Chicago*, 717 F. Supp. 1314 (N.D. Ill, 1989), the warrantless search of fire station lockers was upheld. The court recognized that public employees may have a reasonable expectation of privacy in the workplace. However, in this case, prior to initiating the locker searches, the department discussed the issue with the union prior to implementing the policy, and issued a general order stating that warrantless locker searches would be conducted. The court held that, in light of the general order, as well as the strictly controlled environment that firefighters work under, and the department's interest in assuring that its personnel are free from drugs and alcohol, Chicago firefighters did not have a reasonable expectation of privacy in their lockers **(Figure 6-3)**.

Figure 6-3

A firefighter may not have a reasonable expectation of privacy in his or her locker.

A third exception that allows police to conduct warrantless searches is the so-called stop-and-frisk scenario. Police may conduct a warrantless, pat-down search of a suspect if they have a reasonable fear for their own safety that the suspect could be armed and dangerous. In order for a police officer to be able to stop a person, the officer must be able to point to "specific and articulable facts which, taken together with rational inferences from those facts," would lead to a reasonable conclusion that possible criminal behavior was at hand and that both an investigative stop and a frisk were required. Such a pat-down search is limited to a search for a concealed weapon, and would not authorize a more invasive search. This type of stop-and-frisk is often referred to as a Terry stop, from the Supreme Court case *Terry v. Ohio,* 392 U.S. 1, 20 (1968).

search incident to arrest
a permissible warrantless search that may be conducted by a police officer upon arresting a suspect; the officer may search the suspect and the area immediately around the suspect without a search warrant

The fourth type of warrantless search is what is called a **search incident to arrest**. Upon making an arrest, a police officer may search a suspect and the area immediately around a suspect. This rule goes back to the common law, and is justified by the circumstances of the suspect being taken into custody.

While the search incident to arrest rule has never seriously been challenged, the scope of the "area immediately around the suspect" has been the subject of many cases. Generally, a search incident to arrest is limited to the area from which the subject might obtain a weapon or destroy evidence. However, protective sweeps of an entire house, apartment, or building may be warranted if officers have a reasonable basis to believe that there is a threat posed by unseen third parties. In addition, if a suspect is arrested in a motor vehicle, the entire vehicle may be searched incident to the person's arrest. During any such search incident to arrest, evidence found in plain view may lawfully be seized.

A fifth type of warrantless search involves searches of automobiles. The Supreme Court has recognized a diminished expectation of privacy in an automobile compared to one's home or business. Due to the mobility of cars and the impracticality of obtaining a search warrant to stop and search a car, police may search cars if they have probable cause to believe the vehicle contains contraband **(Figure 6-4)**. A determination of whether or not the officers actually had probable cause to search the car can be made after the fact by a court during proceedings prior to or during the defendant's trial.

Analogous to the Terry stop, police may stop a car if they have an articulable suspicion of traffic, safety, or criminal violations, and may conduct a quick, protective search of the vehicle and occupants for weapons. However, to conduct a thorough search of the vehicle, police must have probable cause.

A sixth exception to the search warrant requirement pertains to searches of pastures, wooded areas, open water, and vacant lots, often referred to as the open fields exception. The Fourth Amendment prohibition

Figure 6-4   *Search warrants are not required for searches of automobiles, provided the police have probable cause. (Photo by James E. Clift.)*

exigent circumstances
an exception to the search warrant rule, holding that police officers may conduct a warrantless search when requiring them to obtain a warrant before acting would have real, immediate, and serious consequences

fire scene exception
an exception to the search warrant requirement, holding that the warrantless entry by firefighters at a fire is lawful

against unreasonable searches and seizures is explicitly limited to "persons, houses, papers, and effects." The Supreme Court has interpreted this to exclude an expectation of privacy for activities conducted out of doors in fields, woods, or similar areas, except in the area immediately surrounding the home. The erection of fences and no-trespassing signs are not sufficient to create a reasonable expectation of privacy in such areas.

The Supreme Court has recognized an additional exception to the warrant requirement that applies to a number of situations that share a common denominator: **exigent circumstances**. Exigent circumstances exist when requiring police officers to obtain warrants before acting would have real, immediate, and serious consequences. Exigent situations include hot pursuit of a fleeing felon; fear of imminent destruction of evidence; the need to prevent a suspect's escape; and a risk of danger to the police or others.

Each of the above situations involves circumstances that preclude the ability of a police officer to obtain a warrant while addressing the immediate needs of the situation. These situations also involve incidents in which the suspect's expectation of privacy is secondary to some other legitimate consideration. For example, if a police officer has a reasonable belief that someone inside a dwelling is in need of immediate aid, that exigency justifies a warrantless entry. Directly related to this "risk of danger" exigent circumstance exception is the **fire scene exception (Figure 6-5)**.

Figure 6-5 *Fires create exigent circumstances, justifying warrantless entries. (Photo by Tim Whalen.)*

Michigan v. Tyler
436 U.S. 499 (1978)
Supreme Court of the United States

MR. JUSTICE STEWART delivered the opinion of the Court. . . .

Shortly before midnight on January 21, 1970, a fire broke out at Tyler's Auction, a furniture store in Oakland County, Mich. The building was leased to respondent Loren Tyler, who conducted the business in association with respondent Robert Tompkins. According to the trial testimony of various witnesses, the fire department responded to the fire and was "just watering down smoldering embers" when Fire Chief See arrived on the scene around 2 a.m. It was Chief See's responsibility "to determine the cause and make out all reports." Chief See was met by Lt. Lawson, who informed him that two plastic containers of flammable liquid had been found in the building. Using portable lights, they entered the gutted store, which was filled with smoke and steam, to examine the containers. Concluding that the fire "could possibly have been an arson," Chief See called Police Detective Webb, who arrived around 3:30 a.m. Detective Webb took several pictures of the containers and of the interior of the store, but finally abandoned his efforts because of the smoke and steam. Chief See briefly "[l]ooked throughout the rest of the building to see if there was any further evidence, to determine what the cause of the fire was."

(*continued*)

(*continued*)

By 4 a.m., the fire had been extinguished and the firefighters departed. See and Webb took the two containers to the fire station, where they were turned over to Webb for safekeeping. There was neither consent nor a warrant for any of these entries into the building, nor for the removal of the containers. . . . .

Four hours after he had left Tyler's Auction, Chief See returned with Assistant Chief Somerville, whose job was to determine the "origin of all fires that occur within the Township." The fire had been extinguished and the building was empty. After a cursory examination, they left, and Somerville returned with Detective Webb around 9 a.m. In Webb's words, they discovered suspicious "burn marks in the carpet, which [Webb] could not see earlier that morning, because of the heat, steam, and the darkness." They also found "pieces of tape, with burn marks, on the stairway." After leaving the building to obtain tools, they returned and removed pieces of the carpet and sections of the stairs to preserve these bits of evidence suggestive of a fuse trail. Somerville also searched through the rubble "looking for any other signs or evidence that showed how this fire was caused." Again, there was neither consent nor a warrant for these entries and seizures. Both at trial and on appeal, the respondents objected to the introduction of evidence thereby obtained.

On February 16, Sergeant Hoffman of the Michigan State Police Arson Section returned to Tyler's Auction to take photographs. During this visit or during another at about the same time, he checked the circuit breakers, had someone inspect the furnace, and had a television repairman examine the remains of several television sets found in the ashes. He also found a piece of fuse. Over the course of his several visits, Hoffman secured physical evidence and formed opinions that played a substantial role at trial in establishing arson as the cause of the fire and in refuting the respondents' testimony about what furniture had been lost. His entries into the building were without warrants or Tyler's consent, and were for the sole purpose "of making an investigation and seizing evidence." At the trial, respondents' attorney objected to the admission of physical evidence obtained during these visits, and also moved to strike all of Hoffman's testimony "because it was got in an illegal manner." . . .

The decisions of this Court firmly establish that the Fourth Amendment extends beyond the paradigmatic entry into a private dwelling by a law enforcement officer in search of the fruits or instrumentalities of crime. As this Court stated in *Camara* . . . the "basic purpose of this Amendment . . . is to safeguard the privacy and security of individuals against arbitrary invasions by governmental officials." . . .

[T]here is no diminution in a person's reasonable expectation of privacy nor in the protection of the Fourth Amendment simply because the official conducting the search wears the uniform of a firefighter, rather than a policeman, or because his purpose is to ascertain the cause of a fire, rather than to look for evidence of a crime, or because the fire might have been started deliberately. Searches for administrative purposes, like searches for evidence of crime, are encompassed by the Fourth Amendment. . . .

(*continued*)

(*continued*)

The petitioner argues that no purpose would be served by requiring warrants to investigate the cause of a fire. This argument is grounded on the premise that the only fact that need be shown to justify an investigatory search is that a fire of undetermined origin has occurred on those premises. . . . In short, where the justification for the search is as simple and as obvious to everyone as the fact of a recent fire, a magistrate's review would be a time-consuming formality of negligible protection to the occupant.

The petitioner's argument fails primarily because it is built on a faulty premise. To secure a warrant to investigate the cause of a fire, an official must show more than the bare fact that a fire has occurred. The magistrate's duty is to assure that the proposed search will be reasonable, a determination that requires inquiry into the need for the intrusion on the one hand, and the threat of disruption to the occupant on the other. . . . Thus, a major function of the warrant is to provide the property owner with sufficient information to reassure him of the entry's legality. . . .

In short, the warrant requirement provides significant protection for fire victims in this context, just as it does for property owners faced with routine building inspections. As a general matter, then, official entries to investigate the cause of a fire must adhere to the warrant procedures of the Fourth Amendment. . . . Since all the entries in this case were "without proper consent" and were not "authorized by a valid search warrant," each one is illegal unless it falls within one of the "certain carefully defined classes of cases" for which warrants are not mandatory. . . .

Our decisions have recognized that a warrantless entry by criminal law enforcement officials may be legal when there is compelling need for official action and no time to secure a warrant. . . . A burning building clearly presents an exigency of sufficient proportions to render a warrantless entry "reasonable." Indeed, it would defy reason to suppose that firemen must secure a warrant or consent before entering a burning structure to put out the blaze. And once in a building for this purpose, firefighters may seize evidence of arson that is in plain view. . . . Thus, the Fourth and Fourteenth Amendments were not violated by the entry of the firemen to extinguish the fire at Tyler's Auction, nor by Chief See's removal of the two plastic containers of flammable liquid found on the floor of one of the showrooms.

Although the Michigan Supreme Court appears to have accepted this principle, its opinion may be read as holding that the exigency justifying a warrantless entry to fight a fire ends, and the need to get a warrant begins, with the dousing of the last flame. . . . We think this view of the firefighting function is unrealistically narrow, however. Fire officials are charged not only with extinguishing fires, but with finding their causes. Prompt determination of the fire's origin may be necessary to prevent its recurrence, as through the detection of continuing dangers such as faulty wiring or a defective furnace. Immediate investigation may also be necessary to preserve evidence from intentional or accidental destruction. And, of course, the sooner the officials complete their duties, the less will be their subsequent interference with the privacy and the recovery efforts of the victims. For

(*continued*)

(*continued*)

these reasons, officials need no warrant to remain in a building for a reasonable time to investigate the cause of a blaze after it has been extinguished. And if the warrantless entry to put out the fire and determine its cause is constitutional, the warrantless seizure of evidence while inspecting the premises for these purposes also is constitutional.

The respondents argue, however, that the Michigan Supreme Court was correct in holding that the departure by the fire officials from Tyler's Auction at 4 a.m. ended any license they might have had to conduct a warrantless search. Hence, they say that even if the firemen might have been entitled to remain in the building without a warrant to investigate the cause of the fire, their reentry four hours after their departure required a warrant.

On the facts of this case, we do not believe that a warrant was necessary for the early morning reentries on January 22. As the fire was being extinguished, Chief See and his assistants began their investigation, but visibility was severely hindered by darkness, steam, and smoke. Thus they departed at 4 a.m. and returned shortly after daylight to continue their investigation. Little purpose would have been served by their remaining in the building, except to remove any doubt about the legality of the warrantless search and seizure later that same morning. Under these circumstances, we find that the morning entries were no more than an actual continuation of the first, and the lack of a warrant thus did not invalidate the resulting seizure of evidence.

The entries occurring after January 22, however, were clearly detached from the initial exigency and warrantless entry. Since all of these searches were conducted without valid warrants and without consent, they were invalid under the Fourth and Fourteenth Amendments, and any evidence obtained as a result of those entries must, therefore, be excluded at the respondents' retrial.

In summation, we hold that an entry to fight a fire requires no warrant, and that, once in the building, officials may remain there for a reasonable time to investigate the cause of the blaze. Thereafter, additional entries to investigate the cause of the fire must be made pursuant to the warrant procedures governing administrative searches. . . . Evidence of arson discovered in the course of such investigations is admissible at trial, but if the investigating officials find probable cause to believe that arson has occurred and require further access to gather evidence for a possible prosecution, they may obtain a warrant only upon a traditional showing of probable cause applicable to searches for evidence of crime. . . .

These principles require that we affirm the judgment of the Michigan Supreme Court ordering a new trial. Affirmed.

Case Name: Michigan v. Tyler

Court: United States Supreme Court

Summary of Main Points: Firefighters do not need a search warrant to lawfully enter a building on fire. Once lawfully present, they may seize evidence in plain view. Firefighters may also remain in a building for a reasonable period after the fire is out to conduct a cause and origin investigation without the need for a search warrant.

Michigan v. Tyler is a critically important case for firefighters and fire investigators. It explains the constitutional issues involved in fire scene investigations, and sets forth the guidelines for cause and origin determinations. These guidelines include:

- The Fourth and Fourteenth Amendment warrant requirements do apply to fire scenes.
- The initial entry by firefighters into a building for the purpose of extinguishing a fire is constitutionally justified as an exigent circumstance exception to the warrant requirement.
- Once lawfully present, firefighters and investigators can seize evidence that is in plain view **(Figure 6-6)**.
- For purposes of the plain view doctrine, the Court does not draw a distinction between government officials who are firefighters, fire inspectors, or law enforcement officers. If any of the governmental agents observe evidence in plain view, the evidence can be lawfully seized.
- Firefighters and investigators may remain on scene without a warrant for a reasonable period of time after the fire has been extinguished in order to conduct their investigation **(Figure 6-7)**. In *Tyler,* the reentry that occurred approximately five hours after the fire had been extinguished was held to be reasonable. The Court considered factors such as the presence of smoke, steam, and lack of lighting in justifying the reentry.
- Post-fire entries that are clearly detached from the initial exigency are subject to the Fourth Amendment warrant requirement. As such, these entries into the building require a warrant. Post-exigency entries to determine cause and origin would require an administrative search warrant, since probable cause may not exist. Entries to search for evidence

Figure 6-6 *Once lawfully present inside a building, firefighters may seize evidence in plain view.*

Figure 6-7 *Fire investigators may remain at a scene for a reasonable period after a fire to investigate the cause.*

of arson or other criminal activity that go beyond the scope of the administrative warrant require a criminal search warrant based upon probable cause.

 **SIDEBAR**

Administrative Search Warrant

When government officials need to enter a person's home or business in order to carry out a legal responsibility, such as conducting a fire code inspection or conducting a cause and origin determination, they may obtain an administrative search warrant. Unlike a criminal search warrant, which requires a showing of probable cause of criminal activity, an administrative search warrant can be issued upon showing the need to conduct a search as part of carrying out a legal responsibility, and a description of the scope of the proposed search. Administrative warrants must be issued in accordance with reasonable legislative or administrative standards, or absent such standards, in accordance with judicially prescribed standards.

Search and Seizure After Tyler

Left unanswered by the *Tyler* case was the question of what constitutes a "reasonable time" after a fire is extinguished, during which fire personnel may legally conduct a cause and origin determination. The *Tyler* case was followed by a second fire scene search and seizure case, also from Michigan, named *Michigan v. Clifford.*

Michigan v. Clifford
464 U.S. 287 (1984)
United States Supreme Court

. . . In the early morning hours of October 18, 1980, a fire erupted at the Clifford home. The Cliffords were out of town on a camping trip at the time. The fire was reported to the Detroit Fire Department, and fire units arrived on the scene about 5:40 a.m. The fire was extinguished and all fire officials and police left the premises at 7:04 a.m.

At 8 o'clock on the morning of the fire, Lieutenant Beyer, a fire investigator with the arson section of the Detroit Fire Department, received instructions to investigate the Clifford fire. He was informed that the Fire Department suspected arson. Because he had other assignments, Lieutenant Beyer did not proceed immediately to the Clifford residence. He and his partner finally arrived at the scene of the fire about 1 p.m. on October 18.

When they arrived, they found a work crew on the scene. The crew was boarding up the house and pumping some six inches of water out of the basement. A neighbor told the investigators that he had called Mr. Clifford and had been instructed to request the Cliffords' insurance agent to send a boarding crew out to secure the house. The neighbor also advised that the Cliffords did not plan to return that day. While the investigators waited for the water to be pumped out, they found a Coleman fuel can in the driveway that was seized and marked as evidence.

By 1:30 p.m., the water had been pumped out of the basement and Lieutenant Beyer and his partner, without obtaining consent or an administrative warrant,

(continued)

(*continued*)

entered the Clifford residence and began their investigation into the cause of the fire. Their search began in the basement, and they quickly confirmed that the fire had originated there beneath the basement stairway. They detected a strong odor of fuel throughout the basement, and found two more Coleman fuel cans beneath the stairway. As they dug through the debris, the investigators also found a crock pot with attached wires leading to an electrical timer that was plugged into an outlet a few feet away. The timer was set to turn on at approximately 3:45 a.m. and to turn back off at approximately 9 a.m. It had stopped somewhere between 4 and 4:30 a.m. All of this evidence was seized and marked.

After determining that the fire had originated in the basement, Lieutenant Beyer and his partner searched the remainder of the house. The warrantless search that followed was extensive and thorough. The investigators called in a photographer to take pictures throughout the house. They searched through drawers and closets and found them full of old clothes. They inspected the rooms and noted that there were nails on the walls, but no pictures. They found wiring and cassettes for a videotape machine but no machine.

Respondents [Cliffords] moved to exclude all exhibits and testimony based on the basement and upstairs searches on the ground that they were searches to gather evidence of arson, that they were conducted without a warrant, consent, or exigent circumstances, and that they therefore were per se unreasonable under the Fourth and Fourteenth Amendments. Petitioner [State of Michigan], on the other hand, argues that the entire search was reasonable and should be exempt from the warrant requirement.

[T]he State does not challenge the state court's finding that there were no exigent circumstances justifying the search of the Clifford home. Instead, it asks us to exempt from the warrant requirement all administrative investigations into the cause and origin of a fire. We decline to do so.

In Tyler, we restated the Court's position that administrative searches generally require warrants. . . . We reaffirm that view again today. Except in certain carefully defined classes of cases, the nonconsensual entry and search of property are governed by the warrant requirement of the Fourth and Fourteenth Amendments. The constitutionality of warrantless and nonconsensual entries onto fire-damaged premises, therefore, normally turns on several factors: whether there are legitimate privacy interests in the fire-damaged property that are protected by the Fourth Amendment; whether exigent circumstances justify the government intrusion regardless of any reasonable expectations of privacy; and, whether the object of the search is to determine the cause of fire or to gather evidence of criminal activity.

We observed in *Tyler* that reasonable privacy expectations may remain in fire-damaged premises. People may go on living in their homes or working in their offices after a fire. Even when that is impossible, private effects often remain on the fire-damaged premises. . . . Privacy expectations will vary with the type of property, the amount of fire damage, the prior and continued use of the premises, and in some cases the owner's efforts to secure it against intruders. Some fires

(*continued*)

(*continued*)

may be so devastating that no reasonable privacy interests remain in the ash and ruins, regardless of the owner's subjective expectations. The test essentially is an objective one: whether "the expectation [is] one that society is prepared to recognize as 'reasonable.'" . . . If reasonable privacy interests remain in the fire-damaged property, the warrant requirement applies, and any official entry must be made pursuant to a warrant in the absence of consent or exigent circumstances.

A burning building of course creates an exigency that justifies a warrantless entry by fire officials to fight the blaze. Moreover, in *Tyler,* we held that, once in the building, officials need no warrant to remain for "a reasonable time to investigate the cause of a blaze after it has been extinguished." . . . Where, however, reasonable expectations of privacy remain in the fire-damaged property, additional investigations begun after the fire has been extinguished and fire and police officials have left the scene generally must be made pursuant to a warrant or the identification of some new exigency.

The aftermath of a fire often presents exigencies that will not tolerate the delay necessary to obtain a warrant or to secure the owner's consent to inspect fire-damaged premises. Because determining the cause and origin of a fire serves a compelling public interest, the warrant requirement does not apply in such cases.

If a warrant is necessary, the object of the search determines the type of warrant required. If the primary object is to determine the cause and origin of a recent fire, an administrative warrant will suffice. To obtain such a warrant, fire officials need show only that a fire of undetermined origin has occurred on the premises, that the scope of the proposed search is reasonable and will not intrude unnecessarily on the fire victim's privacy, and that the search will be executed at a reasonable and convenient time.

If the primary object of the search is to gather evidence of criminal activity, a criminal search warrant may be obtained only on a showing of probable cause to believe that relevant evidence will be found in the place to be searched. If evidence of criminal activity is discovered during the course of a valid administrative search, it may be seized under the "plain view" doctrine. . . . This evidence then may be used to establish probable cause to obtain a criminal search warrant. Fire officials may not, however, rely on this evidence to expand the scope of their administrative search without first making a successful showing of probable cause to an independent judicial officer.

The object of the search is important even if exigent circumstances exist. Circumstances that justify a warrantless search for the cause of a fire may not justify a search to gather evidence of criminal activity once that cause has been determined. If, for example, the administrative search is justified by the immediate need to ensure against rekindling, the scope of the search may be no broader than reasonably necessary to achieve its end. A search to gather evidence of criminal activity not in plain view must be made pursuant to a criminal warrant upon a traditional showing of probable cause.

The searches of the Clifford home, at least arguably, can be viewed as two separate ones: the delayed search of the basement area, followed by the extensive

(*continued*)

(*continued*)

search of the residential portion of the house. We now apply the principles outlined above to each of these searches.

The Clifford home was a two-and-one-half story brick and frame residence. Although there was extensive damage to the lower interior structure, the exterior of the house and some of the upstairs rooms were largely undamaged by the fire, although there was some smoke damage. The firemen had broken out one of the doors and most of the windows in fighting the blaze. At the time Lieutenant Beyer and his partner arrived, the home was uninhabitable. But personal belongings remained, and the Cliffords had arranged to have the house secured against intrusion in their absence. Under these circumstances, and in light of the strong expectations of privacy associated with a home, we hold that the Cliffords retained reasonable privacy interests in their fire-damaged residence, and that the post-fire investigations were subject to the warrant requirement. Thus, the warrantless and nonconsensual searches of both the basement and the upstairs areas of the house would have been valid only if exigent circumstances had justified the object and the scope of each.

As noted, the State does not claim that exigent circumstances justified its post-fire searches. It argues that we either should exempt post-fire searches from the warrant requirement or modify *Tyler* to justify the warrantless searches in this case. We have rejected the State's first argument, and turn now to its second.

In *Tyler*, we upheld a warrantless post-fire search of a furniture store, despite the absence of exigent circumstances, on the ground that it was a continuation of a valid search begun immediately after the fire. The investigation was begun as the last flames were being doused, but could not be completed because of smoke and darkness. The search was resumed promptly after the smoke cleared and daylight dawned. Because the post-fire search was interrupted for reasons that were evident, we held that the early morning search was "no more than an actual continuation of the first, and the lack of a warrant thus did not invalidate the resulting seizure of evidence.". . .

As the State conceded at oral argument, this case is distinguishable for several reasons. First, the challenged search was not a continuation of an earlier search. Between the time the firefighters had extinguished the blaze and left the scene and the arson investigators first arrived about 1 p.m. to begin their investigation, the Cliffords had taken steps to secure the privacy interests that remained in their residence against further intrusion. These efforts separate the entry made to extinguish the blaze from that made later by different officers to investigate its origin. Second, the privacy interests in the residence—particularly after the Cliffords had acted—were significantly greater than those in the fire-damaged furniture store, making the delay between the fire and the midday search unreasonable absent a warrant, consent, or exigent circumstances. We frequently have noted that privacy interests are especially strong in a private residence. These facts—the interim efforts to secure the burned-out premises and the heightened privacy interests in the home—distinguish this case from *Tyler*. At least where a homeowner has made a reasonable effort to secure his fire-damaged home after

(*continued*)

(*continued*)

the blaze has been extinguished and the fire and police units have left the scene, we hold that a subsequent post-fire search must be conducted pursuant to a warrant, consent, or the identification of some new exigency. So long as the primary purpose is to ascertain the cause of the fire, an administrative warrant will suffice.

Because the cause of the fire was then known, the search of the upper portions of the house, described above, could only have been a search to gather evidence of the crime of arson. Absent exigent circumstances, such a search requires a criminal warrant.

Even if the midday basement search had been a valid administrative search, it would not have justified the upstairs search. The scope of such a search is limited to that reasonably necessary to determine the cause and origin of a fire, and to ensure against rekindling. As soon as the investigators determined that the fire had originated in the basement and had been caused by the crock pot and timer found beneath the basement stairs, the scope of their search was limited to the basement area. Although the investigators could have used whatever evidence they discovered in the basement to establish probable cause to search the remainder of the house, they could not lawfully undertake that search without a prior judicial determination that a successful showing of probable cause had been made. Because there were no exigent circumstances justifying the upstairs search, and it was undertaken without a prior showing of probable cause before an independent judicial officer, we hold that this search of a home was unreasonable under the Fourth and Fourteenth Amendments, regardless of the validity of the basement search.

The warrantless intrusion into the upstairs regions of the Clifford house presents a telling illustration of the importance of prior judicial review of proposed administrative searches. If an administrative warrant had been obtained in this case, it presumably would have limited the scope of the proposed investigation, and would have prevented the warrantless intrusion into the upper rooms of the Clifford home. An administrative search into the cause of a recent fire does not give fire officials license to roam freely through the fire victim's private residence.

The only pieces of physical evidence that have been challenged on this . . . appeal are the three empty fuel cans, the electric crock pot, and the timer and attached cord. Respondents also have challenged the testimony of the investigators concerning the warrantless search of both the basement and the upstairs portions of the Clifford home. The discovery of two of the fuel cans, the crock[pot], the timer and cord—as well as the investigators' related testimony—were the product of the unconstitutional post-fire search of the Cliffords' residence. Thus, we affirm that portion of the judgment of the Michigan Court of Appeals that excluded that evidence. One of the fuel cans was discovered in plain view in the Cliffords' driveway. This can was seen in plain view during the initial investigation by the firefighters. It would have been admissible whether it had been seized in the basement by the firefighters or in the driveway by the arson investigators. Exclusion of this evidence should be reversed.

It is so ordered.

Case Name: Michigan v. Clifford

Court: United States Supreme Court

Summary of Main Points: Reaffirming Tyler, the Court ruled that a post-fire cause and origin determination constitutes a search that requires a warrant, unless one of the exceptions to the warrant requirement applies. There is no exception for conducting a cause and origin determination. The fire inspectors should have obtained an administrative search warrant.

In many ways, *Clifford* reinforced and clarified some of the details of the *Tyler* case. The key points of the *Clifford* case were:

- There is no exception to the Fourth Amendment search warrant requirement to conduct a post-fire cause and origin determination. If such a cause and origin determination must be made remote in time from the initial entry, an administrative search warrant must be obtained. Administrative search warrants need not be issued on probable cause, but rather to comply with a legal responsibility. All that is required is a showing of a fire of undetermined origin, and that the search to be conducted will be reasonable in scope and time.

- It is possible, based on fire damage to a building, that no reasonable expectation of privacy can remain in the premises **(Figure 6-8)**. In such a case a warrantless entry would not violate the Fourth or Fourteenth Amendment.

- Where reasonable expectations of privacy remain in a fire-damaged premises, any search into the cause and origin of a fire is subject to the warrant requirement of the Fourth and Fourteenth Amendments, in the absence of consent or exigent circumstances.

- An administrative search warrant will suffice for a cause and origin determination. However, once a search extends beyond cause and origin, and into a search for criminal evidence, a search warrant issued upon probable cause is constitutionally required.

- "An administrative search into the cause of a recent fire does not give fire officials license to roam freely through the fire victim's private residence."

The Court distinguished the *Clifford* case from *Tyler*, noting that

- *Clifford* involved a residence, while *Tyler* involved a business. People's expectation of privacy is greater in their residences than in business property, and thus residences are entitled to greater protection from warrantless searches **(Figure 6-9)**.

- The Cliffords had taken steps to secure their premises after the fire, which increased their expectation of privacy. No such efforts were made in *Tyler*.

Figure 6-8 *A building may be so badly damaged by fire that no reasonable expectation of privacy may exist.*

- In *Tyler*, it was the state's position that the additional entries later in the morning were continuations of the initial entry to fight the fire. The state in *Clifford* did not argue that the warrantless entry at 1:00 p.m. was a continuation of the initial entry of firefighters at 5:00 a.m. to extinguish the fire. The state did not argue that the search had to be delayed due to smoke, steam, or darkness, but rather agreed that the 1:00 p.m. entry was a separate search made without exigent circumstances.

Together, *Tyler* and *Clifford* give the fire service some important guidance in terms of our authority to conduct post-fire cause and origin determinations, and what search and seizure issues arise at fire scenes.

Figure 6-9 *Left unanswered by the Supreme Court is at what point a person loses his or her expectation of privacy in a fire-damaged building. (Photo by Rick Blais.)*

State Law Searches

Besides the United States Constitution, state constitutions also have language pertaining to the rights of citizens, and to search and seizure. State constitutions often have analogous or identical language to the Fourth and Fourteenth Amendments regarding the need for search warrants and due process. While a state constitution cannot provide less protection from warrantless searches than the United States Constitution, states can provide greater protection. As such, state and local law enforcement officials may need to go beyond the scope of *Tyler* and *Clifford* in their states, based on state constitutional law.

Additional Fire Scene Issues

What happens when a firefighter finds evidence of arson, or even of criminal activity unrelated to the fire, in plain view? Must the firefighter bring the evidence to a police officer? Can the firefighter summon a police officer to come into the area where the items were found? Must the police officer first obtain a search warrant?

When operating at a fire scene, a firefighter is a governmental official acting in his or her official capacity. Courts draw no distinction between firefighters, fire investigators, or police officers who find evidence in plain view. The initial exigency authorizes the firefighter to be there lawfully. Once lawfully present on scene, and once having observed the evidence in plain view, there is no further infringement upon a person's expectation of privacy by calling in additional personnel to assist in collecting and documenting the evidence. The reality of the firefighter observing the evidence in plain view (as the first governmental official) extinguishes any expectation of privacy the defendant may have had. (*State v. Eady*, 249 Conn. 431, 1999.)

In the *Eady* case, firefighters found marijuana present in a bedroom during a fire. Firefighters then summoned a police officer who was outside directing traffic. The defendant sought to have the evidence excluded from trial. The court found no meaningful distinction between a police officer finding evidence in plain view and summoning a detective with greater expertise in evidence collection and preservation, and a firefighter finding evidence in plain view and summoning a police officer. The analysis should be on whether the person's expectations of privacy are somehow injured by the additional entry. Once a governmental official has observed the evidence in plain view, there is no longer a reasonable expectation of privacy.

An additional matter that comes up from time to time in arson cases is the admissibility of evidence obtained by warrantless searches conducted by private individuals such as private investigators and fire scene investigators who work for insurance companies. The Fourth and Fourteenth Amendments' warrant requirement applies to governmental action. The exclusionary rule does not apply to evidence seized during a search by a private party, provided the search was not the result of collusion between investigators and law enforcement officials attempting to circumvent the warrant requirement.

Standing

One final matter concerning search and seizure law that is often overlooked is that of **standing**. The exclusionary rule is not an absolute rule that prohibits all use of evidence seized in warrantless searches. Before a criminal defendant can raise an objection to the use of evidence seized in an illegal search, he must have standing to object. Standing comes from having a reasonable expectation of privacy in the place to be searched.

standing
the legal right to bring a case or make a legal claim; for a person to have standing to object to a warrantless search, the person must have had a reasonable expectation of privacy in the place to be searched

When an arsonist breaks into someone else's building and sets a fire, the arsonist has no legitimate expectation of privacy in that building, and thus no standing to object to evidence seized illegally. Of course, in the early stages of a fire scene investigation, it may be impossible to know if the perpetrator has a privacy interest in the premises or not. Thus it is always advisable to secure the appropriate search warrant when necessary. Consider this example.

SIDEBAR

Able owns a commercial building. Baker is a neighbor who has a fight with Able and decides to burn his building down. He hires Charlie to burn the building. Immediately after the fire, the Metro City Fire Department conducts a cause and origin determination, which points toward arson as the cause. Two days later, the inspectors go back without a warrant and without Able's consent. The investigators obtain additional evidence. A week later they return and take some additional samples, again without a warrant or consent.

In a criminal trial for arson, neither Baker nor Charlie has standing to object to the admissibility of the evidence seized in any of the illegal searches, because it was seized in violation of Able's constitutional rights, not theirs. Neither Baker nor Charlie had a legitimate expectation of privacy in Able's building.

CHAIN OF CUSTODY

chain of custody
a law enforcement agency's documented proof of possession of physical evidence, from the time the evidence was seized until its introduction at trial

In order for physical evidence to be admissible at trial, law enforcement agencies must be capable of documenting an unbroken **chain of custody** throughout its possession, from the moment of seizure until the evidence is introduced at trial. Establishing the chain of custody at trial requires the testimony of the investigator who seized the item, and potentially of each person who had possession of the item between the time it was seized and the trial **(Figure 6-10)**.

The chain of custody testimony must establish that the evidence that is sought to be introduced at trial is the same item that was actually seized, and that the evidence was not tampered with, altered, or substituted while in that person's custody. The fewer individuals who are in the chain of custody, the easier it is to have the seized item introduced into evidence.

At fire scenes, securing access to evidence prior to its being seized may be important to its admissibility and value at trial. When fire personnel find evidence, possible evidence, or the point of origin of a fire, the area should be secured and preserved until the arrival of investigators. Access to the evidence by fire personnel should be limited to the greatest extent possible. Access to the area by third parties should be prohibited. Under no circumstances should the area be left unsecured or unattended **(Figure 6-11)**. Evidence should be seized only by trained personnel who are prepared to follow the procedures required by the local courts for documenting chain of custody.

Figure 6-10
Evidence that is seized must be properly handled and documented to establish a chain of custody. (Photo by Ralph Costantino.)

Figure 6-11 *After a fire is knocked down, access to the point of origin and other evidence must be secured. (Photo by Rick Blais.)*

CUSTODIAL INTERROGATIONS AND MIRANDA WARNINGS

The Fifth Amendment to the United States Constitution provides that no person shall be compelled in any criminal case to be a witness against himself. This provision is commonly referred to as the privilege against self-incrimination. The Sixth Amendment provides that people accused of a crime have a right to counsel.

In a 1966 case, *Miranda v. Arizona,* the United States Supreme Court considered these two constitutional rights, and concluded that in order for these rights to be respected and protected, law enforcement officials must advise people in their custody of these rights prior to interrogating them. [384 U.S. 436, 86 S.Ct. 1602, (1966).] Evidence obtained in violation of *Miranda* is subject to the exclusionary rule discussed above in regard to search and seizure. The *Miranda* case was a landmark case in 1966 when it was decided, and it remains of vital importance today.

The *Miranda* decision pertains to **custodial interrogations (Figure 6-12)**, which are interrogations conducted by police officers of suspects who are in their custody. There is a two-part test of the term "custodial interrogation" that must be satisfied for the *Miranda* warning to be required.

Custodial means that the suspect is in custody, under arrest, or reasonably believes himself to be under arrest. The test of whether someone

custodial interrogation
an interrogation conducted by law enforcement officers of suspects who have been taken into custody or who otherwise have been deprived of their freedom in any significant way

Figure 6-12
Custodial interrogation triggers the need for Miranda warnings. (Photo by James E. Clift.)

is in custody is not based upon the arresting officer's subjective intent to arrest. The focus is on whether or not a reasonable person in the suspect's situation would believe himself to be free to leave. A variety of factors must be taken into account, including where the questioning takes place (in a police station, in a police car, at the scene, at a suspect's house), whether the questioning is conducted in a hostile environment, whether the suspect is handcuffed during or was handcuffed prior to the interrogation, whether the suspect approached the police voluntarily, and what the officers have told the suspect. Certainly, if the suspect has been informed that he or she is under arrest, or is not free to leave, the suspect is considered to be in custody.

Interrogation means questioning a person about something that goes beyond general questions about what happened, and into accusatorial or specific questions. Interrogations often involve pressure or persuasion applied to the suspect to convince her to share information with the police.

Before law enforcement officials can interrogate a suspect who is in custody, the suspect has the right to have her constitutional rights explained. These rights are commonly explained in the following manner:

- You have the right to remain silent.

- Anything you say can and will be used against you in a court of law.

- You have the right to talk to a lawyer and have him present with you during questioning.

- If you cannot afford a lawyer, one will be appointed to represent you, if you wish.

Before the police may continue questioning the suspect, they must obtain a knowing and intelligent waiver of these rights. This waiver can be obtained by two additional questions:

- Do you understand each of these rights as I have explained them to you?

- Having these rights in mind, do you wish to talk to us now?

The failure to explain these *Miranda* rights prior to a custodial interrogation can result in the evidence obtained during the interrogation being excluded from trial. In addition, if information gained during an illegal interrogation leads to additional evidence being uncovered, that evidence may also be excluded as "fruit from the poisoned tree." Law enforcement officials who seek to use statements taken in a custodial interrogation must also demonstrate that the defendant's statements were voluntary. Most police departments will obtain the suspect's waiver of rights in writing prior to conducting a custodial interrogation, or anything that may come close to a custodial interrogation.

There are literally thousands of cases interpreting *Miranda* and the myriad of sub-issues that arise: Was the suspect really in custody? Was this just general questioning, or was it an interrogation? Did the suspect intelligently waive his rights? Did the suspect really understand his rights?

Law enforcement officials receive extensive training on the proper methods of obtaining and documenting a waiver of *Miranda* rights. Does *Miranda* have any applications for fire personnel?

George Harvey MEEK, Appellant, v. The STATE of Texas, Appellee
747 S.W.2d 30 (1988)
Court of Appeals of Texas, El Paso

FULLER, Justice.

This an appeal from a conviction for the offense of arson. The court assessed punishment at seven years' imprisonment. We reverse and remand for new trial.

[Defendant, who is the Appellant] . . . asserts the court erred in admitting into evidence the written statement of Appellant, over the objection that it resulted from a custodial interrogation in the absence of advisement to Appellant of his constitutional rights. On April 6, 1986, a fire broke out at 1304 Wedgewood, a residence owned by Mrs. Carole DeWees. On the same date, she gave a statement to Fire Investigator Hector Zubia who was conducting an arson investigation. In this statement, she denied any involvement with regard to the fire. However, she stated that the Appellant, her then estranged husband, had stated he intended to "torch" or set fire to the house. Investigator Zubia made arrangements to interview the Appellant and did so two days later at the Central Fire Station. At the pretrial hearing on Appellant's motion to suppress the statements, the Appellant testified, that before he gave his witness statement, he was taken to an office where he was placed in handcuffs and was generally intimidated. The handcuffs were removed and he was taken to a desk where a secretary transcribed his statement. During the interview he went unaccompanied to his car to retrieve some papers. He left the fire station after the interview; he was not arrested. Investigator Zubia testified that he never placed the Appellant in handcuffs. He also stated that he never considered the Appellant as a suspect during the interview. He testified that he had forgotten about the allegation contained in DeWees' [sic] statement. Shortly after the interview began, the Appellant made the following statement which was transcribed into his witness statement:

> She also told me that she would like to have the house leveled. My interpretation of this statement of hers was that she wanted the house destroyed. I told her that I would help her to make the arrangements and I would tell her what to buy to do it.

(continued)

(continued)

The Appellant proceeded to relate, in a rather contradictory and disjointed manner, the nature of his efforts to get her assistance with regard to her request. At trial, DeWees testified she started the fire and the Appellant, while not present, aided and assisted her in this endeavor. The Appellant's two statements, given that day at the fire station, were read to the jury and were admitted into evidence.

A formal arrest is not a prerequisite before Miranda rights arise. . . . Custodial interrogation means questioning initiated by a law enforcement officer after a person has been taken into custody or otherwise deprived of his freedom of action in any significant way. . . . One primary factor to be considered in analyzing if there was a custodial interrogation is whether or not the focus of the investigation has finally centered on the defendant. . . . Other factors to be considered are probable cause to arrest, subjective intent of the officer and subjective belief of the defendant. . . . However, the courts cannot be expected to determine cases solely on the basis of self-serving statements by the defendant or the interrogating officer. . . .

Investigator Zubia testified that the interview was a question-and-answer session. It defies belief that, when the statement of DeWees regarding the Appellant's threat and the above mentioned statement of the Appellant are taken in conjunction, the focus of the investigation had not centered upon the Appellant, thereby, invoking the rule in Miranda. We hold that the Appellant's statements were inadmissible and that they were the result of a custodial interrogation in the absence of the requisite constitutional warnings. . . .

The judgment of the trial court is reversed, and judgment is hereby remanded for new trial.

Case Name: Meek v. State

Court: Court of Appeals of Texas, El Paso.

Summary of Main Points: Fire personnel are governmental agents, and like police officers, may be subject to constitutional limitations on questioning suspects. In particular, where a fire investigator conducts a custodial interrogation, the suspect being questioned must be given his or her Miranda warnings.

As the *Meek* case points out, a person's constitutional rights can just as easily be violated by a firefighter or fire inspector as by a police officer. *Meek* happened to involve a fire investigator, but the issue of *Miranda* rights could arise if a suspect is questioned by a fire chief, fire officer, or firefighter. Fire personnel are considered to be governmental agents acting in their official capacity. Firefighters, particularly those who may have to question suspects or investigate the cause of a fire, need training in this regard, comparable to that

principal
where more than one person commits a crime, the person who actually commits the physical act of the crime

accessory
someone who helped the principal to commit a crime, but was not actually present on the scene during the commission of the crime

aider and abettor
anyone who knowingly assists someone in the commission of a crime, or who helps them evade capture or cover it up

given to police officers. The *Meek* case was reversed upon appeal, and the Defendant's conviction ultimately reinstated. The text of that decision is in Appendix B, Chapter 6.

ACCOMPLICE LIABILITY: PARTIES TO A CRIME

Often a crime is committed through the efforts of several people. There may be several accomplices, each of whom assists the primary actor or actors in the commission of the crime. We refer to the primary actor or actors as the principals. A **principal** is the person who actually commits the physical act of the crime.

Traditional legal theory called someone who helped the principal to commit a crime, but was not actually present on the scene during the commission, an **accessory** to the crime. Someone who assisted the principal prior to the crime was called an accessory before the fact. Someone who helped the principal escape from the scene, evade capture, or who helped cover up the crime was an accessory after the fact. The penalties for accessories were generally not as severe as those for principals.

The more modern trend is to dispense with the notions of accessory before or after the fact, and treat anyone who knowingly assists someone in the commission of a crime, or who helps them evade capture or cover it up, as an **aider and abettor**. An aider and abettor can be charged as a principal.

📖 **SIDEBAR**

Don decides he is going to burn a building for profit. He discusses his plan with Ed, who offers to supply Don with a timing device and accelerant. Don is arrested shortly after lighting the fire and is charged with arson. Ed is also charged with arson for being an aider and abettor.

CONSPIRACY

conspiracy
an agreement to commit a crime or break the law

When multiple parties are involved in the commission of a crime, in addition to aiding and abetting, the issue of a criminal conspiracy comes up. A **conspiracy** is an agreement to commit a crime or break the law. The conspiracy to commit a crime is a crime in and of itself. In other words, a conspiracy charge can be filed against someone in addition to the underlying charge. Furthermore, each of the co-conspirators can be charged with each crime committed by any of the other co-conspirators that is committed in furtherance of the conspiracy.

SIDEBAR

Mike, Larry, and Kyle conspire to burn down Mike's tailor shop, which has been losing money for years. They agree to split the insurance proceeds equally. Mike calls his insurance company and increases the coverage. He mails the insurance company fraudulent documents stating that he has refurbished the shop and bought new equipment.

Larry buys the gasoline and some illegal explosives to help make sure the building is a total loss. Larry plants the explosives in the shop so they will go off when the fire reaches them, then removes all the valuables from the shop. Kyle agrees to light the fire. The fire is a total loss.

All three perpetrators can be charged with:

- conspiracy to burn the tailor shop
- arson
- mail fraud
- defrauding an insurer
- illegal possession of explosives

It doesn't matter that Larry and Kyle were unaware that Mike would commit mail fraud, nor that Mike and Kyle did not know that Larry would purchase illegal explosives. Each co-conspirator is responsible for the crimes committed by the other co-conspirators in furtherance of the conspiracy.

ATTEMPTS

The attempt to commit a crime, even though uncompleted, may give rise to criminal liability. An attempt requires an act in furtherance of the commission of the crime, committed with the same mental state as required for the underlying crime. However, how much of an act is enough to constitute an attempt? If a person sets a timing device to start a fire in a building, but the timing device fails to go off, is that an attempt? What if the police intervene and stop the device before it is set to go off? What if the police arrest the perpetrator as he arrives at the building with the timing device in hand? What if the police arrest him in his garage as he is constructing the timing device? What if the police sold the person the timing device as part of a sting operation, and they knew it would not go off?

Finding criminal activity in an attempted crime can be quite easy in some cases, but a challenge in others. The easy type of an attempt case is one in which the defendant has done everything necessary to commit the crime, but due to some outside factor beyond her control has not been successful. The harder cases involve situations where the act is more of a preliminary to an attempt, such as purchasing an accelerant or supplies to manufacture a bomb. All jurisdictions share a common requirement that, to be guilty of an attempt, the defendant must commit some act in furtherance of the attempt, which goes beyond mere preparation.

Not all states punish attempt crimes in the same way. Some states punish an attempt to commit a crime as to the same extent as if the person had committed the underlying crime. Some states have statutes that make an attempt to commit any felony punishable as a separate offense. In a few cases, an attempt to commit a crime such as arson may be its own specific criminal offense (see RIGL § 11-4-6 in Appendix B, Chapter 5).

CRIMINAL DEFENSES

There are several categories of defenses to criminal offenses that go beyond a defendant denying that he committed an offense. These defenses may be applicable to any crime, but some are more likely to be used for certain crimes than others.

Self-Defense

Self-defense is a defense that can be used to justify otherwise criminal conduct, up to and including the killing of another person. The defense is valid only when one is confronted with an immediate threat of serious bodily injury. A person is entitled to use reasonable force to defend himself from an attack. The use of deadly force is allowed, but must be reasonable under the circumstances, commensurate with the danger, and used as a last resort.

Consistent with the use of deadly force as a last resort, the common law recognized a duty to retreat rather than use deadly force. This duty to retreat required a person to retreat from someone rather than resorting to deadly force, if the retreat could be made safely. The duty to retreat has been a particular concern when people are in their own homes or businesses and confront an intruder. Many states have abolished the duty to retreat rule in one's home or business. Some states have gone so far as to create a presumption that a person in her own home or business is justified in using deadly force against someone who unlawfully breaks into the home or business. (See RIGL 11-8-8.)

Self-defense does not justify preemptive strikes against someone who a person fears will attack him in the future. For self-defense to be applicable, the threat must be imminent. In the same vein, a punitive attack on someone who previously gave a person grounds to use self-defense is not privileged. Jurisdictions differ on the treatment of someone who was mistaken about the need to use force for self-defense. Some states hold that, so long as the defendant's belief that he or she was being attacked was "reasonable," the use of self-defense was appropriate. In other states, a person who uses force to defend himself acts at his own peril, and if he was mistaken about the need to use force, self-defense cannot be used as a defense.

Defense of Others

Coming to the defense of others is also a valid defense to criminal charges. The basic rules and limitations that apply to self-defense are equally applicable when defending others.

Defense of Property

A person is entitled to use reasonable force to protect and preserve her own property or the property of others. However, deadly force cannot be used merely to protect property.

Insanity

The insanity defense is a highly touted defense on television, but in reality it plays a minor role in criminal prosecution. While states differ on exactly what constitutes insanity, the general rule is that, at the time of the conduct, as a result of mental disease or defect, the perpetrator must have lacked substantial capacity either to appreciate the wrongfulness of his conduct, or conform his conduct to the requirements of the law.

Examples of the two categories are:

1. *Unable to appreciate the wrongfulness of his conduct:* someone who, due to age or infirmity, believed that he was lighting a candle when in reality he was lighting a curtain on fire.

2. *Unable to conform his conduct to the requirements of the law:* someone who acted on an "irresistible impulse."

The term "mental disease or defect" does not include a mental abnormality that is manifested only by repeated criminal or antisocial behavior. [*State v. Johnson,* 399 A.2d 469, 476 (RI, 1979).] Establishing a mental disease requires medical testimony and documentation to survive the type of skepticism that most people inherently feel when the insanity defense is raised. Contrary to popular belief (and television depiction), the insanity defense is regularly rejected by juries, as it was for the arsonist who set the Happy Land Social Club fire on March 25, 1990 in New York City that killed 87 people **(Figure 6-13)**. See *People v. Gonzalez,* 625 N.Y.S.2d 844; 163 Misc. 2d 950 (1995).

States differ on the procedure for filing an insanity defense. Some states consider the insanity defense to require a guilty but insane verdict, while

Figure 6-13 *The arsonist who set the Happy Land Social Club fire that killed 87 people on March 25, 1990 in New York City attempted to use the insanity defense. The jury rejected the defense and found him guilty. (Photo by Corbis.)*

others permit a not guilty by reason of insanity verdict. Still other states, such as Michigan, provide for the jury to have a choice between verdicts of guilty but mentally ill, and not guilty by reason of insanity. See *People v Ramsey,* 422 Mich. 500, 375 N.W.2d 297 (1983).

Another problem with the insanity defense is that it is a double-edged sword. The defense, if successful, absolves the defendant from criminal liability for her action, but she will then be institutionalized indefinitely. Certainly for a charge of murder, such a trade-off may be acceptable. However, for minor charges for which there may be little or no jail time, the insanity defense would not commonly be used, even where it may be appropriate based on the facts.

Entrapment

Entrapment is a defense that calls into question the fairness of police conduct in precipitating the defendant's conduct. It can be used as an affirmative defense to a crime in which the conduct of law enforcement personnel was likely to induce a normally law-abiding person to commit the offense. An example would be a situation in which an undercover police officer asks someone to deliver a package of marijuana to a certain person who needs it for medical reasons, and offers to pay him a large sum of money. Jurisdictions differ over whether the defendant's propensity to commit crimes of this nature is admissible. Some states focus more on the defendant's propensity to commit a crime as a concern that overrides any entrapment by the government, while other states focus more on the propriety of the government's behavior.

Goolsby v. State of Georgia
184 Ga. App. 390, 361 SE2d 684 (GA, 1987)

POPE, Judge.

Defendant Goolsby was employed as an Atlanta firefighter. He was the recipient of several citations for bravery and saving lives. Unfortunately for defendant, in September 1985 he ran into Paul Lester, an old acquaintance who was known by defendant to have a criminal record of convictions and arrests for selling drugs. What defendant did not know was that Lester was "working off" his sentences by serving as an informant for both the City of Atlanta and Clayton County police. From September through November, Lester called defendant between twenty and thirty times asking defendant to help him sell cocaine. Lester appealed to defendant's sympathy by claiming he needed to make money by selling drugs in order to pay for medical care for his mother who was dying of cancer. After repeatedly refusing Lester's requests, defendant finally agreed to cooperate.

On the afternoon of November 25, 1985, defendant met Lester and an undercover police officer, posing as a drug dealer, at the parking lot of a drive-in restaurant. The officer got into defendant's vehicle and defendant sold him one-eighth ounce of a substance which later proved to be cocaine. According to the officer's testimony, after he had already returned to Lester's automobile defendant approached that vehicle and offered to sell the officer twenty tablets of Valium. He asked the officer if he was interested in purchasing a greater quantity and arranged a meeting later on that evening to consummate the deal. Later in the evening the officer and Lester met defendant at the fire station where he worked. Defendant sold the officer 100 additional tablets of Valium. According to the officer, defendant boasted about his own cocaine habit and claimed he had obtained a second mortgage on his house to support the habit. He also boasted about driving the fire engine while under the influence of Valium. Defendant then offered to sell the officer more cocaine. The following day the three men met and defendant sold the officer two additional ounces of cocaine. Defendant was arrested at this third meeting.

At trial the jury acquitted defendant of the first instance of selling cocaine on the ground of entrapment. However, he was found guilty on three counts of violating the Georgia Controlled Substances Act for selling Valium and for the second sale of cocaine. Defendant appeals the convictions claiming the trial court erred in failing to direct a verdict of acquittal on the ground defendant had established the defense of entrapment and the prosecution failed to disprove it.

(continued)

(*continued*)

Here, as in *Robinson* . . . there was no evidence that the accused had been regularly engaged in the illegal sale of drugs. As in *Robinson,* defendant's testimony that he was induced to arrange the sale of contraband by persistent solicitation and undue persuasion of the informant was not rebutted and the informant did not testify. However, there exists no "per se rule that a defendant is entitled to a directed verdict where the informant is not called to rebut the defendant's testimony of entrapment. . . . [A] defendant's testimony as to entrapment, even if unrebutted by any other witness to the alleged misconduct, will not entitle him to a directed verdict of acquittal unless that unrebutted testimony, together with all reasonable deductions and inferences therefrom, demands a finding that entrapment occurred." . . .

Here, the undercover officer testified all sales subsequent to the initial exchange of cocaine were initiated by the defendant himself and not by the informant. Defendant denies he initiated the additional sales and claims he was instructed by the informant to act as if he was conversant with drugs and could provide any amount or type of illegal drug to the undercover officer. Nevertheless, the officer's testimony "was sufficient to raise a jury question as to whether [defendant] was entrapped into the commission of the crime." . . .

"This is one of those cases in which a question of fact was presented as to entrapment for determination by the jury. The evidence did not, however, demand a finding that defendant . . . was entrapped into the commission of a crime." *State v. Royal* . . . "In this case, the issue of whether the [S]tate impermissibly encouraged the [defendant] to evil was properly submitted to the jury, and the evidence authorized a rational trier of fact to find beyond a reasonable doubt that the [defendant] had not been entrapped." *Pierce.*

Judgment affirmed.

Case Name: Goolsby v. State of Georgia

Court: Georgia Court of Appeals

Summary of Main Points: The question of whether a defendant is "impermissibly encouraged" to commit a crime by police is a question of fact for the jury to decide. In this case, the jury found against the defendant, and the appellate court will not overturn the jury's decision.

Statute of Limitations

There is a set period of time after a crime has been committed during which the perpetrator can be charged. This time period is set by the legislature and is referred to as the statute of limitations. After this period of time has elapsed, a person cannot be charged with the crime, and if charged may raise the statute of limitations as a defense.

States differ in the length of time allowed for prosecution of criminal charges under their statutes of limitations. Typically, petty offenses may have a short statute of limitations of three to five years, while more serious crimes may have statutes of limitations from five to 20 years. Some crimes are of such a nature that there is no statute of limitations. These crimes typically include murder, rape, robbery, and arson.

Necessity

The defense of necessity is a common-law defense that can be used when a person has no alternative in a situation but to violate the law. The defense must show that the act was committed in response to an imminent public or private harm that would have occurred had the law been followed. It has been called the "justification" or "choice of evils defense."

In order to invoke the necessity defense, the circumstances cannot be the result of the defendant's own conduct. Some natural phenomenon or the act of some other individual must have brought about the situation.

SIDEBAR

A mountain search and rescue team is searching for a lost mountain climber. During the course of the search, a winter storm arrives sooner than expected. The rescue team initially establishes a tent camp, but the weather is so severe that the tents are destroyed. The team locates a house, and forces entry to seek shelter. In the event that the homeowner sought to have members of the team charged with breaking and entering, the defense of necessity would be applicable.

SUMMARY

Arrests can be made by sworn law enforcement officials, often termed peace officers, as well as by ordinary citizens, when a misdemeanor is committed in their presence, or when there is probable cause to believe that a felony has been committed and the person being arrested committed it. The major difference is that sworn peace officers have immunity from false arrest charges if a mistake is made, while citizens do not. When firefighters make an arrest, they are considered to be citizens and thus have no immunity protection.

The exclusionary rule excludes the use of evidence seized or obtained in violation of a defendant's constitutional rights. A search warrant is constitutionally required prior to most searches. Among the exceptions that exist is the fire scene exception, which allows firefighters and investigators a reasonable period of time after a fire to conduct a cause and origin determination without having to obtain a search warrant. During the fire, and for a reasonable period of time after the fire, personnel may seize evidence that is in plain view.

The law allows a person other than the principal parties to a crime to be charged under one of several theories, including as an accessory, as an aider and abettor, or as a co-conspirator. The law also allows the prosecution of someone who attempts to commit a crime.

REVIEW QUESTIONS

1. Describe the difference between the authority of a peace officer to arrest a person and that of a firefighter.

2. Explain the difference between a criminal complaint, an indictment, and information charging.

3. What must the police establish in order to obtain a criminal search warrant?

4. What must be established in order to obtain an administrative search warrant?

5. Identify and explain six exceptions to the search warrant requirement.

6. Define chain of custody and explain how it relates to firefighters who find evidence of a crime at the scene of a fire.

7. What are the Miranda instructions, and when does a person have a right to be given Miranda warnings?

8. Explain the charges that can be brought against a person who sets a timing device intended to start a fire in a building, if the device malfunctions and does not go off.

9. Define the term "conspiracy" and explain how it can be used to charge someone with a crime that he did not personally commit.

10. Identify and explain five criminal defenses.

DISCUSSION QUESTIONS

1. In the process of extinguishing a fire in an apartment building, a firefighter is part of a search team conducting a primary search of the floor above. In an apartment directly above the fire apartment, she observes what appears to be drug paraphernalia and marijuana on a table in plain view. If the firefighter informs a police officer about the drugs, can the officer legally seize the evidence without a search warrant?

2. The Supreme Court in both *Tyler* and *Clifford* ruled that fire officials have a reasonable period of time after the fire is extinguished to conduct an investigation into the cause and origin of the fire. How long is a reasonable time? What factors will bear on what is a reasonable time?

3. In *Chicago Firefighters IAFF Local 2 v. City of Chicago,* 717 F. Supp. 1314 (N.D. Ill, 1989), fire station locker searches were upheld as constitutional, because the firefighters did not have a reasonable expectation of privacy in the contents of their lockers. What if the fire department did not inform the firefighters—through discussions with the union and the issuance of general orders—that locker searches would be conducted? Would the case come out the same way?

Chapter 7

CIVIL LIABILITY ISSUES

Learning Objectives

Upon completion of this chapter, you should be able to:

- Define the intentional torts of battery, assault, false imprisonment, intentional infliction of severe emotional distress, trespass, trespass to chattels, conversion, misrepresentation, and bad faith.
- Explain how consent is a defense to battery, assault, and false imprisonment.
- Define implied consent and informed consent.
- Explain that a competent adult has an absolute right to decline medical care.
- Identify the factors that are involved in determining if a person lacks capacity to consent to, or decline medical care.
- Explain what should be done to document refusals of care against medical advice.
- Explain the difference between slander, slander per se, and libel.
- Identify the four invasion of privacy torts.

INTRODUCTION

The engine company arrived on the scene for a man passed out. As the crew exited the apparatus, they observed a man lying on the sidewalk, starting to raise his head in acknowledgment of their arrival.

"Hey boys," he slurred, waiving a hand as if to say keep going. "I'm fine, just trying to get some sleep."

"What happened, sir?" the captain asked.

"Just leave me alone, I just wanna sleep," he said, laying his head back down on the sidewalk.

"Sir, can we check you out? We want to be sure you are okay," the captain said as he approached. The man smelled of alcohol, urine, and vomit.

The man struggled to sit up. "I said leave me alone. Don't touch me!" the man barked, still slurring his words. He raised his hand and made a fist in a threatening manner toward the captain.

"Engine Twenty-Two to dispatch. We're going to need the police here for assistance."

Civil law is generally defined as that body of law that addresses the rights between private individuals. However, civil law can also be viewed as including everything in the law that is not considered part of criminal law. Civil law encompasses a variety of legal actions, including suits for breach of contract, intentional torts, negligence, product liability, strict liability, malpractice, domestic relations, and probate.

As we have discussed, the burden of proof in civil actions is the "fair preponderance of the evidence," or the "more likely than not" standard. This civil standard is a significantly lesser burden than the criminal standard of beyond a reasonable doubt. As we will see, many actions that are criminal on the one hand may also give rise to civil actions. The lesser burden of proof makes it easier for someone who has been harmed to recover from a wrongdoer in a civil case than it is to convict the wrongdoer in a criminal case.

tort
a general term for a civil wrong; a tort is an act committed by one or more parties that causes injury to another, for which the law allows a remedy of monetary damages

The term **tort** is a general term for a civil wrong. A tort is an act committed by one or more parties that causes injury to another, for which the law allows a remedy of monetary damages. The purpose of tort law is to compensate the victim of wrongdoing, at the expense of wrongdoer. We will discuss three categories of torts:

1. Intentional torts

2. Negligence (discussed in Chapter 8)

3. Strict liability (discussed in Chapter 8)

INTENTIONAL TORTS

Intentional torts are civil actions that one party may have against another for some intentional act that causes an injury. The origins of intentional torts go back to our common-law foundations, and include assault, battery, false imprisonment, intentional infliction of severe emotional distress, trespass, conversion, fraud, and defamation.

You will notice that many of our discussions in this area of intentional torts are similar to the discussions we had in Chapter 5 on criminal law. While many of the underlying principles are similar, they are not identical. It is important to keep the concept of the civil action for a tort such as battery separate from the criminal offense of battery, and so on.

BATTERY

Battery is an intentional unpermitted contact with another person. The intent required for battery is not a hostile intent nor an intent to do bodily injury. Rather, it is intent to cause contact with another person in a way that is unlawful and unpermitted. Someone who throws a rock at a person intending to strike them, which actually strikes them, commits a battery even if their aim is not good.

Knowledge to a substantial certainty that contact will occur is enough to establish intent. Someone who throws a rock at a large crowd can be liable for damages for battery to the person who is struck, even though the rock thrower lacked an intent or even a knowledge that the particular person, or any person, would be struck. In a similar way, intent may be transferred from a target to an actual recipient through the principle of transferred intent.

📖   SIDEBAR

QUESTIONS: Mike tries to strike Larry with a slap, but Larry ducks, and Mike's slap strikes Chris. Mike had intent only to slap Larry, not Chris. Can Mike be held liable for battery to Chris, even though he did not intend to strike Chris?

ANSWER: Yes. The principle of transferred intent would make Mike liable to Chris.

Battery is considered to be a violation of a person's dignity. As such, damages can be presumed, so that even when there are no actual monetary damages, a jury can award damages. The victim of a battery can also recover all damages that flow from the contact, whether they were foreseeable or not. Damages would include medical bills, property damage, pain and suffering, and lost wages. Someone who intends to push someone out of the way, and ends up knocking the victim down and causing serious bodily injury, can be held liable for all the injuries that result from the push, even though the actor did not intend to cause the serious injuries. The focus is on whether the person intended to make contact, not on whether they intended the end result.

For purposes of battery, contact can include contact with any part of the person's body, or anything closely associated with the victim, including his/her clothing, eyeglasses, hat, cane, a chair in which the victim is sitting, even a horse or a car in which the victim is riding. The contact can also be the result of an object or other person that the perpetrator places in motion. Thus, if a person pushes another person into a third person, the original pusher could be liable to the third person, as well as the second, for battery.

If the perpetrator sets in motion some device or prop that results in unlawful contact, it is still battery. Thus a person who places poison in someone's food could be liable for battery, as would someone who digs a pit and covers it over, intending someone to fall into it.

Battery: Permitted Contact and Consent

One of the key issues with regard to battery is differentiating between permitted contact and unpermitted contact. Physical contact between people takes place every day **(Figure 7-1)**. To the extent that contacts are acceptable by society at large, these types of contact are considered permitted. These would include a light tap on someone's shoulder to get their attention, or shaking someone's hand. However, even these socially acceptable types of

Figure 7-1 *Someone who plays basketball impliedly consents to the type of contact normally associated with basketball.*

consent

the willingness of a person to allow certain physical contact or other conduct that might otherwise be objectionable; consent is a defense to a charge of battery, as well as assault and false imprisonment

express

directly and distinctly stated or expressed, rather than implied or left to inference

implied consent

consent to medical treatment that is implied by law for a patient in need of medical treatment but who lacks the capacity to consent or decline treatment; implied consent is limited to treatment necessary to protect life and limb

informed consent

consent that is informed; for consent to be valid, the person must understand what they are consenting to

contact could constitute a battery where the perpetrator knows that the victim does not want to be touched.

Another category of permitted contact is where the person consents to the contact. **Consent** is the willingness of the recipient to allow the conduct. Consent to bodily contact can be **express** or implied. When a person boards a crowded subway train, he is impliedly consenting to some degree of physical contact with other riders. When someone engages in a hockey or football game, he impliedly consents to physical contact occurring during the course of the game that is in accordance with the rules. These are examples of **implied consent**, where the person's consent to contact can be implied by virtue of his having voluntarily boarded the subway or engaged in a game.

Consent can also be express, for example when a boxer signs a consent form before a boxing match, or when a patient signs a consent form prior to being treated by a doctor.

Battery: Consent and Patient Treatment

The issues of battery and consent have important implications for firefighters and emergency personnel. Whether we realize it or not, issues of consent arise every time we treat a patient. In order for consent to touch or treat someone to be valid, it must be informed and voluntary. Consent that is induced by fraud, mistake, or duress will not be valid.

Informed consent means that in order for consent to be valid, the person must understand what he is consenting to. Someone who believes that he is consenting to a relatively risk-free medical procedure, when he is actually the subject of experimental research with unknown possible outcomes, cannot give informed consent. Individuals who would subject someone to such a procedure would be liable for battery, even if no harm came to the patient.

Informed consent in a hospital setting and in an emergency setting are two entirely different things. The medical community has established standards for informed consent. In fact, some states consider it to be negligence, rather than battery, when a medical provider fails to obtain informed consent. The standards are much less well defined in a pre-hospital setting, due to the exigencies normally associated with treating patients. Nevertheless, emergency responders do have to consider the issue of informed consent when treating patients.

Capacity to Consent

For consent to be valid, the person must be capable of understanding what it is that he is consenting to. Even more problematic is the fact that if the person lacks the capacity to consent, he also lacks the capacity to decline medical treatment. Emergency responders are frequently confronted with the issue of trying to determine whether a patient has the capacity to consent to, or decline,

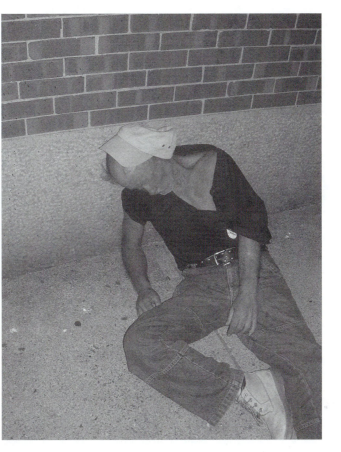

Figure 7-2 *Consent issues arise every time we treat a patient.*

medical treatment **(Figure 7-2)**. A variety of factors come into play in determining a person's capacity to consent, including the person's age, intoxication, medication (prescribed and recreational), and the effect of illness or injury.

Doctors are confronted with the issue of a patient's capacity every day, and frequently differ in their opinions. This reality places emergency responders in the field in a difficult situation. Some authorities recommend using a standard that the patient must be alert and oriented to person, place, and time, not intoxicated or subject to mind-altering medication, not subject to some mental disease or defect, and capable of understanding the consequences of his decision making. State agencies that oversee emergency medical responders often establish their own standards for competency, and should be relied upon by personnel who are subject to their authority.

Where a patient is in need of medial treatment but lacks the capacity to consent to or decline treatment, the law recognizes implied consent to treat the patient for injuries and illnesses that threaten life or limb. This type of implied consent does not go beyond treatment necessary to protect life and limb.

Treatment of Minors

Minors pose a difficult dilemma for emergency responders. As a general rule, minors do not have the capacity to consent to or decline treatment. The age of consent varies from state to state, with some states saying a person as young as 14 may consent to medical treatment, and others saying the person must be 19 years old **(Figure 7-3)**. The following are some examples of state laws governing medical treatment of minors.

EXAMPLE

Indiana Code 16-36-1-3 Consent to own health care; minors

Sec. 3. (a) Except as provided in subsections (b) and (c), unless incapable of consenting under section 4 of this chapter, an individual may consent to the individual's own health care if the individual is:

  (1) an adult; or
  (2) a minor and:
      (A) is emancipated;
      (B) is: (i) at least fourteen (14) years of age;
              (ii) not dependent on a parent for support;
              (iii) living apart from the minor's parents or from an individual in loco parentis; and
              (iv) managing the minor's own affairs;
      (C) is or has been married;
      (D) is in the military service of the United States; or
      (E) is authorized to consent to the health care by any other statute.

(b) A person at least seventeen (17) years of age is eligible to donate blood in a voluntary and noncompensatory blood program without obtaining parental permission.
(c) An individual who has, suspects that the individual has, or has been exposed to a venereal disease is competent to give consent for medical or hospital care or treatment of the individual.

EXAMPLE

Rhode Island General Laws

§ 23-4.6-1 Consent to medical and surgical care. Any person of the age of sixteen (16) or over or married may consent to routine emergency medical or surgical care. A minor parent may consent to treatment of his or her child.

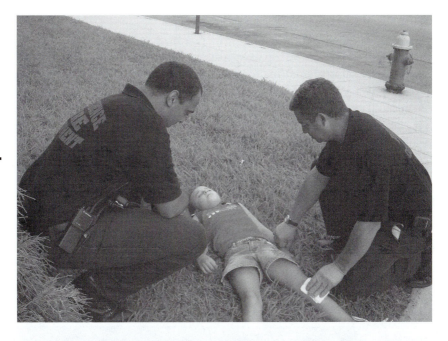

Figure 7-3
Firefighters and emergency medical personnel have implied consent to treat minors in the absence of a parent or guardian to the extent necessary to protect life and limb.

EXAMPLE

Massachusetts General Laws

c.112, §12F Emergency treatment of minors. No physician, dentist or hospital shall be held liable for damages for failure to obtain consent of a parent, legal guardian, or other person having custody or control of a minor child, or of the spouse of a patient, to emergency examination and treatment, including blood transfusions, when delay in treatment will endanger the life, limb, or mental well-being of the patient.

Any minor may give consent to his medical or dental care at the time such care is sought if (i) he is married, widowed, divorced; or (ii) he is the parent of a child, in which case he may also give consent to medical or dental care of the child; or (iii) he is a member of any of the armed forces; or (iv) she is pregnant or believes herself to be pregnant; or (v) he is living separate and apart from his parent or legal guardian, and is managing his own financial affairs; or (vi) he reasonably believes himself to be suffering from or to have come in contact with any disease defined as dangerous to the public health pursuant to section six of chapter one hundred and eleven; provided, however, that such minor may only consent to care which relates to the diagnosis or treatment of such disease.

As is evident from these statutes, emergency personnel need to know the laws of their particular state with regard to age of consent for minors. As with adult patients who lack the capacity to consent to or decline treatment, the law recognizes implied consent to treat minors for injuries and illnesses that threaten life or limb. However, implied consent does not go beyond treatment necessary to protect life and limb.

In addition, issues may come up in regard to who has the authority to make decisions for minors and others who lack capacity. Some states specify by statute who has the legal authority to make medical decisions for those unable to make them for themselves. Most authorities find a priority order of consent as follows:

- spouse
- a parent
- an adult child
- an adult sibling
- an uncle or aunt
- a grandparent
- a person in loco parentis, (in place of the parents), with legitimate responsibility or authority, such as a teacher, camp counselor, or other with a medical authorization form signed by a parent.

EXAMPLE

Arizona Revised Statutes

44-133. Emergency consent for hospital care, medical attention or surgery by person in loco parentis. Notwithstanding any other provision of the law, in cases of emergency in which a minor is in need of immediate hospitalization, medical attention or surgery and after reasonable efforts made under the circumstances, the parents of such minor cannot be located for the purpose of consenting thereto, consent for said emergency attention may be given by any person standing in loco parentis to said minor.

A parent with legal custody of a child has the ultimate legal authority to consent or decline medical aid. Where there are two parents who each have legal custody, and one consents while the other refuses, courts hold that the consent of the willing parent is adequate consent. Where only one parent has legal custody, it is the custodial parent who has the legal right to consent.

Powers of Attorney

Many states have laws that allow people to appoint someone to make important health care decisions for them, in the event they unable to make such decisions for themselves. The names of these documents include power of attorney, durable power of attorney for health care decisions, and living will

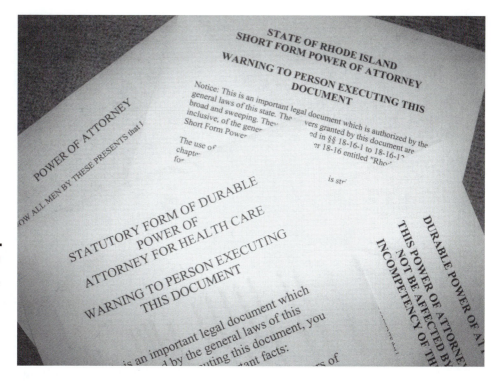

Figure 7-4 *There are a variety of powers of attorney, some of which may give a person the right to make health care decisions for another person.*

(Figure 7-4). Once executed in accordance with state law, these documents give the person named the legal authority to make health care decisions for the patient to the same extent that the patient could, if the patient was able.

Consent Though Fraud or Duress

To be valid, consent must be free from duress or fraud. As we discussed in Chapter 5 on the criminal law of battery, consent that is induced by fraud is invalid. Typical cases in this area involve people who pass themselves off as doctors, and who examine and in some cases treat patients. Another line of cases involves doctors who obtain consent from their patients to perform a medical procedure, but proceed to take advantage of them sexually once they are incapacitated. See *Jeffreys v. Griffin*, 301 A.D.2d 232, 749 N.Y.S.2d 505 (1st Dept., 2002). See also *People v. Griffin*, 242 A.D. 2d 70 (1st Dept 1998), appeal dismissed 93 NY2d 955 [1999], *People v. Teicher*, 422 N.E.2d 506; 52 N.Y.2d 638 (1981).

Obtaining Consent

In a hospital setting **(Figure 7-5)**, consent is usually obtained in writing prior to the initiation of treatment. Obtaining consent in an emergency setting

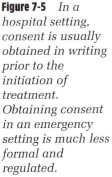

Figure 7-5 *In a hospital setting, consent is usually obtained in writing prior to the initiation of treatment. Obtaining consent in an emergency setting is much less formal and regulated.*

is much less formal and less regulated than obtaining consent in a hospital setting. As a general rule, in a pre-hospital setting there is no specific requirement that personnel ask "May I examine you?" prior to initiating patient care, although such a question is certainly good practice and highly advisable.

Consent to patient contact can be implied from the person submitting to care, and the patient's lack of an objection to contact. With a conscious patient, personnel should explain to a patient exactly what steps are being taken, and if appropriate, explain why. Any indication by the patient that he does not want care, or does not consent to a particular type of care, must be honored if the patient has legal capacity. A patient has a right to consent to some types of care, but not others. A patient may consent to first aid, but decline an intravenous line (IV), oxygen, or transport. It is critical to maintain good communications with the patient right from the start.

Obtaining consent in a pre-hospital setting is usually a function of the patient recognizing who the responders are, and acquiescing to the care that is offered. When personnel respond on fire or emergency vehicles and arrive in uniform, the recognition factor is present. In some communities, volunteer personnel may arrive on scene in their personally owned vehicles and in civilian clothing prior to the arrival of emergency apparatus. In such a case, it is certainly advisable that such personnel inform the patient of their identity and intent to treat prior to initiating patient contact.

Refusal of Consent and Competency

Whenever a patient's capacity to consent is an issue, emergency responders can be placed in a difficult situation. Competent adults have the absolute right to decline medical treatment, even if it means they will die or suffer serious injuries. However, persons who are not competent do not have the capacity to decline treatment, and responders who do not treat such persons properly can be held liable for abandonment and negligence.

In addition to this dilemma over capacity and the right to refuse treatment is the issue of informed consent. Just as a patient must understand what it is he is consenting to for the consent to be valid, the issue of informed consent becomes an issue with someone who declines help. If the person does not understand the possible consequences of declining aid, the decision to decline help is not knowingly made. Thus, emergency personnel have an obligation to inform competent adult patients who choose to decline aid of the possible consequences of that refusal.

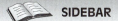 **SIDEBAR**

Patient Abandonment

Abandonment occurs when a medical provider who has assumed care and control of a patient in need of medical attention either stops providing care, leaving the patient unattended, or leaves the patient under the care of someone with lesser qualifications. While some authorities identify abandonment as an intentional tort, most courts that have dealt with the subject view it as a breach of the standard of care, which will be addressed in Chapter 8 on Negligence.

Many books and lecturers on emergency medicine discuss the subject of abandonment authoritatively, stating rigid and absolute requirements that must be adhered to avoid committing abandonment. While certainly recommendations against committing abandonment are well advised and no doubt well intentioned, the truth is that the law surrounding abandonment is quite unsettled, and the cases are relatively few. Most of the cases pertaining to abandonment involve doctors in a clinical environment, and the law in this area is evolving slowly. For those interested in further information on abandonment, the following cases are recommended:

Meiselman v. Crown Heights Hospital Inc., 285 NY 389, 34 N.E. 2nd 367 (1941)

Miller v. Dore, 154 Me. 363, 148 A. 2d 692 (1959),

Crowe v. Provost, 52 Tenn. App. 397, 374 S.W. 2d 645 (1963)

Katsetos v. Knowland, 170 Conn. 637, 368 A. 2d 172 (Conn. 1976)

Schliesman v. Fisher, 158 Cal. Rptr. 527 (1979)

Johnston v. Ward, 288 S.C. 603, 344 S.E. 2d 166 (1986)

Allison v. Patel, 211 Ga. App. 376, 438 S.E. 2d 920 (1993)

Manno v. MacIntosh, 519 N.W. 2d 815 (1994)

Documentation of Refusal of Treatment

When a competent adult patient refuses to accept medical treatment and/or transportation against the advice of the personnel on scene, it is vitally important that this refusal be documented. Where possible, the patient's signature should be obtained on a refusal of aid form or other suitable document. The patient's signature on this form should be witnessed, and the form should document the information provided to the patient regarding the consequences of such a refusal.

In situations where the patient refuses to sign the refusal of aid form, this fact should be documented on the form, and the form signed by all witnesses who observed the refusal to sign. The incident report should reference the patient's refusal of aid, refusal to sign, and again reference what information was given to the patient regarding the possible consequences of the refusal.

Scope of Consent

Consent to allow contact or treatment is not an unlimited license to permit any type of contact with a person. Consent has its limits. When consent is given, consent is to an act, not a result. Consent to having one's blood pressure taken is not consent to an IV.

In terms of consent to other types of battery, if someone consents to being struck with a ball, as may be the case in a game of dodgeball, the fact that being struck by the ball unforeseeably results in death or serious injury does not invalidate the consent. However, consenting to being struck with a ball is not consent to being struck with a punch, even if the injury that results is of the same type that could have been caused by a ball. The consent is exceeded when a person commits acts that are of a different nature from what the victim has consented to.

Cases involving the scope of implied consent may be complicated by the lack of a clear understanding of the scope of actions that the victim has consented to. Someone who decides to play in a tackle football game impliedly consents to the kind of contact normally associated with football. Conduct outside the bounds of the rules, such as being struck over the head with a pipe during a timeout, is clearly beyond the scope of consent. Yet other acts, such as late hits or unnecessary roughness, are violations of the rules, but may be of a nature that is generally tolerated by the sport. In such cases, the scope of consent is a jury question. The defendant in a battery case is entitled to rely upon what any reasonable person would understand plaintiff's consent to be.

ASSAULT

Assault is intentionally placing another in apprehension of imminent bodily contact. It has also been described as an attempted battery that is not successful. As with most common-law intentional torts, damages are presumed in an assault because it is a violation of the person's dignity that triggers liability. Thus, it is not necessary to prove that the victim sustained actual damages in order to recover for an assault. Of course, if damages do result, the perpetrator of the assault would be liable for all damages that result.

FALSE IMPRISONMENT

False imprisonment, sometimes called false arrest, occurs when a person without lawful right intentionally restrains the free movement of another against that person's will. The victim need only be restrained momentarily for the tort to be committed.

False imprisonment does not require imprisonment in a jail cell, but does require a total confinement or limitation of the right of the victim to move about freely within boundaries set by the perpetrator. The imprisonment may be enforced through use of a physical enclosure, by force, the threat of force, or the threat of harm to the victim or the victim's family or property. The area of confinement may be small, such as a jail cell, or it may be large, such as a house, building, or even a city, but the confinement must be total. The area need not have discernible boundaries, for example when a perpetrator points a gun at a victim and tells him not to move. However, it is not false imprisonment if a person merely prevents another person from having access to a place that he wants to go.

A person is not totally confined if there is a reasonable means of escape or exit. In order for a means of exit to be reasonable, it must be known or obvious, and not expose the person to any threat of harm. Furthermore, a means of escape that exposes the person to embarrassment, unreasonable discomfort, or requires him to be heroic is not reasonable.

The victim of false imprisonment may recover damages from the perpetrator for time lost, physical discomfort, any damage to health or well-being, lost wages, and lost business opportunities, as well as harm to the victim's reputation and humiliation that may have accompanied the false arrest. Punitive damages may also be allowed. Like battery and assault, false imprisonment is a violation of a person's dignity, and as such actual damages are not required. A jury may presume damages, and award an appropriate amount.

As an intentional tort, false imprisonment requires that the restraint on a person's movement be made intentionally. Accidentally or negligently causing a false imprisonment is not actionable, although if actual damages occur it may be actionable as negligence. For example, assume that a janitor at closing time fails to check to see that everyone has left an office building before turning off power to the elevators, as he is supposed to do. The janitor turns off power to the elevators, thereby trapping someone in an elevator car. If the janitor realizes that he forgot to check for occupants in the elevator and restores power, resulting in only a momentary imprisonment, there is no liability. However, if the janitor fails to realize his mistake and the patron has a medical condition such as diabetes and is unable to take his medication, the janitor would be liable for the harm that results under the negligence theory.

One common type of false imprisonment case occurs when a law enforcement officer or a private citizen arrests an individual on criminal charges, but is mistaken about either the actual identity of the perpetrator or about whether a crime has actually been committed. This kind of case is commonly referred to as **false arrest**. When a duly authorized law enforcement officer, also referred to as a **peace officer**, makes an arrest in good faith with probable cause to believe that the person has committed a felony, or that the person committed a misdemeanor in their presence, the officer has a "privilege." The privilege effectively provides immunity from liability to the officer for false arrest if it later turns out that the officer was mistaken. Occasionally, a case comes up in which a police officer arrests someone without probable cause, or in bad faith. In such cases, the officer has no privilege, and an action for false arrest can be brought.

A private citizen who acts in good faith and with probably cause is authorized to arrest the perpetrator of any felony, or the perpetrator of a misdemeanor, provided it is committed in his presence. However, someone who makes a citizen's arrest but is mistaken is not entitled to the privilege afforded to peace officers. In such a case, the person who is wrongly arrested may sue the person who made the arrest for false arrest, even though the citizen acted in good faith and with probable cause. As discussed in Chapter 6, firefighters and emergency medical personnel are not peace officers, and do not have the benefit of any privilege when making an arrest. However, fire department personnel who are sworn in as peace officers (in some jurisdictions, arson squad investigators are peace officers) do have the privilege **(Figure 7-6)**.

Another situation of concern to firefighters and emergency responders in regard to false imprisonment involves patients who have an impaired level of consciousness and thereby lack the capacity to decline medial treatment. If the patient lacks the capacity to decline aid, personnel may use reasonable force to restrain and take a patient to a medical facility. However, using restraint to forcibly take a person to a medical facility could be considered false imprisonment if the patient actually has the capacity to decline aid.

In any situation in which a patient appears to lack the capacity to refuse aid, and particularly when the patient is violent or combative, police assistance is an absolute necessity. Police officers can take the patient into protective custody, at which point the officer may authorize the patient's treatment and transportation. In addition, police officers can help document the patient's condition, and the need for restraint.

false arrest

a type of false imprisonment, in which a law enforcement officer or private citizen arrests an individual on criminal charges, but is mistaken about either the actual identity of the perpetrator or about whether a crime has actually been committed

peace officer

another term for law enforcement officer in many states

Figure 7-6 *Many fire departments have personnel who are sworn peace officers.*

Right of Refusal and Responder Liability

It is important for fire and emergency personnel to fully understand the inter-relationships among assault, battery, and false imprisonment, and the issues of consent and refusal of aid. Emergency responders should not be intimidated by fear of lawsuits from making sound professional judgments. When dealing with a patient who lacks the mental capacity to consent or decline medical treatment, it is always better to err on the side of treating such a patient.

A patient trying to prove battery and false arrest against fire and emergency medical personnel who render treatment and transportation in a borderline situation would have a very difficult case to make. First, the person would have the burden to prove that the emergency personnel were mistaken about his mental condition and capacity. Most juries would look favorably upon emergency personnel who appeared to care more about the patient's well-being than the patient did. Second, if the victim were able to convince a jury that the emergency personnel were wrong, it is still unlikely that a jury would award presumed damages in the absence of bad faith. Thus, the patient would have to prove that he suffered actual damages by being treated and taken to the hospital against his will. Fire and emergency personnel who act in good faith in such a situation would be unlikely to be found liable.

However, personnel who wrongfully allow a patient who lacks capacity to decline medical treatment could find themselves on the receiving end of a lawsuit, against which they have no defensible argument **(Figure 7-7)**.

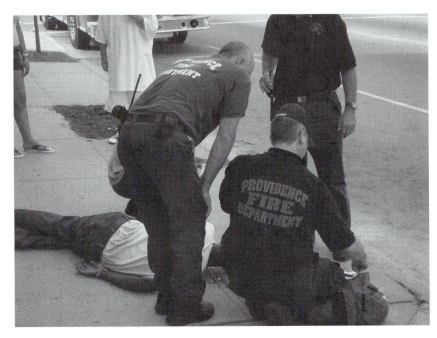

Figure 7-7 *When in doubt about a patient's capacity to consent to, or decline medical aid, it is always advisable to err on the side of treating and transporting.*

See *State v. Wheeler,* 496 A.2d 1382, 1388 (R.I. 1985) and *Patino v. Suchnik,* 770 A.2d 861 (R.I. 2001).

Consent and False Imprisonment

The issue of consent, discussed above with regard to battery, applies equally to the tort of false imprisonment. Someone who has legal capacity can consent to being detained or imprisoned. Issues such as the voluntariness of the consent, fraud, and duress apply equally with regard to false imprisonment as they do to battery. The consent issue frequently arises with regard to shoplifting cases, when there is a question of whether or not the person suspected of shoplifting voluntarily agreed to go with store security personnel to "clear their name," or whether they were under arrest. The same issue may arise when firefighters detain a person while waiting for the police.

INTENTIONAL INFLICTION OF SEVERE EMOTIONAL DISTRESS

Intentional infliction of severe emotional distress is a tort, in which someone through extreme and outrageous behavior intentionally or recklessly causes severe emotional distress to another. The perpetrator can be liable for damages for causing such emotional harm, as well as damages for any physical harm that results.

The type of conduct required must involve more than threats, indecencies, vulgarities, annoyances, and petty oppressions. It must be outrageous and extreme beyond all bounds of decency. Examples of such cases include a person who told a woman that her husband had been seriously injured in an accident; spreading a rumor that a person's son had committed suicide; mutilation of a deceased loved one's body; and disinterment of bodies.

While some cases of intentional infliction of severe emotional distress can be the result of a single action, more often they are the result of a pattern of conduct that is repeated over a prolonged period. Bullying, hounding, repeated threats, public embarrassment, harassment—even repeated invitations directed toward an unwilling woman to have sexual intercourse—can give rise to liability.

Intentional infliction of severe emotional distress can also occur when a perpetrator injures a third party in the victim's presence, resulting in emotional harm. Cases include, for example, a wife who is present as her husband is shot, or a mother who sees her child intentionally run over.

The *Garrison* case is a fire service example of an intentional infliction of severe emotional distress case. The court gives an excellent analysis of the legal issues in the context of a real-life case.

Garrison v. Bobbitt
134 Ohio App.3d 373 (1999)
Court of Appeals of Ohio, Second District, Montgomery County

BROGAN, Judge.
. . . The [Defendants] in this case were Robert Bobbitt, Chief of the Miamisburg Fire Department, and Dennis Lutz, a captain in the department. Bobbitt and Lutz were sued for defamation and for allegedly intentionally inflicting emotional distress on appellee, Orpheus Garrison, who took a medical retirement from the department in 1997. . . .

The claims against Bobbitt and Lutz arose from events occurring during the employment of Orpheus Garrison (known as Gary Garrison) with the Miamisburg Fire Department. Construing the evidence in Garrison's favor, the followings facts, and factual disputes, where applicable, appear in the record. Garrison was first scheduled to start employment with Miamisburg as a firefighter in February 1991. However, because Garrison was a Captain in the Marine Corps, he was instead mobilized for action in Desert Storm in January 1991. At that time, Garrison was told by Miamisburg that his job would be held until he returned. Subsequently, Garrison returned from Desert Storm in June 1991, but Miamisburg was not able to bring him into the department until August. When Garrison started work, Miamisburg assigned him an August seniority date. Garrison disputed the date, claiming that a February 1991 seniority date should be used because federal law protected his job

(continued)

(*continued*)

during mobilization. Miamisburg disagreed, and Garrison was forced to contact the Department of Labor. After the Department of Labor intervened, Miamisburg eventually agreed to give Garrison the February 1991 seniority date. Because a clause in the law also required Garrison to complete probation before Miamisburg had to take action, Garrison was not supposed to receive the correct seniority date until the end of his one-year probationary period. When the probationary period ended, however, the city claimed to have forgotten the terms of the agreement. This resulted in the need for a second intervention by the Department of Labor. Eventually, the matter was resolved and Garrison received the February 1991 seniority date. After this unfavorable beginning, relations did not improve. Garrison testified that he was harassed throughout his employment and was treated differently from other employees. These incidents included denial of overtime pay, charging Garrison with vacation time when he attended training, denial of access to special duty teams formed by the department (scuba, hazardous materials, tactical entry, etc.), refusal to pay expenses Garrison had when similar payments were authorized for other employees, ostracism from the fire chief's softball team, denial of promotions, and unequal treatment when disciplinary matters arose. Garrison was labeled as a troublemaker and was also used as an example in union meetings of how management could make an employee's life miserable if management did not like the employee. Additionally, Garrison was called a "slumlord" by his supervisor because Garrison owned a number of rental properties in Miamisburg. Derogatory references were also made to Garrison's military background and demeanor.

In August 1995, an emergency call came in to the fire department about Garrison's father, either as a non-breather or a DOA ("dead-on-arrival"). Garrison was the first person on the scene, and was quite upset. After this incident, some employees commented to Bobbitt that Garrison was not the same person he had been before. During the same time period, Garrison struggled over a breakup with his girlfriend, who had left him about a month before his father's death, i.e., in July 1995. A friend and co-worker, Randy Botts, discussed the break-up with Garrison's supervisor, Lutz. Botts told Lutz that he had talked to Garrison off-duty about the situation and was concerned that Garrison might be suicidal.

In November 1996, Garrison had a recreational fire at one of his rental properties. At the time, departmental policy was that the fire department should be notified of a recreational fire. The department would then send someone out to inspect the fire. Although Garrison called about 9:00 a.m. to report the fire, no one was sent to inspect. Around 3:00 p.m., a fire inspector appeared and told Garrison to extinguish the fire because it was outside the context of the recreational burn ordinance. Garrison extinguished the fire, as directed. Bobbitt was notified about the incident and told the inspector to make out a report. When the inspector contacted dispatch for a number, the dispatcher said she had received a number of calls on the fire.

Later that day, the fire department received more calls about a fire at the same property. Around 7:00 p.m., the duty officer at the downtown fire station called Bobbitt to tell him about the calls. At that point, Bobbitt told the officer to respond to the scene. Bobbitt also went because one of his employees was involved. When

(*continued*)

(*continued*)

they arrived on the scene, another fire was burning in the back yard, apparently at the location of the earlier fire.

On the day of the fire, Garrison was taking out plastic and putting in drywall at the rental property where the fire was located. His procedure was to take trash cans full of plastic to a dumpster about a half mile away and then come back. Garrison repeated this process all day. When the chief showed up with the duty officer, Garrison went out back to see what had happened. The big pile of coals was burning again. Garrison believed the fire had either rekindled or was restarted by kids who were around. Despite these potential explanations, the chief had Garrison arrested for having a recreational fire. To Garrison's knowledge, he was the only person who had been arrested for having a recreational fire since he had come to work for the department. In fact, the fire department went to one individual's house about three times a year to put out recreational fires, but had never arrested the individual. To Garrison, this incident confirmed what the people in the department had said was true, i.e., Bobbitt was out to get him.

Following his arrest, Garrison did not return to work. He stayed in his house for a couple of weeks and sent his sister to court to plead no contest because he was not in shape to leave the house. Garrison remained off work, went through the Employee Assistance Program ("EAP"), and eventually took a medical retirement on July 2, 1997. During the time that Garrison was on medical leave, rumors occurred about the possibility that Garrison would come back to the fire department and harm people. No real basis was given during anyone's testimony for the origin of these beliefs, and the testimony was conflicting as to where the rumors originated.

Bobbitt denied ever having a conversation about Garrison's mental well-being with anyone other than at a staff level meeting. Rumors were brought to the chief's attention after Garrison failed to return to work. There were concerns that Garrison was not mentally healthy, that he had a lot of trauma in his life, and that he had never gotten over his father's death and a breakup with a young lady. Concerns also existed about the safety of the employees. Bobbitt testified that two different people told him that Garrison had threatened his life. [sic] He also heard that Garrison had threatened Lutz's life. Inexplicably, Bobbitt could not remember the names of the people who had told him this, nor was he aware of any names of specific people who were concerned about their safety. Bobbitt discussed the safety concerns with the assistant city manager, who then sent Garrison a letter barring Garrison from city property until he was released for duty. A second letter allowed Garrison to come on city property to retrieve items from his personal locker, but only if accompanied by an official.

Lutz blamed the rumors on the co-workers on Garrison's shift (Randy Botts, Jack Ikerd, and Bob Robinson). According to Lutz, these workers approached him after Garrison went on leave and expressed concern that Garrison might come back and harm them. Before being approached by these employees, Lutz did not have any concerns about safety. Lutz was not able to recall any comments by specific employees, but did say that the employees knew Garrison's hobbies were guns and military paraphernalia. Lutz also knew Garrison was going to EAP. Lutz

(*continued*)

(*continued*)

denied being aware of any threats that Garrison made on anyone's life. His safety concerns were based on Garrison's suicide talk, the gun collection, and the EAP treatment. After talking to the employees, Lutz went to Bobbitt and told him about the safety concerns.

According to Botts, Garrison's co-worker and friend, Lutz said immediately after Garrison's sick leave that he was concerned about what Garrison would do next. Lutz did not give any basis for his feelings or concern for personal safety other than Lutz's observation as a supervisor. Lutz did not mention concern over the gun collection, Garrison's reaction to his father's death, or the EAP treatment. Botts told Lutz he did not think there was anything to worry about, because he did not believe Garrison was capable of harming anyone. Botts also heard comments from other people about concerns for their safety, but did not think they were completely serious.

As was mentioned, Garrison never returned to work at the department. While on medical leave, he consulted a psychologist and was also sent to a psychologist by the disability board. Garrison was told the work environment was hostile and he should not go back. Accordingly, Garrison took medical retirement.

II

We have previously noted that to establish a claim for intentional infliction of emotional distress, the plaintiff must show "(1) that the actor either intended to cause emotional distress or knew or should have known that actions taken would result in serious emotional distress to the plaintiff, (2) that the actor's conduct was so extreme and outrageous as to go, 'beyond all possible bounds of decency' and, was such that it can be considered as 'utterly intolerable in a civilized community,' * * * (3) that the actor's actions were the proximate cause of plaintiff's psychic injury, and (4) that mental anguish suffered by plaintiff is serious and of a nature that 'no reasonable man could be expected to endure it.'"

In *Yeager* . . . the Ohio Supreme Court explained the nature of this tort by way of the following description, which has been quoted many times in varying factual situations:

"'It has not been enough that the defendant has acted with an intent which is tortious or even criminal, or that he has intended to inflict emotional distress, or even that his conduct has been characterized by "malice," or a degree of aggravation which would entitle the plaintiff to punitive damages for another tort. Liability has been found only where the conduct has been so outrageous in character, and so extreme in degree, as to go beyond all possible bounds of decency, and to be regarded as atrocious, and utterly intolerable in a civilized community. Generally, the case is one in which the recitation of the facts to an average member of the community would arouse his resentment against the actor, and lead him to exclaim, "Outrageous!"*

"'The liability clearly does not extend to mere insults, indignities, threats, annoyances, petty oppressions, or other trivialities. The rough edges of our society are still in need of a good deal of filing down, and in the meantime plaintiffs must necessarily be expected and required to be hardened to a certain amount of rough*

(*continued*)

(continued)

*language, and to occasional acts that are definitely inconsiderate and unkind. There is no occasion for the law to intervene in every case where some one's [sic] feelings are hurt. There must still be freedom to express an unflattering opinion, and some safety valve must be left through which irascible tempers may blow off relatively harmless steam. * * *'" . . .*

The resolution of these kinds of cases is fact-intensive and often depends on distinctions between conduct that seems outrageous and conduct that is undesirable, but not quite offensive enough to be called outrageous. . . .

Had the present case simply involved the events leading up to Garrison's citation for violation of the burn ordinance, we would have concluded . . . that the defendants' alleged conduct was not "outrageous," as that term has been interpreted. The incidents occurring during Garrison's employment, while perhaps annoyances or petty oppressions, were not extreme or outrageous. On the other hand, the events that took place beginning with the burn violation, construed in Garrison's favor, are similar to conduct found objectionable in various cases. Specifically, there was evidence indicating that the rumors and comments that Garrison would physically harm other firemen, including Lutz, originated with Lutz. Although Lutz claimed that such thoughts never entered his mind before the members of Garrison's shift, including Botts, brought the safety issue to his attention, Botts contradicted this account. In this regard, Botts said Lutz was the one who raised the issue of Garrison's potential threat to others, immediately after Garrison's sick leave started. Moreover, Lutz gave Botts no basis for his comments, other than his observations as a supervisor.

However, the record contains no evidence of observations that would reasonably have caused Lutz to believe Garrison would harm others. More important, the record offers no logical explanation why Lutz or anyone else would make inappropriate comments to that effect. Instead, the evidence indicates that before Garrison took sick leave, he had been a competent employee with few discipline problems in five years of service. Furthermore, while Garrison had been depressed during the previous year over the death of his father and a break-up with a girlfriend, the record is devoid of a single instance of aggression on Garrison's part toward anyone. As an additional point, while Lutz claimed to have liked Garrison, his references to Garrison as a "slumlord" and to "Gary's ghettos" do not exactly convey an impression of fondness.

Likewise, the record reveals some evidence that Bobbitt's actions (including those taken in connection with the arrest and the exclusion from city property) may have been motived by ill will. In this regard, Bobbitt's denial that he considered Garrison a troublemaker is contradicted by Botts's testimony. According to Botts, there was a consensus that Garrison was a troublemaker. Furthermore, Botts said some people at the station felt Garrison did not have a chance of being promoted because he was on the Chief's "shit list." Botts also testified that he felt Bobbitt disliked Garrison.

The presence of ill will is also implied by evidence that Garrison was treated differently than others in connection with the violation of the burn ordinance and the exclusion from public property. In this context, Garrison testified that to his knowledge, no one had ever been arrested for a recreational fire before, even

(continued)

(*continued*)

though the department put out a number of recreational fires, including several fires for the same person. Similarly, Garrison also said no one had ever been banned from a public place in Miamisburg before. Bobbitt did not comment on this latter point, but did say that a citation is automatically issued if the fire department is informed of a fire and has to go out a second time. If, in fact, others were cited for recreational fires, the implication of ill will could be offset. However, no such evidence was given to the trial court.

Additional conflicts bearing on credibility appear in the evidence submitted to support summary judgment. For example, Bobbitt said he contacted Lutz after hearing rumors around the station. By contrast, Lutz said he was the one who contacted Bobbitt, upon hearing the safety concerns expressed by the co-workers on Garrison's shift. Bobbitt also testified that he feared for his safety and that he had never had conversations with any member of the department about Garrison's mental well-being, other than at a staff meeting level. Again, by contrast, Botts (who was not a supervisor), said he had talked to Bobbitt a couple of times about Garrison's mental well-being. Botts further said that Bobbitt never expressed any fears about his own safety.

Compared to acts found outrageous in other cases, the defendants' acts in implying that Garrison would physically harm others, in arresting Garrison for a violation that was bound to humiliate him, and in banning Garrison from public property (also a source of humiliation to a reasonable person), would be outrageous if Garrison did nothing to warrant such treatment. . . .

As an additional matter, we reject the defendants' contention that they cannot be held liable because a reasonable person in Garrison's situation would not have suffered mental anguish. . . . In the present case, both Lutz and Bobbitt were in positions of power over Garrison, with actual or apparent authority to affect his interests. Moreover, both men knew or should have known that Garrison's mental condition was already fragile. Bobbitt first had concerns about Garrison's mental health when Garrison's father died. Garrison was distraught at the scene and Bobbitt was aware that some employees felt Garrison "just wasn't the same person" when he returned to work after his father's death. Lutz also specifically knew from talking to Botts before Garrison's medical leave that Garrison might be suicidal over a break-up with a girlfriend. Again, these facts do not indicate a potential threat of harm to others. To the contrary, the only threat Garrison appears to have posed was to himself. On the other hand, the facts do reveal that Garrison may have been peculiarly susceptible to emotional distress at the time. Under the circumstances, unfounded accusations, humiliation, and exclusion would have been unjustifiable, and could have caused even a reasonable person in Garrison's position to suffer mental anguish. Knowledge of Garrison's susceptibility also satisfies the first prong of the standard used to evaluate these types of cases, i.e., "'that the actor either intended to cause emotional distress or knew or should have known that actions taken would result in serious emotional distress to the plaintiff.'" . . . For these reasons, enough evidence was presented . . . and the judgment of the trial court is affirmed.

Judgment affirmed.

Case Name: Garrison v. Bobbitt

Court: Court of Appeals of Ohio, Second District

Summary of Main Points: For the tort of intentional infliction of severe emotional distress, the conduct must be intentional; extreme and outrageous; the proximate cause the of plaintiffs injuries; and of a nature that no reasonable person could be expected to endure. Whether the defendants committed intentional infliction of severe emotional distress is a question of fact for the jury to decide.

Intentional infliction of severe emotional distress is commonly included in discrimination and sexual harassment lawsuits as additional grounds for liability. The victim commonly alleges that in addition to the defendant's conduct constituting unlawful discrimination and harassment, it also constituted intentional infliction of severe emotional distress.

TRESPASS TO LAND

The tort of trespass to land occurs when someone unlawfully enters the land of another, or causes some object, substance, or third person to enter the land of another. Trespass to land may also occur when someone unlawfully remains on the land of another, or fails to remove something that he is under a legal duty to remove from the land of another. The law of trespass has a long and varied history going back to the earliest days of the common law, and continues to evolve to this day. Trespass theories have been used in toxic tort cases, in which hazardous materials have chemically polluted someone's property, and have recently been used in cyberlaw cases **(Figure 7-8)**. See *Intel Corp. v. Hamidi* 30 Cal.4th 1342 (2003).

Someone who enters the property of another without permission is technically trespassing. However, a cause of action for trespass exists only if the trespasser is warned, asked to leave, and refuses, or if the trespass causes damages. A trespass that causes damage is actionable without the need for a warning.

Some examples of trespassing that go beyond a technical trespass because they involve damage include: dumping debris or chemicals on another's property, cutting down trees on another's property, building a fence or wall on another's property, and constructing a building that hangs over a neighbor's property.

A landowner's rights include the air above the real estate, as well as the ground below the surface. Thus, flying a model airplane over someone's property or shooting a gun across the property is trespassing. If no damages ensue, a warning to the trespasser would be required before the property owner could file suit for such trespasses. Building an underground tunnel or

Figure 7-8 *The ancient laws of trespass can apply when a hazardous materials release damages a neighbor's property.*

installing underground pipes that cross onto a neighboring property would also be trespassing.

TRESPASS TO CHATTELS

Trespass to chattels, more commonly known today as trespass to personal property, involves some sort of intentional interference or damage to the personal property of another. The tort of trespass to chattels would occur when the item is damaged, as opposed to stolen or taken away. The perpetrator would be liable for the damages caused.

CONVERSION

The tort of conversion applies to personal property that someone has intentionally, and without legal justification, taken away. Someone who steals personal property would be liable to the owner for conversion. For an action of conversion, the perpetrator must have intended to permanently deprive the owner. The perpetrator would be liable for the full value of the item that was taken.

MISREPRESENTATION

Misrepresentation or fraud involves a false representation of a material fact, intended to deceive the recipient, which actually does deceive the recipient, resulting in damages to that person.

In some ways, fraud overlaps several other torts, such as battery, trespass, and false imprisonment, because by these torts can be committed using false statements to obtain a person's consent. The key factors in fraud cases are whether the perpetrator knew that what he was saying was false, whether he intended to deceive the victim, and whether the victim's reliance on the deception was justifiable.

Unlike torts such as battery, assault, or false imprisonment, damages for fraud are not presumed, and must be proven. Damages may include loss of money or property, economic losses, or in some cases personal injury. When the fraud is deliberate and wanton, courts may award punitive damages.

BAD FAITH

Bad faith is a relatively new tort that developed out of modern contract law. Bad faith involves the intentional failure to comply with the terms of a contract. Bad faith actions usually involve contracts for insurance policies, but have in some cases been extended to employment contracts.

Implied in every contract is a duty of good faith and fair dealing upon both parties. Breach of the duty has traditionally been considered to be a breach of contract. However, many states now recognize that in addition to a breach of contract, certain breaches of the duty of good faith and fair dealing constitute a separate tort action. Insurance companies in particular have been scrutinized in this regard when making decisions about paying on claims.

Zoppo v. Homestead Ins. Co.
71 Ohio St.3d 552, 644 N.E.2d 397 (1994)
Ohio Supreme Court

In the early morning hours of October 13, 1988, Windy's Bar Restaurant, an establishment owned and insured by Donald Zoppo, was destroyed by fire. The fire was incendiary in nature and was started with kerosene, a liquid accelerant. At the time of the fire, Zoppo had an insurance contract in effect issued by Homestead Insurance Company. The coverage on the Homestead policy was $50,000 for the building and $65,000 for its contents.

Following its investigation, Homestead denied coverage to Zoppo. Homestead concluded that there was sufficient evidence that Zoppo had participated in setting the fire. Further, Homestead found that Zoppo had made material misrepresentations regarding his whereabouts on the night of the fire and regarding whether he had additional insurance.

(continued)

(*continued*)

Zoppo then brought suit against Homestead for breach of the insurance contract and for the tort of bad faith refusal to settle. Zoppo sought punitive damages in connection with the bad faith claim. The case went to trial before a jury. . . .

The jury returned verdicts for Zoppo in the sum of $80,000 in compensatory damages on the breach of contract claim and $187,800 on the bad faith claim ($122,800 in compensatory damages plus $65,000 in attorney fees). The jury also found that Zoppo was entitled to punitive damages. [T]he trial court set the amount of punitive damages at $50,000.

Homestead appealed and Zoppo cross-appealed. . . . The court of appeals held that there was no evidence of wrongful intent on the part of Homestead in denying Zoppo's claim. . . . The court of appeals also vacated the award of punitive damages and attorney fees. . . .

I - Bad Faith

. . . Our review of the record indicates the trial court correctly instructed the jury on the law of bad faith using the reasonable justification standard. There was ample evidence to support the jury's finding that Homestead failed to conduct an adequate investigation and was not reasonably justified in denying Zoppo's claim.

From the outset, Homestead's inquiry focused primarily on Zoppo, who claimed that he was in Pennsylvania hunting at the time of the fire. Homestead's investigators did not seriously explore evidence that other individuals, who were previously ousted from the bar by Zoppo, had threatened to burn the bar down. In fact, there was a previous attempt made to set the bar on fire. Two of the ousted men bragged in public that they were responsible for the attempted fire and one said he would be back "to finish the job." Following the actual fire, which occurred only three weeks after the attempted arson, one of the ousted men told a group of bar patrons that he had set the fire.

Despite these leads, and despite the fact that there appeared to have been a robbery and break-in (machines were broken into and one of the windows was broken), there was evidence at trial that the Homestead investigators failed to locate certain key suspects, verify alibis, follow up with witnesses or go to Pennsylvania to determine Zoppo's whereabouts on the morning of the fire. In fact, evidence was presented that when interviewing some of the alleged perpetrators, the investigators did little more than ask cursory questions such as whether they were responsible for the fire. When they answered negatively, their questioning ceased.

The investigators instead focused on the inconsistencies in Zoppo's statements concerning the sequence of events the morning of the fire and on the statement of a bar patron, Dave Pogue. Pogue initially corroborated the theory that the ousted men were responsible for the fire, but he later implicated Zoppo. However, there was evidence that he was paid for this later statement.

Part of Homestead's denial of the claim was based upon its belief that Zoppo had a motive for destroying his building, namely, financial gain. Homestead

(*continued*)

(*continued*)

argued that the bar was overinsured and that it was losing money. However, there was evidence to the contrary. Although Zoppo had purchased the bar six months prior to the fire for $10,000 and had insured it for $50,000, Homestead, in its initial underwriting report, had stated that the building had a market value of $95,798. Furthermore, Zoppo had no debts and had actually made improvements to the bar prior to the fire. Moreover, before the denial of the claim, Zoppo attempted to prevent demolition so that he could rebuild the bar.

Finally, Zoppo's expert, a claims consultant, testified that the Homestead investigation was inadequate and that Homestead was not justified in denying the claim.

Hence, based on the foregoing, we reinstate the trial court's finding of bad faith.

II - Damages

. . . We must next determine whether there were sufficient facts presented for the jury to consider an award of punitive damages. In *Staff Builders* . . . we stated that: "Punitive damages may be recovered against an insurer that breaches its duty of good faith in refusing to pay a claim of its insured upon proof of actual malice, fraud or insult on the part of the insurer." In this case, since Homestead did not act fraudulently in denying Zoppo's claim, the question becomes whether Homestead acted with actual malice. "Actual malice" is defined as "(1) that state of mind under which a person's conduct is characterized by hatred, ill will or a spirit of revenge, or (2) a conscious disregard for the rights and safety of other persons that has a great probability of causing substantial harm." . . .

There is no evidence here of hatred, ill will or a spirit of revenge. Thus, the trial court had the obligation to determine that there was sufficient evidence that Homestead consciously disregarded Zoppo's rights. . . . The record reveals a one-sided inquiry by Homestead investigators as to who was at fault. They did not adequately question suspects or follow up on leads. Homestead breached its affirmative duty to conduct an adequate investigation. The award of punitive damages was justified.

Finally, regarding the issue of compensatory damages and attorney fees, we hold that an insurer who acts in bad faith is liable for those compensatory damages flowing from the bad faith conduct of the insurer and caused by the insurer's breach of contract. . . .

Judgment reversed and cause remanded.

Case Name: Zoppo v. Homestead Ins. Co.

Court: Ohio Supreme Court

Summary of Main Points: An insurer who is not "reasonably justified" in denying a claim may be held liable to the insured for the tort of bad faith. When the insurer consciously disregards an insured's rights by conducting an inadequate investigation, an award of punitive damages is warranted.

DEFAMATION

defamation

an intentional tort that involves damage to a person's reputation through the publication of false, harmful, and unprivileged statements made to others; there are two types of defamation: slander and libel

slander

defamation that is committed orally through a false spoken word or gesture; the general rule with slander is that the victim must prove an actual monetary loss in order for the case to be actionable

libel

defamation through written or printed falsehoods about a person

slander per se

a particular type of slander involving certain categories of falsehoods that are considered to be more serious than others; under slander, actual damages must be proven, while under slander per se, damages will be presumed without the need to show a monetary loss

Defamation is a tort that involves damage to a person's reputation through false, harmful, and unprivileged statements made to others. For defamation to be actionable, the false statements must be "published." To be considered published, all that is required is that the perpetrator tells someone. Any communication to a third party constitutes publication.

Defamation can be broken down into two separate torts, **slander** and **libel**. Slander is defamation that is oral. Libel is defamation that is written.

Slander is a false spoken word or gesture. The general rule with slander is that the victim must prove an actual monetary loss in order for the case to be actionable. However, there are certain categories of slander that are considered to be more serious than others, and are termed **slander per se**. Under slander per se, damages will be presumed without the need to show a monetary loss. The categories for slander per se include:

- The person committed a crime.
- The person has a venereal disease or other communicable disease.
- The person is unfit for his or her business or trade.
- The person has committed serious sexual misconduct.

Libel is written or printed falsehoods about a person. With libel, damages are presumed without the need for showing actual monetary loss.

Modern technology has raised some interesting issues. For example, if someone makes a defamatory statement on television, is it slander or libel? Most states, succumbing to a strong lobby by the television industry, have by statute taken the position that defamation on television is slander. However, some states take the position defamation on television is libel.

Truth is an absolute defense to any action for defamation. In addition, certain types of communications are privileged and do not constitute defamation. There are two type of privilege in this regard, absolute and qualified.

Those with an **absolute privilege** include legislators while in session; judges, attorneys, witnesses, and jurors in court; the president and his cabinet members; and paid political broadcasts, provided equal time is provided for contrary views.

A **qualified** or **conditional privilege** exists when reasonably necessary to inform public officials of perceived wrongdoing, and in other cases where a person is working in an official capacity. In addition, a qualified privilege is granted to someone who has been defamed, in order to defend his legitimate interests. Such a privilege is limited to what is reasonably necessary to defend his own reputation from a prior defamation. For example, if Jim says, "Ron cheated on his taxes," and Jim is mistaken, Ron has a qualified privilege to reply, "Jim is a liar." However, Ron cannot say, "Jim is an arsonist."

absolute privilege
a privilege that arises in the law of defamation that protects certain categories of people from liability for false statements made without regard to whether they have spoken in good faith; those with absolute privilege include judges, attorneys, witnesses, and jurors in court; members of a lawmaking body (such as Congress) for statements made on the floor; the president and his cabinet members; and paid political broadcasts, provided equal time is provided for contrary views

qualified privilege
a privilege that arises in the law of defamation when there are facts justifying the written or spoken statement, but that may not be available when the words are uttered with malice or without good faith; for example, a qualified privilege is granted to someone who has been defamed, in order to defend his legitimate interests

Cross v. Fire Dept. Pension Fund of the City of Gary
537 N.E.2d 1197 (1989)
Court of Appeals of Indiana, Fourth District

Cross was employed as a firefighter with the Fire Department of the City of Gary, Indiana from 1973 to 1985. During 1984, Cross developed asthma which was aggravated by smoke, dust and extreme temperatures. This asthmatic condition prevented Cross from performing his normal duties as a firefighter.

On February 15, 1985, Cross filed an application for disability benefits with the Local Board. On April 24, 1985, Cross brought Bobby Joiner, Chief of the Gary Fire Department, a note purportedly written by Dr. Jatinder Kansal, Cross' [sic] personal physician. The note was written in two different inks. On April 26, 1985, Richard Gilliam, Secretary of the Pension Fund, and Joiner had a conference call with one of Dr. Kansal's employees regarding the authenticity of the note; the employee said that the note would not have been prepared and signed in two different colored inks. On the basis of that conversation, Gilliam wrote a letter to the Board of Trustees of the Public Employees' Retirement Fund ("PERF"):

Dear Sir,

The Local Pension Board of the Gary Fire Department met on April 26, 1985 and reviewed the records of James E. Cross, 316–58–4990. He entered the Gary Fire Pension plan on November 6, 1973 as Albert E. McCain. In 1981 he changed his name to James Earl Cross. He converted to the P.E.R.F. in 1980. Mr. Cross states that he has asthma which is aggravated on exposure to smoke, dust, cold air, sudden change in weather. His own Dr. along with the Local Board's Dr. agree that he can do some sort of desk job. He was offered a job as dispatcher, or in fire prevention by the Fire Chief. He refused and claimed he had another attack the very same day. After talking with his doctor we came to the conclusion that Mr. Cross is putting on an act. Dr. Kansel who is his Dr. states that he can work the dispatcher or fire prevention job. He also tried to falseify [sic] a doctor's slip which is included in his file.

The Gary Fire Pension Board has therefore denied him a disability pension. We are forwarding his records to the P.E.R.F. Board along with this determination. We respectfully await your determination.

Respectfully Yours,
/s/ Richard E. Gilliam
Pension Board Secretary

On January 16, 1986, Cross filed suit for libel based on the above communication. . . .

(continued)

(*continued*)

Discussion and Decision

. . . Whether a statement is protected by a qualified privilege is a question of law, unless the facts are in dispute. . . . The defense of qualified privilege will protect a communication made in good faith on any subject matter in which the party making the communication has an interest or in reference to which he has a duty either public or private, either legal, moral, or social, if made to a person having a corresponding interest or duty. . . . Gilliam as Secretary of the Local Board and P.E.R.F. had a common interest in the reason for denying Cross' disability benefits. Thus, the Local Board has made a prima facie showing that the communication between them was protected by a qualified privilege.

The protection of a qualified privilege may be lost if the plaintiff shows that the publisher was primarily motivated by feelings of ill will toward the plaintiff, if the privilege was abused by excessive publication of the defamatory statement, or if the statement was made without belief or grounds for belief in its truth. . . . While the good faith of the speaker is generally a question of fact, . . . in order to avoid summary judgment, Cross must set forth specific facts showing that there is a genuine issue for trial. . . .

Cross contends that Gilliam was primarily motivated by feelings of ill will toward him in writing the letter to P.E.R.F. However, Cross fails to offer any support of that contention. Cross stated in his Affidavit in Support of Response to Motion for Summary Judgment that "[T]here occurred on several occasions previous to November, 1984, run'ins and arguements [sic] between Plaintiff and Richard Gilliam; and that there was no love lost between them." Cross' bald assertion that Gilliam was primarily motivated by feelings of ill will will not make it so.

Further, Cross argues that the letter was written without belief or grounds for belief in its truth. Cross has failed to offer any support that the doctor's slip referred to in Gilliam's letter was in fact written by Dr. Kansel. Likewise, Gilliam and Joiner spoke with an employee of Kansel who indicated that the note could not have been written by Kansel. There is simply no proof that Gilliam or the Local Board did not have a good faith belief that Cross falsified the doctor's slip.

Because Gilliam's publication is protected by a qualified privilege and Cross has produced no evidence from which ill will or bad faith can be inferred, summary judgment was properly granted in favor of Gilliam and the Local Board. Affirmed.

Case Name: Cross v. Fire Dept. Pension Fund of the City of Gary

Court: Court of Appeals of Indiana, Fourth District

Summary of Main Points: A qualified privilege for statements that may otherwise be considered defamatory may be lost if "the plaintiff shows that the publisher was primarily motivated by feelings of ill will toward the plaintiff, if the privilege was abused by excessive publication of the defamatory statement, or if the statement was made without belief or grounds for belief in its truth."

Figure 7-9 *A public figure is someone who, through his or her accomplishments, fame, mode of living, profession, or calling, has placed him- or herself in the public spotlight. This includes fire chiefs and other high-ranking governmental officials.*

Defamation of Public Figures

A public figure is someone who, through his or her accomplishments, fame, mode of living, profession, or calling, has placed him- or herself in the public spotlight as a celebrity or as a person upon whom public attention is focused **(Figure 7-9)**.

Defamation of public figures is treated differently under the law than defamation of ordinary persons. Public figures can be defamed only when the defamer acts with **actual malice**. Actual malice means that the speaker or writer either knew that what he was saying was false when he said it, or spoke with reckless disregard of whether it was true or false. This is a substantially higher standard than what is applied to defamation of an ordinary person.

There are hundreds of cases that analyze who is a public figure. Someone may be a public figure for a brief period following some event, or while holding a particular position, and others may be public figures their entire lives. Certainly, elected officials are considered public figures, as are police and fire chiefs. In some cases, union presidents and others who have placed themselves in positions in the public eye are considered public figures.

actual malice

a verbal or written statement made by a person who either knew the statement was false, or acted with reckless disregard as to whether it was true or false

INVASION OF PRIVACY

Prior to 1890, no court had ever recognized a right of privacy. Invasion of privacy arose out of defamation and breach of contract cases. Courts saw the need for some sort of remedy to help protect people who were being injured through relatively modern advances in photography and mass print media.

Today there are four distinct torts that fall under the umbrella of invasion of privacy.

1. *Unreasonable intrusion on the seclusion of another.* Intruding upon another in his or her home, or other areas where there is a recognized right to seclusion, including through the use of wiretaps, microphone, or cameras.

2. *Appropriation of another's name or likeness.* Using someone's identity for commercial advantage, such as to advertise a product.

3. *Unreasonable publicity given to another's private life.* The release of private information about a person to the public that the ordinary person would consider offensive and objectionable.

4. *Publicity placing another in a false light.* Publicizing another person in such a way that he or she is falsely associated with something. The false light need not reach the level of defamation, but must involve the person being wrongly associated with something that an ordinary person would consider offensive and objectionable.

Damages for invasions of privacy may include amounts for any resulting illness, harm to the victim's business interests, and any profits that the perpetrator may have recognized from his or her misconduct. However, like other intentional torts such as battery, assault, and false imprisonment that are an affront to a person's dignity, damages may be presumed.

SUMMARY

Intentional torts provide a mechanism for someone who has been injured or harmed by the intentional conduct of another to seek private vindication of his rights, and recover damages to make himself whole again. Most intentional torts have historical roots dating back to the common law, yet all have found a place in modern society, addressing matters pertaining to hazardous materials, the Internet, and mass media. The intentional torts of battery, assault, and false imprisonment have an important connection to fire and emergency personnel when it comes to providing emergency medical care.

REVIEW QUESTIONS

1. Define the tort of battery and explain how consent is a defense.

2. Explain why it is not a battery when one football player tackles another, when neither player signed a consent form, nor even discussed the issue of consent.

3. Why is informed consent important when a person declines medical aid against medical advice in a pre-hospital setting?

4. At what age may persons in your state give consent for medical care?

5. In determining if an adult patient has the capacity to consent to medical care, what factors would you take into account?

6. Explain how the tort of trespass can be used in cases of hazardous materials spilling onto another person's property.

7. Define intentional infliction of severe emotional distress, and give an example of a type of conduct that would probably constitute it.

8. Explain the relationship among the terms defamation, libel, and slander.

9. What are the four categories of slander that do not require proof of actual damages?

10. Explain the difference between defamation of a public figure and defamation of an ordinary person.

DISCUSSION QUESTIONS

1. Consider the tort of battery and sports. How far outside the rules does the contact have to be to constitute battery? Consider a football game, where a player makes a late hit, diving on top of an already tackled player, seconds after the play was over. What if the contact occurred 10 seconds after the play was over? What if the contact was not related to a play, but rather occurred during a timeout?

2. You open up a fire magazine, and in an advertisement for fire hose you see a photo of yourself working at a building fire last year. The photo clearly shows your face in unmistakable detail. Do you have any recourse against the magazine and the advertiser? What if the photo appeared in a newspaper as part of a story about the fire, not an advertisement? Assume the same scenario, except that the photo is of you trying to handle a burst hose. The photo is captioned "Does your hose make you feel incompetent?" Since the photo came out you have been harassed mercilessly by your fellow firefighters. Is there any additional recourse you may have against the magazine or advertiser?

3. How do firefighters and emergency medical personnel determine if a person who appears to be impaired from alcohol or drugs has the legal capacity to consent or decline medical assistance? What precautions are warranted in borderline cases?

Chapter 8

NEGLIGENCE

Learning Objectives

Upon completion of this chapter, you should be able to:

- Define negligence and identify the elements of negligence.
- Explain the concepts of duty, standard of care, breach, damages, and proximate cause as these terms relate to negligence.
- Explain what the standard of care is for professionals and those with specialized training.
- Identify the types of evidence that can be used to establish the standard of care for a professional.
- Explain the defenses to negligence, including assumption of risk, contributory negligence, comparative negligence, last clear chance doctrine, and the rescue doctrine.
- Explain the Fireman's Rule.
- Define gross negligence and recklessness.
- Explain joint and several liability.
- Identify the most common types of activities for which strict liability is imposed.

INTRODUCTION

The firefighters gathered in the classroom for the post-fire critique. The battalion chief began by saying "I know this will be difficult, but we needed to get though it. Charlie Ross, the injured firefighter, is doing okay and is recovering from his injuries." The chief pushed the play button on the tape recorder, and it was as if the incident was happening again.

"Mayday, mayday, mayday – This is Ladder 15. We are on the fifth floor. One of our members was searching and appears to have fallen down the elevator shaft."

"Command to Ladder 15, your mayday is acknowledged. Command to RIT Team One, respond to the base of the elevator shaft and locate the missing member. Command to dispatch, transmit a second alarm."

At the conclusion of the tape, the chief again spoke. "We have determined that an elevator repair company was in the building at the time the fire started, and that they were working on the fifth floor at the time. They said they closed the hoistway door when the alarm went off, and then exited the building, but we're not so sure."

In the back of the room one firefighter muttered "Let's hope Charlie's got a good lawyer."

When it comes to civil liability, the biggest area of concern to fire and emergency personnel is **negligence**.

Most civil suits against fire departments, firefighters, and emergency medical personnel will involve accusations of negligence. Unlike the torts discussed in Chapter 7, negligence is a non-intentional tort.

Negligence is a complex topic, and is impossible to fully define in a sentence or two. As a result, you may come across a variety of definitions of negligence, which differ in varying degrees. Some definitions of negligence emphasize certain facets of negligence, while other definitions focus on other aspects.

Black's Law Dictionary defines negligence as "the omission to do something which a reasonable man, guided by those ordinary considerations which ordinarily regulate human affairs, would do, or the doing of something which a reasonable and prudent man would not do." A more concise definition that captures the most important issues involved in negligence is "the failure to exercise the care that the reasonably prudent person would have exercised under the circumstances which causes damages to another."

Despite any differences in the definitions of negligence from one source to another, there are some basic underlying principles that pertain to negligence that transcend the definitional differences. These principles include: duty, standard of care, breach of the standard of care, damages, and causation. We will discuss these principles in the context of an examination of the elements of negligence.

negligence (civil tort)
the failure to exercise the care that the reasonably prudent person would have exercised under the circumstances that caused damages to another

ELEMENTS OF NEGLIGENCE

In order for an action for negligence to exist, the following three elements must be present:

1. An act or omission
2. Damages to the plaintiff
3. Breach of the standard of care

An Act or Omission

In order for there to be negligence, there must be an act or an omission. The act or omission requirement for negligence is very similar to the act requirement for a crime, which we discussed in Chapter 5. For negligence, the act can be any affirmative act, such as driving a fire truck **(Figure 8-1)**, operating a piece of machinery, or cooking a meal. When people act, they are under an affirmative duty to protect others from foreseeable, unreasonable risks associated with, or created by, their actions.

An omission, or the failure to act, can be grounds for negligence only when one is under a legal duty to act. As we learned in Chapter 5, at common law no one is under an affirmative duty to act in the absence of a relationship or law that creates such a duty. The recognition of a duty to act has historically been based upon the relationship between the individuals. A parent has a duty to come to the aid of a child, and a lifeguard has a duty to come to the aid

Figure 8-1 *Any affirmative act, such as driving a piece of apparatus, creates a duty to protect others from foreseeable, unreasonable risks.*

Figure 8-2 *A duty to act is created by the relationship between the parties. (Photo by Robert A. Gread.)*

of a swimmer in distress **(Figure 8-2)**. In the same way, an on-duty firefighter would have an affirmative duty to render aid when called upon to do so. The creation of the duty is based on the relationship between the parties.

Some states have created statutory duties, such as a duty on the part of the driver of a car to render assistance to anyone who may have been injured in a motor vehicle accident that the driver was involved in. A few states have gone even further and created a generalized duty to render aid to another. The failure to act when one is under a duty to act can give rise to negligence.

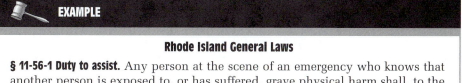

EXAMPLE

Rhode Island General Laws

§ 11-56-1 Duty to assist. Any person at the scene of an emergency who knows that another person is exposed to, or has suffered, grave physical harm shall, to the extent that he or she can do so without danger or peril to himself or herself or to others, give reasonable assistance to the exposed person. Any person violating the provisions of this section shall be guilty of a petty misdemeanor and shall be subject to imprisonment for a term not exceeding six (6) months, or by a fine of not more than five hundred dollars ($500), or both.

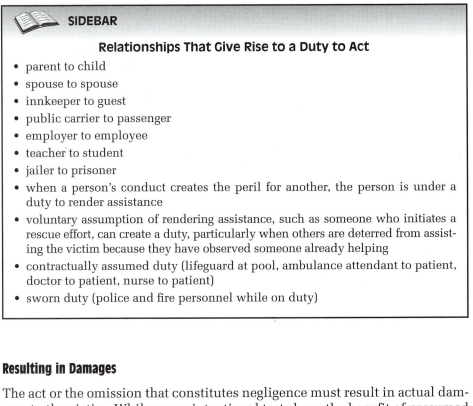

SIDEBAR

Relationships That Give Rise to a Duty to Act

- parent to child
- spouse to spouse
- innkeeper to guest
- public carrier to passenger
- employer to employee
- teacher to student
- jailer to prisoner
- when a person's conduct creates the peril for another, the person is under a duty to render assistance
- voluntary assumption of rendering assistance, such as someone who initiates a rescue effort, can create a duty, particularly when others are deterred from assisting the victim because they have observed someone already helping
- contractually assumed duty (lifeguard at pool, ambulance attendant to patient, doctor to patient, nurse to patient)
- sworn duty (police and fire personnel while on duty)

Resulting in Damages

The act or the omission that constitutes negligence must result in actual damages to the victim. While many intentional torts have the benefit of presumed damages, there is no such presumption of damages with negligence. In order for someone to establish a claim for negligence, he or she must be able to prove actual damages. Damages may be property damage, personal injury, medical bills, pain and suffering, or loss of income. However, being wronged in principle is not sufficient to constitute negligence.

In addition, the negligent act must be the proximate cause of the damages or harm. **Proximate cause** is a legal term referring to the fact that the act in question was the legal cause of the harm that resulted. Proximate cause requires a greater analysis of the circumstances than merely finding a "but for" connection. Harm that is remote or not proximately related to the conduct of the defendant is not actionable.

proximate cause
a legal term referring to the fact that the act in question was the legal cause of the harm that resulted

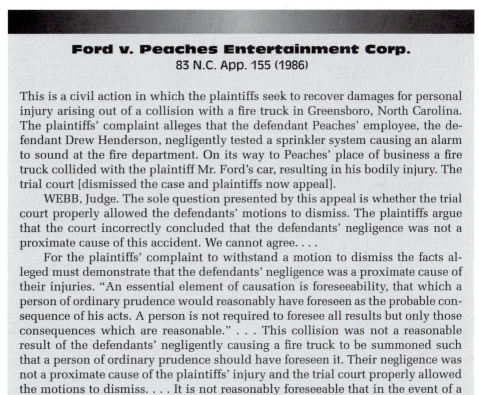

> ## SIDEBAR
>
> 1. Able sells Baker a 50-foot sailboat. Two years later, Baker is killed when the boat sinks at sea during a hurricane. "But for" the fact that Able sold Baker the boat, Baker would not have died. However, the sale of the boat cannot reasonably be called the "proximate cause" of Baker's death. The proximate cause was the hurricane.
> 2. Able takes Baker out on his 50-foot sailboat. At sea, fog rolls in and Able fails to utilize the onboard radar and satellite navigation system to navigate through the fog. The boat runs aground and Baker is killed. Able's failure to utilize the available instrumentation is a proximate cause of Baker's death.

Ford v. Peaches Entertainment Corp.
83 N.C. App. 155 (1986)

This is a civil action in which the plaintiffs seek to recover damages for personal injury arising out of a collision with a fire truck in Greensboro, North Carolina. The plaintiffs' complaint alleges that the defendant Peaches' employee, the defendant Drew Henderson, negligently tested a sprinkler system causing an alarm to sound at the fire department. On its way to Peaches' place of business a fire truck collided with the plaintiff Mr. Ford's car, resulting in his bodily injury. The trial court [dismissed the case and plaintiffs now appeal].

WEBB, Judge. The sole question presented by this appeal is whether the trial court properly allowed the defendants' motions to dismiss. The plaintiffs argue that the court incorrectly concluded that the defendants' negligence was not a proximate cause of this accident. We cannot agree. . . .

For the plaintiffs' complaint to withstand a motion to dismiss the facts alleged must demonstrate that the defendants' negligence was a proximate cause of their injuries. "An essential element of causation is foreseeability, that which a person of ordinary prudence would reasonably have foreseen as the probable consequence of his acts. A person is not required to foresee all results but only those consequences which are reasonable." . . . This collision was not a reasonable result of the defendants' negligently causing a fire truck to be summoned such that a person of ordinary prudence should have foreseen it. Their negligence was not a proximate cause of the plaintiffs' injury and the trial court properly allowed the motions to dismiss. . . . It is not reasonably foreseeable that in the event of a false alarm a fire truck will cause an accident in responding to the alarm.

Affirmed.

Case Name: Ford v. Peaches Entertainment Corp.

Court: North Carolina Court of Appeals

Summary of Main Points: The test of proximate cause relates to forseeability. "A person is not required to foresee all results but only those consequences which are reasonable."

   **SIDEBAR**

Do you agree with the court in *Ford?* Do you think that if the judges who decided the case had been firefighters, their perspective on the forseeability of a fire truck being involved in an accident responding to a false alarm would change? How "foreseeable" does a result have to be? Forseeability also bears on the issue of duty when acting, our first element in negligence.

intervening act
the act of a third person that breaks the chain of causation and eliminates liability between an original wrongdoer and an inured party

standard of care
a measure of what the community expects and demands from a person in a given situation; the ordinary standard of care is the degree of care that the reasonably prudent person would have exercised under the circumstances; see also the professional standard of care

Another interesting case involving proximate cause is *Westbrook v. Cobb*, 105 N.C. App. 64 (1992). Plaintiff's house caught fire when a car struck a utility pole nearby, creating an electrical surge. While the fire department was on the scene extinguishing the fire, plaintiff entered the house to retrieve some personal items and injured his back. Plaintiff sued the driver of the automobile that collided with a utility pole for causing his back injury. The court ruled that plaintiff's entry into the house was an **intervening act**, so that his injuries did not "naturally flow" from defendant's negligence. In other words, the car negligently striking the pole was not the proximate cause of the plaintiff's injuries. It was plaintiff's running into his burning house that was the proximate cause of his injuries.

Breach of the Standard of Care

The third element for the action of negligence is a breach of the **standard of care**. Normally, when engaging in any activity, a person is expected to exercise the care that the reasonably prudent person would exercise under the circumstances. If a person's conduct fails to live up to this reasonably prudent person standard, and someone else is injured or harmed, the injured party can sue for negligence.

The reasonably prudent person standard is sometimes referred to as the ordinary standard of care. The standard has also been equated with the reasonable man, the reasonable man of ordinary prudence, and the man of ordinary sense using ordinary care and skill. In each case of negligence, it is a question of fact for the jury to decide whether the defendant's conduct met or failed to meet the standard of care. Viewed a different way, the standard of care is a measure of what the community expects and demands from everyone.

In a trial involving negligence, the judge will instruct the jury about the reasonably prudent person standard. The jury then considers whether or not the defendant's actions met or failed to meet the reasonably prudent person standard. In making the determination of what the reasonably prudent person would have done under the circumstances, the jury takes into account the collective life experiences of each juror.

professional standard of care

the standard of care that a person with professional skills and training is required to exercise, namely, the care that the reasonably prudent professional of like training and experience would have exercised under the circumstances

malpractice

a common term for professional acts of negligence

Professional Standard of Care How does a jury evaluate the conduct of professionals in regard to negligence? Are firefighters automatically negligent when they respond to a fire, because the reasonably prudent person would not run into a building that is on fire? How about a surgeon? Would the reasonably prudent person cut another person open? This potential dilemma is solved by the establishment of a separate standard of care for people with special expertise and professional training, often referred to as the **professional standard of care**. A person with professional skills and training will be held to exercise the care that the reasonably prudent professional of like training and experience would have exercised under the circumstances **(Figure 8-3)**. Professional acts of negligence are often referred to as **malpractice**.

Jurisdictions differ somewhat on the exact formulations of the standard of care for professionals, but the general principles are universal. An EMT would not be judged by the standards of an emergency room doctor, nor even a paramedic; nor would an emergency room doctor be held to the standard of a neurosurgeon. A paramedic in a given state would be evaluated according to how the reasonably prudent paramedic in that state would have acted under the circumstances.

Evidence of the Professional Standard of Care While defining the standard of care for a professional is one thing, *proving* what is the standard—in the context of an actual case—is another matter entirely. When dealing with the reasonably prudent person standard, the jury members are entitled to rely upon their own life

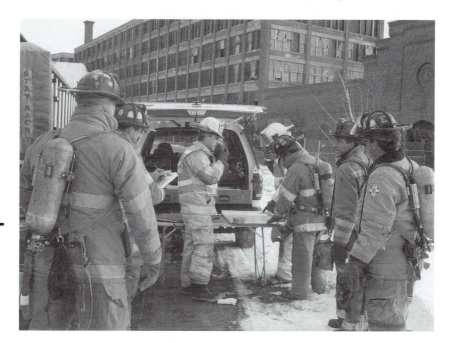

Figure 8-3 *Those with professional skills and training are held to the standard of the reasonably prudent professional.*

experience in determining what the reasonably prudent person would have done. Such a reliance makes sense when the jury is deciding matters such as whether the reasonably prudent person would have driven at 70 miles per hour in a 25 mph zone, or whether the reasonably prudent person would have made sure that his front steps were free of snow and ice four days after a blizzard. However, how does a jury decide whether or not a patient needed a certain type of medical procedure, or a certain medication for a certain condition?

In a negligence case in which the professional standard of care is an issue, the jury must be educated about what the reasonably prudent professional of like training and experience would have done under the circumstances. This educational process involves the introduction of four categories of evidence: expert witnesses, learned treatises, laws, and industry-wide standards.

- *Expert witnesses.* Witnesses who have professional expertise in a given subject area would be permitted to testify to explain their conclusions about how a person with like training and experience should have acted under the circumstances, and what the standard of care should have been.

- *Learned treatises.* This term is a fancy way of saying authoritative books, journal articles, and other writings that explain what the standard of care should be.

- *Laws and regulations.* There is an assumption in the law that the reasonably prudent person or professional would comply with applicable laws. While some jurisdictions give compliance with laws more weight than other jurisdictions, the indisputable fact is that laws and regulations are evidence of what the reasonably prudent professional would have done. In particular, noncompliance with OSHA regulations, even in non-approved plan states, can be used as evidence of negligence in many fire service cases.

- *Industry-wide standards.* Where applicable, industry-wide standards provide evidence of what the reasonably prudent professional would have done under the circumstances. Obviously in the fire service, the National Fire Protection Association standards are a major source of evidence about the standard of care for the reasonably prudent firefighter, fire officer, fire chief, or fire department.

The jury will take the evidence presented, and make a determination about what the standard of care should have been, and whether or not the defendant's conduct met or fell short of the appropriate standard of care.

FIRE SERVICE NEGLIGENCE CASES

The following three cases are provided as examples of negligence suits against firefighters and fire departments. Understanding the facts of each case will help us comprehend some of the important issues related to fire service negligence.

> ### EXAMPLE
>
> In *Kenavan v. New York,* 523 NYS 2d 60 (1987), an engine company arrived at the scene of a car fire. Smoke was obscuring the scene, so the engine company drove through the smoke and pulled past the burning car. As the crew stretched a line, the captain took a handlight and went down the road to warn oncoming traffic. A car drove past the captain through the smoke, striking four firefighters, and killing one of them. The widow of the deceased firefighter and the injured firefighters sued the fire department, the captain, and the driver of the apparatus, among a number of other defendants. The plaintiffs alleged negligence in the placement of the apparatus, and the failure to establish a "fire line." They also alleged that the city was negligent for allowing abandoned cars to remain in the location of the fire. The jury found for the plaintiffs. The case was later reversed on appeal for the driver and captain, based on immunity.

> ### EXAMPLE
>
> In *McGuckin v. Chicago* 191 Ill.App.3d 982, 548 N.E.2d 461 (1989), a fire was discovered in the basement of Chicago's Union Station. The fire department responded and extinguished the visible fire. The fire was believed to have been limited to some trash near an electrical chase. Companies left the scene, without fully checking the upper floors, because of locked doors. Shortly thereafter, a fire erupted on an upper floor, directly above where the basement trash fire had been. The fire department made several heroic rescues, but one occupant of the building succumbed to his injuries. His widow sued the fire department and the building owners. The jury found both the fire department and the building owners negligent. On appeal, the verdict against the fire department was reversed, based on immunity.

> ### EXAMPLE
>
> In *Harry Stroller v. City of Lowell* 587 NE 2d 78 (1992), the plaintiff owned a sprinklered building that caught fire. The fire department arrived, but did not supplement the sprinkler system. Instead, the fire department used water from nearby hydrants to supply their handlines and master streams. The fire spread to five of the plaintiff's buildings. After the fire, the owner sued the fire department, alleging that it was negligent by diverting water away from the sprinklers. The jury found that the fire department was negligent and awarded an $850,000 judgment. On appeal the verdict was upheld, but reduced to meet a statutory damages cap.

In all three of these cases, juries found that fire departments and fire personnel had committed negligence, and held them liable for damages. We will revisit each of these cases in Chapter 9, when we discuss immunity. For now, it is important to recognize that fire departments and firefighters can be sued for negligence in the performance of their duties. Cases involving fireground decision making are well beyond the general knowledge of the average juror. Evidence must be presented to help the jury establish the appropriate standard of care. Consider the type of evidence that must have been used in each of the three cases to establish the standard of care.

EMERGENCY MEDICAL CARE AND NEGLIGENCE

Negligence actions arising out of firefighting are but one aspect of the liability problem facing firefighters and fire departments. Lawsuits alleging negligence in the rendering of emergency medical treatment raise some areas of particular concern. For one thing, states have adopted comprehensive regulations governing pre-hospital emergency care. As should be apparent from our discussions, these regulations establish strong, if not conclusive, evidence of the standard of care for paramedics and emergency medical technicians **(Figure 8-4)**. Deviation from these pre-hospital protocols makes it

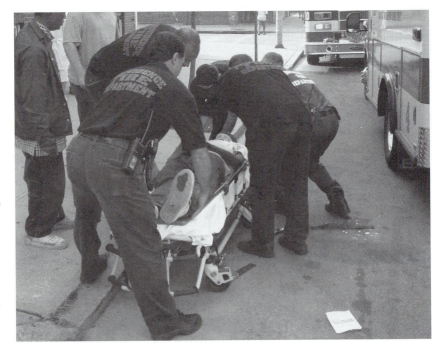

Figure 8-4 *Pre-hospital care protocols issued by state or local authorities are strong evidence of the standard of care for emergency medical personnel.*

relatively easy for an injured patient to establish negligence on the part of emergency medical technicians and paramedics.

Consent to Treat, Battery, and Negligence

In Chapter 7 we discussed the issue of battery and consent in regard to medical treatment. Many jurisdictions now consider issues of patient treatment, consent, informed consent, and implied consent to be matters better suited to consideration as negligence. In other words, rather than analyze a case by determining if the patient gave his or her informed consent to treatment, or whether the medical provider committed battery, courts will look instead at the standard of care of the reasonably prudent medical provider, and determine if the provider was negligent in (A) not providing enough information to the patient; or (B) not obtaining valid consent. Some states refuse to even consider informed consent cases as intentional tort/battery cases, and consider them strictly as cases of the breach of the standard of care.

Patient Abandonment

abandonment
the intentional stopping of medical care without legal excuse or justification

Another important issue related to rendering patient care involves patient **abandonment**. As discussed in Chapter 7, abandonment is the intentional stopping of medical care without legal excuse or justification. A patient's informed refusal of care provides a medical provider with a legal justification for stopping care, as does the transfer of care from a paramedic to a hospital emergency room.

Concerns over abandonment in a pre-hospital setting arise from cases against physicians for the abandonment of patients. These cases arose when physicians left patients in the care of less qualified medical personnel, or failed to properly monitor patients they had been treating. The legal concepts surrounding abandonment have been transposed onto the pre-hospital environment, seemingly in the absence of case law, by well-intentioned writers and lecturers on the subject.

There is a split of opinion on whether abandonment is a type of negligence, in which the analysis focuses on whether or not the provider rendered the appropriate standard of care, or whether abandonment is an intentional tort. Most authorities now seem to conclude that abandonment is based on negligence, and have identified specific areas of concern. For emergency medical personnel, these areas include:

- stopping care and leaving the scene
- leaving a patient temporarily unattended or unsupervised
- allowing a patient who lacks capacity to decline medical treatment
- transferring responsibility for the care of a patient to a lower level of care (for example, if a paramedic transfers a patient to an EMT or first responder)

Another common scenario occurs when a doctor gratuitously attends to a patient at the scene, and then passes care on to paramedics or EMTs upon their arrival. There is no case law on this subject. Given that the weight of authority applies a negligence analysis to abandonment cases, the proper focus is on what the reasonably prudent doctor, EMT, or paramedic would have done under the circumstances. State and local pre-hospital protocols will be an important consideration in establishing the appropriate standard of care.

BEYOND NEGLIGENCE

gross negligence
an aggravated form of negligence that involves an extreme departure from the ordinary standard of care

Between ordinary negligence on the one hand, and intentional torts on the other, are two additional types of conduct that can also create liability. These two types of conduct are gross negligence and recklessness.

Gross negligence is an aggravated form of negligence that involves an extreme departure from the ordinary standard of care. States differ upon an exact formulation for gross negligence. Some courts have described gross negligence as willful misconduct, recklessness, or such utter lack of care as to be evidence of either willful misconduct or recklessness. However, most jurisdictions draw a sharp distinction between gross negligence and recklessness, finding that gross negligence differs from negligence in degree, whereas recklessness differs from negligence in kind. In most cases, the question of whether an act constitutes negligence or gross negligence is left to a jury to decide. Gross negligence is an important issue for firefighters and emergency personnel, as we will see when we discuss sovereign and statutory immunity.

recklessness (civil tort)
an aggravated form of gross negligence involving willful, wanton conduct where the actor had knowledge that harm was likely to result from his behavior, and consciously chose to act despite the risk

Recklessness, which is also referred to as willful, wanton, and reckless behavior, is an aggravated form of gross negligence. The key focus of recklessness is that the actor had knowledge that harm was likely to result from his behavior, and consciously chose to act (or refused to act when under a duty to act) despite the risk. It is the knowledge that harm was likely to result that separates recklessness from gross negligence. In many cases, reckless conduct is considered the equivalent of intentional conduct, and the perpetrator may be liable to the same extent as one who acted intentionally.

Persons guilty of gross negligence or recklessness can be held liable for damages to the same extent as someone who is negligent, and in some jurisdictions may be liable for punitive damages. In addition, there are certain situations in which ordinary negligence is not enough to create liability in the absence of gross negligence or recklessness. In Chapter 9, we will discuss sovereign and statutory immunity protection. As a general rule, where immunity protection is available, it applies only to conduct that constitutes negligence, but offers no protection for gross negligence or recklessness.

Crouch v. Regional Emergency Medical Services

State of Michigan, Court of Appeals

Genesee Circuit Court

LC No. 99-064580-NI No. 238010 (May 22, 2003)

Viewed in the light most favorable to plaintiff, the evidence indicates that defendant, a sheriff's deputy, was dispatched to the Crouch home following a phone call to 911 by plaintiff concerning her husband, Chad Crouch (hereinafter "Crouch"). Following defendant's arrival, plaintiff and a family friend expressed their concerns about Crouch's irrational behavior and his statements suggesting that he was contemplating suicide. Plaintiff wanted him transported to Hurley Hospital. Defendant talked to Crouch, and he calmed down and cooperated with defendant. Defendant handcuffed Crouch and escorted him from the house to the back of the patrol car. Defendant arranged for an ambulance to transport Crouch to the hospital because he was being "totally cooperative." When the ambulance arrived, defendant informed the attendants that Crouch had a history of depression, was talking about suicide, and had been drinking, but was being cooperative. One of the attendants, a licensed paramedic, and defendant helped Crouch into the back of the ambulance. Defendant removed the handcuffs. Crouch was quiet and cooperative and agreed to be transported for evaluation. On the way to the hospital, Crouch opened the door, stepped on the bumper and then fell or stepped off of the moving ambulance, sustaining injuries.

Plaintiff brought this action against defendant and alleged that his conduct fell under the "gross negligence" exception to governmental immunity. . . . Plaintiff alleged that defendant was grossly negligent because he released Crouch from protective custody to the ambulance attendants rather than transporting him to the hospital in the patrol vehicle.

Plaintiff presented testimony from two experts in support of her position. . . .

The trial court ruled that when the evidence was viewed in the light most favorable to plaintiff, the best one could say was that defendant made a wrong choice. However, the court reasoned that "turning Mr. Crouch over to an experienced EMT professional does not rise to the level of what this Court would determine to be reckless to the point of demonstrating a lack of concern for Mr. Crouch's welfare." In addition, the court ruled that plaintiff's proof of proximate causation was deficient because the incident occurred after Crouch was in the care of the paramedic.

Plaintiff argues that the court erred in granting summary disposition because reasonable minds could differ concerning whether defendant's conduct was grossly negligent. . . . We disagree.

The governmental immunity statute defines "gross negligence" as "conduct so reckless as to demonstrate a substantial lack of concern for whether an injury results." . . . Evidence of ordinary negligence does not create a question of fact regarding gross negligence. . . . As the Michigan Supreme Court emphasized in that case, "[T]he Legislature limited employee liability to situations where the contested conduct was substantially more than negligent." . . .

(continued)

(*continued*)

Reasonable minds could not differ in concluding that defendant's conduct in turning Crouch over to the care of the ambulance attendants was not "gross negligence" as defined in the statute. Defendant released Crouch to the care of a paramedic for transportation to the hospital in an ambulance. This conduct did not demonstrate "a substantial lack of concern for whether an injury results." . . .

Under the gross negligence exception, a governmental employee may be held liable if his gross negligence is "the proximate cause" of the plaintiff's injury. . . . In *Robinson v. Detroit*, 462 Mich 439, 459, 462; 613 NW2d 307 (2000), the Court explained that "the proximate cause" means that the employee's gross negligence must be "the one most immediate, efficient, and direct cause" of the injury or damage. We agree with the trial court that reasonable minds could not differ in concluding that defendant's actions do not meet this standard, in light of defendant's having given Crouch over to the paramedics' care and control. . . .

Affirmed.

Case Name: Crouch v. Regional Emergency Medical Services

Court: State of Michigan, Court of Appeals

Summary of Main Points: Gross negligence involves conduct that demonstrates a lack of concern for the consequences. When a defendant's conduct is at worst negligence, there is no question of fact upon which a jury could find gross negligence.

DEFENSES TO NEGLIGENCE

The law recognizes several defenses to negligence. These defenses include: assumption of risk, contributory and comparative negligence, rescue doctrine, last clear chance, and the Fireman's Rule.

Assumption of Risk

The defense of assumption of risk applies to situations in which the injured party knew of the danger or peril, understood the risks, and freely and voluntarily chose to act. Assumption of risk serves to limit the liability of a person who negligently creates a hazard. For example, a novice skier who chooses to ski down a trail that is clearly marked "Warning: Steep Trail—Experts Only" assumes the risk that he or she may be injured in a fall because the ski trail is so steep and/or difficult. In many ways, assumption of risk falls back on the idea of consent that was discussed with regard to battery, in which a person knows of the risk and voluntarily agrees to accept the consequences.

contributory negligence
a defense to a negligence action in which, if the plaintiff was shown to be in any way contributorily negligent in causing his or her own injuries, the defendant could not be held liable; the contributory negligence rule was an absolute defense to a suit for negligence; it has been abolished in favor of comparative negligence

comparative negligence
procedure whereby a jury is responsible for apportioning fault among the various parties to a lawsuit; comparative liability is assigned on a 100 percent scale, with each party receiving a percentage of fault as determined by the jury

Contributory and Comparitive Negligence

Often a person's injuries or property damage are caused primarily by the defendant's conduct, but are also caused in some small part by his or her own conduct. The law was traditionally rather harsh in this regard. At common law, if the plaintiff was shown to be in any way contributorily negligent in causing his or her own injuries, the defendant could not be held liable. This **contributory negligence** rule was an absolute defense to negligence cases. Over the years, all states have abolished the contributory negligence rule in favor of a system called **comparative negligence**.

Under comparative negligence, the jury is responsible for apportioning fault among the various parties to a lawsuit. Comparative liability is assigned on a 100 percent scale, with each party receiving a percentage of fault as determined by the jury. Thus, in a car accident in which the plaintiff sustains $100,000 in damages, if the jury determines that the defendant was 80 percent at fault, and the plaintiff 20 percent at fault, the plaintiff could recover $80,000. States differ on some of the finer points of comparative negligence, but the major principles of comparative negligence operate in much the same way.

Because the contributory negligence rule was so harsh, a variety of exceptions developed, including the Rescue Doctrine and the Last Clear Chance Rule. Despite the implementation of comparative negligence, many of the doctrines that developed under contributory negligence still remain.

Rescue Doctrine   The Rescue Doctrine is a principle that developed under contributory negligence to ensure that someone who came to the rescue of another could recover from the person who negligently caused the situation if the rescuer was injured or suffered damage in the course of effecting the rescue **(Figure 8-5)**.

Figure 8-5   *The rescue doctrine is based on the recognition that danger invites rescue. A rescuer who comes to the aid of another is not prohibited by contributory negligence or assumption of risk from suing for negligence. (Photo by Rick Blais.)*

The rationale for the rescue doctrine was that courts should not discourage people from helping others who are in distress. The rescue doctrine served to limit the ability of a defendant to claim that a rescuer was contributorily negligent or had assumed the risk in situations where the defendant's negligence created the peril that "invited" a rescuer.

The *Ouellette* case shows the way in which some courts still handle the Rescue Doctrine under modern comparative negligence systems.

Ouellette v. Carde
612 A.2d 687 (1992)
Supreme Court of Rhode Island
OPINION

MURRAY, Justice.

This civil action in negligence is before the court on the defendants' appeal from a judgment entered in favor of the plaintiff. We affirm.

On March 18, 1986, defendant Orin V. Carde attempted to change the muffler and tailpipe of his 1979 Mercury Cougar. The defendant parked his car in the closed garage connected to his home and elevated the back of the car with a hydraulic jack. . . .

He began pulling on the muffler and exerted such pressure in attempting to wiggle it free that the right side of the car fell off the bumper jack onto the stanchion jack and trapped defendant underneath the car. Because of the angle at which the car fell, the gas tank landed on the right stanchion jack puncturing the tank and releasing approximately ten gallons of gas onto the garage floor. The defendant remained trapped under the car in a semiconscious state for an unknown period. He eventually recovered and worked himself free, and called the plaintiff, Beverly Ouellette, from the garage telephone. In the middle of the conversation he passed out, and plaintiff, a long time friend and neighbor, immediately drove to defendant's house. She entered the front door of the home and made her way to the garage through the laundry room. She nearly slipped in a puddle of gasoline on the garage floor as she entered the garage and found defendant lying on the ground beneath the dangling phone. She attempted to call a rescue squad but was unable to get a dial tone, and defendant became agitated. He told plaintiff that they should leave through the garage door and directed her to press the electric door opener. When the door was one-half to three-quarters open, the gas ignited in an explosion. Both plaintiff and defendant escaped but were severely burned. The plaintiff was taken directly to the emergency room where she was treated for third-degree burns to her left ankle and to both feet. She was released later that day but was readmitted one week later. She stayed in the hospital for fifteen days during which time she had a series of operations and a skin graft. She was subsequently released and entered a home-care program through which a nurse would visit and change

(*continued*)

(*continued*)

her bandages three times a day. In addition to the physical injuries, plaintiff experienced extreme anxiety and panic attacks. She received treatment for anxiety over a three-year period and was still taking medication at the time of the trial. . . .

[T]he jury returned a verdict in favor of plaintiff for $85,000 plus interest and costs. On appeal defendant raises six issues. We address each issue separately.

I

. . . The defendant argues that Rhode Island's comparative-negligence statute incorporates the public-policy principles of the rescue doctrine and that the jury should have been instructed to apply standards of comparative negligence to plaintiff's case. The plaintiff responds that the rescue doctrine survives the adoption of comparative negligence because comparative negligence inadequately promotes the public policy of encouraging a person under no duty to rescue to save the life of a human being in peril.

The rescue doctrine is a rule of law holding that one who sees a person in imminent danger caused by the negligence of another cannot be charged with contributory negligence in a nonreckless attempt to rescue the imperiled person. . . . The doctrine was developed to encourage rescue and to correct the harsh inequity of barring relief under principles of contributory negligence to a person who is injured in a rescue attempt which the injured person was under no duty to undertake. . . . In practice the doctrine may be used either to establish a plaintiff's claim that the defendant was guilty of actionable negligence in creating the peril which induced the rescue attempt or to eliminate the defenses of contributory negligence and assumption of risk. The instant case, however, raises the question of whether Rhode Island's adoption of the comparative-negligence doctrine . . . requires that rescue-doctrine cases be adjudicated under standards of comparative negligence.

Comparative fault removes the harsh consequences of contributory negligence because a rescuer is not barred completely from recovery for negligently performing a rescue. Under a comparative-negligence standard the trier of fact apportions fault among responsible parties, and the negligent rescuer is entitled to recover only that percentage of total damages for which the party creating the peril is responsible. The comparative-negligence doctrine, therefore, arguably incorporates this policy consideration of the rescue doctrine, but there is a split in authority whether the doctrine of comparative negligence fully addresses the other policy considerations of the rescue doctrine. . . . Most courts addressing this issue focus on the fact that comparative negligence removes the harsh consequences of contributory negligence and have ruled that a plaintiff who is negligent in performing a rescue should recover only a pro rata share of the damages sustained attributable to the defendant. . . .

We are of the opinion, however, that the comparative-negligence doctrine does not fully protect the rescue doctrine's underlying policy of promoting rescue. No common-law duties changed as a result of the enactment of Rhode Island's comparative-negligence statute, and there is nothing other than an individual's moral conscience to induce a person under no legal duty to undertake a rescue attempt. The law places a premium on human life, and one who voluntarily

(*continued*)

(*continued*)

attempts to save a life of another should not be barred from complete recovery. Only if a person is rash or reckless in the rescue attempt should recovery be limited; accordingly we hold that the rescue doctrine survives the adoption of the comparative-negligence statute and that principles of comparative negligence apply only if a defendant establishes that the rescuer's actions were rash or reckless. . . . In adopting this reasoning we recognize that the oft quoted words of Justice Cardozo apply now as they did in 1921:

"Danger invites rescue. The cry of distress is the summons to relief. The law does not ignore these reactions of the mind in tracing conduct to its consequences. It recognizes them as normal. * * * The risk of rescue, if only it be not wanton, is born of the occasion. The emergency begets the man. The wrongdoer may not have foreseen the coming of a deliverer. He is accountable as if he had." *Wagner v. International Railway, Co.,* 232 N.Y. 176, 180, 133 N.E. 437, 437–38 (1921).

In the instant case, defendant was not entitled to a jury instruction on comparative negligence unless plaintiff's rescue was rash or reckless. Because defendant did not assert that plaintiff acted recklessly, the trial justice did not err in denying defendant's requested jury instruction on comparative negligence. . . .

Accordingly, the defendant's appeal is denied and dismissed. The judgment in favor of the plaintiff is affirmed, and the case is remanded to Superior Court.

Case Name: Ouellette v. Carde

Court: Rhode Island Supreme Court

Summary of Main Points: The Rescue Doctrine applies to cases under comparative negligence, and use of comparative negligence as a defense will be allowed only when a rescuer has acted in a "rash or reckless" way.

Last Clear Chance

The Last Clear Chance Doctrine comes into play when two parties have both been negligent, but the second party has the opportunity to avoid the harm and fails to do so. The typical Last Clear Chance case occurs when the plaintiff's negligence creates a situation rendering the plaintiff helpless. The defendant discovers the peril that the plaintiff is in, and could avoid injury to plaintiff, but negligently fails to do so. The purpose of the Last Clear Chance Rule was to allow a plaintiff to recover from a defendant who had a last clear chance to prevent an injury simply because the plaintiff had been negligent in creating the situation.

After comparative negligence laws were adopted, some states abolished the Last Clear Chance Rule, choosing instead to have the jury apportion fault. Some states require that for the plaintiff to be able to recover, the defendant's conduct must involve a higher degree of fault than mere negligence, such as gross negligence or recklessness. Some states will also approach this issue from the perspective of proximate cause, finding the defendant liable only if the defendant's negligence was the proximate cause of the plaintiff's injuries.

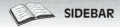

SIDEBAR

Jones is driving his pickup truck on a highway late at night. He dozes off and his truck strikes a barrier. Jones is shaken up but otherwise uninjured. Soon a fire truck, ambulance, and police cruiser arrive on the scene. Jones is seated in the police cruiser while waiting for a tow truck. Shortly thereafter, a tractor trailer driven by Smith at an excessive rate of speed slams into the police cruiser and fire truck, injuring Jones. While Jones's negligence may have created the situation leading to his own injuries, Smith had the last clear chance to avoid injury to Jones. The case could also be analyzed from the perspective of proximate cause. Smith's negligence could be viewed as the proximate cause of Jones's injuries, since Jones's injuries did not "naturally flow" from his own negligence, but rather resulted from the "intervening act" of Smith.

Fireman's Rule

The Fireman's Rule is a defense to lawsuits filed by firefighters and police officers who are injured in the line of duty, against persons who (1) negligently caused the incident to which they responded; or (2) negligently created a dangerous condition at the scene that caused their injuries **(Figure 8-6)**.

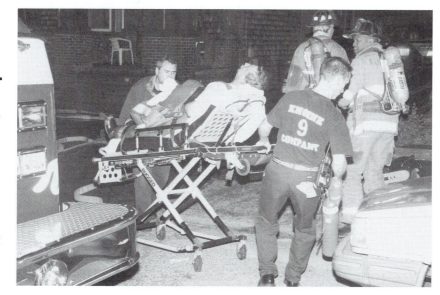

Figure 8-6 *The fireman's rule prohibits an injured firefighter from suing the person who negligently causes a fire or emergency, or negligently maintains the property on which a fire or emergency occurs. (Photo by Rick Blais.)*

The Fireman's Rule has been considered to be an exception to the Rescue Doctrine, as well as being founded on both the assumption of risk and Last Clear Chance Doctrines. The rationales for limiting the ability of firefighters and police officers to sue based on the Fireman's Rule are:

- Concerns about placing too heavy a burden on property owners to maintain their premises in a prepared and safe condition at all times in the event that firefighters or police officers may respond
- Concerns that citizens may be discouraged from calling for help if they think they may be subject to liability, resulting in delayed responses and an increase in civilian casualties and property damage
- Firefighters and police officers know the risks and voluntarily assume them.
- Firefighters and police officers are adequately compensated through sick leave and disability programs paid for by taxpayers.
- The cost of injuries to firefighters and police officers should be spread among all the taxpayers in a community, not just those who have a fire or emergency.

Hack v. Gillespie
74 Ohio St.3d 362, 658 N.E.2d 1046 (1996)
Ohio Supreme Court

Appellant Stephen Hack was a fire fighter [sic] for the city of Lakewood, Ohio. On March 1, 1989, Hack responded to a fire at 1589 Larchmont Avenue. He gained access to the residence by entering a porch located on the second floor. While on the porch, Hack leaned over a decorative railing to retrieve some equipment. The railing, however, gave way, causing Hack to fall to the ground. As a result, Hack suffered a broken hip and elbow. . . .

The issue presented for our consideration concerns the liability of an owner of private property to a fire fighter who enters the premises and, while performing his official duties, suffers harm as a result of the condition of the premises. Specifically, we are asked to reexamine the rule in Ohio regarding a landowner's liability to police officers and fire fighters . . . generally referred to as Ohio's "Fireman's Rule."

The term "Fireman's Rule," which is used to include fire fighters and police officers, refers to a common-law doctrine . . . [which] classified fire fighters as licensees entering upon property for their own purposes and with the consent of the property owner or occupant. . . . Thus, the landowner or occupant owed no

(continued)

(*continued*)

duty to the fire fighter unless the firefighter's injury was caused by the owner's or occupier's willful or wanton misconduct. . . .

The rule was originally created to apply to fire fighters, but it has evolved and has been extended to include police officers. . . . It appears that a vast majority of our sister states have adopted or have retained some form of the Fireman's Rule. The rule, however, is by no means a uniform rule. Rather, those jurisdictions which have adopted or retained some vestige of the rule have done so by applying various legal theories and principles, resulting in several different versions. . . .

In the case at bar, appellants ask this court to . . . hold that a landowner owes a duty of reasonable care, in all instances, to fire fighters who enter upon the private premises in the exercise of their official duties. In this regard, appellants suggest that fire fighters [sic] who enter upon private premises should be classified as invitees and, accordingly, may recover for personal injuries suffered as a result of the possessor's ordinary acts of negligence. Alternatively, appellants contend that . . . a fire fighter can recover against a negligent landowner where, as here, the dangerous condition that caused the injury was in no way associated with the emergency to which the fire fighter responded. . . .

We concede that this court has, previously, determined that the duty of care owed by a landowner to a fire fighter (or police officer) stems from common-law entrant classifications, i.e., licensees or invitees. However, Ohio's Fireman's Rule is more properly grounded on policy considerations, not artificially imputed common-law entrant classifications. Indeed, persons such as fire fighters or police officers who enter land pursuant to a legal privilege or in the performance of their public duty do not fit neatly, if ever, into common-law entrant classifications. . . .

First, fire fighters [sic] and police officers can enter the premises of a private property owner or occupant under authority of law. Hence, fire fighters [sic] and police officers can be distinguished from ordinary invitees. . . . Second, because a landowner or occupier can rarely anticipate the presence of safety officers on the premises, the burdens placed on possessors of property would be too great if fire fighters and police officers were classified, in all instances, as invitees to whom a duty of reasonable care was owed. . . . Third, the rule has been deemed to be justified based on a cost-spreading rationale through Ohio's workers' compensation laws. In this regard, this court has recognized that all citizens share the benefits provided by fire fighters and police officers and, therefore, citizens should also share the burden if a fire fighter or police officer is injured on the job. . . .

We believe that many of the reasons supporting the rule . . . are well founded and are still sound and valid in our society today. Fire fighters and police officers assume risks by the very nature of their chosen profession. The risks encountered are not always directly connected with arresting criminals or fighting fires. Members of our safety forces are trained to expect the unexpected. Such is the nature of their business.

(*continued*)

(*continued*)

The risks they encounter are of various types. A fire fighter, fighting a fire, might be attacked by the family dog. He or she might slip on an object in the middle of a yard or on a living room floor. An unguarded excavation may lie on the other side of a closed doorway, or the fire fighter might be required to climb upon a roof not realizing that it has been weakened by a fire in the attic. Fortunately, Ohio has statutory compensation schemes which can temper the admittedly harsh reality if one of our public servants is injured in the line of duty.

Further, appellants argue that fire fighters [sic] and police officers are treated unfairly in Ohio because they are not entitled to the same protection as other individuals/employees who enter a landowner's or occupier's premises. However, unlike water, electric and gas meter readers, postal workers and others, fire fighters can enter a homeowner's or occupier's premises at any time, day or night. They respond to emergencies, and emergencies are virtually impossible to predict. They enter locations where entry could not be reasonably anticipated, and fire fighters often enter premises when the owner or occupier is not present. We believe that under these circumstances abrogation of Ohio's Fireman's Rule, as suggested by appellants, would impose too great a burden on Ohio landowners and occupiers and their insurers. . . .

We are aware that a few jurisdictions have abolished or modified their original rule. . . . We are also cognizant that the Fireman's Rule has been the subject of considerable commentary. . . . However, we believe that the principles set forth . . . strike an appropriate balance between the interests of a possessor of land and the right of a fire fighter or police officer to avoid exposure to unlimited or unreasonable risks of injury.

Accordingly, we hold that an owner or occupier of private property can be liable to a fire fighter or police officer who enters premises and is injured in the performance of his or her official job duties if (1) the injury was caused by the owner's or occupier's willful or wanton misconduct or affirmative act of negligence; (2) the injury was the result of a hidden trap on the premises; (3) the injury was caused by the owner's or occupier's violation of a duty imposed by statute or ordinance enacted for the benefit of fire fighters or police officers; or (4) the owner or occupier was aware of the fire fighter's or police officer's presence on the premises, but failed to warn them of any known, hidden danger thereon.

Based on the foregoing, we hold that summary judgment was properly granted in favor of Gillespie. We affirm the judgment of the court of appeals.

Judgment affirmed.

Case Name: Hack v. Gillespie

Court: Ohio Supreme Court

Summary of Main Points: The Fireman's Rule prohibits a firefighter who is injured in the line of duty from suing the owner or occupier of land for negligence. However, certain exceptions are recognized.

Some states that recognize the Fireman's Rule consider it to be an extension of property and trespass laws, and limit the application of the Fireman's Rule to circumstances in which firefighters or police officers come upon a defendant's property. Such states would not apply the Fireman's Rule to injuries to a firefighter that occur on public streets or locations anywhere other than the defendant's property. Other states focus more on the assumption of risk aspect, and apply the Fireman's Rule to all emergency scenes, including motor vehicle accidents.

As the court in the *Hack* case indicated, there are a number of exceptions to the Fireman's Rule that permit injured firefighters to sue those responsible for causing the fire or creating the dangerous conditions at the scene that caused their injury. These exceptions include:

- The person's conduct was willful, wanton, or intentional.
- The injury was the result of a hidden trap.
- The injury was the result of a violation of law enacted to protect firefighters or police officers.
- The owner/occupier was aware of a hidden danger on the property, and failed to warn the firefighters or the police officers of its presence.
- The rescuer was off duty, stopped voluntarily at an accident scene to offer assistance, and was injured by another driver.
- Manufacturers are subject to strict liability for any defective products they manufacture.

Some states, such as Oregon, have completely abolished the Fireman's Rule by case law. See *Christensen v. Murphy*, 296 Or. 610 (Or., 1982). Other states, such as New Jersey, have abolished it by statute. See New Jersey Public Statutes 2A:62A-21 in Appendix B—Chapter 8. However, most states continue to recognize and apply the Fireman's Rule to firefighters and police officers.

strict liability
the legal doctrine that makes an actor responsible for any and all harm that may occur without regard to fault; in a civil sense, strict liability refers to liability for damages in tort without regard to negligence, gross negligence, recklessness, or intentional conduct

STRICT LIABILITY

In addition to intentional torts and negligence, there is a third type of civil liability known as strict liability. Throughout most of the study of law, we have seen a principle that equates liability with fault. Implicit in our justice system is the belief that before someone should be liable, whether criminally or civilly, there should have to be some degree of fault. It offends our notions of justice to think that someone, through no fault of their own, could be held liable.

However, some activities are of such a nature that society has demanded that responsibility not be limited to situations where fault must be proven. These situations result in what we call **strict liability**. If harm results, the actor will be responsible for any and all harm without regard to fault. Whether the actor exercised due care, or even exceeded due care, he or she will be liable for any damage that occurs.

The categories for which strict liability is applied are few, namely:

- *Keepers of dangerous animals*. The owners of dangerous animals are strictly liable for any damage done by the animal. In the case of normally dangerous animals, such as lions, tigers, bears, and wolves, the owners are strictly liable for any and all damages that their animals cause. In the case of animals that are normally harmless, such as dogs, cats, cattle, or sheep, if the owner knows of their propensity to be dangerous, the owner is strictly liable for any damages. This latter rule is often referred to as the "one free bite" rule. If the owner of a dog is unaware of its propensity to bite, normal rules of negligence will apply. Once the owner is aware that the dog has bitten someone, strict liability attaches. The rule has been applied to other animals, such as cattle, if they have a known propensity to escape and go upon the roadways.

- *Workers' compensation*. While not a common-law doctrine, statutory workers' compensation systems have created, in effect, strict liability systems whereby employers are automatically liable for workplace injuries to employees, even when the injuries are the result of the employees' negligence. Workers' compensation systems will be discussed in detail in Chapter 10.

- *Strict product liability*. The sellers of goods have been held strictly liable for defects in their goods that injure users. This form of strict liability includes foods as well as other items that cause harm. Liability attaches even when manufacturer and seller exercised reasonable care in the manufacture and sale of the product. The case of *Rucker v. Norfolk & Western Ry. Co.,* 64 Ill. App.3d 770, 381 N.E. 2d 715 (Ill.App. Dist., 1978) arose out of the massive explosion in Crescent City, Illinois of a railroad tank car full of liquefied petroleum gas (propane) in June, 1970 **(Figure 8-7)**. The court specifically rejected the defense by the manufacturer of the tank car that it was "state of the art," finding such a defense to be irrelevant.

- *Dram shop liability*. The seller of liquor to an intoxicated patron can be held strictly liable for damages to a third party under laws passed by many states, commonly referred to as "dram shop" laws. Liability is strict, and no showing of negligence on the part of the seller is required.

- *Abnormally dangerous activities*. Those who engage in an abnormally dangerous activity, also termed an ultrahazardous activity, will generally be held strictly liable for any damage that results. The challenge is in defining abnormally dangerous, or ultrahazardous. The most common example of an abnormally dangerous activity is blasting, which is almost universally recognized as a strict liability activity. The flying of airplanes at one time fit the definition of abnormally dangerous, making the aircraft owner liable for all ground damage resulting from a crash. While air travel fortunately no longer fits the definition of ultrahazardous, the

Figure 8-7 *Liquefied petroleum gas tank car explosion, Crescent City, Illinois, at 6:53 a.m. on June 21, 1970. (Photo courtesy of Scheiwe's Print Shop, Crescent City, Illinois.)*

concept has been incorporated into Federal legislation, which imposes strict liability on aircraft owners and pilots.

The rationales for imposing liability on abnormally hazardous activities include encouraging those who engage in such activities to use all possible efforts (not just reasonable efforts) to prevent harm, and an economic approach that supposes that those who create a hazard and intend to profit from it should pass the cost of any harm that occurs on to their customers as a cost of doing business.

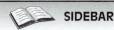 **SIDEBAR**

Fire and Strict Liability

According to one noted authority on tort law, Prof. William L. Prosser, author of *Handbook of the Law of Torts, 4th edition* (St. Paul, MN: West Publishing, 1971), at common law landowners were held to strict liability for fires that started on their land and damaged the property of another. English case law recognized some exceptions, such as for acts of God or the intervening act of a stranger. This policy was changed by statute in 1707 to prevent liability against a landowner for accidental fires. Thereafter, English courts held landowners liable only for fires that were caused by negligence or were intentionally set.

The American courts have taken the approach that, absent negligence, there is no liability on the part of a landowner for fire that damages a neighbor's property. However, there are statutes in some states that recognize that some uses of fire are very dangerous, and apply strict liability. Some examples include strict liability for people who start outdoor fires during a specified dry season, and for railroads for fires started by trains along the tracks.

The fire problem in the United States is significantly higher than in any other industrialized country in the world. Given our ability to put men on the moon and conquer a host of previously fatal illnesses, it stands as a paradox of American society that fire remains such a problem. In many countries, people who have accidental fires are considered to be at fault and liable for any damage that occurs. In some countries, those who have an accidental fire can be charged criminally. One has to wonder what the effect would have been on the fire problem in the United States if strict liability for fire had remained the common-law rule. Would owners and occupiers of land have been encouraged to take extra precautions? Would our overall approach to the fire problem have changed?

RESPONDEAT SUPERIOR

respondeat superior
a legal doctrine that holds an employer liable for the torts of its employees, provided the torts are committed within the scope of the workers' employment

Respondeat superior is a legal doctrine that holds an employer liable for the torts of its employees, provided the torts are committed within the scope of the workers' employment. Respondeat superior is a form of strict liability for the employer, because it does not matter that the employer was not negligent. The negligence of the employee is imputed to the employer. This type of liability is often referred to as vicarious liability.

The biggest limitation on the application of respondeat superior is that the tort must be committed within the scope of the employee's employment. The *Thorn* case below thoroughly discusses this issue.

Thorn v. City of Glendale
28 Cal.App.4th 1379, 35 Cal.Rptr.2d 1 (1994)
Court of Appeal, Second District, Division 2, California

GATES, Acting P. J. James Thorn and his business, Glendale Spa City, Inc. (Spa City), appeal from a judgment upon demurrer entered in favor of the City of Glendale (Glendale) in their action against Glendale and its employee and fire marshal,

(*continued*)

(*continued*)

John Orr, for fire damage to Spa City. The complaint alleges that Orr set a fire at Spa City while acting in his official capacity and that Glendale is liable for the ensuing loss both under respondeat superior principles and for negligently supervising Orr. . . .

The complaint alleges the following. On February 22, 1991, Orr entered Thorn's premises under color of authority to conduct a fire inspection. He then committed arson by setting incendiary devices which destroyed the premises and the business conducted thereon. Since Orr was acting within the scope of his employment, Glendale is liable for the resulting damage. Moreover, Glendale knew or should have known that the fire marshal was an arsonist and negligently failed to supervise him.

Appellants urge that Glendale is liable under the doctrine of respondeat superior because Orr's alleged acts were committed within the scope of employment. . . .

Historically, the scope of employment doctrine has been limited to acts which are directly or indirectly in furtherance of the employer's purpose, precluding vicarious liability for criminal acts not related to the employer's enterprise. . . .

[E]mployers have been held liable for the wrongful and unauthorized acts of their employees where they were committed in the course of a series of acts of the agent which were authorized by the principal. . . .

More recently, our Supreme Court stated, "'A risk arises out of the employment when' in the context of the particular enterprise an employee's conduct is not so unusual or startling that it would seem unfair to include the loss resulting from it among other costs of the employer's business. [Citations.] In other words, where the question is one of vicarious liability, the inquiry should be whether the risk was one 'that may fairly be regarded as typical of or broadly incidental' to the enterprise undertaken by the employer." . . .

Policy reasons suggested for imposing vicarious liability include that it will tend to (1) provide a spur towards accident prevention; (2) provide greater assurance of compensation for accident victims; and (3) assure that accident losses will be broadly and equitably distributed among the beneficiaries of the enterprise that entail them.

None of the foregoing tests favor liability in the present case. A fire marshal's entering a building and setting an incendiary device for the purpose of burning it down is so startling and unusual an occurrence as to be outside those risks which should fairly be imposed upon the public employer. The alleged act did not arise from the pursuit of the employer's purpose but was rather the result, we must assume, of a personal compulsion.

While Orr's ability to request access to private areas of a building arose from his employment, that ability is not unique. Similar permissive access is available to security guards, repairpersons, and utility workers. Glendale would have no greater reason to guard against and deter the alleged acts than would employers of other workers whose duties entail their entering private premises. Moreover, property damage resulting from fire, as distinguished from personal injury and

(*continued*)

(continued)
trauma, is commonly insurable by the business enterprise victim. In truth, property owners would appear far better able to insure against the loss than would the public entity, particularly where the conduct in question is felonious in nature. . . .

The judgment is affirmed.

Case Name: Thorn v. Glendale

Court: Court of Appeal, Second District, Division 2, California.

Summary of Main Points: For purposes of respondeat superior, the scope of employment does not include acts that are well outside those that the reasonable employer would expect to be "typical of or broadly incidental" to their business.

Respondeat superior must be distinguished from cases in which an employer is sued directly for the employer's own negligence arising out of something that an employee did. Employers may be sued for:

- negligent hiring of an employee
- negligent retention of an employee
- negligent supervision of an employee
- negligent training of an employee

In each of the above situations, the employer is sued for its own negligence, not the negligence of the employee.

JOINT LIABILITY

When the negligence of two or more persons combines to cause damage to a victim, each of the negligent parties can be held liable for the entire amount of damages.

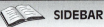 **SIDEBAR**

Able and Baker decide to drag race their cars on a public street. During the course of the race, Able's car strikes a pedestrian, causing serious injuries. The pedestrian can sue both Able and Baker, or either one individually. If Baker has insurance and Able does not, the pedestrian can collect the full amount of damages from Baker.

tortfeasor

the legal term for someone who commits a tort

The legal term for someone who commits a tort is **tortfeasor**. In the case of two or more individuals who commit a tort, the term used is joint tortfeasors. The liability of joint tortfeasors is called "joint and several." In other words, as in the Able and Baker drag racing example, either tortfeasor can be held personally liable for the full amount. This rule does not permit the injured party to collect more than the actual amount of damages, but rather avoids the situation in which the injured party is limited to collecting a prorata share from each joint tortfeasor. Once the victim is compensated, a joint tortfeasor may seek an appropriate contribution from the other joint tortfeasors for the damages paid to the victim. In the drag racing example above, if Baker has to pay the entire amount of damages to the pedestrian, Baker can then seek payment from Able for one-half of the amount paid.

AUTHOR'S COMMENTARY

Concluding thought's on liability: SOPs and SOGs

Somewhere along the line in the fire service, someone created a controversy by recommending that fire departments rename all their standard operating procedures (SOPs) standard operating *guidelines* (SOGs). The theory was that the term "procedure" implied a mandatory prescription for action, much like an airline pilot's preflight checklist. As such, use of the term "procedure" left no room for discretion. If firefighters failed to follow one item on the SOP checklist, they would automatically be negligent. According to the SOG theory, the use of the term "guideline" implied that there was some flexibility involved.

The theory went on to conclude that if a fire department simply changed the name of all its standard operating procedures to standard operating guidelines, its liability would automatically be lessened.

But there is no legal happenstance that occurs when you change the name of SOPs to SOGs. The same perils that await someone who violates a "procedure" still await someone who violates a "guideline." The solution lies not in renaming, but in clearly defining. Fire departments need to define their SOPs or SOGs, so as to make it clear that they provide some degree of flexibility to officers and firefighters. Officers and firefighters should be explicitly authorized by department rules and regulations to deviate from SOPs/SOGs when and where, in their professional experience and training, such deviation is warranted.

On the other hand, SOPs or SOGs that are so absolute that they cannot be violated under any circumstance should clearly be identified. If, for example, personnel are absolutely forbidden from attempting a winter water rescue without an exposure suit, the SOP/SOG should clearly state that policy. If personnel are absolutely forbidden from entering a structure fire without a self-contained breathing apparatus, the SOP/SOG should clearly state this. At the same time, where there are tactical options and choices, those options should be identified as well.

Most important, SOPs or SOGs should be defined as tools that explicitly anticipate and require experienced personnel to deviate from them when and where appropriate.

SUMMARY

Negligence is the biggest area of civil liability for fire departments, firefighters, and emergency medical personnel. The law of negligence is evolving as new issues emerge. However, at the heart of negligence remains a consideration of what the reasonably prudent person would have done under the circumstances. An analysis of the standard of care for fire service or emergency medical personnel cannot escape consideration of NFPA standards and OSHA regulations. The torts of gross negligence, recklessness, and strict liability also have important application to fire and emergency responders.

REVIEW QUESTIONS

1. Define negligence.
2. When can the failure to act satisfy the act requirement for negligence?
3. What is proximate cause?
4. In negligence, what constitutes damages?
5. What are the four categories of evidence used to establish the standard of care for a professional?
6. What is the difference between contributory negligence and comparative negligence?
7. Explain the difference between negligence, gross negligence, and recklessness.
8. Explain the Rescue Doctrine and the Fireman's Rule.
9. List four exceptions to the Fireman's Rule.
10. Define the term "joint and several liability."

DISCUSSION QUESTIONS

1. Consider the *Thorn v. Glendale* case. Was the court correct in ruling that the fire inspector was acting outside the scope of his employment? What is the difference between a fire inspector who negligently knocks over a can of gasoline during an inspection—which leads to a fire—and a fire inspector who sets a fire while on an inspection? Why should the fire department be liable in one instance and not the other?

2. Is there a difference between the reasonably prudent person standard for a person driving a car and the standard of care expected of a firefighter driving a fire truck? Should the standard be the same, or should a special standard of care apply for the firefighter?

3. What is the rationale for applying strict liability when someone engages in an abnormally hazardous activity?

Chapter
9

IMMUNITY FROM LIABILITY

Learning Objectives

Upon completion of this chapter, you should be able to:

- Describe the history and current status of sovereign immunity as it applies to the American fire service.
- Explain the purpose and role of tort claims acts.
- Explain statutory immunity and how it differs from sovereign immunity.
- Explain the difference between discretionary acts and functionary acts, and governmental function and proprietary function, and how the difference impacts immunity protection.
- Describe whom the Volunteer Protection Act of 1997 applies to, and the immunity protection afforded.
- Explain the public duty doctrine, the special duty exception, and the insurance waiver doctrine.
- Explain the limitation on immunity protection for acts that constitute gross negligence, recklessness, or intentional acts.

INTRODUCTION

Mr. and Mrs. Smith entered the lawyer's office, and took a seat at the end of the conference table. It had been years since they had last been there for their wills, but the office looked familiar.

"Our house burned down and the fire department did an awful job," said Mrs. Smith. "Took them forever to get there."

"I read about the fire in the newspaper," replied the attorney. "I am very sorry to hear about your tragedy. Thank god you are both okay. Did you bring the documents that I requested?

"Certainly. Here are the fire reports and here is our insurance policy. As you can see, our insurance coverage has clearly lapsed," said Mr. Smith.

"Hummm," said the attorney, studying the documents carefully. "I see on the fire report that the fire department got on scene in four minutes."

"Really," replied Mrs. Smith, rather surprised. "It seemed much longer than that. But anyway, they broke all the windows, destroyed our ceilings, and put a huge hole in the roof."

"Well," replied the attorney "Cases like this are difficult and expensive. First we'll need an expert witness to help us determine if the fire department was negligent. We have to be sure that our expert is better than their expert for us to win. In fact, we may need several experts and they can be expensive."

"Oh," sighed Mrs. Smith.

The lawyer continued, "And then there's the problem of immunity."

"What's that?" asked Mrs. Smith.

sovereign immunity
a common-law rule that the government is immune from liability for actions in tort; the immunity arose from the old English principle that "The king can do no wrong"

At common law, the government was completely immune from liability for actions in tort. The immunity arose from the old English principle that "The king can do no wrong." This immunity was referred to as **sovereign immunity**.

The principle of sovereign immunity was adopted in the United States through our adoption of the English common law, despite the fact that we had no king. It served to prevent anyone from successfully suing any level of the government—Federal, state, or local—in tort. Over time, courts and state legislatures began to recognize the inequities of sovereign immunity to those who were injured by governmental actions. Some states abolished sovereign immunity by statute, while others did so through court decisions. As a result, sovereign immunity has been abolished or severely limited in all jurisdictions.

TORT CLAIMS ACTS

As sovereign immunity has been abolished, legislatures have adopted statutes called "tort claims acts." These laws were created to give private parties a mechanism for recourse in the event they were injured by some governmental activity **(Figure 9-1)**.

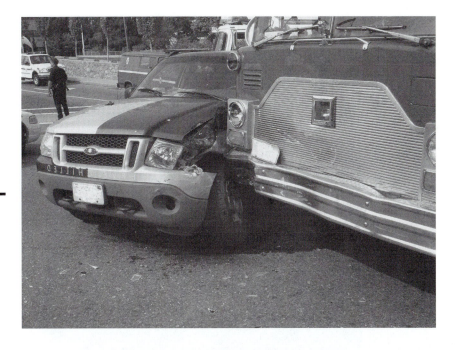

Figure 9-1 *Tort claims acts give private parties a mechanism for recourse in the event they are injured by some governmental activity.*

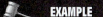 **EXAMPLE**

28 USC Sec. 2674 Liability of United States

§2674. The United States shall be liable, respecting the provisions of this title relating to tort claims, in the same manner and to the same extent as a private individual under like circumstances, but shall not be liable for interest prior to judgment or for punitive damages. . . .

§ 2680. Exceptions. The provisions of this chapter and section 1346 (b) of this title shall not apply to—

(a) Any claim based upon an act or omission of an employee of the Government, exercising due care, in the execution of a statute or regulation, whether or not such statute or regulation be valid, or based upon the exercise or performance or the failure to exercise or perform a discretionary function or duty on the part of a federal agency or an employee of the Government, whether or not the discretion involved be abused.

Tort claims acts have served three separate functions. First, in states that had not previously abolished sovereign immunity, tort claims acts formally established governmental liability. Second, tort claims acts established a procedure by which an injured party could file a damage claim against a governmental entity. Third, tort claims acts reserved some limited exceptions, in which immunity protection would remain. Under the typical tort claims act, a person who is injured by a tort committed by governmental action must initially file a claim with the government. If the claim is denied, the person is then entitled to file suit against the government and the persons responsible as if they were private entities. In addition, many jurisdictions place a monetary cap on the amount of damages that a municipality or the state could be liable for under the tort claims acts.

While the principle of abolishing sovereign immunity and the adoption of tort claims acts seems straightforward, its application has created considerable chaos for those trying to understand governmental liability. Many authorities have expressed frustration at the present state of the law

Figure 9-2 *Fires in New England's old mills make for spectacular blazes, and occasionally give rise to lawsuits. (Photo by Rick Blais.)*

regarding governmental immunity. There are 50 state legislatures, plus Congress, who have written separate tort claims acts. There are 50 state supreme courts plus the Federal court system that have interpreted these tort claims acts. On top of that, we have what some legal commentators have concluded are judges struggling to reach a just result in a case, and in doing so stretching the language of tort claims acts to find ways to protect firefighters, fire departments, and municipalities from liability **(Figure 9-2)**. This problem may best be understood by examining the *Harry Stoller* case in detail.

Harry Stoller & Co. v. Lowell
412 Mass. 139 (1992)
Massachusetts Supreme Judicial Court

WILKINS, J.

On April 23, 1978, five brick buildings in Lowell and their contents were destroyed by fire. The fire started on the sixth floor of one of the buildings. Three of the buildings, including the one in which the fire started, had sprinkler systems. In this action, the owner of the premises sought recovery against the city of Lowell . . . based on the claimed negligence of the city's firefighters in combating the fire.

The jury in a 1990 trial returned a verdict of $850,000 for the plaintiff. Because of the statutory limitation on the amount for which a municipality may be liable . . . a judgment for $100,000 was entered against the city. The city, which had moved unsuccessfully for a directed verdict, sought and obtained the entry of a judgment in its favor notwithstanding the verdict. The judge concluded that the city was exempt from liability under the discretionary function exception set forth in . . . the Massachusetts Tort Claims Act. . . . We reverse the judgment for the city.

We are concerned solely with the question whether the city is entitled to immunity from liability by application of the discretionary function exception to governmental tort liability. The city does not argue that it owed no duty to the plaintiff or that the evidence did not warrant a finding that the city negligently violated that duty. We must, however, discuss the conduct on which liability was based, because it is that conduct that must have involved a discretionary function. . . .

The theory of the plaintiff's case was that the city negligently failed to use the building's sprinkler systems to fight the fire. The jury would have been warranted in finding the following facts. The sprinkler systems had been tested two days before the fire, and they worked satisfactorily. Water pressure adequate to allow the sprinkler system to work properly on the sixth floor of the building in which

(*continued*)

(*continued*)

the fire started was not maintained during the fire. During the early stages of the fire, water was coming out of the sixth-floor sprinkler system. A pumper initially attached to the sprinkler system was disconnected shortly thereafter. The fire department hoses and the sprinkler system used the same water source, and use of the hoses reduced the pressure in the sprinkler systems. Accepted practice in fighting a fire high in a building of the type involved here required the use of the sprinkler system in the circumstances. It would be rare if a sprinkler system properly supplied with water pressure did not put out such a fire, or at least contain it until it could be put out by manual means.

The first step in deciding whether a plaintiff's claim is foreclosed by the discretionary function exception . . . is to determine whether the governmental actor had any discretion at all as to what course of conduct to follow. Quite obviously, if the governmental actor had no discretion . . . a discretionary function exception to governmental liability has no role to play in deciding the case.

The second and far more difficult step is to determine whether the discretion that the actor had is that kind of discretion for which . . . provides immunity from liability. Almost all conduct involves some discretion, if only concerning minor details. If allegedly tortious conduct were to be immunized from causing liability simply because there was some element of discretion in that conduct, the discretionary function exception would go a long way toward restoring the governmental immunity that [the Tort Claims Act] was designed to eliminate. As we shall show, however, the discretionary function exception, both under our Act and under the Federal Tort Claims Act, is far narrower, providing immunity only for discretionary conduct that involves policy making or planning. Because of the limitation of the exception to conduct that is policy making or planning, the words "discretionary function" are somewhat misleading as a name of the concept. . . .

In the *Whitney* opinion, we said that the dividing line should be between those functions that "rest on the exercise of judgment and discretion and represent planning and policymaking [for which there would be governmental immunity] and those functions which involve the implementation and execution of such governmental policy or planning [for which there would be no governmental immunity]." . . . We added that, when the conduct that caused the injury has a "high degree of discretion and judgment involved in weighing alternatives and making choices with respect to public policy and planning, governmental entities should remain immune from liability." . . . But, when that conduct "involves rather the carrying out of previously established policies or plans, such acts should be governed by the established standards of tort liability applicable to private individuals or entities." . . . We granted that the general rule, as stated, was not "a model of precision and predictability" because "the performance of all functions involves the exercise of discretion and judgment to some degree." . . .

In an anticipatory attempt to assist the process of differentiation between functions that are discretionary and those that are not, the court identified certain considerations as relevant. If the injury-producing conduct was an integral part of

(*continued*)

(*continued*)
governmental policymaking or planning, if the imposition of liability might jeopardize the quality of the governmental process, or if the case could not be decided without usurping the power and responsibility of either the legislative or executive branch of government, governmental immunity would probably attach. . . . The general rule, however, should be one of governmental tort liability. . . .

The Federal Tort Claims Act and cases interpreting it underlie the discussion of the discretionary function exception. . . . The path of the opinions of the United States Supreme Court concerning the discretionary function exception has been neither straight nor clear. . . . The important lesson from the opinions of the Supreme Court is that governmental immunity does not result automatically just because the governmental actor had discretion. Discretionary actions and decisions that warrant immunity must be based on considerations of public policy. . . . Even decisions made at the operational level, as opposed to those made at the policy or planning level, would involve conduct immunized by the discretionary function exception if the conduct were the result of policy determinations. . . .

Cases involving alleged governmental negligence in fighting fires provide us with some help in deciding this case under the discretionary function test that we have established. . . . In *Defrees v. United States,* 738 F. Supp. 380 (D. Or. 1990), the judge concluded that discretionary function immunity applied to bar liability where forest service employees had had to make social and economic policy decisions in assigning firefighting personnel and equipment and in considering the property interests and endangered species to be protected. . . . One court has said, on the other hand, that shutting down a sprinkler system during a fire is not necessarily a discretionary or policymaking decision. *Industrial Risk Insurers v. New Orleans Pub. Servs., Inc.,* 735 F. Supp. 200, 204 (E.D. La. 1990) (Louisiana law).

There are aspects of firefighting that can have an obvious planning or policy basis. The number and location of fire stations, the amount of equipment to purchase, the size of the fire department, the number and location of hydrants, and the quantity of the water supply involve policy considerations, especially the allocation of financial resources. In certain situations, firefighting involves determinations of what property to attempt to save because the resources available to combat a conflagration are or seem to be insufficient to save all threatened property. In such cases, policy determinations might be involved, and application of the discretionary function exception would be required.

The case before us is different. The negligent conduct that caused the fire to engulf all the plaintiff's buildings was not founded on planning or policy considerations. The question whether to put higher water pressure in the sprinkler systems involved no policy choice or planning decision. There was a dispute on the evidence whether it was negligent to fail to fight the fire through the buildings' sprinkler systems. The firefighters may have thought that they had a discretionary choice whether to pour water on the buildings through hoses or to put water inside the buildings through their sprinkler systems. They certainly had discretion in the sense that no statute, regulation, or established municipal practice required the firefighters to use the sprinklers (or, for that matter, to use hoses exclusively).

(*continued*)

(*continued*)

But whatever discretion they had was not based on a policy or planning judgment. The jury decided that, in exercising their discretion not to use the buildings' sprinkler systems, the Lowell firefighters, were negligent because they failed to conform to generally accepted firefighting practices. When the firefighters exercised that discretion, policy and planning considerations were not involved. Therefore, the discretionary function exception does not shield the city from liability.

The judgment notwithstanding the verdict entered in favor of Lowell is vacated. Judgment shall be entered in favor of the plaintiff in the amount of $100,000. So ordered.

Case Name: Harry Stoller & Co. v. Lowell

Court: Massachusetts Supreme Judicial Court

Summary of Main Points: Sovereign immunity has been abolished in Massachusetts, and replaced with a tort claims act. The tort claims act recognizes a narrow exception, in which immunity still exists for acts that involve a "discretionary function." A broad interpretation of "discretionary function" would essentially reinstitute the sovereign immunity that the tort claims act sought to eliminate. It would be inappropriate for courts to apply such a broad definition of discretionary function to acts that do not involve making public policy, but rather involve carrying out public policy. It is up to the legislature, not the courts, to create an additional exception to protect fire departments and firefighters from such liability.

discretionary act

an act for which there is no fixed requirement for a course of action, such as would eliminate the exercise of discretion; in relation to tort claims acts, discretionary acts are often considered to be decisions of a policymaking nature made by elected and appointed officials that go to the heart of our democratic form of government

functionary act

an act of carrying out established policy, as opposed to exercising discretion

DISCRETIONARY VERSUS FUNCTIONARY ACTS

As the Massachusetts Supreme Judicial Court recognized in *Stoller*, many jurisdictions draw an important distinction between discretionary acts on the one hand, and functionary acts on the other hand. In these jurisdictions, acts deemed to be discretionary are entitled to immunity, whereas acts that are functionary are not.

Courts have struggled with how to differentiate between discretionary acts and functionary acts. For purposes of our discussions we will define **discretionary acts** as those of a policymaking nature that go to the heart of our democratic form of government: Should a community have a paid fire department or should it rely on volunteers? Where should the fire stations be located? Should taxpayer funds be spent on a new aerial ladder or a new pumper? These types of questions are fundamental matters of public policymaking in which the exercise of discretion by elected and appointed officials should rightfully be protected from being second-guessed haphazardly through the court system. **Functionary acts**, or ministerial acts, are acts that carry out established policy, and are not entitled to immunity protection. Between the two extremes are a variety of actions and decisions than can go either way, depending upon one's perspective.

 SIDEBAR

The Massachusetts legislature responded to the *Stoller* decision with an amendment to the Massachusetts Tort Claims Act, creating a special exception for fire departments. Under this law, fire departments would have immunity protection by virtue of being excluded from the Tort Claims Act.

Massachusetts General Law c.258, §10 Application of sections 1 to 8

The provisions of [the Massachusetts Tort Claims Act which give a person injured by a governmental action the right to sue the government], shall not apply to: . . .

(g) any claim based upon the failure to establish a fire department or a particular fire protection service, or if fire protection service is provided, for failure to prevent, suppress or contain a fire, or for any acts or omissions in the suppression or containment of a fire, but not including claims based upon the negligent operation of motor vehicles or as otherwise provided in clause (1) of subparagraph (j).

Note: Would the Massachusetts law provide immunity protection to a fire department that was handling an emergency that was not fire-related, such as a water rescue, or vehicle extrication?

WHAT IS DISCRETION?

In *Kenavan v. New York,* 523 NYS 2d 60 (1987), which we examined in Chapter 8, the jury returned a verdict in favor of a firefighter's widow and injured New York City firefighters, finding that the captain, the driver of the engine company, and the city were liable for injuries sustained when a motorist struck four firefighters operating at a car fire. The accusations against the captain and driver were that their fireground decision making on where to park the apparatus and whether to establish a fire line were negligent. On appeal, the court found that New York law provided immunity protection for the captain and driver, because the decision on where to park the apparatus, and the establishment of a fire line, involved the exercise of "judgment and discretion" **(Figure 9-3)**.

In *Norton v. Hall,* 2003 ME 118 (2003), a police officer driving her car to the scene of an emergency was determined to be exercising a "discretionary

Figure 9-3 *When a company officer decides where to stretch the initial attack line, is he exercising discretion or carrying out policy? (Photo by Ray Taylor.)*

function," and thus entitled to immunity. In another case from Maine, *Roberts v. State,* 731 A.2d 855, (ME, 1999), the act of a prison guard in closing a jail cell door on the fingers of an inmate was determined to be a "discretionary function."

Virtually every act by a governmental official involves some level of discretion. If carried to its logical conclusion, such an analysis would completely insulate the government from liability despite the adoption of tort claims acts. Some legal scholars believe that what the courts are saying is that it is unfair to scrutinize the split-second decision making of these emergency service professionals in court.

Consider the following quotation from *Wilcox v. City of Chicago* 107 Ill. 334, 339, (1883):

> "'If liable for neglect in this case the city must be held liable for every neglect of that [fire] department, and every employee connected with it when acting within the line of duty. It would subject the city to the opinions of witnesses and jurors whether sufficient dispatch was used in reaching the fire after the alarm was given; whether the employees had used the requisite skill for its extinguishment; whether a sufficient force had been provided to secure safety; whether the city had provided proper engines and other appliances to answer the demands of the hazards of fire in the city; and many other things might be named that would form the subject of legal controversy. To permit recoveries

to be had for all such and other acts would virtually render the city an insurer of every person's property within the limits of its jurisdiction. It would assuredly become too burdensome to be borne by the people of any large city, where loss by fire is annually counted by the hundreds of thousands, if not by the millions. . . . To allow recoveries for the negligence of the fire department would almost certainly subject property holders to as great, if not greater, burdens than are suffered from the damages from fire. Sound public policy would forbid it, if it was not prohibited by authority.'"

The mechanism that many courts use to reach the conclusion that emergency scene decision making warrants liability protection stretches the definition of discretionary function. Consider the following case, *Chandler Supply v. Boise,* and how the *Stoller* case would have been decided had the fire occurred in Idaho.

Chandler Supply Co., Inc. v. City of Boise
104 Idaho 480, 660 P.2d 1323 (1983)
Supreme Court of Idaho

. . . Chandler had brought suit against the appellant, City of Boise, under the Idaho Tort Claims Act . . . alleging negligence on the part of Boise's fire department. The facts appear . . . as follows.

On December 22, 1976, at about 5:34 p.m., the Boise Fire Department responded to the activation of a building fire alarm. On arrival at the reported location, the firefighters discovered a grass fire of unknown origin burning between a set of railroad tracks and a fence enclosing a warehouse. The firefighters extinguished the fire using burlap bags, shovels, and buckets of water. Before leaving the scene at approximately 5:55 p.m., members of the fire department checked the exterior of the warehouse and found no evidence that the building was involved in the fire. At approximately 6:12 p.m., in response to a telephone report of a building on fire, the fire department returned to the location of the earlier grass fire. The firefighters discovered that the previously checked warehouse was on fire. The fire was fought and extinguished, but resulted in substantial damage to property owned by Chandler. Both Chandler and his insurer filed timely claims with the city. A jury trial was held which resulted in a special verdict finding Boise 75% negligent and Chandler 25% negligent. Total damages amounted to $116,331.31. The city of Boise now appeals.

(continued)

(*continued*)

The primary question raised by appellant is whether the trial court erred in refusing to hold the city of Boise immune from liability under . . . the "discretionary function" exception to the Tort Claims Act. . . .

Although the Tort Claims Act which our legislature enacted is very similar to that passed by Congress and many other states, our review . . . of the "melange of decisions" interpreting such acts yields the conclusion that the decisions from other jurisdictions provide little guidance for defining the scope of governmental liability under our own Tort Claims Act. . . .

Interpretation of the discretionary function exception must begin with a review of the status of sovereign immunity in Idaho immediately preceding the enactment of the Tort Claims Act. In *Smith v. State*, . . . this Court held the following:

> "[W]e hereby hold that the doctrine of sovereign immunity is no longer a valid defense in actions based upon tortious acts of the state or any of its departments, political subdivisions, counties, or cities, where the governmental unit has acted in a proprietary as distinguished from a governmental capacity."

The Court in Smith also invited the legislature to exercise its own prerogative in the area of sovereign immunity and therefore delayed the effect of the *Smith* decision until "60 days subsequent to the adjournment of the First Regular Session of the Forty-First Idaho State Legislature" . . . .

The significance of the *Smith* decision in the present case is the fact that while this Court abolished its prior court made rule of sovereign immunity for proprietary functions of a governmental unit, that abolition did not extend to traditional governmental functions. The forty-first legislature, however, subsequently waived sovereign immunity for tortious acts with respect to not only proprietary but also to governmental functions of a governmental unit which have a parallel in the private sector. . . . Thus, the legislature was more expansive in allowing relief from governmental torts than was this Court in *Smith*. Of course, the legislature enacted certain exceptions to governmental liability, including the discretionary function exception. . . . The question is, what was the intent behind establishing those exceptions, particularly the discretionary function exception.

In abolishing its prior rule of sovereign immunity for tortious acts arising from proprietary functions of the state, this Court in *Smith* did not provide for any exceptions to such immunity. The unstated but obvious reason underlying such unrestricted tort liability arising from a governmental unit's proprietary functions was that such liability does not impinge upon the ability of the government to supply the services for which it has traditionally been responsible. . . . . It seems apparent that a basic purpose behind the legislature's creation of a list of exceptions to governmental liability was to limit the effect of its waiver of sovereign immunity with respect to governmental functions. Such is particularly true with reference to the discretionary function exception. In our view, the purpose behind the discretionary function exception is to preserve governmental immunity from tort liability for the consequences which arise from the planning and

(*continued*)

(*continued*)

operational decision-making necessary to allow governmental units to freely perform their traditional governmental functions. A review of this Court's decision in *Ford* . . . helps to illustrate this principle.

In *Ford,* a young boy was injured when he was visiting the city [sic] of Caldwell's fire station. The boy was in the ready room on an upper floor of the fire station. He had with him a rocket toy which he showed to two firemen who were present. While the two firemen played with the toy, the boy fell through the hole in the floor of the ready room onto the concrete floor below and was seriously injured. Suit was brought against the city, but the trial court dismissed it. On appeal, this Court affirmed the dismissal. This Court stated that in the absence of a statute providing otherwise, a municipality was "not liable for the torts of its officers and employees occurring in the exercise of a governmental function . . . ." . . . It was then held that the maintenance of a fire department by a municipal corporation is a governmental function, and that the city was therefore immune from suit for the negligence of the firemen.

Ford illustrates the type of result that the legislature most likely intended to change when it waived governmental tort immunity for governmental as well as proprietary functions. Although the firemen in *Ford* were working within the scope of a traditional governmental function, i.e., a publicly maintained fire department, they were not engaged in the planning and operational decision-making necessary to the fulfillment of the primary function of the fire department, i.e., fighting fires and providing emergency assistance. Planning and operational decision-making, as it relates to a fire department's primary function of fighting fires, is an example of the sort of decision-making we believe was intended by the legislature to remain protected under the cloak of governmental immunity through the enactment of the discretionary function exception. To hold otherwise would be to open the door to such governmental liability as that evidenced in *Downs v. United States*, 522 F.2d 990 (6th Cir. 1975).

In *Downs,* the federal government was found to be potentially liable under the federal tort claims act for murders perpetrated by a skyjacker when FBI agents shot out one of the aircraft's engines to prevent a takeoff. The court there held that the FBI agents' actions did not constitute a discretionary function because the "agents were not involved in formulating governmental policy." . . . This Court clearly disapproved of *Downs* . . . and certainly our legislature did not intend such a narrow interpretation to be placed upon the term "discretionary function." Public officers engaged in preserving the peace and safety of a community are called upon to exercise their judgment in a manner which often means life or death to themselves and others. Decisions in such areas as law enforcement and firefighting must often be made in an instant. Surely, by enacting the discretionary function exception, the legislature recognized that discretion in making such judgments is entitled to deference at least equal to that given to legislators and judges who have the luxury of time, debate and a comparatively safe and comfortable place to ponder and decide the ways in which governmental business should be conducted.

(*continued*)

(*continued*)

We therefore hold that the discretionary function exception . . . shields governmental units from tort liability for the consequences arising from the planning and operational decision-making necessary to the performance of traditional governmental functions. Since the action in the present case is based upon a claim of negligence with regard to the operational decisions of city firemen in fighting a fire, a traditional governmental function, the action is barred. . . . The judgment is reversed. Costs to appellants.

Case Name: Chandler Supply Co., Inc. v. City of Boise

Court: Supreme Court of Idaho

Summary of Main Points: When the legislature enacted the tort claims act, it reserved immunity protection for governmental actors who are exercising a discretion function. The operational decision making of city firefighters at a fire scene is a discretionary function and therefore the fire department is immune from liability.

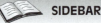 **SIDEBAR**

The Court in *Chandler* acknowledged the difficult decisions that firefighters and emergency responders must make on a daily basis. The concern that many legal scholars have with cases such as *Chandler, Kenevan, Norton,* and *Roberts,* which apply a broad interpretation of discretion, is that one would be hard pressed to identify a decision by a firefighter that could be considered nondiscretionary. If acts such as deciding where to park a fire truck at a scene, whether or not a fire has been extinguished adequately, or when to close the door to a jail cell are considered to be discretionary functions, are any decisions functionary? This situation also leaves firefighters and emergency personnel to wonder whether the next lawsuit challenging fireground tactical decision making will result in continued immunity, or in a court taking notice of the problem as did the court in *Stoller,* and imposing liability. Jurisdictions such as Massachusetts have addressed this problem directly, and have created clear statutory immunity protection for fire departments and fireground decisions without having to stretch the definition of "discretionary function."

THE OTHER DISTINCTION: GOVERNMENTAL VERSUS PROPRIETARY

In addition to the discretionary-functionary distinction made by some courts and tort claims acts, some jurisdictions draw a distinction between a government agency acting in a governmental capacity, and a government

governmental function
an activity undertaken by a governmental agency or actor that is of a type that is generally provided by the government, such as police, fire, highways, and public health; some jurisdictions that provide immunity protection for governmental actors provide it only when the actor is involved in a governmental function

proprietary function
a governmental activity that is commonly performed by private sector entities, such as running hospitals, golf courses, swimming pools, parking garages, and utilities

agency acting in a proprietary capacity. Where this distinction is recognized, governments acting in a truly **governmental function** are entitled to immunity protection, while governments acting in a **proprietary function** are subject to liability.

Jurisdictions differ about how to distinguish between governmental and proprietary functions. The distinction usually comes down to some sort of comparison between what services governments provide compared with what services the private sector provides. If the government provides a service that, generally, only the government provides (police, fire, highways, public health), immunity protection is available. Such a service would be termed a governmental function. If the government is engaged in an activity that private businesses also provide, it will be subject to liability just as a private sector entity would. Examples of proprietary functions would include running hospitals, golf courses, swimming pools, parking garages, and utilities **(Figure 9-4)**.

The State of Ohio has a tort claims act that specifically declares certain governmental actions to be either "governmental functions" or "proprietary functions," Ohio Revised Code §2744.01. It provides immunity protections for governmental functions and establishes liability for proprietary functions. Ohio law designates police, fire, emergency medical, ambulance, and rescue activities as governmental functions (see Appendix B, Chapter 9).

Figure 9-4 *In many jurisdictions, immunity protection is not available for governmental activities that are proprietary functions, such as running a parking garage.*

Buchanan v. Littlehales
606 A. 2d 567 (1992)
Commonwealth Court of Pennsylvania

Joseph Buchanan (Buchanan) appeals an order of the Court of Common Pleas of Berks County (trial court) granting summary judgment in favor of Volunteer Fire Company No. 1 of Grill and LTL Lounge (fire company). Buchanan's complaint alleges that he was injured in an automobile accident which occurred on November 22, 1986. He admits that his car struck two vehicles while driving north on Pennsylvania Route 568, while he was intoxicated. After hitting the two vehicles, he alighted from his vehicle and attempted to walk off the roadway and was struck by Paul R. Littlehales' [sic] vehicle. His sole allegation against the fire company is that it served intoxicating beverages to him while he was already in an intoxicated state, thereby violating . . . the Liquor Code. . . .

The trial court held that the fire company was immune from suit pursuant to what is commonly referred to as the 1980 Immunity Act (Act), articulated in the Judicial Code. . . . The trial court criticized opinions denying volunteer fire companies total immunity from suit, reasoning that these opinions which make a distinction between governmental and proprietary functions are in derogation of the Act which articulates all of the exceptions to governmental immunity. The trial court reasoned that volunteer fire companies are local government agencies and, therefore, pursuant to the Act, they have immunity unless one of the eight exceptions articulated in the Act applies.

The issue to be decided in this case concerns whether volunteer fire companies, when serving alcoholic beverages for profit, are local government agencies immune from liability pursuant to the Act. While the Judicial Code defines volunteer firemen as local government "employees" . . . it does not define volunteer fire companies as local government agencies. Hence, this issue is subject to judicial interpretation. . . .

Buchanan argues that the trial court "erred in construing Pennsylvania law to require that the volunteer fire company was immune from suit in the present case, thereby entitling it to summary judgment." He contends that case law has previously addressed the issue of whether a volunteer fire company is included within the definition of "local agency." Buchanan asserts that a volunteer fire company is a local government agency with immunity only when the fire company is in the performance of public fire fighting duties. Hence, according to Buchanan, because the fire company was not performing a public fire fighting function when it served him alcohol, it is not entitled to immunity, and summary judgment should not have been granted.

We begin our analysis with the case of *Radobersky v. Imperial Volunteer Fire Department,* 368 Pa. 235, 81 A.2d (1951), in which the Supreme Court held that a volunteer fire company is a charity, as well as an agency, which performs "governmental functions." As such, the Supreme Court stated that a volunteer fire company has immunity "while acting in furtherance of the defendant's [fire company] corporate purpose to 'engage in the prevention and control and extinguishment of

(*continued*)

(*continued*)

fires in the town of Imperial, Pennsylvania, and in the surrounding vicinity'." . . . Furthermore, the court stated that

> such immunity from liability did not attend the fire company while it was returning from participation in a firemen's parade at a point beyond the territory of its corporate purpose and that, in such instance, the company was subject to the same liability with respect to its fire truck as applies to other motor vehicles while being operated upon a public highway.

. . . The law is well settled that a volunteer fire company has immunity only when in performance of public fire fighting duties. The *Wilson* Court definitively applied the public fire fighting duties analysis and held that a volunteer fire company had immunity from liability for pollution, which the fire company caused when chemicals which it was using to clean up a diesel fuel spill on a highway, polluted a lake. . . .

In the present case, the volunteer fire company, obviously, was not performing a public fire fighting duty when it allegedly served Buchanan the intoxicating beverages resulting in his driving in an intoxicated state. Therefore, we conclude that the volunteer fire company was not acting as a local government agency entitled to immunity. . . .

Pursuant to the above analysis, we hold that the trial court erred in granting the volunteer fire company immunity, because the volunteer fire company was not acting as a local government agency, performing a public fire fighting function when it allegedly served alcoholic beverages to Buchanan in this matter.

Accordingly, the order of the trial court is reversed, and this matter is remanded for further proceedings consistent with the foregoing opinion.

Case Name: Buchanan v. Littlehales

Court: Commonwealth Court of Pennsylvania

Summary of Main Points: A volunteer fire company is a local governmental agency for purposes of immunity protection only while it is performing a public firefighting function. When it is serving alcoholic beverages, it is not subject to immunity protection.

STATUTORY IMMUNITY

In place of sovereign immunity, or in some cases in addition to sovereign immunity, states have created statutory immunity protection for emergency responders. Some statutory immunity protection arises directly from the language of the tort claims acts, as discussed above, while other statutes create specific types of immunity for specific actors or situations.

Statutory immunity protection that is incorporated into tort claims acts commonly contains qualifying language that limits its coverage to activities that involve a discretionary act while engaged in a governmental function, as opposed to a proprietary activity. Immunity statutes that exist independently of tort claims acts commonly provide a more blanket immunity for certain types of activities, without drawing a distinction between governmental and proprietary, or discretionary and functionary. Immunity statutes may provide liability protection for firefighters, fire departments, emergency medical technicians, pre-hospital emergency medical services (EMS) providers, Good Samaritans, and even those who manufacture, purchase, teach, and use automatic external defibrillators (AEDs), to name a few (see Appendix B—Chapter 9). Good Samaritan laws provide immunity protection for anyone who voluntarily stops to render aid to anyone.

The wording of these immunity statues is critical for determining the scope of protection provided. Some immunity statutes apply only to municipal fire departments, while others are broad enough to include private nonprofit volunteer fire companies and emergency medical providers. Firefighters should learn the various statutory protections that exist in their state, as well as the case law interpreting these statutes.

VOLUNTEER PROTECTION ACT

In an effort to encourage more people to volunteer their time for charitable causes, Congress passed the Volunteer Protection Act of 1997, 42 USC 14501. The Volunteer Protection Act is a Federal law that provides immunity protection for people who volunteer their time to work for a state or municipal government, or who volunteer for a charitable organization **(Figure 9-5)**. The act defines a volunteer as anyone who receives no compensation, other than reasonable reimbursement or allowance for expenses, nor anything else of value in excess of $500 per year.

The Act provides individuals who volunteer with immunity from negligence suits, but does not include protection for gross negligence, recklessness, or intentional acts, nor does it offer protection for suits arising out of the operation of a motor vehicle. The Volunteer Protection Act provides protection only for the person who volunteers, not for the charitable organization. A copy of the act is included in Appendix B, Chapter 9.

PRIVATE NONPROFIT VOLUNTEER FIRE COMPANIES

As we discussed in Chapter 3, private nonprofit volunteer fire companies are not municipal agencies, and as such may not be entitled to sovereign immunity protection for fireground actions. In addition, private volunteer fire companies may not be eligible for statutory immunity protection where the immunity statute specifically provides immunity protection to municipal, county, or state fire departments. The ultimate determination on immunity or liability must be based on an analysis of state law. Some states have specific

Figure 9-5 *Congress passed the Volunteer Protection Act of 1997 to provide immunity protection to those who volunteer their time for charitable causes.*

immunity protection that extends to volunteer fire companies, while in other states the matter depends upon the interpretation of existing statutes. The *Buchanan* case is one example where immunity protection can extend to private volunteer fire companies while engaged in firefighting operations, and the *Catlett* case is another example **(Figure 9-6)**.

National Railroad Passenger Corporation v. Catlett Volunteer Fire Company, Incorporated, et al.
241 Va. 402, 404 S.E.2d 216 (VA, 1991)
Supreme Court of Virginia

CARRICO, Chief Justice.
 Following a railway crossing collision between a passenger train and a fire truck, National Railroad Passenger Corporation (Amtrak), owner of the train, filed
(continued)

Figure 9-6 *The cab of the fire truck that collided with the Amtrak train in Catlett, Virginia. (Photo courtesy of* The Fauquier Citizen.*)*

(*continued*)
a complaint in the United States District Court for the Eastern District of Virginia against Catlett Volunteer Fire Company, Incorporated, owner of the fire truck (Catlett), and the estate of Mark Jay Miller, a volunteer fireman who was operating the fire truck and who was killed in the collision (Miller). In the complaint, Amtrak sought recovery of the sum of $910,000 for property damage allegedly sustained by the train in the collision. . . .

The [District] court granted Catlett's motion for summary judgment . . . [and ruled] that [Miller's] estate could be held liable only if Miller's conduct constituted gross negligence.

Upon appeal to the Fourth Circuit, Amtrak filed a motion for certification of state law questions to this Court. . . . Accordingly, the Fourth Circuit certified to this Court the following questions:

   a. [Is] Catlett . . . immune from suit pursuant to Virginia Code . . . ?
   b. [Does] Miller [have] qualified immunity and can only be liable if his actions are found to constitute gross negligence? . . .

On September 28, 1989, members of Catlett responded to the scene of a car fire on private property adjacent to Route 28 in Fauquier County, Virginia. Miller, the volunteer fireman who was driving the lead fire truck, discovered through
(*continued*)

(*continued*)

radio communications that he had driven past the driveway leading into the property where the car fire was burning. He then turned around and proceeded back on the same highway towards the driveway entrance. Again, Miller overshot the driveway. He stopped the fire truck just beyond the driveway entrance to his right and backed up in order to make a right turn. He then made a wide right turn into the driveway. The driveway is unpaved and crosses over the railroad tracks. After Miller turned onto the driveway, he drove slowly up the drive's incline towards the train tracks and proceeded to cross. The truck's emergency lights and headlights were operating at all times during the response to this call. As the fire truck was crossing the railroad tracks, it was hit by a southbound Amtrak train.

Prior to the accident, Catlett required its member drivers, including Miller, to undergo extensive in-house training as to the safe operation of the fire company's vehicles. Clyde M. Lomax, the fire company's chief, instructed the volunteer firemen of the company that they must obey all traffic laws without exception when responding to a call. Lomax specifically recalls instructing his drivers, including Miller, to stop and look both ways before crossing any railroad tracks.

Catlett is a non-profit corporation that exists independently of Fauquier County (the "County"). Catlett's members are volunteers, whom neither Catlett nor the County compensates for firefighting. Catlett owns its station house and the vehicles that it uses. The County does not control or supervise the daily operations of Catlett and the County is not involved in the selection of Catlett's members, officers or drivers. The County provides no direct funding to Catlett. The only cash disbursement that Catlett receives from the County is a one-thirteenth (1/13) share of a lump sum grant that the County makes available to the Fauquier County Fire and Rescue Association (the "Association"), an unincorporated association that is comprised of appointed members from the thirteen volunteer fire and rescue companies that operate within the County. The County neither directs the Association as to how this grant should be distributed nor requires the Association to account for how the grant is disbursed. The remaining expenditures that the County makes on behalf of the Association are for the payment of bills that the Association submits to the County.

I. Is Catlett Immune from Suit Pursuant to Code § 27-23.6¹

In pertinent part, Code § 27-23.6 provides as follows: *Any county may contract with any volunteer fire-fighting* [sic] *companies or associations in the county or towns therein for the fighting of fire in any county so contracting. If any contract be entered into by a county the fire-fighting company shall be deemed to be an instrumentality of the contracting county and as such exempt from suit for damages done incident to fighting fires therein.*

Amtrak contends that Code § 27-23.6 does not exempt Catlett from liability because the statute requires the existence of a contract between Catlett and the County and no contract was entered into between these parties. Amtrak also contends that the statute only applies when property is damaged incident to the

(*continued*)

(*continued*)

actual fighting of a fire and, hence, does not apply to damage caused by the negligent operation of a fire truck en route to the scene of a fire.

A. The Contract Question

In granting summary judgment in favor of Catlett and Miller's estate, the District Court held that there was no express contract between Catlett and the County but that an implied contract existed between them. Amtrak argues that the District Court correctly found that no express contract existed between Catlett and the County yet "erred in imposing an implied-in-fact contract between them to the detriment of Amtrak."

In finding the existence of an implied-in-fact contract, the district judge stated: *Where an express contract is not made, but might have been, or in equity and good conscience should have been made, the law will impose the duty and infer the necessary promises to effect a contractual relation between the parties. . . .*

After making this statement, the district judge surveyed the Virginia statutes concerning volunteer fire companies, reviewed the evidence supplied by Catlett concerning "the ongoing [close working] relationship between Catlett and [Fauquier County]," and noted the requirement . . . that "the Act shall be liberally construed." Concluding, the judge said: "Liberally construing Section 27-23.6 in light of the evidence before this court regarding the relationship between Fauquier County and Catlett, the Court finds an implied contract does indeed exist between these two parties."

We think the District Court correctly stated Virginia law on the subject of implied contracts. . . . We also think the Court properly applied that law to the evidence before it. Accordingly . . . we will adopt the District Court's views as our own and find the existence of a contract between Catlett and Fauquier County.

B. The Meaning of "Incident To"

One of the remaining questions relating to the application of Code § 27-23.6 is whether the exemption from suit for "damages done incident to fighting fires" encompasses damages resulting from the operation of a fire truck en route to a fire. Amtrak argues that the statute "only exempts a volunteer fire company from suit for property damages 'done incident to' the actual act of fighting a fire." . . .

We think that both definition and common sense compel the conclusion that the operation of a fire truck en route to the scene of a fire is incident to fighting the fire. Accordingly, we hold that Catlett is entitled to the exemption from suit for damages provided by Code § 27-23.6, and we answer the first certified question in the affirmative.

II. Does Miller Have Qualified Immunity?

Amtrak contends that Miller was not entitled to sovereign immunity because the exemption from suit contained in Code § 27-23.6 is provided to firefighting companies but not to individual members of those companies. Amtrak also contends that even if the exemption provided by Code § 27-23.6 is applicable, Miller is not

(*continued*)

(*continued*)

entitled to immunity because driving across railroad tracks without stopping is not the sort of discretionary act upon which immunity may be based.

A. Nature of the Exemption

It is true that Code § 27-23.6 does not in specific terms extend its exemption from suit to individual members of fire-fighting companies. But that does not end the inquiry. If a contract exists between a fire-fighting company and a county, Code § 27-23.6 makes the fire-fighting company "an instrumentality of the contracting county and as such exempt from suit for damages done incident to fighting fires." To this extent, the company would be entitled to the cloak of the county's sovereign immunity and, in turn, the cloak may be available to the company's members.

Amtrak argues, however, that the exemption provided fire companies by Code § 27-23.6 is something less than the immunity enjoyed by the county under general law. Amtrak says that Code § 27-23.6 "is an 'exemption from suit' statute, the effect of which is to limit the remedy of a property owner whose property is destroyed in the course of actual firefighting [and for] this reason, the statute does not bestow sovereign immunity on Catlett."

We disagree with Amtrak concerning the effect of Code § 27-23.6. While it does limit a property owner's remedy, it also has the wholesome effect of encouraging the provision of fire protection services on a voluntary basis in those areas of the Commonwealth where such services might not otherwise be made available. These public policy considerations no doubt prompted the General Assembly to include in the statute the clause granting exemption to fire companies from suits for damages done incident to fighting fires. . . .

We have used the terms "exemption" and "immunity" interchangeably in several of our decisions. . . . And a court in a sister state has said that "[a]fter a search of several standard dictionaries, legal dictionaries and other definitions found in numerous cases, the Court finds that the most common synonym for the word "exemption" is the word "immunity." . . .

We too, think that the terms "exemption" and "immunity" are synonymous. We hold, therefore, that the exemption provided Catlett as an "instrumentality" of Fauquier County is the equivalent of sovereign immunity to the extent of damages done incident to fighting fires. . . .

Miller was a member of a fire company that was made an "instrumentality" of the county with which the company had contracted for the fighting of fire. Further, on the occasion in question, Miller was the "officer . . . in charge." Given the nature of Miller's position, we see no legal distinction between his status and that of the police officer involved in *Colby*, at least insofar as the availability of the defense of sovereign immunity is concerned.

B. Use of Judgment and Discretion

Amtrak argues that . . . Miller is entitled to invoke the defense of sovereign immunity only if "the act complained of involved the use of judgment and discretion." . . . Amtrak says that "the act complained of [here] is Miller's act of driving

(*continued*)

(*continued*)

a fire truck across a railroad crossing without first stopping as required by state law, Catlett's By-Laws, and Catlett's internal safety policies." Hence, Amtrak concludes, "Miller's crossing the railroad tracks without first stopping was a ministerial act, not a discretionary act to which sovereign immunity attaches."

We disagree with Amtrak. "Such situations involve necessarily discretionary, split-second decisions balancing grave personal risks, public safety concerns, and the need to achieve the governmental objective.". . . We think [such] acts involve the exercise of judgment and discretion.

We hold, therefore, that Miller is entitled to invoke the defense of sovereign immunity and is liable only for gross negligence. Accordingly, we answer the second certified question in the affirmative.

First Certified Question Answered in the Affirmative. Second Certified Question Answered in the Affirmative.

Case Name: National Railroad Passenger Corporation v. Catlett Volunteer Fire Company, Inc.

Court: Supreme Court of Virginia

Summary of Main Points: In Virginia, a volunteer fire company that operates under an agreement with a county to provide fire protection services is entitled to immunity protection to the same extent as a governmental entity.

Question: Does the court in Catlett apply a narrow or broad interpretation of "discretion" in reaching a decision?

LIMITATIONS ON IMMUNITY

Immunity protection, whether from sovereign immunity or statutory immunity, is not an absolute protection from all lawsuits. There are a number of limitations upon immunity protection. These limitations vary from state to state and statute to statute, but generally include gross negligence, recklessness, and intentional acts; the doctrine of insurance waiver; and the special duty exception.

Gross Negligence, Recklessness, or Intentional Act Limitation

As a general rule, immunity protection is only applicable for acts of negligence. Acts that constitute gross negligence, recklessness, or that are intentional, are exempt from most immunity protection. Recall that in Chapter 8 we discussed the distinction between negligence and gross negligence. Immunity protection is one of the reasons why the distinction between negligence and gross negligence is so important.

Insurance Waiver Limitation

Many states recognize that immunity protection exists, but will find the immunity to be waived when a governmental entity has purchased liability insurance that would cover a particular claim by an injured third party. Generally, the immunity waiver would allow liability only to the extent of the insurance. This is called the insurance waiver doctrine, and is discussed in the following case.

Luhman v. Hoenig and Cape Carteret Volunteer Fire and Rescue Department, Inc.
358 N.C. 529, 597 S.E.2d 763 (2004)
Supreme Court of North Carolina

WAINWRIGHT, Justice.

On 26 February 2000, a brush fire started in plaintiff Luhmann's neighborhood. Defendant Cape Carteret Volunteer Fire and Rescue Department, Inc. responded to the fire with several vehicles. Two of the vehicles, a tanker truck and a pumper truck, were connected to one another by a fire hose.

While the fire was being extinguished, plaintiff approached the trucks to speak with a fireman. Plaintiff was not asked to leave the area. As plaintiff was speaking with the fireman, Fire Chief Harold Henrich instructed defendant, fireman Billy Hoenig ("Hoenig"), to leave the scene and replenish the water supply in the tanker truck. Contrary to standard procedures, Hoenig failed to walk around the truck to check for connected hoses. As Hoenig backed away in the tanker truck, the hose connecting the tanker truck to the pumper truck tightened and pinned plaintiff's legs against the pumper truck. Plaintiff felt his leg breaking as someone yelled for the truck to stop.

Plaintiff suffered a fractured tibia, tears in his meniscus cartilage and ruptures in his anterior cruciate ligaments. Plaintiff had two surgeries and underwent physical therapy. As a result of his injuries, plaintiff was forced to sell the auto repair business that he owned. Plaintiff wears a leg brace and has developed a chronic pain syndrome called reflex sympathetic dystrophy. Plaintiff will likely never again climb, stoop, kneel, or crouch. He can occasionally walk. Pain remains a significant part of plaintiff's life. At some point, it is likely that plaintiff will need further treatment, including a possible knee replacement.

On 14 June 2000, plaintiff filed suit against Hoenig and the Cape Carteret Volunteer Fire and Rescue Department, seeking damages for their alleged negligence. On 5 February 2002, following summary judgment motions by both parties, the trial court ruled that defendants were negligent as a matter of law. On 2 May 2002, a jury awarded plaintiff $950,000 in damages.

The critical issue in the present case is whether defendants are entitled to the statutory immunity in N.C.G.S. § 58-82-5 or the sovereign immunity in N.C.G.S. § 69-25.8.

(continued)

(*continued*)

N.C.G.S. § 58-82-5 states in pertinent part:

(a) For the purpose of this section, a "rural fire department" means a bona fide fire department incorporated as a nonprofit corporation which under schedules filed with or approved by the Commissioner of Insurance, is classified as not less than Class "9" in accordance with rating methods, schedules, classifications, underwriting rules, bylaws, or regulations effective or applied with respect to the establishment of rates or premiums used or charged pursuant to Article 36 or Article 40 of this Chapter and which operates fire apparatus of the value of five thousand dollars ($5,000) or more.

(b) A rural fire department or a fireman who belongs to the department shall not be liable for damages to persons or property alleged to have been sustained and alleged to have occurred by reason of an act or omission, either of the rural fire department or of the fireman at the scene of a reported fire, when that act or omission relates to the suppression of the reported fire or to the direction of traffic or enforcement of traffic laws or ordinances at the scene of or in connection with a fire, accident, or other hazard by the department or the fireman unless it is established that the damage occurred because of gross negligence, wanton conduct or intentional wrongdoing of the rural fire department or the fireman.

N.C.G.S. § 69-25.8 states in pertinent part:

Any county, municipal corporation or fire protection district performing any of the services authorized by this Article shall be subject to the same authority and immunities as a county would enjoy in the operation of a county fire department within the county, or a municipal corporation would enjoy in the operation of a fire department within its corporate limits. . . . Members of any county, municipal or fire protection district fire department shall have all of the immunities, privileges and rights, including coverage by workers' compensation insurance, when performing any of the functions authorized by this Article, as members of a county fire department would have in performing their duties in and for a county, or as members of a municipal fire department would have in performing their duties for and within the corporate limits of the municipal corporation.

. . . [T]he facts of the present relationship between the County and the fire department are consistent with a fire protection district within the meaning of Chapter 69. N.C.G.S. § 69-25.5(1) authorizes a board of county commissioners to provide fire protection services for a district by contracting with an incorporated nonprofit volunteer fire department. . . . Under N.C.G.S. § 69-25.4(a), a board of county commissioners is authorized to fund its fire protection services by levying and collecting taxes for that purpose. N.C.G.S. § 69-25.4(a) (2003).

In the present case, the Carteret County Board of Commissioners entered into a contract with the fire department on 13 October 1997, whereby the fire department agreed to provide continuing fire protection within the Cape Carteret Fire and Rescue Service District in exchange for compensation from Carteret County funded by

(*continued*)

(*continued*)

the levy and collection of an ad valorem property tax not to exceed ten cents per one hundred dollars valuation on all taxable property within the district. This contractual arrangement generated approximately $850,000 a year for the fire department, accounting for approximately 98% of its annual budget, and transforming the department from one staffed by volunteers to one staffed by paid professionals.

Thus, based on defendants' own representations to the trial court, as well as our fact-specific examination of the relationship between the Cape Carteret Volunteer Fire and Rescue Department and Carteret County, we are satisfied that the fire department in this case constitutes a fire protection district within the meaning of Chapter 69. As such, the fire department is entitled to the same immunities as a county or municipal fire department under N.C.G.S. § 69-25.8.

The well-established common law [sic] principle of sovereign immunity referenced in Chapter 69 precludes a county, as a recognizable unit of the state, from being sued except upon its consent or waiver of immunity. . . . Under N.C.G.S. § 153A-435(a), the purchase of liability insurance "waives the county's governmental immunity, to the extent of insurance coverage, for any act or omission occurring in the exercise of a governmental function." . . .

Here, the Cape Carteret Volunteer Fire and Rescue Department was covered by liability insurance in excess of one million dollars. Therefore, to the extent of this insurance coverage, the Fire Department has waived its sovereign immunity pursuant to N.C.G.S. § 153A-435(a) and is liable for damages.

Accordingly, the opinion of the Court of Appeals is reversed as to the issue of sovereign immunity.

Case Name: Luhman v. Hoenig and Cape Carteret Volunteer Fire and Rescue Department, Inc.

Court: Supreme Court of North Carolina

Summary of Main Points: When an entity that otherwise would be entitled to immunity protection purchases liability insurance, the immunity protection is waived to the extent of the insurance.

Case Note: Many who learn about the insurance waiver exception for immunity come to the conclusion that purchasing liability insurance is a bad idea for fire departments. Liability insurance provides a number of benefits in addition to payment for damages to third parties, not the least of which is the cost of defense. A determination about whether or not liability insurance should be purchased should only be made only after consulting with competent local counsel who is familiar with all of the issues.

Special Duty Exception

Many states recognize an exception to immunity protection when the public entity owes the plaintiff a duty that is somehow special compared to the duty owed to the average member of the public. This "special duty" can arise in a number of ways. The *Odie* case discusses all the major cases in the area and provides solid reasoning for this special (also called "private") duty exception.

City of Gary v. Odie
638 N.E.2d 1326 (1994)
Court of Appeals of Indiana, Fourth District

RILEY, Judge.

In this suit for damages brought by the Estate of Eddie Odie, the jury found the Defendant-Appellant the City of Gary, Indiana (Gary), liable for negligent response to Katie Odie's calls for emergency assistance made on the 911 number established and serviced by Gary. Gary brings this appeal and raises one restated issue for our review: Did the trial court err in determining that the City of Gary owed a private duty to Eddie Odie?

FACTS   The facts most favorable to the judgment reveal that shortly after 4:00 a.m. on August 14, 1988, Eddie Odie got out of bed to let his dog into the house. When he returned to bed he experienced breathing difficulties. At 4:07 a.m., Katie Odie dialed 911 and told the dispatcher that her husband could not "catch his breath" and asked for an ambulance to be sent. (R. 269). The dispatcher assured her that the ambulance was on its way. Katie Odie then called her neighbor, her sister, brother-in-law and niece, and Eddie Odie's daughter and grandson.

Shortly after her neighbor arrived, Katie Odie called 911 for the second time. Again she was assured that the ambulance was coming. When her sister arrived, Katie Odie called for a third time. A fourth call was made by Katie Odie's brother-in-law. Katie Odie made the fifth call and was assured that the ambulance would arrive momentarily. When she told the dispatcher that the ambulance still had not come, the dispatcher informed her in a discourteous tone that the ambulance was probably outside already. The ambulance arrived at 4:49 a.m., 42 minutes after Katie Odie's first call.

Katie Odie testified that she relied on the assurances of the dispatcher and had she known the length of time it would take for the ambulance to arrive, she would have taken her husband to the hospital by car.

At all times relevant to this claim, the City of Gary (Gary) was the exclusive operating authority of the Gary Fire Department Ambulance Service (Ambulance Service). On August 14, 1988, the Ambulance Service was short-staffed with only two of the four regular ambulance crews working. That night the supervisor ordered Ambulance Crew # 404 to station themselves on the northwest side of Gary at Fire Station # 9. Crew # 404 disregarded the order, stationing themselves on the south east side of Gary at Fire Station # 10. The Odies lived about a mile and a half from Fire Station # 9.

The dispatcher with whom Katie Odie spoke chose Crew # 404 to respond to the Odie call; however, when the Odie call first came in, the dispatcher could not find Crew # 404. When he located them at Firehouse # 10, they were asleep. One member of the crew answered the call and logged the response time as 4:35 a.m.; however, it took some time before the team was prepared to leave the station. At the time of Katie Odie's call, other ambulance crews were available to respond.

(continued)

(*continued*)

The average response time for Gary and Northwest Indiana reporting region in 1988 was approximately six minutes; cardiac care time parameters for basic life support was four to six minutes and for advanced life support was eight minutes.

When the ambulance arrived at the Odies' home, the EMTs failed to promptly assess Eddie Odie's condition, failed to promptly call for the assistance of a paramedic who could provide more definitive care, and failed to promptly initiate CPR. Eddie Odie arrived at St. Mary's Medical Center at 5:29 a.m.; he died at 5:45 a.m. of congestive heart failure, pulmonary edema, and cardiac arrest. . . .

DISCUSSION

In light of the elimination of governmental immunity pursuant to the Indiana Tort Claims Act as a defense, the Estate was required to prove the traditional elements of actionable negligence in order to prevail. . . . In order to recover damages, the Estate had the burden of establishing: (1) a duty owed by the defendant to conform its conduct to a standard of care necessitated by its relationship with the decedent; (2) a breach of that duty; and (3) an injury proximately caused by the breach. . . .

Gary contends that it did not owe a duty to Eddie Odie beyond the general duty owed to all members of the public. Specifically, it argues that there was insufficient contact between Eddie Odie and the Ambulance Service to create a private duty.

The duty to exercise care for the safety of another arises as a matter of law out of a relationship which exists between the parties, and it is the province of the court to determine whether a relationship gives rise to a duty. . . . Factual questions may be interwoven with the determination of the existence of a relationship, rendering the existence of a duty a mixed question of law and fact, ultimately to be resolved by the fact-finder. . . .

Generally two types of duty, public and private, can arise in situations similar to the case at bar. . . . If the duty which an official authority imposes upon its agent is a duty to the public, a failure to perform it, or an inadequate or erroneous performance is a public injury and must be redressed in some form of public prosecution. . . . However, to ensure responsibility and the utmost protection possible within limited means, a municipality must be accountable for its negligence to some degree. . . . Thus, if the duty is a duty to an individual, then the neglect to perform it, or to perform it properly, is an individual wrong, and may support an individual action for damages. . . .

A private duty must be particularized to an individual. . . . In specific circumstances, a governmental entity or agent can, by its conduct, narrow an obligation which it owes to the general public into a special duty to an individual. . . . Thus, before Gary can be held liable for negligence, the Estate must show that the Ambulance Service or its dispatcher owed a private duty to Eddie Odie. Absent a duty, there can be no breach of duty, and no negligence or liability based on the breach of duty. . . .

Courts throughout the United States have had difficulty defining the exact nature of the private duty and when Indiana courts have considered the concept,

(*continued*)

(*continued*)

they generally have not found it. . . . *City of Hammond v. Cataldi* (1983), Ind. App., 449 N.E.2d 1184, 1188 (fire department did not owe special duty to victims of a fire in its attempt to extinguish the fire because it was made in response to its general duty to protect the safety and welfare of the public). . . . However, the concept of a private duty owed by a public entity to individuals has eroded the doctrine of immunity. Indeed, in a discussion about exceptions to the governmental doctrine where a private duty exists, Judge Sullivan stated: "The presence of the amorphous and ill-defined duty owed by government to the public as a predominating factor which insulates government from tort responsibility is giving way to consideration of a more fundamental duty owed to private individuals." . . .

The cases [in Indiana] stand for the majority position that liability to an individual for damages will not lie where the officer or city simply owes a duty to the public generally. . . . A common thread through all of these cases is the lack of direct, personal contact between the plaintiffs and the government entity or agent upon which the special duty relationship depends.

This absence of personal contact was emphasized in another case from this court . . . in which evidence did not reveal that the deceased had any special relationship with the city giving rise to an individual duty towards the deceased. . . . The deceased's wife dialed 911 three times but the calls were not answered. . . . A fourth call, made by another person, did reach the emergency service. . . . The estate contended that the ten minute delay in medical assistance contributed significantly to the decedent's death three months later. . . .

In a tragic case, the New York Courts found a county and city did owe a special duty to a person using a 911 system. In *DeLong v. Erie County* (1982), 89 A.D.2d 376, 455 N.Y.S.2d 887, affirmed in *DeLong v. Erie County* (1983), 60 N.Y.2d 296, 469 N.Y.S.2d 611, 457 N.E.2d 717, the appellate division and the court of appeals found Erie County and the City of Buffalo owed a special duty to a murder victim who had called 911 for assistance. The 911 operator failed to ask the caller's name and to verify her address. The operator then dispatched police to the wrong house number of a street with almost the same name as that where the victim resided. The 911 operator also told the victim that police would arrive "right away" and the victim waited. At the time of the call, the murderer had not yet entered the victim's home. The victim lived one-half block from a police station, and had she called the station assistance could have arrived in less than 60 seconds. After the victim was stabbed, a neighbor called the local police station, and police did indeed arrive within 60 seconds.

In affirming the appellate division's decision, the court of appeals relied on Judge Cardozo's statement that: *[i]f conduct has gone forward to such a state that inaction would commonly result, not negatively merely in withholding a benefit, but positively or actively in working an injury, there exists a relation out of which arises a duty to go forward. (citations omitted). . . .*

Gary urges us to adopt specific criteria, like that of Illinois, for determining the existence of a private duty. In Illinois, the court looks to see if (1) the municipality

(*continued*)

(*continued*)

is uniquely aware of the particular danger or risk to which plaintiff is exposed; (2) there are allegations of the municipality's specific acts or omissions; (3) the specific acts or omissions are either affirmative or willful in nature; and (4) the injury occurred while the plaintiff was under the direct and immediate control of the employees or agents of the municipality. . . . Applying this analysis to the case at bar, a private duty could not arise until after the ambulance had arrived at the Odie home.

The Illinois criteria and the cases in which it is employed reflect the Illinois governmental tort immunity act and the Illinois decisional law regarding the "special relationship" exemption to the immunity afforded therein. It was not written with an eye toward the Indiana Tort Claims Act or our controlling case law, and Gary offers no compelling reason why we should adopt the Illinois standard. Further, Gary fails to note that the Illinois tort immunity act and the four-part "special relationship" test was found to be inapplicable to a city's 911 emergency system because that system is not a police protection service, instead a city is liable under a standard of willful and wanton misconduct under another Illinois statute. . . . Thus, we decline to adopt the Illinois criteria; however, we agree with Gary that the adoption of some criteria would serve as a useful guide both for our review and for trial courts.

A perusal of the law of other state jurisdictions reveals that the New York courts have considerable experience in applying both the public duty rule and its special duty exception. The New York Court of Appeals has delineated a narrow class of cases in which they have recognized an exception to the general rule of municipal immunity based on the "special relationship" between the municipality and the claimant. . . . In order to find a special relationship, a plaintiff must prove: (1) an explicit assurance by the municipality, through promises or actions, of an affirmative duty to act on behalf of the injured party; (2) knowledge on the part of the municipality's agents that inaction could lead to harm; (3) some form of direct contact between the municipality's agents and the injured party; and (4) that party's justifiable reliance on the municipality's affirmative undertaking. . . .

Our review of Indiana case law indicates that in order to prove the existence of a private duty there must be both personal contact, beyond a simple request for protection or aid, and some additional factor. One of the additional factors suggested by our decisions is the reliance upon the promise of aid which lulls the victim into inaction. In order to facilitate the analysis of a special duty claim in both the trial court and on review, we adopt the principles enunciated in [New York] for application in cases in Indiana wherein a special duty is alleged. We believe that these principles are well-suited to the Indiana Tort Claims Act and our case law.

Considering the case at bar, the record reveals that Katie Odie called the 911 emergency number four times from 4:07 a.m. until 4:35 a.m. on the morning of her husband's death, and told the dispatcher that her husband was experiencing trouble breathing. This satisfies the third element of the analysis, that of direct

(*continued*)

(*continued*)

contact, and the second element of the analysis, that the dispatcher knew that inaction could lead to harm. Each of the calls were [sic] answered by the dispatcher on duty who gave assurances to Katie Odie that an ambulance was on its way. This satisfies the first element of the analysis that the municipality assumed an affirmative duty to act on Eddie Odie's behalf. Katie testified that had she known that help was not immediately available either she or one of her relatives, who arrived at the house before the ambulance, could have transported Eddie Odie to the hospital. Thus, Katie was lulled into inaction by the 911 dispatcher's continued assurances that the ambulance would arrive and those assurances deprived Eddie of assistance that reasonably could have been expected from other sources. The average response time for Gary at that time was six minutes, 36 minutes less than the time it took for the ambulance to arrive at the Odie's home. This satisfies the final element of the analysis, that of reasonable reliance of the victim.

The trial court was correct in determining that the Estate demonstrated a relationship between the decedent and Gary giving rise to a special duty. Thus, we affirm the judgment of the trial court.

Judgment affirmed.

Case Name: City of Gary v. Odie

Court: Court of Appeals of Indiana, Fourth District.

Summary of Main Points: Even where a fire department has immunity protection, if a special duty exists between a particular victim and the fire department, and that duty is breached, the fire department may be liable for damages.

Recall that in Chapter 8 we discussed *McGuckin v. Chicago* 191 Ill.App.3d 982, 548 N.E.2d 461 (1989), in which a trash fire was discovered in the basement of Chicago's Union Station. The fire department responded, extinguished the visible fire, but apparently failed to check the upper floors for extension prior to leaving the scene. Shortly thereafter, a fire erupted on an upper floor directly above where the basement trash fire had been, claiming the life of one occupant. Illinois had a Tort Immunity Act which provided immunity for fire departments for fireground operations. At trial, the jury determined that under the circumstances the fire department owed the victim a special duty, and was negligent. On appeal the verdict against the fire department was reversed. The appellate court applied a four-prong "special duty" test similar to the test used in the *Odie* case. The court concluded that the deceased victim was not under the direct and immediate control of the fire department, and thus the special duty exception did not apply.

The Flip Side of the Special Duty Exception: the Public Duty Doctrine

Under the "special duty exception," the duty that a governmental actor owes to the public in general is known as a "public duty." When a "special duty" arises, a governmental actor who otherwise would be entitled to immunity protection may be found liable for breaching the special duty. As the *Odie* case discusses, a special duty may arise from the relationship between the parties, and the reliance of the victim upon the promise of assistance.

public duty doctrine

doctrine that when a governmental actor is engaged in a governmental function involving the exercise of discretion, the actor is not subject to liability for breaches of his or her public duty; rather, liability only attaches if the actor breaches a duty to a particular plaintiff to whom a special duty is owed

There is a flip side to the special duty exception, known as the **public duty doctrine**. The public duty doctrine has been used with increasing frequency in some states to create de facto immunity protection where the tort claims acts would seem to have imposed liability on all governmental entities, including fire departments.

The public duty doctrine holds that when a governmental actor is engaged in a governmental function involving the exercise of discretion, the actor is not subject to liability for breaches of his or her public duty. Rather, liability attaches only if the actor breaches a duty to a particular plaintiff to whom a special duty is owed. Some examples of cases in which negligence suits were barred by a finding of a public duty include:

- Wrongful issuance of a driver's license to a person who should not have been qualified to drive, and who later caused an accident injuring a third party

- Improper issuance of an entertainment permit in light of problems in the past, when someone was injured during the event

- The release of a prisoner on parole by the parole board, who, days after being released, shot a police officer during a robbery

The public duty doctrine does not provide governmental actors with immunity per se. Rather, it serves to block liability for negligence through a judicial determination that a public official lacks a legal duty to the public at large. Recall that in Chapter 8, we discussed that in order for there to be negligence, there must be a legal duty that is breached. A duty arises to all who may foreseeably be injured by one's affirmative acts, but for an omission a duty arises only if the parties have a special relationship.

The *Grogan* case arose out of the Beverly Hills Supper Club fire, which killed 165 people **(Figure 9-7)**. The court follows the rationale of the public duty doctrine to provide liability protection to the City of Southgate and the Commonwealth of Kentucky, without recognizing it as the public duty doctrine. The essence of the court's ruling is that for there to be liability for negligence on the part of a public official, there must be a duty that is greater than the duty that a public official owes to the public at large.

Figure 9-7 *The aftermath of the Beverly Hills Supper Club fire. (Photo by Corbis.)*

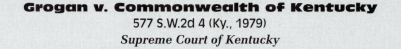

Grogan v. Commonwealth of Kentucky
577 S.W.2d 4 (Ky., 1979)
Supreme Court of Kentucky

On May 28, 1977, a fire at the Beverly Hills Supper Club in the City of Southgate, Kentucky, resulted in a great number of deaths and personal injuries. Shortly thereafter numerous injured parties and personal representatives of those who had lost their lives filed damage suits in the Campbell Circuit Court. In all of these actions, which in due course were consolidated, the City of Southgate and the Commonwealth of Kentucky were named as defendants. The plaintiffs now appeal from separate judgments dismissing the actions, on the pleadings, as to the city and the Commonwealth. . . .

[T]he doctrine of sovereign immunity no longer protects municipal corporations in this state from tort liability. . . . [H]owever, . . . cities are different creatures from natural persons, private corporations, and other suable entities, and that the fundamental bases for tort liability are not necessarily the same for all of them in all situations.

As observed in . . . our most recent opinion on the subject, "a city's relationship to individuals and to the public is not the same as if the city itself were a private individual or corporation, and its duties are not the same. When it undertakes

(continued)

(*continued*)
measures for the protection of its citizens, it is not to be held to the same standards of performance that would be required of a professional organization hired to do the job. If it were, it very well might hesitate to undertake them. . . . A city cannot be held liable for its omission to do all the things that could or should have been done in an effort to protect life and property."

Broadly speaking, the theory on which the city's liability is premised is that it failed to enforce laws and regulations, including its own, establishing safety standards for the construction and use of buildings within its corporate limits, and that its failures in this respect were a substantial factor in causing the tragedy. In other words, the charge is that the city did not enforce a law or laws designed for the safety of the public and that its taxpayers must therefore bear a loss occasioned by someone else's failure to comply with the law.

Though appellants indulge the facile assumption that under similar circumstances a private individual would be liable at common law, we do not believe that the common law, as applied to individuals, offers any reasonably comparable analogy. There is, of course, the familiar principle that one who undertakes the care of another, or of his property, even though it be voluntary and without consideration, owes him the duty of reasonable care. But in the enactment of laws designed for the public safety a governmental unit does not undertake to perform the task; it attempts only to compel others to do it, and as one of the means of enforcing that purpose it may direct its officers and employes [sic] to perform an inspection function. The failure of its officers and employes [sic] to perform that function "does not constitute a tort committed against an individual who may incidentally suffer injury or damage, in common with others, by reason of such default." . . .

The law that applies to this case has been carefully considered, clearly enunciated, and firmly settled . . . . These precedents do not restore the doctrine of sovereign immunity, nor do they evince a retreat toward it. . . . [T]he arbitrary distinction between governmental and proprietary activities [has been erased], thus extending a city's exposure to tort liability into the realm of "governmental functions," but it did not purport to create new torts. As it happens, the existence of the sovereign-immunity doctrine served to prevent a normal development of common-law tort principles in the field of municipal liability until recent times, and the subject cannot reasonably be covered by fitting it out with a ready-made suit of clothes borrowed from the law of torts as it applies to individuals and private corporations. They simply are not the same animals. . . .

What we have said thus far with regard to the city's responsibility applies with equal force to the Commonwealth. There being no basis for tort liability, the question of governmental immunity becomes irrelevant. The answer to the argument that federal constitutional rights are violated by any result that has the effect of shielding governmental bodies from liability that would attach except for their status is that the principles articulated in our previous cases and applied here do not have that effect. To the contrary, that status would be the only basis for holding a city or state liable, because only a governmental entity possesses the

(*continued*)

(*continued*)
authority to enact and enforce laws for the protection of the public. Hence it is that in delineating the areas and extent of public responsibility we are dealing with a subject quite apart and different from the world of individual and corporate relationships. There being no reasonable basis for comparison, there can be no discrimination.
[A]ffirmed.

Case Name: Grogan v. Commonwealth of Kentucky

Court: Supreme Court of Kentucky

Summary of Main Points: A governmental actor is not liable to members of the general public for the failure to enforce laws.

Another public duty doctrine case was *Stone v. North Carolina Department of Labor, Occupational Safety and Health Division,* 347 N.C. 473, 495 S.E.2d 711 (N.C., 1998). On September 3, 1991, 25 workers were killed and 56 were injured in a fire in a chicken processing plant in Hamlet, North Carolina. Out of 90 people in the plant at the time of the fire, 81 were killed or injured due to locked doors and a variety of unsafe conditions **(Figure 9-8)**. Survivors and relatives of the deceased sued the state, alleging negligence in the failure of state OSHA to ever have inspected the premises and discovered the numerous safety violations that led to the tragedy. The court applied the public duty doctrine, and ruled that the state was not liable because the victims were not owed a "special duty." Incidentally, Imperial Foods was fined $808,150 by the State of North Carolina for OSHA violations, and the owner of the plant was sentenced to 19 years and 11 months in prison.

AUTHOR COMMENTARY: CONCLUDING THOUGHTS ON CIVIL LIABILITY, SOVEREIGN IMMUNITY, AND TORTS CLAIMS ACTS

The area of civil liability, sovereign immunity, and tort claims acts is a very difficult and complex area of the law. The approach of each state to immunity varies tremendously in so many ways that it is difficult to draw hard and fast conclusions about similar cases from different states. However, there are two basic principles that tend to transcend jurisdictional differences. They are:

1. The more a governmental activity looks like a discretionary act, the more likely it will be to have some immunity protection. The more a governmental activity looks like a functionary act, the more likely it will be that there is liability.

Figure 9-8   *Exit doors were locked at the Imperial Foods fire in Hamlet, NC, on September 3, 1991, leading to the deaths of 25 employees. (Photo courtesy of* The News and Observer.*)*

2. The more an activity looks like a governmental function, the more likely it will be that there will be immunity, and the more an activity looks like a propriety act the more likely it will be that there is liability.

Utilizing this approach, it is possible for fire service leaders to be proactive in minimizing an organization's liability exposure. Consider the following scenario:

It is 2:00 a.m. on a Sunday. A fire alarm comes in from a commercial building, and a full first-alarm response is dispatched. Upon arrival, the first-in company officer discovers that the building is locked up tight, and there are no visible signs of a problem. Personnel diligently check the exterior of the building for any signs of a problem, or a way in. No keys to the building are available, and the dispatch center is unable to contact a responsible party. Companies have been on scene for nearly 20 minutes. Confronted with a choice of doing damage to the building to gain access, or checking as best they can and leaving, the officer decides:

Scenario (A): To order the members to force entry. This entry causes about $500 worth of damage to an aluminum window and casing. The alarm is determined to be accidental.

Scenario (B): Not to do any damage, and places all units back in service. Two hours later, crews are called back to the scene and the building is heavily involved in fire.

In either of the preceding scenarios, the officer's decision was a functionary decision. Admittedly, some states would find it to be a discretionary decision, but the officer was attempting to carry out the policy of the fire department, not establish a fire department policy. Company officers at 2:00 on a Sunday morning are not making fundamental policy decisions that go to the heart of our democratic form of government. In this case, the building owner may come back and sue the officer and the department, alleging that the officer's decision was negligent. In the absence of immunity protection, the officer and the fire department may have some exposure.

Now let's assume that the fire department has a written policy on how company officers are expected to handle situations like this. The policy dictates that attempts will be made to access the building, contact a responsible party, and determine if a fire condition exists. If, after 20 minutes' time, a responsible party cannot be contacted, no visible signs of fire appear, the entire perimeter has been checked, and the only way to access the building is by doing physical damage to enter the structure, companies may leave the scene without entering. (Note: Some fire departments may choose the alternative policy.)

In either event, the building owner now finds himself challenging a policy decision made at the highest levels of a governmental agency. He is no longer challenging the decision of a company officer, because the officer was merely carrying out established policy. The owner must challenge the department's policy and allege that the department was negligent in adopting such a policy. Policy decisions made at the highest levels of a fire department are much more likely to have the benefit of immunity protection than one made at 2:00 a.m. on a Sunday by a company officer.

Just as in firefighting, there are no absolutes in this area of the law. However, the more something looks like a policy decision, the more likely it is that courts will find there to be immunity.

Having addressed all the issues in the past three chapters, some final insights on liability are warranted. Fire chiefs, fire officers, and firefighters can help to limit their liability by taking the following steps:

1. Comply with national standards, particularly the NFPA standards, to the greatest extent possible.
2. Comply with all laws, including OSHA regulations.
3. Provide up-to-date policies and procedures for personnel to follow.
4. Train and document training for all personnel on policies and procedures.
5. Embrace the concept of risk management.
6. Join with your brother and sister firefighters and fire chiefs in lobbying your state legislature for statutory immunity protection for fireground and emergency scene actions.

SUMMARY

Immunity is a difficult and challenging subject, which is dynamic, evolving, and critically important to firefighters and fire departments. Sovereign immunity has been abolished or limited in all states. Each state and the Federal government have enacted tort claims acts, which allow suits against governmental actors, but often provide specific exceptions for governmental actions that involve a discretionary function. Part of the difficulty in understanding immunity law arises because the law varies tremendously from state to state, making it difficult to apply cases from one state to another.

Several general exceptions exist, in which immunity protection is not provided to public agencies and governmental officials. These exceptions include cases of gross negligence, recklessness, and intentional conduct; where the governmental agency has purchased insurance protection; and where a special duty is found to exist between the governmental actors and the victim. Some states recognize the public duty doctrine, which provides the functional equivalent of immunity by requiring that, for a public official to have liability for negligence, there must be a duty owed to the plaintiff that is greater than the duty a public official owes to the public at large. In the absence of this special duty, a governmental actor cannot be held liable for negligence.

REVIEW QUESTIONS

1. Identify two ways in which sovereign immunity was abolished.
2. Explain the insurance waiver doctrine.
3. Explain why private volunteer fire companies can be entitled to some immunity protection.
4. What is the difference between sovereign immunity and statutory immunity?
5. Distinguish between a discretionary act and a functionary act.
6. Distinguish between a governmental function and proprietary function.

7. Why is there so much confusion relating to fire department liability and immunity?

8. What three functions do tort claims acts serve?

9. Define volunteer as that term is used in the Volunteer Protection Act.

10. Does the Volunteer Protection Act provide immunity protection for a volunteer firefighter who causes a motor vehicle accident while driving a fire truck negligently? Would it matter if the volunteer firefighter was driving recklessly?

DISCUSSION QUESTIONS

1. Do you agree with the reasoning behind the special duty exception to immunity? What public policy concerns does it address, and does it raise any public policy concerns? Can you give an example of a situation in which the special duty exception would be applicable?

2. How can the public duty doctrine create de facto immunity protection for a fire department in a state that has enacted a tort claims act that offers no exceptions for fire departments?

3. In a state that recognizes the insurance waiver limitation to immunity, why is buying liability insurance still a good idea for most fire departments and emergency organizations?

Chapter

10

CONTRACT LAW AND EMPLOYMENT ISSUES

Learning Objectives

Upon completion of this chapter, you should be able to:

- Define the terms contract, offer, acceptance, consideration, and promissory estoppel.
- Explain the difference between actual and apparent authority, and void and voidable contracts.
- Identify the types of contracts commonly associated with the fire service.
- Explain why insurance companies are so highly regulated.
- Identify the important issues associated with mutual aid agreements.
- Define employee at will and explain the requirements of due process as it relates to the discipline of a firefighter.
- Explain how workers' compensation systems operate in general, and the variations commonly associated with firefighters.
- Explain how the principle of exclusivity functions to provide immunity protection to employers.

INTRODUCTION

The captain spoke firmly but authoritatively to the homeowner, as if he had said the same words a hundred times before. "There is definitely a problem with the electrical wiring to this light fixture. We have shut off power to the circuit that supplies power to this light. You need to get an electrician to check the wiring before power is turned back on."

The homeowner answered, "Uh, huh."

"Do you understand?" asked the captain.

"Yeah, I guess so," replied the homeowner.

"Well, let me just add one more thing. I have advised you to get an electrician, and not to turn the electricity back on to this circuit until the electrician checks the wiring. If you choose to ignore this instruction, you may start a fire. Do you understand?"

"Yes," replied the homeowner, now listening more intently.

The captain continued. "If you turn the power back on before an electrician checks out the wiring, and it starts a fire, your insurance company may not provide coverage. Do you understand?"

"Really? Wow. Okay, I understand," the homeowner replied, somewhat concerned by what the captain had said.

Back on the apparatus, Gary asked Jack about the exchange they had overheard. "Was the captain right in what he said? Will an insurance company really deny a claim just because the homeowner ignores the captain's instruction?"

"Kid, the captain is right on. There's a provision in insurance policies that requires people not to intentionally do anything to increase the hazard to their property, and ignoring the captain's instruction does exactly that."

A **contract** is an agreement between two or more people that contains mutual promises that are enforceable at law. Contracts may be oral, written, or in some cases implied by the conduct of the parties. For a contract to be enforceable, each party to the contract must provide consideration. **Consideration** is a legal term for the inducement to enter into the agreement. Consideration can be an act or a promise. Consideration must involve either conferring a benefit upon another, or the acceptance of a recognizable detriment. For example, paying someone money is conferring a benefit, as well as being a recognizable detriment to the person who pays. Promising to pay someone money is a recognizable detriment.

A person's promise to do something that he or she is already required to do does not constitute consideration. No benefit is conferred, and no detriment is incurred because the person making the promise is doing no more than already required.

As a general rule, one-sided promises without consideration from the other party are not enforceable. Such one-sided promises are often referred to

---

contract

an agreement between two or more people that contains mutual promises that are enforceable at law

consideration

a legal term for the inducement that a party to a contract must provide in order for the agreement to be binding; it involves either conferring a benefit upon another, or accepting a recognizable detriment; for example, paying someone money is conferring a benefit upon that other person, as well as being a recognizable detriment to the person who pays, while promising to pay someone money is a recognizable detriment

Figure 10-1 *A contract is an agreement between two or more people that contains mutual promises that are enforceable at law.*

as unilateral promises, since only one party has committed to do something. From a legal perspective, a promise to do something that is unsupported by consideration from the recipient is in the nature of a promise to give a gift, and is not enforceable. In order for an enforceable contract to be created, there must be consideration from both parties **(Figure 10-1)**.

📖 SIDEBAR

- *Holly promises to paint Chelsea's house. Chelsea agrees.* This is simply a promise by Holly, not a binding contract, because Chelsea has not promised to do anything in return. Chelsea has provided no consideration. Holly has made a unilateral promise to paint Chelsea's house. Holly's promise is not enforceable at law.

- *Holly promises Chelsea that she will paint Chelsea's house in exchange for Chelsea paying her $250. Chelsea agrees to pay $250.* There is a binding contract. Holly has agreed to paint the house, and Chelsea has agreed to pay her. Both parties have provided consideration. It is the mutual promises that give rise to an enforceable agreement.

- *Chelsea says, "Holly, if you paint my house, I will pay you two hundred fifty dollars." Without verbally promising to do anything, Holly paints Chelsea's*

> *house.* Holly's execution of her obligations under Chelsea's offer constitutes acceptance of the offer. Holly's consideration was painting the house. An enforceable agreement exists between Holly and Chelsea.
>
> - *Chelsea is a firefighter. Holly is a police officer. At the scene of a car accident, someone steals Chelsea's helmet from the truck. Chelsea promises to pay Holly $250 if she can find the thief and return the helmet. Holly investigates the theft, arrests the perpetrator, and returns Chelsea's helmet.* Chelsea's promise to pay Holly is not enforceable. Holly already has a preexisting duty to investigate the theft and apprehend the wrongdoer. Aside from the obvious public policy issues, Holly's actions were nothing more than that which she is already required to do. See *Gray v. Martino,* 91 NJ Law 462, 103 A. 24 (1918).

In contract law, the requirement that both parties provide consideration is called mutuality. Without mutuality, a contract is not created, and any promises that are made cannot be enforced.

OFFER AND ACCEPTANCE

A contract is created when one party makes an offer, and the other party accepts the offer. In the absence of an offer and an acceptance, there is no contract. Once the offer is accepted, the agreement is binding upon both parties. The conduct of the person making the offer and the person making the acceptance must be intentional, voluntary, and knowing.

The law of offer and acceptance can become quite complicated. However, for our purposes, a clearly defined offer, combined with an unambiguous acceptance, supported by mutual consideration, creates a contract.

statute of frauds
a state law that requires that certain types of contracts have at least some written memorandum of their existence; the statute of frauds varies somewhat from state to state, but as a general rule, a statute of frauds will apply to contracts that take over one year to complete, contracts for the sale of real estate, and contracts for the sale of goods over $500 in value

CONTRACT FORMALITY

Contracts may be in writing, oral, or implied. An oral contract or an implied contract is just as valid and enforceable as is a written contract. Implied contracts can arise by the conduct of the parties. For example, a person walks into a convenience store and without saying a word picks up a candy bar, and places money on the counter. The clerk picks up the money and places it in the cash register. An implied contract was created and fully executed even though there was no writing and no words were spoken.

Some types of contracts are of such a nature that the law requires them to be in writing in order to be enforceable **(Figure 10-2)**. Each state has its own **statute of frauds**, which requires that certain types of contracts have at least

Figure 10-2 *There are a variety of written contracts, including sales agreements, leases, and insurance policies.*

some written memorandum of their existence. The statute of frauds is based on public policy concerns that certain types of contracts are so important that they need to be in writing in order to be enforceable. The statute of frauds varies somewhat from state to state, but as a general rule a statute of frauds will apply to contracts that take over one year to complete; contracts for the sale of real estate; and contracts for the sale of goods over $500 in value **(Figure 10-3)**. For contracts that fall under the statute of frauds, a verbal offer and acceptance do not create an enforceable agreement.

BREACH AND DAMAGES

When one party fails to live up to the promises made in a contract, or fails to properly execute his or her part of the agreement, the contract is said to have been breached. The party who breached the contract is then liable to the other party for damages. The subject of damages is a complex topic. However, the basic principle in contract damage cases is to put the non-breaching party in as good a position as they would have been had the other party not breached the contract. Thus, someone who is injured by a breach of contract can generally recover the costs associated with having a third party complete

Figure 10-3 *The statute of frauds requires contracts for the sale of goods worth more than $500 to be in writing to be enforceable.*

the contract (called the cost of "cover"), plus incidental damages. If the non-breaching party performed a service or provided goods, he or she would be entitled to recover the contractually agreed-upon price. In some instances, it may be possible to get a court order to compel the other party to comply with the terms of the agreement, but such a case is the exception. Courts prefer not to use that power except in rare instances in which monetary damages would not be adequate.

AUTHORITY TO CONTRACT

A person who has reached the age of majority under state law has the legal ability to enter into a contract. The age of majority differs from one state to another, but most consider 18 to be the age at which a person may enter into contracts. The analysis regarding capacity to contract is very much like the determination of the age of consent in regard to medical treatment. Those under the age of majority, or those who are mentally impaired, lack the capacity to enter into an enforceable contract.

Another issue that arises with regard to authority to contract pertains to the authority of an employee or agent to bind a corporate entity to a contract. The formal legal authority to bind another to a contract is referred to as

actual authority
the formal legal
authority of a person
to bind another to a
contract

due diligence
the requirement that,
before entering into a
contract with a person
(agent) who purports
to be acting on behalf
of another (principal)
the seller determines
if the agent has the
actual authority to
bind the principal to
the contract

apparent authority
the authority of a
person (called an
agent) to legally bind a
principal based upon
the conduct of the
principal in holding
the agent out as
having the authority;
a principal who
knowingly allows an
agent to hold himself
out as having the
authority to bind the
principal can be
bound by the acts of
the agent

actual authority. Actual authority to enter into a contract for a state or municipal entity is usually spelled out by statute, ordinance, or charter. For a private corporation, the authority to contract is usually specified by the organization's charter and/or bylaws.

Does this mean that before a vendor sells a pair of gloves on account to a volunteer fire company, the vendor must examine the company's bylaws and charter? The answer is yes and no. The only way a vendor can know for sure who in a fire company has actual authority to contract is to research the corporate documents. Before selling an expensive item, such as a piece of fire apparatus or land, to a fire company, the seller would have a responsibility to exercise **due diligence** to make sure that the person signing the purchase and sales contract has actual authority to bind the corporation. However, for routine purchases, when a corporation holds a person out as having the authority to bind the corporation, or knowingly allows a member to hold himself out as having the authority, then absent knowledge to the contrary by the vendor, the corporation can be bound. This is called **apparent authority**.

📖 **SIDEBAR**

Chief Jones of the Smithville Volunteer Fire Company has routinely purchased gloves, helmets, and boots from ABC Fire Equipment for the past five years. Each time Chief Jones has ordered equipment, ABC Fire Equipment has delivered the order and mailed the fire company an invoice. Each time the fire company has paid the invoice without question. According to the fire company's bylaws, only the company's president has the authority to enter into contracts for the Smithville Volunteer Fire Company.

The chief places an order for five pairs of gloves, five helmets, and five pairs of boots, which ABC Fire Equipment promptly delivers. When the invoice is received, the fire company refuses to pay, claiming that the chief did not have the legal authority to purchase the equipment.

The fire company was aware that the fire chief had been holding himself out as having the authority to make purchases on behalf of the fire company by having paid the bills incurred by the chief for the past five years. In the absence of knowledge by ABC Fire Equipment that the fire chief lacked the authority to make such purchases, the chief will be deemed to have had the apparent authority to bind the fire company, thus making the fire company liable to pay for the equipment.

Note: If the transaction had been for the purchase of a new $400,000 pumper, the fire company could raise the question of due diligence in refusing to honor the agreement. The issue will focus on whether the vendor had a duty to investigate the fire chief's actual authority to enter into the agreement before proceeding with the transaction, and whether the vendor met that duty.

void contract

a contract that is unenforceable even when both parties to the contract desire it to be valid; contracts may be declared void because they are for an illegal purpose, violate a law, or are against public policy

voidable contract

a contract that can be rescinded by one or both parties; if a party lacks the capacity to enter into a contract, enters into a contract out of duress, or if both parties are mistaken about a material factor, the agreement is voidable

VOID OR VOIDABLE AGREEMENTS

Contracts that are invalid or unenforceable for one reason or another can be classified as either void or voidable. A **void contract** is unenforceable even where both parties to the contract desire it to be valid. Contracts may be declared void because they are for an illegal purpose, violate a law, or are against public policy. A **voidable contract** is one that can be rescinded by one or both parties. A voidable agreement may be enforceable depending on the circumstances. If a party lacks the capacity to enter into a contract, enters into a contract out of duress, or if both parties are mistaken about a material factor, the agreement is voidable.

Dover Professional Fire Officers Association v. Dover

124 N.H. 165, 470 A.2d 866 (1983)

New Hampshire Supreme Court

PER CURIAM. . . . At issue is the validity and constitutionality of a contract entered into by the City of Dover with a private corporation, Wackenhut Services, Incorporated, for the provision of fire protection and ambulance services in the city of Dover. . . .

On June 22, 1983, the defendant, the City of Dover (the city), by vote of the Dover City Council, executed a contract with Wackenhut Services, Incorporated (Wackenhut), under which Wackenhut was to provide fire protection and ambulance services for the city. Performance of these services was scheduled to begin on August 1, 1983, and to continue for a five-year period.

On June 24, 1983 . . . the Dover Professional Fire Fighters Association and the International Association of Fire-fighters, Local 1312, filed a petition in the Strafford County Superior Court seeking to restrain the city from contracting out its fire protection services. . . .

On May 24, 1983, the legislature enacted chapter 101 of the Laws of 1983, effective that same date, mandating that "[n]otwithstanding any other provision of law to the contrary, no city council shall contract with any private firefighting unit unless said unit has been certified by the State fire marshal pursuant to RSA 153:4-a." Laws 1983, 101:1.

The city entered into its contract with Wackenhut for fire protection services on June 22, 1983. Both the Dover City Council and Wackenhut were fully aware, at the time of contracting, of the existence and provisions of Laws 1983, 101:1. They were also cognizant of the fact that Wackenhut, a private corporation, had never been certified by the State fire marshal's office.

The city and Wackenhut contend . . . that their contract, although entered into prior to Wackenhut's certification by the State fire marshal, does not violate

(continued)

(continued)

the statutory requirements of Laws 1983, 101:1. Both parties maintain that the underlying purpose of this statute is to prevent the performance and implementation of any written agreement between a city and a private firefighting unit, prior to that unit's having obtained certification from the State fire marshal. They contend that it is not the *signing* of the agreement, but its *performance* and *implementation*, which the statute is intended to prohibit. Wackenhut explicitly noted in its brief that it fully intended, pursuant to the terms of its contract with the city, to obtain the necessary certification from the State fire marshal's office prior to initiating its performance under the terms of the contract. By its recent motion, dated October 27, 1983, Wackenhut notified the court that it had in fact obtained certification from the State fire marshal's office on October 26, 1983. . . .

In the instant case, the language of Laws 1983, 101:1 is clear and concise: *"No city council shall contract* with any private firefighting unit *unless said unit has been certified. . . ."* (Emphasis added.) The use of the word "shall" is generally regarded as a command . . . and the use of the negative as a prohibition. In enacting Laws 1983, 101:1, the legislature explicitly mandated that *"no* city council *shall contract . . ."* (emphasis added) prior to certification. . . .

The parties, in this instance, cannot ignore the plain language of Laws 1983, 101:1, and simply infer that the ultimate intent of the legislature was not to prohibit a city council from entering into a contract with a private firefighting unit prior to certification but, rather, to prevent the city from implementing that contract prior to certification. "Where the language of a statute is plain, we will give the words their usual and customary meaning." . . .

We conclude, based upon the plain meaning of the language of Laws 1983, 101:1 and its legislative purpose, that the contract executed by the City of Dover and Wackenhut Services, Incorporated, is void for having been entered into in direct contravention of State law. The fact that Wackenhut has since obtained certification from the State fire marshal does not affect our judgment as to the validity of the contract.

Case Name: Dover Professional Fire Officers Association v. Dover

Court: New Hampshire Supreme Court

Summary of Main Points: A void contract cannot be enforced, despite the fact that both parties to the contract may desire that it be enforced.

While contracts that violate a law or violate public policy are void, contracts in which one party lacks the capacity to contract, contracts signed under duress, or contracts based upon a mutual mistake of both parties are voidable. The *Fox* case below discusses the issue of duress and voidability in the context of the validity of a firefighter's resignation. Issues of duress and voidability apply equally to a firefighter's resignation as they do to a contract, because in essence these issues involve the fundamental capacity to contract and reach an enforceable agreement between the parties.

Fox v. Piercey
119 Utah 367, 227 P.2d 763, (1951)

CROCKETT, Justice.

This appeal presents the question: Do the findings of the court below support its judgment that the resignation of the plaintiff from the Salt Lake City Fire Department was obtained by duress? . . .

On the evening of August 5, 1948, Harold Fox, who was then a first-grade fireman in the Salt Lake City Fire Department, engaged in a party and some drinking at his home. Considerable disorder developed, during which he allegedly struck a neighbor woman, and his son fired a 22-rifle [sic] into the floor in an attempt to scare him into behaving himself. He was arrested, placed in the city jail and booked on a charge of drunkenness. Information concerning his conduct reached the newspapers that night and was published the next day. J. K. Piercey, Chief of the Fire Department, visited him while he was in jail and requested that he appear at the Chief's office the next morning, which he did after some delay. Piercey had summoned his four assistant chiefs to be present. The Chief informed Fox that he would have to discharge him or that he would be given the opportunity to resign if he so desired. Fox claims that upon his refusal to resign Piercey threatened him by stating: "If you do not resign, I will blast you and smear you in every newspaper in Salt Lake City. I will make it so miserable you can't get a job in the city." Piercey and the four assistant chiefs who were all present denied that any such threats were made.

Fox left the office of the Chief about 10:30 a.m., and was directed to return at 1:30. Upon his return at 1:30, the Chief handed him a letter of discharge. He left the office with this letter but returned in a few minutes and asked if he could still have the privilege of resigning. Piercey thereupon had a letter of resignation prepared, which Fox signed. The next day, after consulting with his attorney, he sent a letter to the Chief stating that he withdrew his resignation. . . . Fox then commenced this action in the district court alleging that his resignation was void because it was obtained by duress. . . .

The trial court expressly rejected plaintiff's contention that Chief Piercey made the threats mentioned. . . . However, the court did find,

"* * * Defendant Piercey requested plaintiff to resign * * * and stated that unless he resigned he would be discharged; and said Piercey further informed plaintiff that his discharge would be accompanied by detrimental publicity and would seriously and detrimentally affect plaintiff's opportunities for obtaining employment in Salt Lake City and vicinity. * * *"

From these findings, the trial court determined that the resignation of plaintiff was procured while the plaintiff was frightened and alarmed and under the influence of duress and that the resignation was void. . . .

There is, of course, no disagreement with the general proposition that a resignation would be voidable for duress the same as any contract. . . .

(continued)

(*continued*)

[T]here have been four distinct phases in the development of the law regarding duress:

1. The ancient rule limiting it to certain specified acts;
2. The enlargement to include any threats, but requiring the "brave man" test;
3. The relaxing of this rule to apply the "man of ordinary firmness" test; and
4. The modern rule that any wrongful act or threat which actually puts the victim in such fear as to compel him to act against his will constitutes duress.

We approve this modern rule. It is obvious that applying this subjective test might theoretically degenerate to a point where a person desiring to avoid a contract might claim that practically any conduct of another put him in fear and overcame his will. It is necessary that there be some objective standard for determining when duress has been practiced. It must appear that the threat or act is of such a nature and made under such circumstances as to constitute a reasonable and adequate cause to control the will of the threatened person. . . .

Notwithstanding the fact that we approve this modern and liberal rule as a test of whether or not duress has been practiced, under all the authorities, ancient and modern, the act or threat constituting duress must be wrongful. . . .

In the Restatement of the Law of Contracts . . . it is stated:

"Acts or threats cannot constitute duress unless they are wrongful, even though they exert such pressure as to preclude the exercise of free judgment. But acts may be wrongful within the meaning of this rule though they are not criminal or tortious or in violation of contractual duty * * * ."

Applying these rules to the conduct of the appellant fire chief as found by the court, we are unable to see how Mr. Fox can avoid the effects of his resignation. It is true that Fox's testimony, as disclosed by the record, would support his claim of duress but we are limited to a consideration of the findings made by the court.

In the first place, the only claim of duress made by Fox . . . was that the threats which he alleges were made by Piercey, as set out at the beginning of this opinion, were what put him in such fear as to overcome his will and constitute duress. As before recited, the court expressly found these threats were not made, which means the court found the cause of duress claimed by Fox did not exist.

The findings the court did make that Piercey informed Fox that he would be discharged and that his discharge would "be accompanied by detrimental publicity and would seriously and detrimentally affect Fox's opportunity for obtaining employment," neither state nor fairly imply that Piercey would do anything beyond the act of discharge to publicize Fox's misconduct or discharge. Assuming he had done so, he would not have been threatening to give the newspapers any information concerning the respondent which they did not already possess, except the fact of his discharge, and the discharge of a public officer would be public information anyway. Piercey merely advised Fox of what the latter knew, quite as well as Piercey did, would have been the inevitable result of his discharge. It is uniformly held that mere persuasion or advice does not constitute duress. . . .

(*continued*)

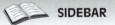

(*continued*)

We now consider whether Piercey's threat of discharge can properly be termed wrongful. . . . Under the circumstances, it could hardly be said in fairness that appellant's threat to discharge respondent was wrongful, such as if it had been wholly capricious and arbitrary or if the threat had been made for some ulterior or improper purpose.

It is not uncommon for an administrative officer who finds it necessary to remove an employee to give the employee an opportunity to resign rather than be discharged. . . . This is indulging a kindness to the employee in protecting him and his work record. It would be a dangerous doctrine to hold that to offer an employee his choice of resigning or accepting a discharge would amount to such compulsion that the employee could avoid his resignation for duress. If such were the law, then any time an employer mentioned the subject of discharge to his employee, he would have to go ahead and discharge him, and could not give the latter the choice of resigning because the resignation would be voidable. . . .

Our conclusion is that the findings of the court are insufficient to support the determination that Fox's resignation was procured by duress. The judgment is reversed and the cause remanded for dismissal. Costs to appellant.

Case Name: Fox v. Piercey

Court: Utah Supreme Court

Summary of Main Points: For a contract to be voidable due to duress, the duress must be wrongful and of a nature sufficient to preclude the ordinary man of the exercise of free judgment.

PROMISSORY ESTOPPEL

promissory estoppel
a unilateral promise made by someone who knows that the recipient will be relying upon the promise to his or her detriment, which courts may enforce when equity requires

As a general rule, a unilateral promise to do something is not legally enforceable. An exception is recognized when the recipient of a unilateral promise relies upon the promise to his or her detriment, and the party who made the promise knew that the recipient would be relying upon the promise. When such a situation arises, the person who relied upon the promise may ask a court to enforce the promise under a theory known as **promissory estoppel**.

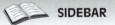   **SIDEBAR**

The Smithville Volunteer Fire Company has the opportunity to purchase a piece of land for a new training facility. The property is for sale at a greatly reduced price from an elderly person who used to belong to the fire company. Upon learning of the transaction, the town manager promises to give the fire company some surplus land the town has in a better location for free. In reliance upon the town manager's promise, the fire company opts not to buy the land. The elderly person's land is then sold to the town manager's brother, who is a real estate developer. Shortly thereafter, the town manager refuses to transfer the surplus land to the fire company.

The Smithville Volunteer Fire Company may have an action for promissory estoppel. The company relied upon the promise of the town manager. The town manager knew that the fire company would rely upon his promise. As a result of the reliance, the company suffered a detriment.

The basis for promissory estoppel is "detrimental reliance." When the recipient of a unilateral promise reasonably relies upon the promise to his or her detriment, and when the maker of the promise knew that the recipient would rely upon the promise, a court may enforce the promise. Promissory estoppel is an "equitable" principle, which in theory "estoppes" or "stops" the person who made the promise from claiming there is no contract. When courts are asked to follow "equitable" principles, the court is required to explore all of the equities involved in the transaction. If the court determines that justice may be done only by enforcing the unilateral promise, the court may do so.

SIDEBAR

Equitable Principles and "At Law"

Suits for breach of contract are said to be cases "at law." In "at law" cases, the court looks solely at the facts and the law, and determines who should prevail. When courts are asked to apply equitable principles such as promissory estoppel, the judge is required to make an additional inquiry into the overall equities of the case. The distinction goes back to old English law, in which there were two entirely different court systems, law courts and chancery (or equity) courts, each with its own procedural rules.

When courts are asked to make "at law" determinations, no additional inquiry into the equities of the case is made. At law, if a party is entitled to win based on a specific legal principle, that party will prevail. When courts are asked to apply equitable principles, but the person seeking the equitable remedy has acted in bad faith, has come to court with "unclean hands," or there are other equitable issues that come into play, the court may decline to rule in that party's favor.

Consider the following example. Able owed a large sum of money to Baker. Able knew that Baker was going to sue him, and to preserve his assets from Baker, Able fraudulently deeded his house and other property to Charlie. Baker sued Able, but due to Able not having any assets, Baker settled the case for a smaller amount than the amount owed. After the case was settled, Able asked

Charlie to deed the property back, and Charlie refused. Able then sued Charlie, seeking an injunction to compel the return of all of Able's assets. In such a case, Able is seeking an equitable remedy, namely an injunction. Able has come into court with unclean hands, having attempted to defraud Baker. Under such a circumstance, the court may decline Able's request for an equitable remedy to order Charlie to return the assets.

FIRE SERVICE CONTRACTS

Among the types of contracts of concern to fire service personnel are:

- mutual aid agreements
- insurance policies
- purchasing
- employment agreements
- collective bargaining agreements (see Chapter 11)

MUTUAL AID AGREEMENTS

Mutual aid agreements are contracts between fire departments, fire districts, and municipalities to provide for assistance in the event of an emergency **(Figure 10-4)**. Mutual aid agreements set forth the rights and responsibilities

Figure 10-4 *Mutual aid agreements need to address a wide range of matters that arise when emergency help is sent or received. (Photo by Ray Taylor.)*

of the parties. For many years, these agreements were informal, and often were not in writing. Over the years, mutual aid agreements have become more and more important, and correspondingly more complex. Some regions have adopted mutual aid pacts, in which several communities enter into a comprehensive multiparty mutual aid agreement that addresses a wide range of community-to-community assistance during emergencies.

Mutual aid agreements should address the numerous matters that arise when emergency help is sent or received, including:

- scope of assistance to be provided
- chain of command and authority
- workers' compensation issues
- reimbursement for costs of disposables
- damage to equipment and apparatus
- repairs to apparatus
- hazardous materials billing issues
- EMS billing issues
- length of time companies and personnel can be committed
- food, housing, and incidentals
- who is responsible for overtime and backfill
- liability, hold harmless, and indemnification issues

Thomas v. Town of Lisbon
209 Conn. 268, (1988)
Connecticut Supreme Court

COVELLO, J.

This is an appeal from a decision of the compensation review division affirming an order by the workers' compensation commissioner requiring the town of Lisbon to pay workers' compensation benefits to two Lisbon volunteer fire fighters [sic]. . . .

The operative facts are not in dispute. The claimants, Wayne Thomas and William Atterbury, were members of the Lisbon volunteer fire department. That department entered into a mutual aid agreement with the Norwich fire department whereby Lisbon fire fighters would come to the aid of the Norwich fire department in an emergency and Norwich fire fighters would reciprocate in Lisbon when needed. On June 25, 1983, a fire occurred in the Taftville section of Norwich. Because the town of Lisbon had pumping equipment specially suited to fight the fire, Taftville fire officials, using a citizens band radio, requested the aid of Lisbon volunteer fire fighters. The claimants arrived at the scene together. Atterbury spoke with the chief of the

(continued)

(*continued*)

Lisbon fire department. The chief told Atterbury to proceed to a nearby fire hydrant to prepare for refilling the Lisbon tanker trucks as they became empty. On the way to the hydrant, the claimants were injured in a motor vehicle accident.

Both parties agree that the claimants were injured and both agree that the claimants are entitled to workers' compensation benefits. The parties disagree, however, as to which town is liable for the payment of that compensation.

The appellee, the city of Norwich, argues that § 7-322a mandates that the appellant, the town of Lisbon, pay the required workers' compensation benefits. We disagree. Section 7-322a provides: "Any active member of a volunteer fire company who offers his services to an officer or person in charge of another fire company which is actively engaged in fire duties, and whose services are accepted by such officer or person, shall be entitled to receive all benefits payable under the provisions of sections 7-314 and 7-314a. Such payments shall be made by the municipality in which the fire company of which such a fireman is a member is located." This provision contemplates a situation where an individual fire fighter offers his or her personal services to another fire department, which thereafter accepts the offer of services. . . .

General Statutes § 7-322a . . . is inapplicable in this situation. Atterbury spoke only to the chief of the Lisbon fire department and never offered his services to anyone from the Norwich fire department. Furthermore, Thomas never offered his services to anyone. Since neither man spoke with anyone from Norwich their services were, obviously, never accepted by Norwich, as required by the statute. Absent the specific factual circumstance articulated in § 7-322a, we agree with the town of Lisbon that § 7-314a is controlling and, therefore, the city of Norwich is responsible for the compensation payments in issue.

Section § 7-314a provides in relevant part: "(a) Active members of volunteer fire departments shall be construed to be employees of the municipality for the benefit of which volunteer fire services are rendered. . . ." This statute allows volunteer fire fighters to receive workers' compensation benefits as if they were employees of the municipality that benefited from their services. While it is arguable that the injured fire fighters benefitted the town of Lisbon by serving as volunteer fire fighters of that town and working in a mutual aid exchange with another town, it is evident that the volunteer fire services in this instance were rendered for the benefit of Norwich. Furthermore, whereas § 7-322a speaks to individuals, § 7-314a speaks to the conduct of towns. That is exactly the situation before us. The town of Lisbon's fire department responded to Norwich's call for help, it was not just the individual claimants who responded. Therefore, § 7-314a is applicable and the city of Norwich is liable for the compensation benefits due the claimants.

Case Name: Thomas v. Town of Lisbon

Court: Connecticut Supreme Court

Summary of Main Points: The financial liabilities associated with mutual aid response are governed by a variety of laws. At times these laws conflict and require the assistance of the courts to determine which laws apply in a given situation.

As is evident from the *Thomas* case, mutual aid agreements may inter-relate with a variety of state laws. These laws may affect issues contained in mutual aid agreements, such as workers' compensation, liability, and reimbursement of expenses.

Mutual aid agreements that involve crossing state lines raise a number of difficult problems. Differences in liability, immunity, workers' compensation, OSHA compliance, and authority vary tremendously from state to state. Most states have adopted laws to address some of the common issues that may arise. However, any mutual aid agreement that involves responses across state lines must be thoroughly researched.

INSURANCE POLICIES

Insurance policies are contracts between an insurance company and an insured that provide for indemnification or payment by the insurer in the event of a loss by the insured. Insurance policies are unique in that they are heavily regulated by state government. Some types of insurance are also subject to Federal oversight. Insurance companies are so heavily regulated that they must submit a sample insurance policy to insurance regulators for review and approval prior to being able to sell that type of policy. The reasoning for such scrutiny comes down to the unique and vital function served by insurers in our society.

When an insurance company breaches its duties or the terms of the policy, policyholders have an action for breach of contract. In addition, policyholders may also have an action in tort for bad faith, as discussed in Chapter 7.

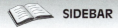 **SIDEBAR**

Insurance Policies

Insurance policies are contracts, but are actually so much more. Experience has shown that without governmental oversight, some insurers would take money from customers and escape liability for claims by skillfully worded policy language incomprehensible to the average layperson. In fact, in many cases, policyholders never even obtain a copy of all the policy language until after coverage has been purchased.

Typically there is a large power imbalance between insurers and policyholders. Insurance companies are usually large corporate structures, while individual policyholders are often individuals who lack the sophistication necessary to understand the meaning of lengthy boilerplate language in the contracts. Insureds pay their premiums and then rely upon the insurer to honor their contracts. Most insureds lack the financial resources to battle with an insurer in the event of claim. Insurers, on the other hand, have a financial motive to deny as many claims as possible based on minor or manufactured technicalities, while making the claims process as difficult and as lengthy as possible.

To protect insureds, courts now recognize a special duty on the part of insurers to exercise good faith and fair dealing. States also scrutinize the financial well-being of insurance companies closely, out of a concern for their remaining solvent and able to pay claims. For all these reasons, insurers and insurance policies are treated uniquely in our society.

Fire Insurance Policies

Firefighters commonly come in contact with insurance policies that provide coverage in the event of a fire. These policies may be separate fire risk policies, or they may be part of a comprehensive homeowner's or commercial property insurance policy that includes insurance for fires as well as a variety of other risks. Insurance policies that provide fire coverage are commonly categorized as property insurance policies.

Most fire insurance policies have specific provisions that either condition coverage on the insured's agreement not to "increase the hazard through any means within the control or knowledge of the insured," or that provide an exclusion to coverage if the insured "increases the hazard through any means within the control or knowledge of the insured." Consider the following case:

Good v. Continental Insurance Company
277 S.C. 569, 291 S.E.2d 198 (1982)
Supreme Court of South Carolina

HARWELL, Justice:
[The Goods, who owned a house and insured it with Continental (appellant), won a jury verdict for $49,600.00 against Continental for a fire they sustained. Continental now appeals claiming that the Goods (referred to as respondants [sic]) voided the insurance policy by increasing the hazard to the house.]

[Continental] insured respondents' house against loss by fire with a standard homeowners policy. The policy contained the following standard increase hazard clause:

> *Unless otherwise provided in writing added hereto this company shall not be liable for loss occurring (a) while the hazard is increased by any means within the control or knowledge of the insured.*

(continued)

(*continued*)

A fire occurring on November 29, 1977 almost totally destroyed respondents' residence. Appellant disclaimed liability upon the ground that after the policy was written and without any written authorization, the hazard was increased by means within the control or knowledge of the respondents.

While extinguishing the fire, firemen discovered an illegal liquor still on the second floor concealed in a false closet under the eaves of the roof. The still, encased by brick and mortar, consisted of a ninety-gallon copper vat over an eight to ten inch butane gas burner. Firemen also found nine fifty-five (55) gallon drums full of mash, three fifty-five (55) gallon drums approximately three-quarters full of mash, twenty-two (22) half-gallons of moonshine liquor, and hundreds of broken bottles and melted plastic containers.

Immediately after the fire was extinguished and while the embers were still smoking, a police department detective found the copper vat three-fourths full of "steaming" mash while the mash in the twelve fifty-five (55) gallon drums was merely "warm." The bricks encasing the copper vat which were exposed to the butane burner were considerably hotter on the inside surfaces than the outside surfaces which were exposed only to the heat of the fire. When the detective dismantled the burner, he found no indication of gas in the gas line. Based upon his observations, he opined that the still was in operation when the fire occurred. However, respondent Sims Good denied that the still had been in operation. He admitted, though, that he had installed the still two years earlier some time after his insurance policy became effective and that he had pled guilty to the criminal offense of unlawfully possessing a distillery. . . . He admitted that he made liquor in his home for presents but not to sell. Appellant had no knowledge of the illegal still's existence. In fact, respondent Sims Good testified that no one else knew of the still's existence, not even his wife, respondent Dorothy Good.

Appellant contends that the only reasonable inference to be drawn from the evidence presented is that the use of the dwelling for the manufacture of illegal liquor and the storage of mash and flammable liquors in large quantities substantially increased the hazard assumed under the policy; that the increased hazard resulted from means solely and exclusively within respondent Sims Good's knowledge or control; that the increased hazard was accomplished after issuance of the policy and its effective date; that the conditions causing the increased hazard were permanent and habitual; that the hazard was materially and substantially increased; and that language and conditions of the policy rendered it immaterial whether the use of the premises for the manufacture of illegal whiskey caused or contributed to the cause of the fire. . . .

[T]his Court [has] held that coverage under an insurance policy was suspended if the increased risk was permanent and continuous even though it did not produce the loss. However, if the increased risk caused the loss, an occasional, temporary increase of risk was sufficient to void a policy.

In addition to these guidelines, both appellant and respondents acknowledge that to void the policy the increase of hazard must be material and substantial such that the insurer could not reasonably be presumed to have contracted to

(*continued*)

(*continued*)

assume it. . . . Also, according to the policy, increased hazard must be accomplished by means solely within the control and knowledge of the respondents after the insurance contract was effected.

The evidence leads unmistakably to the following: the distillery was permanently installed; it was regularly used at least during the holiday season; respondent Sims Good installed the still after the insurance policy became effective; and, only he operated the still.

We conclude that the only reasonable inference from the evidence is that there had been an increase of hazard, thereby voiding the policy. The use to which respondents put their dwelling was so foreign to the normal uses of a dwelling as to become beyond the contemplation of the insurer. . . . Surely the insurer did not contemplate providing insurance coverage on a residence which concealed an illegal still and the fruits of its operation.

Reversed.

Case Name: Good v. Continental Insurance Company

Court: Supreme Court of South Carolina

Summary of Main Points: Standard homeowner's policies carry an exclusion of coverage for acts by the owner that increase the hazard to the property.

Fire insurance cases that involve "increasing the hazard" frequently arise when the sprinkler systems are shut off, or the property is allowed to remain vacant for an extended period. While both situations may be considered "increasing the hazard," many policies also have specific provisions that exclude coverage in the event that sprinklers are shut off, or the building is vacant for a designated period such as 30 days, unless the insured notifies the insurer. The shutting off of sprinkler systems, in particular, is a huge concern to insurers. Insurance rates for sprinklered buildings are considerably less than for unsprinklered buildings. Policies usually require the insured to immediately notify the insurer if the sprinklers are shut down. If the insured fails to notify the insurer, and the sprinklers are out of service when a fire occurs, the insurer can deny a claim.

Subrogation

subrogation
the right of a person to assume the legal claims of another party; the right of an insurer who pays out a claim to an insured, to file suit against whoever was responsible for causing the damage to the same extent that the insured could have sued

Subrogation is the right of a person to assume the legal claims of another party. Subrogation commonly arises in an insurance context when an insurer pays out a claim to an insured. When an insurance company pays out such a claim, the insurer may then file suit against whoever was responsible for causing the damage to the same extent that the insured party could have sued.

> ### 📖 SIDEBAR
>
> Mrs. Smith's house is insured by the ABC Insurance Company. Mrs. Smith purchases a coffee maker made by XYZ Corp. The coffee maker is defective and starts a fire that damages Mrs. Smith's house. The ABC Insurance Company pays Mrs. Smith for all her damages, and then sues XYZ Corp. to recoup all amounts that they paid to Mrs. Smith.

Insurance companies have used subrogation to sue fire departments for negligence in fighting a fire. Recall the *Stoller* case discussed in Chapters 8 and 9, in which the fire department was sued for negligently diverting water away from the sprinklers in order to supply their attack lines. It was the insurance company, not the property owner, that brought the lawsuit under the insurer's right of subrogation.

PURCHASING

The buying and selling of apparatus, equipment, supplies, and real estate all involve contract law. Every sales transaction creates a contractual relationship between the buyer and seller. In state and municipal environments, the purchasing process is typically a complex procedure aimed at preventing fraud and abuse, while securing a cost savings to taxpayers. Issues of actual authority, apparent authority, and due diligence frequently arise. Only officials designated by law may bind the state or a municipal entity to a contract. The purchasing process is typically overseen by those officials in accordance with statutes, ordinances, and charters.

At times, government purchasing procedures seem to work to hamper legitimate efforts to get the best price, but must nonetheless be complied with. Jurisdictions vary in regard to the consequences for violations of purchasing procedures. In some jurisdictions, the transaction may be voidable at the option of the municipal or state agency, while in others there may be civil liability on the part of public officials who violate the procedures.

EMPLOYMENT AGREEMENTS

The employment of a person is a contractual arrangement **(Figure 10-5)**. The employer agrees to pay the employee a wage and provide certain benefits, while the employee agrees to work. Some employment agreements are complex written contracts with a variety of terms, obligations, and contingencies

Figure 10-5 *The employment of a person is a contractual agreement.*

on each party. Other employment agreements may be oral or implied contracts, with some of the terms left vague or poorly defined.

Unless both parties specifically agree otherwise, employment relationships are considered to be "at-will," which means that the employee may be terminated at any time, without notice, for any or no reason. The employee may also quit at any time without consequence or penalty. When terminated from employment, an "at-will" employee has no right to a hearing, and there is no need for the employer to show just cause.

In the public sector, the at-will status of an employee may be abrogated by agreement of the parties, or by law. A contract provision, civil service regulation, charter provision, ordinance, or statute stating that an employee may be terminated only for "cause" or "just cause" changes the status of the employee from "at-will."

While an employer may terminate an at-will employee for any reason or no reason, the employer cannot do so in violation of law. For example, an employer cannot fire an at-will employee due to race, religion, sex, age, or some other prohibited discriminatory reason, or in retaliation for a protected activity, such as complaining about an unsafe or illegal practice. We will discuss this subject in greater detail in Chapters 12 and 13.

Due Process

When it comes to public employment, an additional consideration that comes into play is due process. States and municipalities are subject to the Fourteenth Amendment, which states:

> *No State shall make or enforce any law which shall abridge the privileges or immunities of citizens of the United States; nor shall any State deprive any person of life, liberty, or property, without due process of law. . . .*

Accordingly, before a state or a political subdivision of a state can take an action that impacts a person's "life, liberty, or property," the person must be afforded due process. Whenever due process is implicated in an employment context, several complex issues arise. One way to approach this difficult area of due process is to consider three distinct questions: *whether, how much,* and *when.* "Whether" refers to whether or not a due process right has been implicated. "How much" refers to how much process is due. "When" refers to the timing of the process. We will discuss each of these questions separately.

Whether   For the due process clause to be applicable to a person's employment, courts have determined that the person first must have a property interest in their continued employment. In order for a property interest to exist, the employee must have an expectation of continued employment. Employees who may be terminated only for cause or just cause are generally considered to have a property interest in their employment.

At-will employees do not have such an expectation of continued employment, because they may be terminated at any time. Probationary employees, as well as employees who have signed a "last chance agreement" may lack an expectation of continued employment, depending upon the facts of the case. If an employee has a property interest in his or her employment, due process is applicable, and conversely, if an employee lacks a property interest, he or she has no right to due process.

Dorf v. Sylvania Twp. Bd. of Trustees
2 Ohio App.3d 196, 441 N.E.2d 275 (1981)
Ohio Court of Appeals

POTTER, J.

The trial court has summarized the facts which led to the appeal to that court. They are, in pertinent part, as follows:

"The plaintiff-appellant was employed by Sylvania Township as a full-time township fireman on July 7, 1978. On November 8, 1979, the Sylvania Board of Township Trustees, acting on the recommendation of the Sylvania Township Fire Chief, David A. Drake, terminated plaintiff-appellant's employment with the fire department. The professed reason for Mr. Dorf's discharge was that Dorf was unable to perform his duties as a township fireman due to an off-duty injury which the plaintiff-appellant suffered sometime during July, 1979. The Board admits that it did not afford Mr. Dorf a hearing to determine the sufficiency of the cause for his discharge.

". . . Although the Board did not officially terminate plaintiff-appellant's employment until November 8, 1979, the evidence adduced at the hearing before the

(*continued*)

(*continued*)

Court shows that on November 7, 1979, Mr. Dorf learned that he would be fired from a fellow firefighters [sic]. On the following day, Mr. Dorf attempted to meet with Fire Chief Drake at the Sylvania Township Hall to discuss the matter. . . .

"The Board and Fire Chief Drake were informed of plaintiff-appellant's presence and his desire to speak with the Fire Chief. The Board and Drake refused to meet with Dorf and instead adjourned the regular meeting of the Board and went into a secret executive session to discuss the question of Dorf's status as a township firefighters [sic]. This executive session lasted thirteen (13) minutes after which the regular meeting was reconvened.

"Fire Chief Drake then recommended to the Board that Dorf be discharged due to the aforementioned off-duty injury. . . . In his testimony before this Court, Drake conceded that he had neither sought nor obtained from plaintiff-appellant any medical reports concerning Dorf's injury before he made his recommendation, but rather, had relied on unverified information which he had received from third parties.

"Township Trustee Lucille Laskey then moved to discharge Dorf from the roster of the township firefighters. This motion was seconded by Trustee Deane Allen and approved unanimously by the Board, effective immediately. Subsequently, Mr. Dorf filed his notice of appeal from the Board's decision."

The trial court . . . ordered the board to reinstate Mr. Dorf to his former position as a township firefighter and gave Mr. Dorf . . . judgment at six percent interest equal to the wages he had lost due to his illegal discharge.

The Board of Trustees of Sylvania Township maintains that . . . when a Sylvania Township firefighter is physically incapable of performing his job, he may be unilaterally dismissed. . . . Appellant relies on the common law right to sever an employer-employee relationship at will. Appellant also maintains that Mr. Dorf had no constitutionally protected property interest in his position; therefore, he had no right . . . to notice and a hearing on the issue of his disability.

Mr. Dorf maintains that the legislature has passed certain laws which entitle him to notice and a hearing before the board may terminate his contract and that he has a protected property interest in his public employment. . . .

Interpreting the statute, we find that a firefighter has a protected right of tenure in his position unless he is removed for cause and after a due process hearing. [T]he statute provides that the firefighter has tenure, the expectation of continued employment, and, therefore, a constitutionally protected property interest in his job. . . . In a detailed and excellent opinion rendered in the appeal below, Judge Franklin reviewed many of the cases which have commented on this protected property interest. We quote, with approval, the following from that opinion:

"The initial question before the Court is whether plaintiff-appellant was entitled to the rudimentary elements of due process, notice and a hearing, prior to the Board's decision to terminate his employ as a township firefighter. The resolution of this question turns on whether the plaintiff-appellant's employment status is a property interest protected by the Due Process Clause of the Fourteenth Amendment of the United States Constitution. . . .

(*continued*)

(*continued*)

"The Fourteenth Amendment provides, in relevant part, that:

"'[N]or shall any State deprive any person of life, liberty, or property, without due process of law * * *.'

"Whether public employment constitutes a protected property interest under the Fourteenth Amendment depends on the sufficiency of the claim to entitlement to continued public employment. This necessarily entails a consideration of state law, because, as the Supreme Court has held,

"'Property interests, of course, are not created by the Constitution. Rather, they are created and their dimensions are defined by existing rules or understandings that stem from an independent source such as state law-rules or understandings that secure certain benefits and that support claims of entitlement to those benefits.' . . .

"In *Bishop v. Wood,* supra, the Supreme Court outlined the proper analytical scheme for evaluating due process claims arising out of public employment:

"'A property interest in employment can, of course, be created by ordinance, or by an implied contract. In either case, however, the sufficiency of the claim of entitlement must be decided by reference to state law * * *. Whether such a guarantee [of entitlement to continued employment] has been given can be determined only by an examination of the particular statute or ordinance in question' . . . (footnotes omitted)."

Although we find that a firefighter can be removed for reason of physical disability and need not permanently be carried on the rolls, this decision cannot be made unilaterally. Mr. Dorf had a protected right and this right could not be terminated without notice and a hearing, or, as stated in Judge Franklin's opinion, without the fundamental guarantees of due process. When these were violated, Mr. Dorf had a right of appeal. . . .

We find . . . that Mr. Dorf possessed a constitutional right to notice and a hearing. . . . The judgment of the Court of Common Pleas of Lucas County ordering Martin Dorf reinstated is affirmed. . . .

Case Name: Dorf v. Sylvania Twp. Bd. of Trustees

Court: Ohio Court of Appeals

Summary of Main Points: A firefighter who has an expectation of continued employment has a protected property interest in his employment, and cannot be terminated without first being afforded his due process rights. These rights include the right to have notice and a hearing.

The leading case on due process requirements for public employees is *Cleveland Board of Education v. Loudermill,* 470 U.S. 532, 105 S.Ct. 1487, 84 L.Ed 2d 949 (1985). While the U.S. Constitution requires that a public employer respect the due process rights of a public employee, the question of whether or not a given state or local employee has a property interest in his or her continued employment requires an inquiry into state law. Any property interest that exists in a state or local employee's employment arises out of state law. Federal law does not create an expectation of continued employment. In other words, whether or not the person is an at-will employee, or has a property interest in his or her continued employment is a matter of state law, not Federal law. In the *Dorf* case, Mr. Dorf's property interest in his employment arose under state law.

How Much If it is determined that an employee has a protected property interest in his or her continued employment, the analysis next turns to "how much" process is due. When considering how much process is due, courts are guided in part by a three-part balancing test outlined in *Mathews v. Eldridge,* 424 U.S. 319, 335 (1976). The *Mathews* test requires consideration of: (1) the property or liberty interest that will be affected; (2) the risk that the procedures followed will result in an erroneous deprivation of rights in light of other possible alternatives; and (3) the interests of the state or municipality, including an analysis of the concerns that the government is trying to address, and the financial and administrative burdens that the alternative procedures might require.

As such, temporary deprivations are considered differently from permanent deprivations, and minor deprivations (such as a brief suspension) are considered differently from major deprivations (such as a termination).

At a minimum, in cases where due process rights are found to exist, the person has a right to be informed of the allegations against him or her, and given the opportunity to explain the circumstances. When appropriate, the person may have a right to a full-blown evidentiary proceeding analogous to a trial, where testimony is taken under oath, the person has the right to confront and cross-examine witnesses, the right to be represented by counsel, and the right to an impartial decision maker.

When In addition to determining how much process is required, the timing of that process is critical in determining whether or not due process has been provided. Some types of deprivations require a pre-termination process, while for other types of deprivations a post-termination process is considered adequate.

Lovingier v. City of Black Hawk, Colorado
198 F.3d 258 (10th Cir., 1999)
U.S. Court of Appeals, Tenth Circuit

Appellee Brady Lovingier, a former firefighter with the City of Black Hawk Fire Department, brought this civil rights action under 42 U.S.C. § 1983, claiming that his termination violated his due process rights. . . .

(*continued*)

(*continued*)

Beginning in January of 1994, the City of Black Hawk employed appellee Lovingier as a firefighter. On April 17, 1996, the Fire Chief, Brian Lesher, acting on the recommendation of the Assistant Fire Chief, Ervin Meacham, terminated Lovingier's employment. Meacham recommended Lovingier's termination in March 1996; on April 17, Lesher terminated Lovingier by handing him a notice of dismissal, "effective immediately." . . . Although disputed by appellants, Lovingier asserts that Fire Chief Lesher gave him no opportunity to respond to the charges in the notice.

Lovingier appealed his termination to the City Manager, Lynette Hailey. He moved for her recusal on grounds of partiality, but she declined to recuse herself. After a postponement of the hearing, Hailey conducted post-termination proceedings at which Lovingier was represented by counsel and had the opportunity to call and cross-examine witnesses. In her findings, Hailey concluded that Lesher "did not give Lovingier an opportunity to respond to the allegations contained in the summary and did not give Brady Lovingier reasonable time to prepare a response to the allegations," in violation of the city's policies. . . . While Hailey ultimately sustained Lovingier's termination, to remedy the injury resulting from the city's failure to give Lovingier a proper termination hearing, she ordered the city to pay him his regular salary from April 17, 1996, the date of termination, to May 30, 1996, the date the post-termination hearing was originally scheduled.

Asserting a violation of his Fourteenth Amendment right to due process, Lovingier thereupon filed suit against the city and defendants-appellants Hailey, Lesher and Meacham. The defendants filed a motion to dismiss . . . the due process claims, which the district court granted as to defendant City of Black Hawk and denied as to defendants Hailey, Lesher, and Meacham. Appellants now appeal the denial of their motion to dismiss Lovingier's due process claims on qualified immunity grounds.

. . . With regard to whether appellants violated Lovinger's federal constitutional or statutory rights, Lovingier insists the defendants violated his Fourteenth Amendment right to due process because they terminated him without an adequate opportunity to be heard. The Fourteenth Amendment provides that there shall be no deprivation of "life, liberty, or property without due process of law." U.S. Const. amend. XIV.

It is undisputed that Lovingier suffered a deprivation of a property interest as a result of his termination because he "possessed a legitimate claim of entitlement to his continued employment as a firefighter with the City of Back Hawk sufficient to invoke due process." . . . The relevant question for our review, therefore, is whether Lovingier was deprived his legitimate entitlement to municipal employment without due process of law.

"Due process requires that plaintiff have had an opportunity to be heard at a meaningful time and in a meaningful manner before termination. . . . 'This requirement includes three elements: 1) an impartial tribunal; 2) notice of charges given a reasonable time before the hearing; and 3) a pretermination [sic] hearing,

(*continued*)

(*continued*)

except in emergency situations.'" . . . Lovingier asserts that he was denied entirely a pretermination hearing, and denied an impartial tribunal at his post-termination hearing.

A. Pretermination Hearing

The Supreme Court has held that the Due Process Clause requires " 'some kind of a hearing' prior to the discharge of an employee who has a constitutionally protected property interest in his employment." . . . "The pretermination 'hearing,' though necessary, need not be elaborate." . . . At a minimum, it must provide the employee notice and an opportunity to respond. . . . Lovingier argues that the pretermination process he received did not satisfy this minimum, and the deficiency cannot be cured by the post-termination procedures afforded him. . . .

Lovingier claims that "when Defendant Fire Chief Lesher met with Plaintiff on April 17, 1996 (date of termination), he merely handed Plaintiff the written Notice of Dismissal without allowing Plaintiff the opportunity to respond to the allegations or to exchange information with the Fire Chief." . . . This states a violation of the requirement clearly established by *Loudermill*. . . . Even the City Manager's decision found that Lesher did not give Lovingier an opportunity to respond to the allegations contained in the summary and reasonable time to prepare a response to the allegations. Thus Lovingier states a claim that Lesher's pretermination actions constituted inadequate procedure even under the relatively lenient standards for a pretermination hearing. . . . A brief, face-to-face meeting with a supervisor can satisfy the pretermination due process requirements of *Loudermill*, provided that it affords some notice of and opportunity to contest the grounds for termination. . . .

Defendants argue that prior notice of the disciplinary infractions on which Lovingier's termination was based afforded him adequate pretermination process. We disagree. Lovinger was terminated at the very moment he was given notice of the charges against him. The transcript of his termination reveals that Lesher terminated him "effective immediately." . . . Thus, this case is clearly unlike those cases in which an employee is given the duration of a meeting, or even several days, to respond to charges before she is terminated. . . . Lovingier merely had ten days to appeal following termination, not an opportunity to respond prior to termination. . . .

B. Post-termination Hearing

"When the pretermination process offers little or no opportunity for the employee to present his side of the case, the procedures in the post-termination hearing become much more important." . . . Lovingier's allegations of constitutionally inadequate post-termination process arise out of the contention that defendants failed to provide him with an impartial tribunal. Impartiality of the tribunal is an essential element of due process. . . . We have held, however, that "a substantial showing of personal bias is required to disqualify a hearing officer or tribunal in

(*continued*)

> *(continued)*
> order to obtain a ruling that a hearing is unfair." . . . A complaint that contains only "conclusory allegations of bias, without alleging factual support" is insufficient to make this showing. . . . The person who terminates a public employee does not [necessarily] constitute an unbiased decisionmaker [sic] for due process purposes. . . .
>
> [T]he allegation that defendant Hailey, in her capacity as City Manager, "was closely involved professionally with the Fire Chief," . . . was insufficient to make the "substantial showing" necessary to disqualify a hearing officer for personal bias. . . .
>
> [W]e conclude that Lovingier has not pleaded sufficient specific facts to overcome the presumption of "honesty and integrity on the part of a tribunal," . . . and to support an inference of bias or abandonment of impartiality on the part of Hailey. . . .

Case Name: Lovingier v. City of Black Hawk, Colorado

Court: U.S. Court of Appeals, Tenth Circuit

Summary of Main Points: "Due process requires that plaintiff have had an opportunity to be heard at a meaningful time and in a meaningful manner before termination." In the context of the termination of a firefighter with a protected property interest in his employment, some sort of pre-termination proceeding must be afforded.

Name-Clearing Hearing

name-clearing hearing
a hearing required by the due process clause of the Fourth and Fourteenth Amendments whenever a public employee is accused of wrongdoing that is of a stigmatizing nature, and the accusations are made public; the interest at stake in a name-clearing hearing is a liberty interest, and the purpose of the hearing is limited to providing the employee with an opportunity to be heard relative to the stigmatizing accusations

In the *Loudermill* case, the United States Supreme Court recognized that public employees who have a property interest in their continued employment have a right to due process before being terminated or disciplined. Public employees who are considered to be at-will employees do not have an expectation of continued employment, and thus are not entitled to a due process hearing based on the deprivation of property.

While the due process clause protects public employees from property deprivations such as terminations and demotions, it also applies to governmental actions that affect a person's life and liberty. A public employee's liberty interests may be affected by an employment decision in which the basis for the decision could be stigmatizing to the employee. Courts have recognized a constitutionally based liberty interest in clearing one's name from a stigmatizing reason for termination or job-related action, when the reason has been publicly released. See *Cox v. Roskelly*, 359 F.3d 1105 (9th Cir. 2004). The liberty interest applies to employees who have a property interest in their employment as well as to at-will employees. Thus even an at-will employee may have a right to a due process hearing, commonly referred to as a **name-clearing hearing**, when accused of wrongdoing.

If an employee has already received a *Loudermill*-type disciplinary hearing, there is no need for a second name-clearing hearing. It is only when the employee has not had the opportunity to be heard relative to the stigmatizing allegations that the need for a name-clearing hearing arises. In addition, the purpose of a name-clearing hearing is not to reverse a proposed termination or employment decision, but rather to ensure that the employee has the opportunity to be heard.

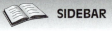 **SIDEBAR**

Davis was the fire chief in Metro City. Davis was appointed by the mayor, and served as an at-will employee. A female firefighter accused Davis of sexual harassment. When the accusation became public, it was accompanied by a great deal of media attention. Davis attempted to contact the mayor to explain the circumstances, but the mayor refused to take his calls. The following day the mayor terminated Davis's employment.

Even though Davis was an at-will employee, and had no expectation of continued employment, he had a right to a name-clearing hearing because the basis for the termination was stigmatizing. At the name-clearing hearing, Davis does not have the right to contest his termination, but rather has the right to an opportunity to clear his name from the stigmatizing accusations.

The procedural requirements for a name-clearing hearing are similar to the hearing requirements under *Loudermill.* A full evidentiary hearing is not necessarily required. Generally, the individual is entitled to notice of the allegations and an opportunity to respond in a meaningful way and at a meaningful time prior to the final decision being made. The right to a name-clearing hearing may apply to situations that do not involve terminations. Some authorities suggest that name-clearing hearings are required whenever potentially stigmatizing documentation is placed into an employee's personnel file. See *Cox v. Boxer,* 359 F.3d 1105 (9th Cir. 2004).

A name-clearing hearing is required only when the allegation would be stigmatizing to an individual in such a way that it seriously damages his or her reputation in the community, or the ability to obtain other employment. The requisite stigma could arise when an employee is accused of committing a crime, dishonesty, immorality or moral turpitude, racism, or similar charges impugning the employee's moral character.

Examples of the kinds of stigmatizing accusations that have been found to implicate a person's liberty interest include employees who are accused of:

- lying on an employment form, or engaging in other forms of dishonesty
- immoral behavior, such as taking drugs or engaging in prostitution

- committing sexual harassment
- using a government position to obtain kickbacks or other special privileges

On the other hand, accusations of incompetence, inability to meet expectations, or negligence are not sufficiently "stigmatizing" to trigger a liberty interest. Liberty interest claims have been rejected in cases in which the employee was accused of:

- being tardy, failing to schedule leave, and engaging in horseplay
- questioning the authority of his supervisor
- failing to meet performance standards
- failure to submit required documentation
- being a poor manager

State Law Due Process Issues

When considering procedural due process requirements for employees of state and local government, state constitutional law must also be considered. Most state constitutions have language that is similar or identical to the Fourteenth Amendment's due process clause, thereby creating state constitutional protection that is in addition to Federal constitutional protections. While states cannot provide less due process than the Federal Constitution requires, they may provide protections that exceed those provided by the Fourteenth Amendment. Some state supreme courts have interpreted state due process rights to exceed the due process rights provided by the U.S. Constitution.

State Action Requirement

It is important to remember that the Fourteenth Amendment's requirement that people be afforded due process applies only to "state action," or action by a state or political subdivision of a state. Public sector fire departments that are a part of state, municipal, or county government, as well as a fire district, would be required to afford employees due process rights, assuming that a property or liberty interest was affected. Private sector employers, including volunteer fire companies, are not subject to the due process requirements, because they are not state actors. However, some states have enacted laws that extend due process requirements to volunteer fire companies, and due process rights to volunteer firefighters. Recall in Chapter 3 the New York case of *Crawford v. Jonesville Board of Fire Commissioners*. New York State law grants volunteer firefighters the right to a due process hearing before being terminated.

In addition to state laws, some volunteer fire companies have charters and internal bylaws that afford members certain rights in the event of disciplinary action. Absent a state statute, local ordinance, charter provision, or fire company provision, members of private volunteer fire companies have no specific right to due process. As such, members are the volunteer equivalent of employees at-will, who may be terminated for any or no reason.

WORKERS' COMPENSATION

At common law, an employee who was injured at work had no right to a continued salary while recuperating, absent a contractual arrangement. The employee's only recourse was to sue the employer, alleging that the employer was somehow negligent in causing the employee's injury. In response, the employer could invoke the common-law defenses of assumption of risk and contributory negligence on the part of the employee. This system was a source of constant friction between employers and employees, leaving many injured workers destitute or reliant upon the government for support.

In 1908, after years of debate and discussion, a no-fault system commonly known as workers' compensation was implemented by the Federal government for employees engaged in hazardous activities and common carriers. In 1911, states began to follow suit, and by 1948 all states had some form of workers' compensation program **(Figure 10-6)**. As we will see, workers' compensation involves concepts of strict liability, no-fault insurance, and immunity.

Figure 10-6 *Workers' compensation is an important consideration for firefighters, given the dangers inherent in their job. (Photo by Ray Taylor.)*

The purpose of the workers' compensation system is to provide a worker who sustained a job-related injury or illness with medical care and a fixed income, without the need to resort to costly and contentious litigation against his or her employer. The system is one of strict liability for the employer. In other words, it does not matter whether the employer was negligent or not, nor does it matter that the employee's own negligence contributed to the injuries. Employers are required to carry workers' compensation insurance. Through the workers' compensation insurance company, the employer is liable for the employee's medical expenses and a stipulated portion of the employee's income, typically two-thirds of his or her normal salary or wage. The cost of caring for injured workers thus becomes a cost of doing business for employers, which can be passed along to their customers.

In exchange for the benefits that employees receive under the workers' compensation program, employers are granted immunity from lawsuits arising out of injuries sustained by employees in the scope of their employment. The immunity protection provided to employers by workers' compensation systems is referred to as **exclusivity**. Exclusivity refers to the fact that workers' compensation insurance is the *exclusive remedy* for an injured or ill worker.

exclusivity
a principle associated with workers' compensation that prohibits an employee from suing his or her employer for injuries sustained in the scope of employment; the employee's exclusive remedy is to accept workers' compensation benefits

Strickland v. Galloway
348 S.C. 644, 560 S.E.2d 448b (2002)
Court of Appeals of South Carolina

GOOLSBY, Judge:
. . . Strickland brought this action against . . . Galloway seeking to recover damages for personal injuries sustained by Strickland in an automobile accident in January 1998. At the time of the accident, both men were serving as volunteer firefighters with Anderson County and were responding to a fire. Strickland had parked his vehicle on the shoulder of the road and was putting on his fire-fighting gear. It was raining heavily. As Galloway pulled off the highway onto the shoulder, his car slid into Strickland, causing him injuries.

Strickland received workers' compensation benefits from the Anderson County Fire Department. He then sought compensation from Galloway individually under a negligence theory. . . .

In circumstances in which the South Carolina Workers' Compensation Act covers an employee's work-related accident, the Act provides the exclusive remedy against the employer. The exclusive remedy doctrine was enacted to balance the relative ease with which the employee can recover under the Act: the employee gets swift, sure compensation, and the employer receives immunity from tort actions by the employee.

(continued)

(*continued*)

The immunity is conferred not only on the direct employer, but also on co-employees. . . . [A] co-employee who negligently injures another employee while in the scope of employment is immune under the Workers' Compensation Act and cannot be held personally liable.

In the present case, if Galloway was acting within the scope of employment, he would be afforded immunity by the Workers' Compensation Act. Having conceded his own status of employee at the time of the accident, Strickland is arguing Galloway was not yet conducting the business of the fire department at the time of the accident. The only apparent distinction is Galloway had just arrived at the scene of the fire, while Strickland had already donned his gear when the accident occurred.

Basing his argument on the "going and coming rule," Strickland maintains Galloway had not yet conducted fire department business at the time of the accident. Under this rule, "an employee going to or coming from the place where his work is to be performed is not engaged in performing any service growing out of and incidental to his employment, and, therefore, an injury sustained by accident at such time does not arise out of and in the course of employment."

South Carolina courts have not addressed this somewhat unique issue of whether or not a volunteer firefighter is acting within the scope of employment while responding to a fire. The issue was addressed in a 1977 Attorney General's opinion:

Since the furnishing of transportation to and from the scene of a fire saves the fire department the expense of transporting the volunteer fireman to and from a fire and also enables the fireman to proceed promptly and directly to and from the scene of the fire, it is self-evident that such a journey represents a substantial part of the volunteer fireman's service to the fire department and the community. Therefore it appears that the volunteer fireman's furnishing of his own transportation directly to and from the scene of the fire is incidental to his duties, and injuries sustained thereby arise out of and in the course of employment so as to be compensable under the South Carolina Workmen's Compensation Act. . . .

The Attorney General's opinion concluded "injuries sustained by a volunteer fireman while on the way directly to . . . a fire are compensable under the South Carolina Workmen's Compensation Act. . . ."

Courts in other jurisdictions have held the going and coming rule does not apply in the context of a volunteer firefighter responding to a fire. The general reasoning followed by these courts is the volunteer firefighter is not "going to work" when responding to the call but is "at work" when responding to the emergency call. Because these volunteers must respond immediately and expeditiously, they are performing the fire department's business when they embark on their response to a fire.

We hold Galloway was conducting the fire department's business at the time of the accident. Thus, the exclusive remedy doctrine of section 42-1-540 bars Strickland from suing co-employee Galloway for his alleged negligence in the accident.

Affirmed.

Case Name: Strickland v. Galloway

Court: Court of Appeals of South Carolina

Summary of Main Points: Workers' compensation is the exclusive remedy for workers injured at work. Volunteer firefighters may be subject to workers' compensation coverage and, if so, are subject to the exclusivity principle. Exclusivity extends not only to the employer but also to coworkers. Coverage begins when the employees begin work, which usually excludes "coming and going" to work. However, in the case of volunteer firefighters it begins when they respond to the call.

As we see in *Strickland,* exclusivity protection is not limited to preventing suits against an employer and managerial personnel, but extends protection to workers from negligence suits by coworkers who are injured. It should also be noted that the applicability of workers' compensation to volunteer firefighters is a matter of state law. Some states such as South Carolina provide coverage for volunteer firefighters, while others, such as Washington, do not. See *Doty v. Town of South Prairie*, 155 Wn.2d 527, 120 P.3d 941 (2005).

Scope of Workers' Compensation

The scope of injuries and illnesses covered by workers' compensation is governed by state law. The typical language in workers' compensation legislation refers to injuries and illnesses "arising out of and in the course of the employment." Each state has developed a large body of case law that interprets the scope of coverage. Most states have a policy to interpret work-related injuries and illnesses broadly so as to provide the maximum amount of workers' compensation coverage with the fewest possible exceptions **(Figure 10-7)**. The *Strickland* case shows how courts will commonly interpret workers' compensation cases.

In managing the caseload of workers' compensation claims, states have created administrative agencies and, in some states, judicial systems to handle workers' compensation cases.

Exceptions to Exclusivity

Some states recognize an exception to exclusivity immunity, when an employee is injured through intentional, reckless, or grossly negligent conduct by an employer or coworker. In *Seymour v. Lenoir County*, 152 N.C. App. 464 (N.C. Ct. App, 2002), a firefighter who was seriously injured in a live burn was allowed to file suit alleging willful and wanton negligence, as well as intentional misconduct, against the department and the instructor who ran the training, under an exception to exclusivity.

Firefighters and Workers' Compensation

In some states, firefighters are specifically exempted from workers' compensation coverage. In other states, firefighters are subject to workers' compensation

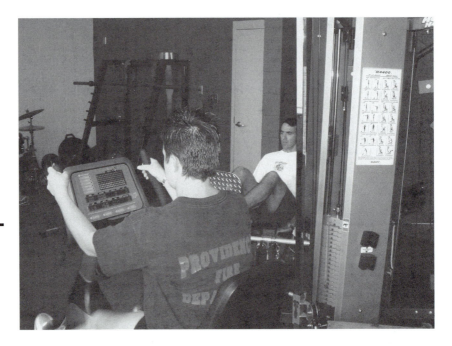

Figure 10-7 *Injuries sustained during on-duty physical training are commonly covered by workers' compensation.*

laws, but are subject to special provisions applicable only to firefighters and other public safety personnel. Many states have statutes that require cities and towns to continue to pay the full salaries of injured firefighters, while other employees receive two-thirds of their regular salary for line-of-duty injuries.

In states where firefighters are expressly exempt from workers' compensation, an important issue arises under the exclusivity (immunity) principle. It would seem that if firefighters are expressly exempt from workers' compensation, they would seemingly be exempt from exclusivity, and thus able to sue their employer and coworkers. The *Kaya* case, while directly pertaining to a police officer, applies equally to firefighters.

Kaya v. Partington
681 A.2d 256 (RI, 1996)
Supreme Court of Rhode Island

WEISBERGER, Chief Justice.
. . . On May 2, 1993, plaintiff, a thirteen-year veteran of the Providence police department who had attained the rank of sergeant, was injured by an unknown assailant while attempting to make an arrest. The plaintiff was dispatched to an

(*continued*)

(continued)

area opposite 187 Benefit Street in Providence's East Side along with several other police officers when Rhode Island School of Design (RISD) security officers requested assistance from the Providence police department. A group of several hundred young people was gathered around a bonfire, singing and chanting, near the entrance to an abandoned railroad tunnel located in the area. The police had reason to believe that these youths, most of whom were costumed and several of whom were filming the event, were participating in a "satanic ritual." Upon arriving at the scene the police, including plaintiff, attempted to disperse the group. The crowd grew hostile and violent. The situation became unruly, with many in the crowd shouting obscenities and throwing debris at the officers. In the midst of this melee, several of the officers were seriously injured, including Kaya, who was struck in the face with a chunk of asphalt while attempting to arrest a female participant.

In accordance with [RIGL 45-19-1], and predicated on his work-related injury, plaintiff received "injured-on-duty" (IOD) benefits. The plaintiff was released by his physician and has returned to the full performance of his duties as a police officer.

Kaya subsequently brought this action for damages, claiming that defendants negligently failed to provide him, as well as other ranking officers on the scene, with blue shirts as part of their uniforms, as opposed to the white shirts issued. Kaya alleges that his white shirt made him a "more susceptible and likely target of violent assault" and that he was "specially targeted for such assault by virtue of his uniform." In addition, plaintiff alleges that he did not have any protective riot gear at the time he approached the disturbance and subsequently sustained his injuries. Kaya specifically contends that defendants acted negligently, willfully, and intentionally by failing to provide a blue shirt and riot gear. . . .

The issue presented in this case is whether a police officer can maintain an action in tort against a municipality and his or her superior officers for injuries sustained in the course of his or her employment and whether IOD benefits constitute the exclusive remedy of a police officer injured in the line of duty.

Police officers in Rhode Island are statutorily entitled to compensation for injuries incurred in the performance of their duties. [RIGL 45-19-1] provides for such compensation and benefits. . . . The IOD statute provides that any police officer in Rhode Island who is injured in the performance of his or her duties may recover the benefits under the statutory scheme. These benefits include a mandate for payment of medical and related expenses for police officers injured in the line of duty as well as the full salary to which they would be entitled had they not been so incapacitated. . . .

Section 45-19-1 is a separate and distinct compensation statute from the Workers' Compensation Act (WCA) which originally covered police officers. Section 45-19-1 is unlike the WCA in that it requires that their full salaries be paid to police officers by the municipality that employs them. Under the WCA injured workers receive only a percentage of their salaries. . . .

(continued)

(*continued*)

In contrast the IOD statute is not optional. It is a mandatory compensation act and must be complied with as the exclusive remedy provided to police officers who become ill or injured in the line of duty. . . .

We hold that § 45-19-1 is the exclusive remedy for police officers injured in the line of duty with *respect to their employers.* The IOD remedy, like workers' compensation benefits, allows a recovery without showing of fault and is not subject to the various tort defenses (contributory negligence, assumption of the risk, lack of foreseeability, lack of a duty or breach of a duty, lack of notice, and intervening causes) that would be available to any alleged tortfeasor, including a municipality, to defeat the claims of a police officer seeking damages under the State Tort Claims Act. . . .

The plaintiff makes much of the exclusivity provision of the WCA and the absence of such an explicit provision in the IOD statute. The WCA was established to provide expeditious relief to injured workers under a nofault [sic] system of benefits replacing a cumbersome and often lengthy tort system. . . . The specific legislative intent was to establish the WCA as the exclusive remedy available to injured workers, completely replacing all other remedies then available. . . . Section 28-29-20, the exclusivity provision of the WCA, states: *"The right to compensation for an injury under chapters 29–38 of this title, and the remedy therefor granted by those chapters, shall be in lieu of all rights and remedies as to that injury now existing, either at common law or otherwise against an employer, or its directors, officers, agents, or employees; and those rights and remedies shall not accrue to employees entitled to compensation under those chapters."*

It is true that when adopted by the Legislature, the IOD statute contained no such exclusivity provision. However, as with the WCA, the remedy provided for in the IOD statute must be exclusive. Although the IOD statute is not explicit on this score, similar sound public policy requires that the exclusivity of the remedy should apply not only with respect to the employer but also with respect to fellow officers, superior officers, and officers of the municipal corporation. The individual defendants are entitled to judgment as a matter of law. . . .

At the time the IOD statute was adopted, police officers and firefighters had no effective right of action against their state or municipal employers because of sovereign immunity. Any right of action that they might have against persons whose conduct inflicted negligent injury upon them would have been gravely circumscribed by the fire-fighters'/police officers' rule. . . .

The IOD statute and the WCA are similar statutes, and though not identical in their provisions, the two are designed to achieve similar goals. In construing the provisions of statutes that relate to the same or to similar subject matter, the court should attempt to harmonize each statute with the other so as to be consistent with their general objective scope. . . .

In our opinion it would create a result not intended by the Legislature for this court to hold that in addition to IOD benefits, police officers, firefighters, and crash-rescue crew-members [sic] should have a right to sue their municipal

(*continued*)

(continued)

and/or state employers. It would be productive of near chaos if we should recognize a right of action for police officers, firefighters, and crash-rescue crewmembers to sue their superior officers and fellow employees. In a paramilitary organization nothing could be more detrimental to good order and discipline than the encouragement of civil actions by police, fire, and emergency personnel against their employers and their superior officers arising out of perceived shortcomings in preparing them for dangerous circumstances that they must encounter on a daily basis.

For the reasons stated, the appeal of the plaintiff is denied and dismissed. The summary judgment entered in the Superior Court is affirmed. The papers in the case may be remanded to the Superior Court.

Case Name: Kaya v. Partington

Court: Rhode Island Supreme Court

Summary of Main Points: Where police officers (and firefighters) are protected by a statutory injury compensation system that is analogous to workers' compensation (although not formally part of workers' compensation), the exclusivity principle shall nonetheless apply. As such, police and fire departments shall not be subject to suits for negligence arising out of workplace injuries to personnel.

The exclusivity immunity afforded by workers' compensation systems provides an important liability protection for fire chiefs, fire officers, and fire departments for injuries arising out of on-duty injuries. The protection afforded by the exclusivity principle avoids consideration of discretionary-functionary and governmental-proprietary distinctions we saw under sovereign and statutory immunity.

Not all states recognize the exclusivity principle for firefighters. Recall the *Kenevan* case, out of New York City, discussed in Chapters 8 and 9. New York law permits career firefighters to sue their employer and coworkers for negligence. Oddly enough, New York law provides for exclusivity for volunteer firefighters who are injured in the line of duty. See *Lima v. State of New York,* 541 N.E.2d 407, 74 N.Y.2d 694 (1989).

SUMMARY

Contracts are created by an offer and an acceptance. Contracts may be oral, implied, or written. Parties entering into an agreement must have the capacity to contract, and must do, or agree to do, something that they are not already required to do. There are a variety of contractual issues that impact the fire service, including mutual aid agreements, purchasing, insurance, and employment agreements. An insurance policy is a type of contract that is subject to considerable governmental regulation due to the important role that insurance plays in our economy.

Workers' compensation systems were established to address the problem of employees who are injured at work. Injured employees are paid a fixed compensation, plus coverage of their medical expenses, as the exclusive remedy for their injuries. In exchange, workers are prohibited from suing their employer, supervisors, or coworkers for negligence in causing the injuries. The treatment of firefighters under workers' compensation varies from state to state. In some states, firefighters are covered under workers' compensation, while in other states firefighters are exempt from workers' compensation under analogous laws that provide comparable coverage. In most states, firefighters who are injured in the line of duty are prohibited from suing their fire department or coworkers for negligence, based on the exclusivity principle.

REVIEW QUESTIONS

1. Define consideration and explain how it relates to mutuality.

2. What is the difference between actual and apparent authority?

3. How does due diligence relate to authority to enter into a sales contract for the purchase of a new fire truck?

4. What are the consequences to a homeowner who "increases the hazard" to her property?

5. Explain how the principle of exclusivity provides the equivalent of immunity protection. Who has the benefit of the immunity?

6. Identify at least six issues that should be addressed in mutual aid agreements.

7. What is the difference between a mutual aid agreement and a mutual aid pact?

8. Does the fire chief of a municipal fire department have the authority to enter into a mutual aid agreement?

9. If a firefighter who resigns later seeks to rescind the resignation, claiming that he resigned under duress, what must he prove?

10. If a full evidentiary hearing is required as part of the disciplinary process for a firefighter in order to satisfy due process, what exactly is required?

DISCUSSION QUESTIONS

1. You are on a routine building inspection and discover that the sprinkler system is shut off. What possible consequences can you explain to the property owner that may motivate him to ensure that sprinkler coverage to the building is restored?

2. Can a firefighter who is an employee at-will be terminated for cause without a hearing? What if there is a city charter requirement that says that firefighters can be terminated only for cause?

3. In the absence of workers' compensation systems, what recourse does an injured employee have to provide an income for his or her family, and pay the medical bills? Explain how workers' compensation systems change this.

Chapter

11

LABOR LAW AND COLLECTIVE BARGAINING

Learning Objectives

Upon completion of this chapter, you should be able to:

- Define collective bargaining, bargaining in good faith, past practice, strike, union shop, closed shop, open shop, dues check-off, agency shop, fair-share agreement, maintenance of membership, and right-to-work.
- Explain the primary differences between private sector labor relations and public sector labor relations.
- Identify the three categories of subjects for collective bargaining (mandatory, prohibited, and permissive), and explain each.
- Explain the various dispute resolution mechanisms commonly used for the three types of impasse disputes: representational, interest, and grievance disputes.
- Explain the duty of fair representation.
- Explain how Weingarten and Garrity Rights serve to protect employees.

INTRODUCTION

The firefighter was escorted into the fire chief's office by his captain. "Firefighter Smith, you know why you are here," the chief said in a stern voice.

"I believe I do, sir," replied the firefighter, "But the rumors going around about what happened are not true."

"Well," said the chief, "You'd best tell me what happened."

"Chief, I believe I have a right to union representation before I talk to you."

"That's fine. The union rep is on the way here. I was hoping we could handle this matter between us."

"Chief, I believe we will be able to resolve this, but I need to consult with my union rep first."

COLLECTIVE BARGAINING

Unions have existed in both the public and private sectors since well before the establishment of the first private sector labor relations acts in the 1930s, or their state law counterparts in the 1950s and 1960s. However, prior to the introduction of labor relations acts, there was no requirement that employers recognize a union, nor were there limitations on what employers could do to rid themselves of union organizers or sympathizers. Dissatisfied workers often resorted to militancy, strikes, and violence when agreements could not be reached. Employers were no less unscrupulous in punishing and intimidating their workforce.

The goal of labor relations legislation was to establish an environment in which parties could bargain collectively, and thereby minimize the disruption to the economy that accompanies labor unrest. **Collective bargaining** refers to the process whereby an employer and duly appointed representatives of the employees negotiate an agreement pertaining to wages, hours, and other terms and conditions of employment. Implicit in collective bargaining is that both parties will negotiate with each other fairly and with open minds, in an effort to reach a mutually satisfactory agreement. The concept of bargaining in good faith remains at the heart of labor relations acts.

Labor relations is a vitally important issue for firefighters. Led by the International Association of Firefighters (IAFF), firefighters have been recognized as the most extensively organized occupation in the United States **(Figure 11-1)**. The IAFF has taken a leadership role in researching and advocating firefighter safety in a way that impacts all firefighters.

collective bargaining
the process whereby an employer and duly appointed representatives of the employees negotiate an agreement pertaining to wages, hours, and other terms and conditions of employment

Figure 11-1 *Fire-fighting has been recognized as the most extensively organized occupation in the United States.*

PUBLIC SECTOR LABOR RELATIONS

The right of firefighters and other public sector workers to bargain collectively is governed primarily by state law. There are many differences from state to state in terms of the rights of firefighters, the scope of bargaining, and the manner in which labor disputes are resolved. As we will see, some of these differences are quite striking.

Public employees have no constitutional right to bargain collectively, except in states such as Florida and New Jersey, where such rights are expressly granted in the state constitution. Authorization for collective bargaining in the public sector is found in statutes, charters, ordinances, and executive orders. In the absence of constitutional, legislative, or executive authorization for collective bargaining, most states hold that a public employer does not have the authority to bargain collectively with its employees, and any contracts that result from such collective bargaining are invalid.

The National Labor Relations Act (NLRA) is a Federal law that addresses private sector labor relations in all 50 states. Passed by Congress in the 1930s, the NLRA created the National Labor Relations Board (NLRB), whose function is to administer and enforce the NLRA.

States have adopted a variety of statutory and organizational structures to address public sector labor relations, with most states modeling their public sector labor laws on the NLRA. Typically the state legislature will establish a state agency analogous to the NLRB. This agency may be called a state labor board, labor relations commission (LRC), or employment relations board (ERB). The purpose of the board is to oversee and administer public sector labor relations, and serve in an adjudicatory capacity in resolving disputes over appropriate bargaining units or allegations of unfair labor practices.

A limited number of states do not have collective bargaining statutes for public employees. In such states, collective bargaining may be authorized by local charter, executive order, or local ordinance. Five states have legislation requiring municipal employers to "meet and confer" with employee representatives, rather than collective bargaining statutes. These meet and confer states require public employers and public employee representatives to meet, and allow them to enter into a nonbinding memorandum of understanding, but do not require collective bargaining.

In a few states, it is illegal for a state or municipal entity to negotiate, collectively bargain with, or enter into an agreement with a public employees' union. Consider the following statute from North Carolina.

EXAMPLE

North Carolina Statutes 95-98. Contracts between units of government and labor unions, trade unions or labor organizations concerning public employees declared to be illegal. Any agreement, or contract, between the governing authority of any city, town, county, or other municipality, or between any agency, unit, or instrumentality thereof, or between any agency, instrumentality, or institution of the State of North Carolina, and any labor union, trade union, or labor organization, as bargaining agent for any public employees of such city, town, county or other municipality, or agency or instrumentality of government, is hereby declared to be against the public policy of the State, illegal, unlawful, void and of no effect.

PRIVATE SECTOR VERSUS PUBLIC SECTOR

Collective bargaining in the public sector differs considerably from that in the private sector. First of all, states and political subdivisions of a state must comply with the due process provision of the Fourteenth Amendment. Similarly, a public employer cannot deny employees their First Amendment right of free speech, the right to peaceably assemble, and the right to petition government for a redress of grievances **(Figure 11-2)**.

Second, the goals of government are politically driven, while the goals of business are financially driven. As a result, political activism on the part of employees for or against one's bosses, or for or against certain issues, is a frequently used tactic in the public sector. Firefighters can engage in political activities in support of or against an elected official, thereby exerting political pressure. On

Figure 11-2 *Public employees have the constitutional right to assemble peacefully, as well as the right to petition government for a redress of their grievances. (Photo by Larry O. Warner.)*

the other hand, politicians can use public sentiment in the political arena to justify taking a hard line against public employees when public opinion dictates. Politics is much less of a tactical factor in the private sector.

Third, in many public sector jurisdictions, labor laws have been superimposed upon an existing civil service system. Civil service systems were initially established to address many of the labor concerns that now clearly fall under the purview of collective bargaining, including hiring, firing, promotions, transfers, and discipline. As a result, many jurisdictions have had to struggle to differentiate which cases should rightfully go through the civil service process and which should go through the collective bargaining process.

Dedham v. Labor Relations Commission
365 Mass. 392, 312 N.E. 2d 548 (1974)
Massachusetts Supreme Judicial Court

KAPLAN, J.
We face again the problem of meshing the new labor rights guaranteed to public employees with earlier provisions of law.

On September 9, 1970, Warren W. Vaughan, a Dedham firefighter . . . engaged in a "heated conversation" with the deputy chief of the fire department, James
(continued)

(*continued*)

Hall, at the Dedham fire station. On September 14 the chief of department, John L. O'Brien . . . notified Vaughan that commencing that day he was suspended for five days with loss of pay for "insubordination" toward a superior officer arising out of the incident with Deputy Chief Hall.

On September 23 Vaughan requested a hearing before a member of the Civil Service Commission . . . as to whether the suspension was for "just cause." . . . [O]n September 30, Vaughan filed a complaint with the Labor Relations Commission charging a prohibited labor practice on the part of the appellees town of Dedham and its fire chief . . . in that they had violated his protected rights . . . to engage in activities on behalf of the firefighters for mutual aid free from interference, restraint, or coercion.

A Civil Service Commissioner held a hearing on October 20 attended by Vaughan and counsel, and on January 13, 1971, the Civil Service Commission notified the fire chief that the suspension was justified but that the penalty should be reduced to a two-day suspension with loss of pay. . . . Meanwhile the Labor Relations Commission, after investigation by its agents, issued its formal complaint on November 5, 1970, against the town and fire chief. At a hearing on January 5, 1971, before a Labor Relations Commissioner, the town and fire chief moved to dismiss the complaint on the ground that the commission lacked jurisdiction of the subject matter. The motion was not allowed. Testimony was taken and recorded, and the following facts as to the September 9 incident appeared, embodied in the "Findings of Fact and Decision" of the commission, made part of its "Decision and Order" contained in the record on appeal.

Vaughan was a member of the executive board and past president of Local 1735, Dedham Firefighters Association. He was off duty on September 9, when he got into the "heated conversation" with Deputy Chief Hall in the presence of another firefighter. The subject was the duties to be performed by firefighters on holidays (such as Labor Day just passed). Vaughan objected to the men's being assigned window washing and similar chores and said this was in violation of the practice of "holiday routine" by which the men were to be excused certain maintenance jobs. Deputy Chief Hall refused to discuss this issue on the ground that "Vaughan was not running the Department." Vaughan told the deputy chief that he was going to bring the matter up at the next union meeting. He advised the men not to do work on holidays in the future beyond the "holiday routine." He then left the fire station.

On evidence going beyond the immediate incident, the Labor Relations Commission also found that Vaughan had an excellent record as a firefighter. In processing grievances and negotiating on labor matters over the previous two years, he had had many heated discussions with the fire chief. Examining the circumstances surrounding the fire chief's decision to suspend Vaughan, the commission found that the chief had ordered the suspension for other than disciplinary reasons. It may be added that the commission found there had in

(*continued*)

(*continued*)

fact been a right to a "holiday routine" which had become vested by practice over a period of years despite "rules and regulations" promulgated by the fire chief.

On the whole case, the commission concluded that the formal complaint it had issued was supported by the testimony. Accordingly, it issued its order in two parts: first, that the appellees, town of Dedham and its fire chief, cease and desist from interfering with their employees in the exercise of their protected rights under the statute; second, that they take affirmative action to "reinstate" Vaughan and make him whole by payment of the withheld salary, make available on request the records as to back pay, post a notice announcing their intention to comply with the directions to cease and desist and to reinstate, and notify the commission as to steps taken to comply with the order. . . . [The town filed suit in Superior court seeking a finding that the Civil Service Commission decision took precedence to the Labor Relations Commission decision.]

If the Civil Service Commission were to administer in such cases only the substantive law which it has historically fashioned under the title of "just cause," and no recourse could be had there or before the Labor Relations Commission to the new labor statute in its material aspects, then a plain perversion of the legislative purpose would occur. . . . [I]f . . . the Civil Service Commission should attempt . . . to apply the labor law as well, with the Labor Relations Commission wholly excluded—a suggestion made by the appellees in their brief, although with little elaboration—then there would still be a plain defiance . . . of the labor statute.

Employees . . . (and their employers as well) are entitled to the specialized services of the Labor Relations Commisssion in the administration of the labor rights, and to the related adjective arrangements. Considering the indissoluble linkage of the character of a tribunal, its procedure, and the substantive law that it enforces, it seems clear that the parties before the Civil Service Commission would not—and in the nature of things could not—secure from that body alone substantive rights equivalent to those assigned by the statute for enforcement to the other commission. So the idea of using the Civil Service Commission to act as a substitute for the Labor Relations Commission in cases involving employees in the civil service would turn out to be quite unsatisfactory. It must, after all, have been a prime legislative purpose in creating the Labor Relations Commission to promote uniformity rather than disuniformity of interpretation and application of the labor law. In this light we need hardly point out that "cease and desist" and "affirmative" remedies, not only available but required in certain cases under the labor statute, could in no event attach to determinations by the Civil Service Commission, and that the nature and course of judicial review of orders of the Civil Service Commission depart from those prescribed for review of orders of the Labor Relations Commission. . . .

Conclusion. [The Labor Relations Commission ruling is valid and enforceable.]

So ordered.

Case Name: Dedham v. Labor Relations Commission

Court: Massachusetts Supreme Judicial Court

Summary of Main Points: A union official was disciplined by his fire chief for matters apparently related to union activities. The union representative sought to contest the discipline through the civil service commission and by filing a prohibited practice complaint with the Massachusetts Labor Relations Commission. Conflicts between collective bargaining laws and other laws, such as civil service laws, will occur. Courts will attempt to interpret potential conflicts in such a way that the laws can be reconciled, so as to avoid a conflict wherever possible. In this case, the court interpreted the labor relations act to require the Massachusetts Labor Relations Commission to have jurisdiction over unfair labor practices, even though the same substantive issue may come up within the context of a civil service commission review of a disciplinary action.

LABOR RELATIONS ACTS

It would be helpful at this point to examine the language used in public sector collective bargaining laws. Some states have a single labor relations act that covers all public employees, while other states have separate acts for municipal workers and state workers. In addition, some states have special collective bargaining laws for special categories of public employees, such as teachers, firefighters, and police officers. In a few states, labor relations laws are enacted at the local level either through charter provisions or ordinances. The following statute is from the State of New York.

EXAMPLE

New York State Consolidated Laws, Article 20, Section 703. Rights of employees. Employees shall have the right of self-organization, to form, join, or assist labor organizations, to bargain collectively through representatives of their own choosing, and to engage in concerted activities, for the purpose of collective bargaining or other mutual aid or protection, free from interference, restraint, or coercion of employers, but nothing contained in this article shall be interpreted to prohibit employees from exercising the right to confer with their employer at any time, provided that during such conference there is no attempt by the employer, directly or indirectly, to interfere with, restrain or coerce employees in the exercise of the rights guaranteed by this section.

The preceding language is analogous to that used in the NLRA, relating to the rights of private sector employees. In addition to defining employees' rights to organize, labor relations acts also define the kinds of actions that are prohibited. Again mirroring the NLRA, the New York statutes categorize these prohibited actions as "unfair labor practices."

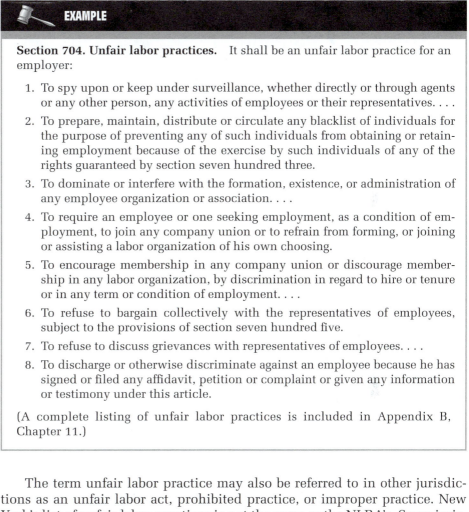

EXAMPLE

Section 704. Unfair labor practices.   It shall be an unfair labor practice for an employer:

1. To spy upon or keep under surveillance, whether directly or through agents or any other person, any activities of employees or their representatives. . . .

2. To prepare, maintain, distribute or circulate any blacklist of individuals for the purpose of preventing any of such individuals from obtaining or retaining employment because of the exercise by such individuals of any of the rights guaranteed by section seven hundred three.

3. To dominate or interfere with the formation, existence, or administration of any employee organization or association. . . .

4. To require an employee or one seeking employment, as a condition of employment, to join any company union or to refrain from forming, or joining or assisting a labor organization of his own choosing.

5. To encourage membership in any company union or discourage membership in any labor organization, by discrimination in regard to hire or tenure or in any term or condition of employment. . . .

6. To refuse to bargain collectively with the representatives of employees, subject to the provisions of section seven hundred five.

7. To refuse to discuss grievances with representatives of employees. . . .

8. To discharge or otherwise discriminate against an employee because he has signed or filed any affidavit, petition or complaint or given any information or testimony under this article.

(A complete listing of unfair labor practices is included in Appendix B, Chapter 11.)

The term unfair labor practice may also be referred to in other jurisdictions as an unfair labor act, prohibited practice, or improper practice. New York's list of unfair labor practices is not the same as the NLRA's. Some jurisdictions have additional protections, and some may not be as comprehensive as New York's. The NLRA, as well as some states, also specifies some unfair labor practices that a union can commit.

When one party alleges that the other party is not bargaining in good faith, or otherwise has committed an unfair labor practice, the procedure is to file a complaint with the appropriate labor board. The labor board then investigates the accusation and issues a ruling aimed at resolving the dispute. Any ruling issued by a labor board may be appealed to a court.

GOOD-FAITH BARGAINING

In 29 USC 159 (d), the NLRA provides:

"For the purposes of this section, to bargain collectively is the performance of the mutual obligation of the employer and the representative of the employees to meet at reasonable times and confer in good faith with respect to wages, hours, and other terms and conditions of employment, or the negotiation of an agreement, or any question arising thereunder, and the execution of a written contract incorporating any agreement reached if requested by either party, but such obligation does not compel either party to agree to a proposal or require the making of a concession."

The preceding section is the primary basis for the requirement of bargaining in good faith. Analogous, if not identical, language exists in most state labor relations acts. Bargaining in good faith does not require a party to make a concession or change a position. It requires both sides to participate actively in deliberations, and to show an intention to find a basis for agreement—it is the cornerstone of collective bargaining **(Figure 11-3)**. Both parties must keep an open mind and make a sincere effort to reach common ground. The employer is prohibited from taking any unilateral action on matters pertaining to wages, hours, or terms and conditions of employment without first offering to negotiate with the union. Whether a party is bargaining in good faith is a factual question that will turn on the totality of the factors present in a given case.

Figure 11-3 *Good-faith bargaining is a cornerstone of collective bargaining.*

IAFF Local 2010 v. City of Homestead
Case No. 72-9285 (1973), aff'd 291 So. 2d 38 (Fla. Ct. App. 1974)
Florida Circuit Court, 11th Judicial Circuit

Plaintiff, Local No. 2010 of the International Association of Firefighters, is a labor organization representing a majority of the persons employed as firefighters by the Defendant, City of Homestead. In January, 1971, Plaintiff sent to the City Manager and the City Councilmen of the City of Homestead a letter seeking recognition as collective bargaining agent for the firefighter employees of the City of Homestead. Subsequently, recognition was granted by the City Council and thereafter the Council designated Homestead City Manager, Olef R. Pearson, as the City's bargaining representative. Negotiations between Pearson and Plaintiff commenced with the mutual understanding that any agreement reached between the parties would be subject to the approval of the City Council.

In November, 1971, after more than 50 hours of negotiations, the City Manager and the Plaintiff reached accord on a collective bargaining agreement. This agreement with the recommendations of the City Manager attached was submitted to the Homestead City Council for its approval. The council met with representatives of Plaintiff on January 10th, 1972 for that purpose. At this meeting, however, the City Council proceeded to renegotiate the contract from the beginning and, in fact, changed every provision of the contract brought up before the meeting terminated. Among the changes made, the Council altered the "bargaining unit" clause, (which had been agreed to by City Manager Pearson and Plaintiff) by excluding certain members of the Union from the contract's coverage, and proposed that the entire negotiated wage provisions be stricken from the contract and in substitution therefore these wages unilaterally established by the City Council in their budget hearings of the year before being inserted in the contract. In addition, notwithstanding the fact that the City had recognized the Plaintiff and had bargained with their representatives, the Council directed the City Attorney to write an opinion concerning the city's duty to further recognize and bargain with the Plaintiff.

In response to these actions of the City Council, Plaintiffs representatives walked out of the meeting, and on January 27th, 1972, sent a letter to the City Manager invoking the arbitration provisions of the Firefighters Collective Bargaining Act. However, Defendant City failed to respond to said letter except by passing a Ordinance . . . designed to supersede the Firefighters Collective Bargaining Law. . . .

This action was brought by Plaintiffs to enforce their constitutional right to bargain collectively and to bring the City once again to the bargaining table.

This litigation presents several issues to the Court for decision. First, the Court must determine whether the City has performed its obligation of negotiating in good faith with its employees under Article I, Section 6 of the Florida Constitution. Second, the Court must determine if the Firefighters Collective

(continued)

(*continued*)

Bargaining Law establishes collective bargaining guidelines in support of the Constitutional obligation and, if so, the respective duties of the parties thereunder.

Article I, Section 6 of the Constitution of the State of Florida grants public employees the right of collective bargaining. The Florida Supreme Court and the Circuit Courts of this State have on several occasions held that this Section imposes a duty upon the public employer to bargain in good faith with their employees through an organization such as Plaintiff Union.

This Court finds that the practices followed by the Defendant City in this case did not constitute good faith collective bargaining. The defendant's conduct in attempting to renegotiate the entire contract after the lengthy negotiations between the Union and the City Manager, indicates that the bargaining between the City Manager and the Union was only surface bargaining and not a good faith effort by the City to reach agreement with Plaintiff Union. This change in the ground rules after the lengthy contract negotiations were completed demonstrates that the City Manager's function was not to negotiate with the Union on behalf of the City, but rather *to induce the Union to compromise some of its demands in the belief that they were reaching an agreement and then present these compromises to the City Council where further concessions from the Union were to be demanded.* The refusal of the City to show any confidence in the preliminary agreement reached by the City Manager (its appointed negotiator) and its attempt to renegotiate the entire agreement and gain further concessions from the Union on almost every provision of the preliminary agreement, is not good faith [sic] bargaining and does not fulfill the duty imposed by the Florida Constitution . . . [making this] a proper case for the award of punitive damages; this is especially so since the compensatory damages awarded for actual losses to the persons deprived by his conduct are small and difficult to ascertain. The denial of a constitutional right . . . even without any accompanying financial loss, is reprehensible. It is the duty of the Court to protect the constitutional rights of our citizens and to punish those who deliberately subvert these rights. To that end, the Court hereby declares that Councilman Rhodes is a violator of the Constitutional rights of the citizens of this State and reprimand [sic] him for his unlawful activities which are unbecoming to any American citizen and are even more unworthy in one who purports to be a public servant. There is no higher duty in one who serves in government than obedience to the law. When an elected official acts as Councilman Rhodes has acted, in deliberate violation of his oath of office (that is, to support the Constitutions of this State and of the nation), he sets a shameful and humiliating example of lawlessness.

Since this is the first instance of judicial action upon one of the many recalcitrant public employers who wish to purposely ignore the new collective bargaining mandates of Article I, Section 6 of our State's Constitution this Court will, in addition to the foregoing reprimand, only impose $1.00 as punitive damages upon Councilman Rhodes. Such judgment should, however, be a warning and an indication of the intent of this Court to compel obedience to our laws and to the Constitution of this state in the future, by whatever means may be required.

Case Name: IAFF Local 2010 v. City of Homestead

Court: Florida Circuit Court, 11th Judicial Circuit

Summary of Main Points: Attempts by a city council to renegotiate an entire agreement after negotiations had reached a tentative agreement, is not bargaining in good faith.

---

impasse
a labor term referring to the situation in which the parties cannot reach an agreement despite bargaining in good faith

Examples of bargaining in bad faith include situations in which the employer sends representatives to the bargaining table who lack the authority to make genuine proposals, as well as refusing to bargain over mandatory subjects for negotiation. It has been held to be bad-faith bargaining when an employer communicates directly with employees in an effort to undermine the union's role in collective bargaining, or if an employer bargains directly with employees rather than the union.

representational impasse dispute
a labor dispute in which the parties cannot reach agreement concerning whether employees desire union representation, who should be the bargaining representative of the employees, or which positions should rightfully be part of the bargaining unit

IMPASSE DISPUTES

In labor terms, an **impasse** occurs when the parties cannot reach an agreement despite bargaining in good faith. There are three basic types of impasse disputes: representational, interest, and grievance. **Representational impasse disputes** involve issues related to whether the employees desire union representation, who should be the bargaining representative of the employees, and which positions should rightfully be part of the bargaining unit.

Interest impasse disputes, also called collective bargaining impasse disputes, are disputes relating to the negotiation of a collective bargaining agreement. **Grievance impasse disputes** are disputes that arise once a collective bargaining agreement is in existence, and involve the interpretation and administration of an existing collective bargaining agreement.

interest impasse dispute
a labor dispute in which the parties cannot reach agreement over the negotiation of a collective bargaining agreement

Each type of impasse has a different mechanism for resolution. In addition, impasse resolution mechanisms differ from state to state, and within a state between different types of employee groups.

I. Representational Impasse—Who Can Join a Union

grievance impasse dispute
a labor dispute in which the parties cannot reach an agreement over the interpretation and administration of an existing collective bargaining agreement

A fundamental principle in labor relations law is that, where employees desire to organize, there shall be one *exclusive* employee representative with whom the employer must negotiate. Collective bargaining statutes establish guidelines for the make-up of appropriate bargaining units. These laws often speak of the need for members of a bargaining unit to have a "community of interest." When representational disputes and impasses arise in the public sector, the state labor board is authorized to resolve in the matter in the first instance, based upon the appropriateness of the bargaining unit.

Representational conflicts commonly arise in one of three ways: first, whether certain categories of employees have a right to organize; second, whether the majority of employees desire representation; and third, in which bargaining unit certain positions should appropriately be placed. Fire service cases typically involve questions over whether certain officers can organize at all, and if so, whether these officers should be part of the firefighters' bargaining unit, as well as whether positions such as a fire department dispatcher or mechanic should be part of the firefighters' union or the municipal workers' union.

In the private sector, under the NLRA, supervisors are considered to be part of management and have no right to organize. In the public sector, however, supervisors do not have the same authority as supervisors in the private sector. In addition, concerns about protecting supervisory employees from political pressure and partisanship have resulted in supervisors being included in bargaining units throughout the public sector, with some limited exceptions.

International Assoc. of Fire-Fighters Local 2905 v. Town of Hartford
146 Vt. 371, 503 A.2d 1143 (1985)

Hayes, J.

This is an appeal by the Town of Hartford (Town) from an order of the Vermont Labor Relations Board (Board). The question presented for our review is whether the Deputy Fire Chief of the Town is a supervisor within the meaning of 21 V.S.A. § 1502(13).

The Vermont Municipal Labor Relations Act . . . specifically excludes from collective bargaining units employees considered to be supervisors. . . .

The Town contends that the Deputy Fire Chief, Richard Taylor, is a supervisor. The term "Supervisor" is defined in section 1502(13) as:

an individual having authority, in the interest of the employer, to hire, transfer, suspend, lay off, recall, promote, discharge, assign, reward or discipline other employees or responsibly to direct them, or to adjust their grievances, or effectively to recommend such action, if in connection with the foregoing the exercise of such authority is not of a merely routine or clerical nature but requires the use of independent judgment.

The Board concluded that the Deputy Chief did not possess the authority set forth in the statute and held that he was properly included in the bargaining unit proposed by the Hartford Career Firefighters Association, Local 2905, IAFF. We agree and affirm.

The findings indicate that the Deputy Chief does not have the authority to hire, transfer, lay off, recall, promote, or discharge employees or to effectively recommend such action. The Town maintains, however, that the Deputy Chief is

(continued)

(*continued*)

personally responsible for directing all activities of the Town ambulance service and exercises substantial supervisory authority in the fire-fighting [sic] function of the Hartford Fire Department.

In performing his duties as a shift commander and ambulance director, the Deputy Chief can suspend an employee for the remainder of the shift. This authority to discipline, however, is the same as that exercised by the lieutenants who are included in the bargaining unit. In the seven years Richard Taylor has been Deputy Chief, he has never suspended or dismissed any ambulance service personnel or any employee of the Fire Department and no employee has been suspended or dismissed as a consequence of his recommendation.

The Board properly held that the Deputy Chief's authority to discipline is extremely limited and that such authority standing by itself does not make an employee a supervisor. . . . "*The statutory test is whether or not an individual can effectively exercise the authority granted him; theoretical or paper power will not make one a supervisor. . . . Nor do rare or infrequent supervisory acts change the status of an employee to a 'supervisor'*" . . . (citations omitted).

In this case, decisions as to which Fire Department employees will receive merit pay raises are made exclusively by the Fire Chief. The Deputy Chief does not have the effective authority to recommend these pay raises. The Deputy Chief may recommend pay increases for ambulance workers, but the Board found no indication that those recommendations are actually followed by Town authorities.

The authority which makes one a supervisor is authority which, when exercised, is not "of a merely routine or clerical nature but requires the use of independent judgment." . . . The Board found no indication that the Deputy Chief's authority to assign and direct firefighters was other than that of a merely routine nature. Moreover, the Board was not persuaded that the Deputy Chief's duties as ambulance director, in assigning and directing employees, require independent judgment. Thus, the Board concluded that his position fell short of being supervisory in nature. We agree with this conclusion.

The next matter we consider is the Deputy Chief's authority to adjust grievances. As to this, the Board found that Fire Department employees have submitted grievances directly to the Chief in the past two years and have not first grieved to the Deputy Chief. Based on all the evidence, the Board held that the Deputy Chief does not have effective authority in this regard.

The Town asserts that, in the absence of the Fire Chief, the Deputy assumes the authority and responsibility of that position. This contention is of little avail. "An employee does not acquire a supervisor's status by reason of temporarily taking over the supervisor's duties in his absence." . . .

The findings of the Board are supported by the evidence below, and the conclusions of law are supported by the findings. It is not enough for the Town to demonstrate that the evidence below might lend itself to a different reading by a different tribunal. We will not substitute our assessment of the evidence for that of a regulatory body. . . . This is particularly so where the subject matter lends itself to Board expertise, as does the area of labor relations in this case. . . .
Affirmed.

Case Name: International Assoc. of Fire-Fighters Local 2905 v. Town of Hartford

Court: Vermont Supreme Court

Summary of Main Points: A deputy chief is not considered to be a "supervisor" as that term is defined in the Vermont Municipal Labor Relations Act, and thus should be included in the firefighters' bargaining unit.

The inclusion of supervisory personnel in a firefighters' bargaining unit is a fairly common practice in many states. Other states allow supervisory employees in the public sector to organize, provided that they have a bargaining unit that is separate from the firefighters. See *Concord v. Public Employee Labor Relations Board,* 119 N.H. 725, 407 A.2d 363 (1979).

II. Dispute Resolution in Interest Bargaining

interest bargaining
a labor term referring to the process of negotiating a collective bargaining agreement

Interest bargaining refers to the process of negotiating a collective bargaining agreement **(Figure 11-4).** In the private sector, once an interest impasse has been reached, management is free to implement its proposals unilaterally, and the union is free to strike, with very limited exceptions. Because such a policy would have detrimental effects on vital public services such as police,

Figure 11-4 *Interest bargaining is the process of negotiating a collective bargaining agreement.*

fire, emergency medical, sanitation, public health, schools, prisons, and public works, a variety of impasse resolution mechanisms have been developed in the public sector. These mechanisms include mediation, fact-finding, voluntary arbitration, and mandatory arbitration.

mediation

a labor impasse tool whereby an independent third party helps to facilitate the parties in reaching an agreement among themselves

Mediation Most states have some mechanism for the **mediation** of collective bargaining disputes. Typically either party can request mediation, although some states require both parties to request it. Mediation is most commonly conducted by one mediator, although some states provide for a panel of three mediators.

The role of the mediator is to help facilitate the parties in reaching an agreement. This includes maintaining communication between the parties, injecting a neutral presence into an otherwise adversarial relationship, identifying and framing the main issues, and proposing solutions acceptable to both sides. Mediators have no power to dictate terms or to compel either side to make concessions.

fact-finding

a labor impasse tool consisting of an independent party or panel appointed to investigate and report the facts related to an impasse

Fact-Finding Fact-finding is an impasse tool available in the private sector under the NLRA to assist with disputes deemed to constitute national emergencies. Most states specifically authorize fact-finding as an impasse resolution mechanism for public employees. Fact-finding is usually conducted by a panel. Generally, fact-finders are prohibited from attempting to mediate the dispute, although some states specifically allow this. Some states delineate what factors the fact-finding panel must look at, such as comparable wages in the private sector, comparable wages in like-sized communities, the interest and welfare of the public, and so on. Fact-finders may be authorized to include recommendations at the conclusion of fact-finding, or fact-finding may be without recommendations.

arbitration

a dispute resolution mechanism whereby the parties submit the disputed issues to a neutral third party, who will render a decision

Arbitration In more than half the states, legislation authorizes some type of **arbitration** as an interest impasse resolution mechanism. Arbitration is a dispute resolution mechanism whereby the parties submit the disputed issues to a neutral third party who will render a decision. There are a variety of methods of arbitrating interest disputes, ranging from voluntary nonbinding arbitration to compulsory binding arbitration.

voluntary arbitration

an impasse resolution mechanism whereby the parties voluntarily agree to submit some or all of their unresolved issues to arbitration

Voluntary arbitration is an impasse mechanism whereby the parties voluntarily agree to submit some or all of their unresolved issues to arbitration. The parties may agree that they will be bound by the arbitrator's decision, or they may agree that the arbitrator's decision will be advisory in nature.

compulsory binding arbitration

a type of arbitration in which unresolved issues are submitted to an arbitrator, or arbitration panel, whose final ruling is binding on all parties

Compulsory binding arbitration refers to arbitration in which unresolved issues are submitted to an arbitrator—or arbitration panel—whose final ruling is binding on all parties. Compulsory binding arbitration is most commonly applicable to "essential" employees, when the prospect of a strike is

Figure 11-5
Compulsory binding arbitration is most commonly available to firefighters and police officers, because a strike can be harmful to public safety. (Photo by Larry O. Warner.)

final offer arbitration (FOA)

a form of labor arbitration adopted in some states that requires each party to submit a final offer to an arbitrator or arbitration panel, who must choose one proposal or the other; a number of variations on final offer arbitration exist, such as allowing the arbitrator to choose the more reasonable solution offered by the parties on each disputed issue, or requiring the arbitrator to select one final package or the other

not tolerable **(Figure 11-5)**. The public employee groups most commonly subject to compulsory binding arbitration are firefighters and police officers. Some states authorize compulsory arbitration for other groups of public employees, but provide that the arbitrator's decision is not binding, but rather merely advisory.

Compulsory binding arbitration is usually referred to an arbitration panel, not to a single arbitrator. There is considerable variation among states in regard to compulsory binding arbitration. Some states authorize all unresolved issues to be submitted to arbitration. Other states place limits on what issues may be submitted to binding arbitration, or more commonly, on which decisions of the arbitration panel will be binding. For example, in some states the arbitrator's decision is binding only on nonmonetary issues, and advisory on matters that require the expenditure of funds.

Some states have adopted a form of arbitration known as **final offer arbitration (FOA)**. FOA involves the submission of a final offer by each party to the arbitrator, who must choose one proposal or the other. A number of variations on FOA exist, such as allowing the arbitrator to choose the more reasonable solution offered by the parties on each disputed issue, or requiring the arbitrator to select one final package or the other. Some versions of FOA also allow the arbitrator to reject both final offers and require the parties to submit new proposals.

In the *Vallejo* case, the municipality refused to submit certain issues to arbitration, requiring the firefighters' union to seek a court order to compel arbitration. At that point, the municipality challenged the authority of the arbitrator to hear the contested matters.

Fire Fighters Union v. City of Vallejo
12 Cal 3d 608, 526 P.2d 971, 116 Cal. Rptr. 507 (Cal.,1974)
Supreme Court of California

. . . In 1971, during negotiations between representatives of the City of Vallejo and the Fire Fighters [sic] Union as to the terms of a new contract, the parties failed to agree on 28 issues. Pursuant to the process prescribed in the city charter, they submitted the disputed matters to mediation and fact finding. When these procedures failed to effect a resolution, the city agreed to submit 24 of the issues to arbitration but contended that four other issues, namely "Personnel Reduction," "Vacancies and Promotions," "Schedule of Hours," and "Constant Manning Procedure," involved the "merits, necessity or organization" of the fire fighting [sic] service and did not come under the arbitrable provisions. The city refused to accept the recommendations of the fact finding panel with respect to these issues or to submit them to arbitration.

On December 22, 1971, prior to the scheduled hearing before the board of arbitrators, the Fire Fighters Union filed a complaint in the Solano Superior Court seeking mandate to compel the city to submit the four disputed issues to arbitration. The court found for the union on all the issues. . . . The court therefore ordered that a peremptory writ of mandate issue directing the city to proceed to arbitration on the disputed issues. The city appeals. . . .

In the instant case, as we have stated, we are called upon to render a preliminary decision as to the scope of the arbitration. The arbitration process, however, is an ongoing one in which normally an arbitrator, rather than a court, will narrow and define the issues, rejecting those matters over which he cannot properly exercise jurisdiction because they fall exclusively within the rights of management. As Professor Grodin has observed: ". . . collective bargaining and issues arbitration are together a dynamic process, in which the positions of the parties and their interaction with the arbitrator is in a state of constant flux. Proposals get modified and non-negotiable positions become negotiable as the parties sort out their priorities, develop understanding of the implications of their positions, and perceive alternative solutions which they may not previously have considered. To determine what is arbitrable and what is not against this changing context is a bit like trying a balancing act in the middle of a rushing torrent." . . .

"Because arbitration substitutes for economic warfare the peaceful adjudication of disputes, and because controversy takes on ephemeral shapes and unforeseeable forms, courts do not congeal arbitration provisions into fixed molds but

(*continued*)

(continued)

give them dynamic sweep." We therefore must be careful not to restrict unduly the scope of the arbitration by an overbroad definition of "merits, necessity or organization." Nor does this cautious judicial approach expose the city to an excessive assertion of the arbitrators' jurisdiction; the city council after the rendition of the award may reject any award that invades its authority over matters involving "merits, necessity or organization" since the charter itself limits the scope of the arbitration decision to that which is "consistent with applicable law."

With this caveat in mind, we approach the specific problem of reconciling the two vague, seemingly overlapping phrases of the statute: "wages, hours and working conditions," which, broadly read could encompass practically any conceivable bargaining proposal; and "merits, necessity or organization of any service" which, expansively interpreted, could swallow the whole provision for collective negotiation and relegate determination of all labor issues to the city's discretion.

In attempting to reconcile these provisions, we note that the phrase "wages, hours and other terms and conditions of employment" in the MMBA was taken directly from the National Labor Relations Act (hereinafter NLRA). . . . The Vallejo charter only slightly changed the phrasing to "wages, hours and working conditions." A whole body of federal law has developed over a period of several decades interpreting the meaning of the federal act's "wages, hours and other terms and conditions of employment."

In the past we have frequently referred to such federal precedent in interpreting parallel language in state labor legislation. . . .

The origin and meaning of the second phrase—excepting "merits, necessity or organization" from the scope of bargaining—cannot claim so rich a background. Apparently the Legislature included the limiting language not to restrict bargaining on matters directly affecting employees' legitimate interests in wages, hours and working conditions but rather to forestall any expansion of the language of "wages, hours and working conditions" to include more general managerial policy decisions.

Although the NLRA does not contain specific wording comparable to the "merits, necessity or organization" terminology in the city charter and the state act, the underlying fear that generated this language—that is, that wages, hours and working conditions could be expanded beyond reasonable boundaries to deprive an employer of his legitimate management prerogatives—lies imbedded in the federal precedents under the NLRA. As a review of federal case law in this field demonstrates, the trepidation that the union would extend its province into matters that should properly remain in the hands of employers has been incorporated into the interpretation of the scope of "wages, hours and terms and conditions of employment."

Thus, because the federal decisions effectively reflect the same interests as those that prompted the inclusion of the "merits, necessity or organization" bargaining limitation in the charter provision and state act, the federal precedents provide reliable if analogous authority on the issue. . . .

(continued)

(*continued*)

We now turn to an analysis of the specific bargaining proposals which are at issue here.

1. Schedule of Hours. The issue of Schedule of Hours by which the union proposed a maximum of 40 hours per week for fire fighters on 8-hour shifts and 56 hours per week for fire fighters on 24-hour shifts is clearly negotiable and arbitrable despite the city's argument that it involves the "organization" of the fire service. The Vallejo charter provides explicitly that city employees shall have the right to bargain on matters of wages, hours and working conditions; furthermore, working hours and work days have been held to be bargainable subjects under the National Labor Relations Act. . . . The city cites no authority to the contrary. Accordingly, we conclude that Schedule of Hours is a negotiable issue.

2. Vacancies and Promotions. The union's Vacancies and Promotions proposal concerns fire fighters' job security and opportunities for advancement and therefore relates to the terms and conditions of their employment. . . .

3. Constant Manning Procedure. An examination of this issue illustrates the wisdom of judicial self-restraint in attempting pre-arbitral definitions of the scope of arbitration. Apparently the union originally sought to add one engine company and to increase the personnel assigned to the existing engine companies. If these union demands required the building of a new fire house or the purchase of new equipment, they could very well intrude upon management's role of formulating policy. In view of the union's counterclaim that such a station and equipment were necessary for the safety of the men, this issue could have presented a complex problem. But the very flow of the proceedings washed away these questions because the union altered its position and accepted the recommendation of the fact finding committee "that the manning schedule presently in effect be continued without change during the term of the new Memorandum of Agreement." Hence we do not face the problem of whether the construction of a new fire house and the purchase of new equipment would intrude upon managerial prerogatives of policy making.

Although the city challenges even the limited status quo version of the manpower issue, contending that the fact finding ruling involves the "merits" and "organization" of the fire department and is therefore excluded from the scope of bargaining, we cannot conclude at this stage that the manpower proposal is necessarily nonarbitrable.

The city argues that manpower level in the fire department is inevitably a matter of fire prevention policy, and as such lies solely within the province of management. If the relevant evidence demonstrates that the union's manpower proposal is indeed directed to the question of maintaining a particular standard of fire prevention within the community, the city's objection would be well taken.

The union asserts, however, that its current manpower proposal is not directed at general fire prevention policy, but instead involves a matter of workload and safety for employees, and accordingly falls within the scope of negotiation

(*continued*)

(*continued*)

and arbitration. Because the tasks involved in fighting a fire cannot be reduced, the union argues that the number of persons manning the fire truck or comprising the engine company fixes and determines the amount of work each fire fighter must perform. Moreover, because of the hazardous nature of the job, the union also claims that the number of persons available to fight the fire directly affects the safety of each fire fighter [sic].

Insofar as the manning proposal at issue does in fact relate to the questions of employee workload and safety, decisions under the National Labor Relations Act fully support the union's contention that the proposal is arbitrable. . . . [T]he courts have recognized rules and practices affecting employees safety as mandatory subjects of bargaining since they indirectly concern the terms and conditions of his employment. . . .

Given the parties' divergent characterizations of the instant manpower proposal, either one of which may well be accurate, we believe the proper course must be to submit the issue to the arbitrators so that a factual record may be established. The nature of the evidence presented to the arbitrators should largely disclose whether the manpower issue primarily involves the workload and safety of the men ("wages, hours and working conditions") or the policy of fire prevention of the city ("merits, necessity or organization of any governmental service"). On the basis of such a record, the arbitrators can properly determine in the first instance whether or not, and to what extent, the present manpower proposal is arbitrable.

Furthermore, the parties themselves, or the arbitrators, in the ongoing process of arbitration, might suggest alternative solutions for the manpower problem that might remove or transform the issue. Indeed, the union in the instant case has already abandoned one position and assumed another. These are the elements and considerations that argue against preliminary court rulings that would dam up the stream of arbitration by premature limitations upon the process, thwarting its potential destination of the resolution of the issues. Hence we hold that the charter provision as to "merits, necessity or organization" of the service does not at this time preclude the arbitration of the union proposal that the manning schedule presently in effect be continued for the term of the new agreement.

4. Personnel Reduction. Finally, the union advanced a Personnel Reduction proposal which would require that the city bargain with the union with respect to any decision to reduce the number of fire fighters. Under the proposal, any reduction would be on a least-seniority basis, and no new employees could be hired until all those laid off were given an opportunity to return. The city objects to that part of the proposal requiring bargaining on a decision to reduce personnel and contends that any such matter is not negotiable because it involves the merits, necessity or organization of the fire fighting [sic] service.

A reduction of the entire fire fighting force based on the city's decision that as a matter of policy of fire prevention the force was too large would not be arbitrable in that it is an issue involving the organization of the service.

(*continued*)

(*continued*)

Thus cases under the NLRA indicate that an employer has the right unilaterally to decide that a layoff is necessary, although it must bargain about such matters as the timing of layoffs and the number and identity of the employees affected. . . . In some situations, such as that in which a layoff results from a decision to subcontract out bargaining unit work, the decision to subcontract and lay off employees is subject to bargaining. . . . The fact, however, that the decision to lay off results in termination of one or more individuals' employment is not alone sufficient to render the decision itself a subject of bargaining. . . .

On the other hand, because of the nature of fire fighting, a reduction of personnel may affect the fire fighters' working conditions by increasing their workload and endangering their safety in the same way that general manning provisions affect workload and safety. To the extent, therefore, that the decision to lay off some employees affects the workload and safety of the remaining workers, it is subject to bargaining and arbitration for the same reasons indicated in the prior discussion of the manning proposal.

Our conclusion that the issues of Personnel Reduction, Vacancies and Promotions, Schedule of Hours and Constant Manning Procedure, except as limited above, involve the wages, hours or working conditions of fire fighters and are negotiable requires in the context of this suit that the City of Vallejo submit these issues to arbitration. We in no way evaluate the merit of the union proposals, but hold only that under the Vallejo charter they are arbitrable.

Such a result comports with the strong public policy in California favoring peaceful resolution of employment disputes by means of arbitration. We have declared that state policy in California "favors arbitration provisions in collective bargaining agreements and recognizes the important part they play in helping to promote industrial stabilization." . . .

We affirm the judgment as herein modified and remand the case to the superior court. . . .

Case Name: Fire Fighters Union v. City of Vallejo

Court: California Supreme Court

Summary of Main Points: A variety of issues can arguably be mandatory subjects for collective bargaining, or management prerogatives. The preferred method for dealing with such disputes is to leave it in the first instance to an arbitrator, rather than have a court block arbitration before the issues are fully developed.

The *Vallejo* case is somewhat unusual, in that the authority to arbitrate was embodied in the Vallejo City Charter. In another case from California, *Firefighters v. San Francisco*, 68 Cal. App. 3d 896, (1977) the California Court of Appeals ruled that an arbitration provision agreed to in a memorandum of understanding between the city and the firefighters was an impermissible delegation of legislative power. The court distinguished the *Vallejo* case, finding that the existence of the arbitration provision in the Vallejo City Charter was a valid delegation of authority to an arbitrator, something that a simple memorandum of understanding could not do.

In a number of states, compulsory binding interest arbitration has been challenged as constituting an unconstitutional delegation of power to an arbitrator. Most states uphold interest arbitration laws, provided there are legislatively defined guidelines to adequately direct arbitrators, and the opportunity for judicial review. See *City of Richfield v. Firefighters Local 1215*, 276 N.W.2d 42, (Minn., 1979). Other states have found such compulsory arbitration laws to be unconstitutional delegations. See *Salt Lake City v. IAFF*, 563 P 2d 786 (Utah, 1977); *City of Aurora v. Aurora Firefighters Protective Association*, 556 P 1356, (Co. 1977); and *City of Sioux Falls v. Sioux Falls Firefighters Local 814*, 234 N.W. 2d 35 (SD, 1975).

Appeal of Interest Arbitration Awards In a public sector collective bargaining case, an arbitrator or arbitration panel plays a very important role, one that implicates important constitutional issues of delegation of power. An arbitrator has the authority to make decisions that will involve the expenditure of taxpayer funds and the administration of public agencies. For this reason, judicial review of an arbitrator's decision must constitutionally be allowed. However, the need for finality in labor disputes dictates that the appellate review be limited to avoid the entire matter being re-litigated on appeal.

States have differing standards by which an arbitration award can be reviewed. Some states uphold arbitrators' decisions unless unsupported by competent, material, and substantial evidence. Other jurisdictions limit the grounds for appeal to allegations of fraud, corruption, arbitrary or capricious conduct, or if the arbitrators exceeded their authority.

III. Grievance Impasses

grievance
an allegation by one party to a collective bargaining agreement that the other side has violated the agreement

A **grievance** is an allegation by one party to a collective bargaining agreement that the other side has violated the agreement. The procedures for grievance handling are negotiated by the parties and included in the collective bargaining agreement. The near universal solution for handling grievances that reach impasse is to submit the matter to grievance arbitration. Use of a single arbitrator is the most common and economical method in grievance arbitrations. Usually the arbitrator's ruling will be binding upon the parties, although the agreement may state that it will be advisory only.

In *Denver v. Denver Firefighters Local No. 858*, 663 P.2d 1032 (Colo., 1983) a firefighter filed a grievance for an alleged contractual violation. The city refused to submit the matter to arbitration, based on a recent Colorado Supreme Court decision, *Greeley Police Union v. City Council*, which held that the delegation of public responsibility to a non-elected arbitrator was unconstitutional. In deciding the *Denver* case, the Colorado Supreme Court upheld the delegation of the grievance to the arbitrator, and distinguished the *Greeley* case, which involved an interest arbitration.

Colorado is not the only state with differing views of the constitutionality of public sector arbitration, depending upon whether it is for interest arbitration or grievance arbitration. Such states uphold the delegation of power to an arbitrator in the context of resolving a grievance, while finding the delegation of authority to an arbitrator to resolve an interest dispute to be unconstitutional.

Judicial Enforcement and Review of Grievance Arbitrations   Unlike a court, an arbitrator has no inherent power to enforce a decision. If one party refuses to comply with a grievance arbitration decision, or seeks to challenge the validity of a decision, the matter must be brought to a court for determination of the issues and enforcement.

Judicial review of grievance arbitrations is generally limited to questions of corruption, fraud, or the arbitrators exceeding their authority. Most courts give substantial deference to arbitrators' rulings in grievance arbitrations. Some courts will make additional inquiries in determining if arbitrators exceeded their authority by analyzing whether the arbitrators' decision "draws its essence" from the collective bargaining agreement.

IAFF Local 669 v. Scranton
1982 PA. 42499, 448 A.2d 114, 68 Pa. Commw. 105 (1982)
Commonwealth Court of Pennsylvania

Local Union No. 669 of the International Association of Fire Fighters, AFL-CIO, appeals a Lackawanna County Court of Common Pleas order, which vacated a portion of a labor arbitrator's award, involving an interpretation of a provision in the collective bargaining agreement between the union and the City of Scranton. The court ruled that the arbitrator improperly ordered the city to maintain a minimum complement of 215 regularly appointed fire fighters [sic].

The disputed provision of the agreement is Article XVIII, Section 1, which provides in relevant part: "The City agrees with the Union that there shall be no arbitrary or capricious changes in job classifications, the transfers of personnel or reduction in force, or the creation of new job classifications. . . ."

The union asserted . . . that the city had violated this provision by reducing the number of fire fighters to under 215 persons, the work force level at the time the agreement became effective. The arbitrator agreed, and directed the city, "[t]o maintain a minimum level of bargaining unit members at 215 for a two-year period first commencing with the date on which the number of fire fighters is in fact returned to 215."

(continued)

(*continued*)

In its appeal to the common pleas court, the city asserted that the arbitrator should be reversed for failing to interpret the disputed provision as requiring a showing that the city's reduction in the work force was arbitrary and capricious.

Although the common pleas court noted that the matter of maintaining a minimum work force was contained within the terms of the agreement . . . [it] concluded that the arbitrator's interpretation was unreasonable, stating:

> We do not believe that the arbitrator's interpretation as to the size of the bargaining unit is reasonable in view of the language of the relevant provision and the evidence presented to him. He could conclude from the "essence" of the agreement that the parties agreed to maintain a force of 215 in the bargaining unit, but he cannot ignore the clear language of the parties that an arbitrary and capricious reduction in force is required before that agreement is violated. There was no evidence that such was the case here. Without such evidence we cannot correct the award. Accordingly, that portion of the award must be set aside.

We must disagree. . . . [J]udicial review of an arbitrator's decision is limited by the "essence test" requiring "a determination as to whether the terms of the agreement encompassed the subject matter of the dispute." . . . A court's inquiry ends "once it is determined that the issue properly defined is within the terms of the agreement." . . .

Given the key point that maintenance of minimum work force is within the term of Article XVIII, Section 1 of the agreement, judicial exegesis of the modifiers in that section is barred. Although the common pleas court understandably concluded that the language of this section lends itself to the alternative interpretation, our Supreme Court, in *Leechburg,* has made clear that it looks unfavorably upon interference by the judiciary, even where the court could reasonably adopt a different interpretation. This limited standard of review reflects the desire of our Supreme Court to treat an arbitrator as a "court of last resort," and a recognition that, in construing particular language, an arbitrator, as here, may rely on additional evidence, derived from the history of the bargaining.

Accordingly, that portion of the decision of the court of common pleas concerning the minimum work force is hereby reversed.

Case Name: IAFF Local 669 v. Scranton

Court: Commonwealth Court of Pennsylvania

Summary of Main Points: "[J]udicial review of an arbitrator's decision is limited by the 'essence test' requiring 'a determination as to whether the terms of the agreement encompassed the subject matter of the dispute.' . . . A court's inquiry ends 'once it is determined that the issue properly defined is within the terms of the agreement.'"

arbitrability
a labor law term used to describe challenges to whether or not an issue can be decided by an arbitrator

substantive arbitrability
a labor law term used to describe challenges to whether or not the collective bargaining agreement allows a particular dispute to be submitted to an arbitrator

procedural arbitrability
A labor law term describing questions about whether the proper procedure was followed in processing and handling a grievance

Arbitrability of Grievances The term **arbitrability** is often used to describe challenges to whether or not a grievance issue can be decided by an arbitrator. The arbitrability of an issue is a matter of law, and frequently is viewed as going to the heart of whether an arbitrator has contractual or jurisdictional authority to even hear a matter. States differ on how a party is allowed to raise the issue of arbitrability, although most states follow the private sector rule originally formulated in three cases decided in 1960 by the United States Supreme Court, referred to as the Steelworkers Trilogy.

These cases recognize two types of arbitrability issues in grievance arbitrations: substantive arbitrability and procedural arbitrability. **Substantive arbitrability** refers to whether or not the collective bargaining agreement allows a particular dispute to be submitted to an arbitrator. The focus in substantive arbitrability cases is whether the parties agreed to submit this type of issue to arbitration. Unless expressly agreed otherwise in the contract, issues of substantive arbitrability are to be made by a court, not an arbitrator. The reviewing court is limited to determining whether or not the matter should be decided by an arbitrator, and must not get into consideration of the actual merits of the grievance.

Procedural arbitrability refers to questions about whether the proper procedure was followed in processing and handling a grievance. Procedural arbitrability issues are to be left to the arbitrator. In cases in which both substantive and procedural issues arise, procedural issues should be resolved by the arbitrator, after the court makes a determination that the underlying dispute is one that the parties have agreed to submit to the arbitrator. The Steelworkers Trilogy cases are listed in Appendix B, Chapter 11.

Whose Grievance Is It? A common issue that arises in grievance cases involves the right of an individual union member to pursue a grievance independently from the union. The general rule is that the grievance belongs to the union, and can only be brought forward by the union. Consider the following case.

City of Pembroke v. Zitnick
792 So.2d 677 (Fla.App. Dist.4, 2001)
Florida Court of Appeals

. . . Zitnick was a firefighter for Pembroke and a member of the International Association of Firefighters Union, Local 2292 (IAFF). Zitnick was fired after an investigation into allegations that he falsified materials in a city-sponsored contest. In response, Zitnick filed a grievance under the CBA seeking backpay and reinstatement. The IAFF conducted an independent inquiry and ultimately declined to pursue the matter to arbitration due to a lack of merit. Zitnick was informed of the IAFF's decision by letter.

(continued)

(continued)

Zitnick then filed directly with Pembroke. However, Pembroke declined to entertain the grievance. . . . As a result of the IAFF's refusal to grieve the matter, Zitnick brought the underlying action. . . . The trial court conducted the hearing [and issued] its order compelling arbitration. . . .

Pembroke claims that the IAFF is a certified bargaining union that has retained control over the grievance procedure; therefore, it is not obligated to submit to arbitration because the union declined to process the grievance because it lacked merit. We agree.

Where a certified bargaining agent retains contractual control over the arbitral step of the grievance procedure and it declines to process a grievance to arbitration because it believes the grievance to be without merit, the public employer is not obligated to arbitrate the dispute if the grievant submits it to arbitration directly. . . .

In the case at bar . . . the IAFF has retained control over the grievance process. Moreover, the correspondence between Zitnick and the IAFF demonstrates that the IAFF determined that Zitnick's claim lacked merit. Therefore, in accordance with the holding of Galbreath, we reverse and vacate the trial court's order compelling arbitration.

Case Name: City of Pembroke v. Zitnick

Court: Florida Court of Appeals

Summary of Main Points: The control of employee grievances rests with the union, which may choose to decline to proceed with a grievance at any point. In such a case, the individual member may not continue to proceed through the grievance process.

The rationale behind the rule that grievances belong to the union, not the individual member, rests on solid policy grounds. Grievance resolution can result in the interpretation of ambiguous contractual language and changes to workplace rules, which impact more than just the member who filed the grievance. Allowing any union member to bring a grievance forward independently of the union could result in changes to workplace rules contrary to the interests of the union and the majority of union members.

SCOPE OF BARGAINING

The scope of subjects that may or must be discussed during collective bargaining in the public sector is defined initially in the labor relations acts, and refined by court decisions. Public sector collective bargaining laws usually mirror the language used in the NLRA to describe the scope of bargaining. The commonly defined scope of bargaining is said to be: "wages, hours and other terms and conditions of employment."

In the private sector, the concept of "management prerogative" serves to keep important managerial decision making, which is at the core of entrepreneurial control, from being subject to mandatory collective bargaining. In the public sector, concerns over management prerogative are magnified and implicate issues of constitutionality.

The concern over management prerogatives has led some states to adopt management rights or management prerogative provisions in their collective bargaining acts, to ensure that such subjects remain exempt from bargaining. Legislatures have gone so far as to curtail the scope of bargaining in response to court decisions that liberally interpreted the scope. See Nev. Rev. Stat. Sec. 288.150 (1976); *Washoe Co. Teachers Assn. v. Washoe Co. School District,* 90 Nev. 442, 530 P. 2d 114 (1974).

There are three distinct categories of subjects for collective bargaining: mandatory, permissive, and prohibited subjects.

mandatory subjects
collective bargaining subjects over which both management and labor are required to bargain

Mandatory subjects are those subjects for which both management and labor must collectively bargain. Mandatory subjects include those issues that clearly fit within the description of "wages, hours and other terms and conditions of employment." The parties can be compelled to bargain over mandatory subjects, and refusing to negotiate over a mandatory subject of negotiation constitutes an unfair labor practice. In addition, neither side is permitted to unilaterally change a mandatory subject without first having to bargain over it.

prohibited subjects
collective bargaining subjects that the parties are prohibited from bargaining

Prohibited subjects are those that the parties are prohibited from bargaining over. If the parties were to reach an agreement over a prohibited subject, the agreement would not be enforceable. Subjects are usually determined to be prohibited because they would violate a law, or would violate a public policy. Management is permitted to unilaterally change a prohibited subject without first having to negotiate.

permissive subjects
collective bargaining subjects that, while not within the scope of the mandatory subjects, are not of the type that are prohibited

Permissive subjects are subjects that, while not within the scope of the mandatory subjects, are not of the type that are prohibited, either. While neither side is required to bargain over permissive subjects—and neither side can be *compelled* to bargain over permissive subjects—bargaining is permitted when both sides agree. Typically, permissive subjects are those that fall within a management prerogative. Management may—but is not required to—bargain over a permissive subject. However, if an agreement is reached on a permissive subject, management will be bound to honor those permissive subjects that become embodied in the agreement.

Examples of each of the three types of subjects, along with the associated cases, are included in Appendix B, Chapter 11. Not all states recognize the three categories of subjects for bargaining. For example, New Jersey recognizes only mandatory subjects and prohibited subjects. Subjects that would otherwise be considered permissive subjects are considered to be prohibited subjects in New Jersey.

Determining whether or not a particular subject is bargainable can be a challenge. Most scope of bargaining disputes involve issues that affect "wages, hours and other terms and conditions of employment" while at the same time they involve matters that have traditionally been considered management prerogatives. To help resolve these disputes, most jurisdictions have developed tests to aid in the determination about whether a subject is bargainable or not.

Some tests look at whether a subject has a "significant relationship" or "direct and intimate relationship" with "wages, hours, and others terms and conditions of employment." Other tests attempt to take into account the competing interests by balancing management prerogatives with employees' right to bargain.

IAFF, Local 314 v. City of Salem
68 Or. App. 793, 684 P2d 605 (1984)

WARREN, J.

. . . The City and the firefighters entered into collective bargaining negotiations for a successor agreement to their 1981–83 contract. The firefighters proposed several items during the negotiations over which the City refused to bargain, claiming that they were only permissive subjects of bargaining. The firefighters filed an unfair labor practice complaint with [the Oregon Employment Relations Board] (ERB) to compel the City to bargain over the items, claiming that they were mandatory subjects of bargaining and that the City's refusal violated its duty to bargain in good faith under ORS 243.672(1)(e). ERB entered an order stating that the firefighters' proposals concerning layoff and safety were mandatory subjects and directed the City to bargain over those proposals. The City appeals that portion of the order compelling bargaining over the firefighters' proposal entitled "Article-21-Safety." The City challenges ERB's reasoning process and claims that there is not substantial evidence to support the conclusion that the proposal is mandatory. . . .

The firefighters contend, and ERB's order concludes, that their proposal is a condition of employment, mandatory for bargaining, because safety is of "like character" to the statutory examples of "employment relations" provided in ORS 243.650(7) and that the firefighters had proved that the preponderant purpose of the specific language of the proposal is to protect employees rather than to affect staffing decisions, which is a management prerogative. Because "employment relations" as . . . are subject to mandatory bargaining, ERB ordered the City to negotiate over the proposal. . . .

[T]he issue in this appeal is whether ERB's interpretation of "other conditions of employment," ORS 243.650(7), to include this proposal is within the range of acceptable interpretation of law ERB is empowered to make and is supported by substantial evidence.

(continued)

(*continued*)

The City claims that ERB's reasoning process placed an inappropriate burden on the employer, ignored the *specific* union proposal and abandoned the "balancing test" normally applied to determine whether a particular proposal is a mandatory or permissive subject for bargaining. It argues that ERB failed to recognize that the minimum personnel requirement . . . engines and truck companies is no more than a back door attempt to infringe on management prerogatives to control staffing and is inappropriately labeled "safety."

ERB analyzed the safety proposal in two stages. It concluded, first, that the general subject of "safety" is a condition of employment of like character to those employment relations enumerated in ORS 243.650(7). Having so decided, it then examined the specific proposal and the evidence presented to determine whether the proposal was in reality a "staffing" proposal rather than a safety proposal.

The evidence presented by the firefighters focused on the interdependent nature of firefighters' tasks. A private consultant in the field of fire protection reviewed the City's firefighting system and testified that, in his opinion, a drop from a three-person engine company to a two-person company would cause an immediate increase in the frequency and severity of injuries to the remaining firefighters, based on the tasks that must be accomplished, at least until employees had been retrained and new safety rules had been developed. A current Salem firefighter described the teamwork under the current three-person engine company. He concluded that a reduction from a three-person to a two-person crew would increase his risk of personal injury. The City's only witness was the Salem Fire Chief, who confirmed that, under the current three-person system, there is a "good possibility" that a reduction of a crew from three to two could be a safety consideration.

Having considered that evidence, ERB explained:

"* * * Essentially, all we are saying in the majority opinion is that the IAFF proved its case; i.e., that any change in the current system of staffing fire engines would have a greater effect on employee safety than it would have on management's prerogative to establish staffing levels. That being the case, we believe the proposal at issue (maintenance of current staffing practices or the bargaining over changes) is an appropriate approach to the safety concerns the IAFF wishes to address. * * *"

ERB determined that the specific proposal has a greater effect on employee safety than on staffing prerogatives. This reasoning comports with the spirit of the "balancing test." The fact that ERB bifurcated its reasoning process takes nothing from the balance. No doubt the proposal affects management prerogatives in the allocation of staff. Although the order to bargain over the proposal would not require the City to bargain over the number of engine companies necessary to meet the public need, it is an attempt by the firefighters to require that three persons be assigned to each engine company that the City determines is necessary. ERB's decision recognizes that, at some point, reductions in personnel levels present safety concerns which are as critical to the conditions of employment as are monetary benefits, hours, vacations, sick leave and grievance procedures. . . . The

(*continued*)

(*continued*)

evidence presented establishes that personnel levels and safety concerns are inter-related. ERB' s decision that safety concerns outweigh management policy in this instance is supported by substantial evidence. . . .

We conclude that the order is within the interpretation of the law ERB is empowered to make and is supported by substantial evidence.

Affirmed.

Case Name: IAFF, Local 314 v. City Of Salem

Court: Oregon Court of Appeals

Summary of Main Points: The balancing test used by the Oregon Employment Relations Board to determine whether a subject is a mandatory subject or a prohibited subject (management prerogative) is within the discretion of the ERB, and is supported by "substantial evidence."

Staffing decisions are among the most hotly contested cases involving the scope of collective bargaining for firefighters. Staffing relates to many different mandatory subjects for firefighters, including working conditions, safety, and workload. Yet staffing also goes to the heart of management prerogative, in terms of what level of service a community will provide, how much funding a community should allocate for fire protection, and how available resources will be distributed around a community.

Another hotly contested issue for firefighters is that of residency.

St. Bernard v. State Employee Relations Board
74 Ohio App.3d 3 (1991)
1st District Court of Appeals of Ohio, Hamilton County

. . . The IAFF is the exclusive representative for the bargaining unit consisting of the city's firefighters and paramedics. In October 1985, the IAFF filed a notice seeking to commence negotiations with the city for an initial collective-bargaining agreement. . . . Among the items for negotiation, the IAFF listed the issue of residency. However, the city repeatedly refused to bargain on this issue, contending that residency was not a subject of mandatory collective bargaining. On January 15, 1986, after the factfinder [sic] . . . failed to make findings as to residency, the IAFF filed an unfair-labor-practice charge with SERB. Eventually, the parties did enter into an agreement which was silent as to residency. Almost immediately thereafter, on April 30, 1986, the city enacted its first-ever residency ordinance,

(*continued*)

(*continued*)

requiring all employees to reside in the city within six months of passage or "forfeit their position."

Following investigation of the IAFF's complaint, SERB found probable cause to believe that the city had committed an unfair labor practice, and directed that a complaint be served. After a hearing, the SERB-appointed hearing officer issued a proposed order finding that the city had committed an unfair labor practice. On March 15, 1989, upon its determination that the residency requirements were a subject of mandatory bargaining, SERB ordered the city to negotiate with the IAFF on the issue. The city appealed to the trial court, which granted the appeal, denied SERB's cross-petition for enforcement, and set aside SERB's order.

Mandatory subjects of collective bargaining are deemed to be matters of immediate concern that vitally affect the terms and conditions of employment of the bargaining-unit employees. . . . R.C. 4117.08(A) provides that the following in Ohio are subjects of mandatory bargaining: "*All matters pertaining to wages, hours, or terms and other conditions of employment and the continuation, modification, or deletion of an existing provision of a collective bargaining agreement are subject to collective bargaining between the public employer and the exclusive representative, except as otherwise specified.*"

As further required by R.C. 4117.08(C), public employers must also bargain in areas that are subjects of management rights and direction of the governmental unit if they "affect wages, hours, terms and conditions of employment * * *." . . . Therefore, a public employer's decision to exercise a management right which affects the terms and conditions of the unit's employment becomes a mandatory subject for bargaining. . . . In this case, we agree with SERB that initial or continued employment for employees within the bargaining unit is contingent upon residence. Residency is said to be a "qualification" for employment under Ordinance No. 17-1986, and therefore is outside the scope of management rights because the city has expressly made it a condition of employment. . . .

The judgment of the trial court is reversed, and SERB's order is reinstated. This cause is remanded to the trial court with instructions to enforce SERB's order.

Case Name: St. Bernard v. State Emp. Relations Bd.

Court: 1st District Court of Appeals of Ohio, Hamilton County

Summary of Main Points: Residency requirements are a mandatory subject for bargaining, because they directly affect the terms and conditions of employment of workers who are subject to them.

The United States Supreme Court has upheld residency rules as constitutional. One of the leading cases on the constitutionality of residency requirements is a firefighter case, *McCarthy v. Philadelphia Civil Service Commission*, 424 U.S. 645 (1976), which will be discussed in Chapter 13. Most successful challenges to residency rules have been based upon state law issues.

IMPACT BARGAINING

impact bargaining

mandatory bargaining over the impact of a decision that involves a management prerogative, and would otherwise be beyond the scope of bargaining

Where a subject is clearly considered to be a management prerogative, and thus beyond the scope of bargaining, but the impact of such a decision affects "wages, hours and other terms and conditions of employment," some states require bargaining on the impact of the decision. This is called **impact bargaining**. One area where impact bargaining has commonly been permitted is the subject of firefighter layoffs. While the initial decision to lay off firefighters for economic reasons is a management prerogative, the impact of that layoff must be bargained in jurisdictions that recognize impact bargaining.

UNILATERAL CHANGES TO MANDATORY SUBJECTS AND PAST PRACTICES

An employer is prohibited from unilaterally implementing changes that involve a mandatory subject for collective bargaining. When a subject is clearly embodied in the collective bargaining agreement, the matter is clear-cut and any unilateral change to that subject would give rise to a grievance. When the matter involves something that the parties did not contemplate at the time the agreement was negotiated, but that clearly involves a mandatory subject, the employer is similarly prohibited from unilaterally implementing the proposed change. See *IAFF Local 1974 v. City of Pleasanton*, 56 Cal. App. 3d 959, 129 Cal. Rptr. 68 (Cal.App.Dist., 1976).

past practice

a labor term relating to a practice that has been followed by both labor and management consistently and over a long period of time, but which has not been incorporated in writing into the collective bargaining agreement

When both the employer and the union have followed a long and consistent practice, but the practice has not been included in the collective bargaining agreement, the practice is commonly referred to as a **past practice**. For example, certain procedures for calculating sick leave accrual or determining seniority that are not embodied in the collective bargaining agreement could qualify as past practices. In most jurisdictions, an established past practice is treated the same as a written contractual provision, and violating or unilaterally changing an established past practice can give rise to a grievance.

The law of past practice is fairly complex and subject to a number of exceptions and variations. Generally, for a past practice to exist, the practice must have been applied consistently and occurred repeatedly over a substantial period of time. It cannot have been a special, one-time practice or intended at the time to have been an exception to the rule. Both the employer and the union must have known that the practice existed and consented or at least allowed it to occur. In addition, the matter cannot involve a management prerogative.

When a past practice contravenes clear contractual language, the contractual language will ordinarily supersede the past practice. However, some states find that where the past practice is widely acknowledged and mutually accepted by both parties, it can be considered to have amended the contract. The party seeking to supersede the contractual language with a past practice must prove that the parties had a meeting of the minds with regard to the past practice, such that there was an agreement to modify the contract. See *Flint Professional Firefighters Union Local 352 v. City of Flint*, Unpublished Decision, No. 244953 (Mich.App. 06/17/2004). Not all states follow such reasoning.

UNION SECURITY PROVISIONS

dues check-off

a form of union security agreement whereby the employer agrees to deduct a designated amount from an employee's paycheck, and pay the amount withheld to the union

closed shop

a form of union security whereby an employer agrees to hire only workers who are already members of a particular union, and requiring that workers maintain their union membership throughout their employment

union shop

a form of union security whereby an employer agrees to mandate that employees become union members within a stipulated period of time after being hired

Union security provisions are clauses in collective bargaining agreements that allow a union to collect dues in a regular and timely manner. Union security is considered vital to the effective existence and administration of unions. There are six basic types of union security provisions: dues check-off, closed shop, union shop, maintenance of membership, agency shop, and fair share.

A **dues check-off** provision is the simplest form of union security. It is an agreement by the employer to deduct a designated amount from an employee's paycheck, and pay the amount withheld to the union. A dues check-off must be specifically authorized by an employee in order to be valid.

In *City of Charlotte v. Local 660, IAFF*, 426 U.S. 283 (1976), the firefighters' union sought to challenge the city's refusal to withhold union dues as a violation of equal protection, because the city allowed check-offs for other purposes. The Supreme Court found no violation of equal protection by a public employer who refuses to honor a union dues check-off simply because it honors requests for check-offs for other purposes.

A **closed shop** provision requires an employer to hire only workers who are already members of a particular union, and furthermore requires that workers maintain their union membership throughout their employment. Under a closed shop provision, membership in the union is a condition of employment. While originally lawful under the NLRA as drafted in 1935, closed shops were expressly prohibited in the private sector in 1947, and are also illegal in the public sector.

Closely related to a closed shop provision is a **union shop** provision, which requires an employer to mandate that employees become union members within a stipulated period of time after being hired. The common time frame for becoming union members is 30 days. In the public sector, a minority of states authorize public employees to agree to union shop clauses in their collective bargaining agreements. The NLRA permits union shop provisions to be negotiated and included in collective bargaining agreements in the private sector, except in so-called "right-to-work" states.

right-to-work states
states that have statutes that expressly prohibit union security agreements, as well as any agreement or requirement that an employee join or pay dues to a union as a condition of employment, whether in the public or private sector

maintenance of membership
a form of union security whereby an employer agrees that once employees becomes members of the union, they must maintain union membership as a condition of employment; however, maintenance of membership provisions do not require employees to join a union

agency shop
a form of union security whereby an employer agrees to mandate that all employees pay union dues—or an amount equal to the periodic union dues—as a condition of employment

Right-to-work states have statutes that expressly prohibit union security agreements, as well as any agreement or requirement that an employee join or pay dues to a union as a condition of employment, whether in the public or private sector. The philosophy behind right-to-work laws is to protect employees from compulsory unionism. The euphemism "right-to-work" is intended to imply that workers in such states have the right to work, unimpeded by a union that they do not support. Research shows an unintended result: on average, workers in **right-to-work states** earn less than workers in other states. Examples of right-to-work laws are included in Appendix B, Chapter 11.

Maintenance of membership provisions are agreements that mandate that once employees becomes members of the union, they must maintain union membership as a condition of employment. However, maintenance of membership provisions do not require employees to join a union. Maintenance of membership provisions are prohibited in right-to-work states, since they would require membership in a union as a condition of employment.

An **agency shop** provision requires an employer to mandate that all employees pay union dues—or an amount equal to the periodic union dues—as a condition of employment. In *Abood v. Detroit Board of Education,* 431 U.S. 209 (1977), the United States Supreme Court determined that the Constitution prohibits a public employees' union from using union dues mandated by agency shop rules to advance political or ideological causes not germane to its duties as a collective bargaining representative. The Court ruled that, when employees do not object to advancing such causes, unions may rightly utilize funds for such a purpose, but a member cannot be forced to contribute any amounts beyond those necessary to cover the costs associated with collective bargaining and contract administration.

After *Abood,* an employer may collect an agency shop fee, provided it represents a fee for collective bargaining and related expenses rendered by the union. It is typically the union dues less an amount spent by the union on political and ideological issues. Agency shop provisions are legal under the NLRA, and in most states for public employees, but are prohibited in right-to-work states.

A **fair share** provision is similar to an agency shop provision except that where in an agency shop an employee is required to pay an amount equal to union dues as a condition of employment, under a fair share provision all employees who do not pay union dues must pay a proportionate share of the cost of collective bargaining activities.

Although not a union security provision, the term **open shop** refers to an employer that does not mandate that employees belong to a union, nor require the payment of fees or dues as a condition of employment. Open shops are most common in "right-to-work" states.

fair share
a form of union security whereby an employer agrees that employees who are not members of the union and thus do not pay union dues, will be required to pay a proportionate share of the cost of collective bargaining activities

open shop
an employer that does not mandate that employees belong to a union, nor require the payment of fees or dues as a condition of employment; open shops are most common in right-to-work states

DUTY OF FAIR REPRESENTATION

A union that is recognized as an exclusive bargaining representative has an obligation to represent the rights of each employee in the bargaining unit fairly. The duty of fair representation extends to all members of the bargaining unit, not just to union members.

The duty of fair representation arises out of the fact that the union is the exclusive representative of the employees. Fair representation does not require the union to take an employee's side in every situation, nor does it require the union to process every grievance presented. The union must make a good-faith determination in each case—based on the merits of the case—in deciding whether it will support an employee's contentions. Decisions that are arbitrary, discriminatory, or made in bad faith violate the duty of fair representation, and may give rise to an employee lawsuit against the union for damages.

Unions are often called upon to make difficult decisions in situations in which the protection of one union member's rights comes at the expense of another member. For example, if one union member feels he or she was improperly passed over for promotion or transfer, filing a grievance can affect the rights of the union member who was selected. In such a situation, the union must make a good-faith decision based upon the merits of the case in deciding to file a grievance.

STRIKES, SLOWDOWNS, PICKETING, AND CONCERTED JOB ACTIONS

strike
any concerted activity by employees that serves to impact an employer in order to make demands or protest a managerial decision; this would include work stoppages, slowdowns, sick-outs, or other organized refusals to work

The term **strike** as defined in the NLRA includes "any strike or other concerted stoppage of work by employees (including a stoppage by reason of the expiration of a collective-bargaining agreement) and any concerted slowdown or other concerted interruption of operations by employees," 29 USC § 142. Any concerted activity that serves to impact work in order to make demands or protest a managerial decision is a strike. This would include work stoppages, slowdowns, sick-outs, or other organized refusals to work.

At common law, all strikes by public employees were considered to be illegal, and courts had the authority to issue restraining orders to enjoin strikes **(Figure 11-6)**. Many states still adhere to this doctrine in regard to strikes by public employees, while some state laws specifically prohibit strikes by some or all public employees. In the private sector, strikes are considered to be economic weapons of labor, while in the public sector strikes are more often tools to exert political pressure on public officials, although in many cases the financial impact cannot be ignored, either.

Figure 11-6 *At common law, strikes by public employees were illegal. (Photo by Larry O. Warner.)*

In *Dover v. IAFF Local 1312,* 322 A 2d. 918 (N.H., 1974), the refusal of firefighters to return to duty for alarms while off duty as part of a bargaining strategy for a new contract, was considered to be a "slowdown," which was enjoined by a court for being an illegal strike. In another case from New Hampshire, *Manchester v. Manchester Firefighters Association,* 120 N.H. 230, 413 A.2d 577 (N.H., 1980), a sick-out was determined to be a strike and enjoined by a court. In upholding the injunction, the New Hampshire Supreme Court further ruled that the failure of the city to bargain in good faith did not excuse or justify the job action by the firefighters.

Not all states follow the common-law rule prohibiting strikes by public employees. A few states grant some public employees a limited right to strike. In *IAFF Local 1494 v. City of Coeur d'Alene,* 586 P. 2d 1346 (ID, 1978), a statute outlawed strikes by firefighters during the term of a collective bargaining agreement. The Supreme Court of Idaho ruled that the law did not apply to a strike that commenced after a contract had expired, but left open the question of whether striking firefighters could be disciplined for their participation in the strike. The court ruled that, when the employer refused to bargain in good faith, the striking firefighters could not be disciplined. In *Garavalia v. City of Stillwater,* 283 Minn. 354, 168 N.W. 2d 336 (MN, 1969), firefighters were terminated for walking off the job in violation on an anti-strike statute that declared strikers to have abandoned and terminated their jobs. The court upheld the termination of the striking firefighters.

Figure 11-7

Firefighting is one of the least likely public services in which a strike can be tolerated.

A difficult question that arises with regard to strikes and concerted activities concerns drawing the line between an illegal strike and a lawful exercise of one's rights. Peaceful picketing that is unrelated to an unlawful strike has generally been upheld. See *Tassin v. Local 832, NUPO,* 311 So. 2d. 591 (LA, 1975). States have adopted a variety of perspectives on this matter, some of which focus on how to define an illegal strike, while others focus more on what remedies are appropriate when a strike occurs. Most courts that have dealt with strikes by public employees have tended to focus on how vital a service the striking employees provide, and how disruptive the concerted activity is to public safety. Because firefighting is one of the most vital public services, it is among the least likely in which job actions will be tolerated **(Figure 11-7)**.

Besides the use of injunctions to stop a strike, courts may also enforce no-strike laws through contempt power, incarceration of union leadership and members, fines, penalties, and damage assessments. Some authorities suggest that damage suits by third parties injured by illegal strikes may be brought against public sector unions.

WEINGARTEN RIGHTS

In one of the most important labor law cases in history, the United States Supreme Court upheld an NLRB determination that it is an unfair labor practice for an employer to deny an employee the right to have union representation at an investigatory meeting or hearing where the employee's answers to questions could result in disciplinary action **(Figure 11-8)**. The case, *NLRB v. Weingarten, Inc.,* 420 U.S. 251 (1975), was a private sector case under the NLRA, but has been recognized universally throughout the United States for both private and public sector employees. The rationale for the Weingarten rule is that employees have a right to engage in concerted action for mutual aid and protection, and that to deny an employee union representation during a disciplinary meeting or hearing violates this right.

Union firefighters and fire service managers should understand several aspects of Weingarten rights:

- The right to have a union representative present applies only in situations where an employee reasonably believes that the investigation will result in disciplinary action.

- The meeting must be investigatory in nature. When the purpose of a meeting is simply to discuss or convey management's complaints about the employee's performance in a non-disciplinary manner, the employee's Weingarten rights have not been denied by not having a union representative present.

Figure 11-8 *Under* Weingarten, *an employee has a right to union representation at any meeting that could result in disciplinary action.*

- The employee must request that a union representative be present, or the right is considered to be waived. *Weingarten* does not require an employer to warn an employee of the right to union representation.

When Weingarten rights are violated, the employer is liable for an unfair labor practice, and the employee may not be disciplined based upon the information obtained from the improper questioning. In some states, specific laws such as a police officers' bill of rights, which specifies how disciplinary procedures are to be handled, may supersede *Weingarten*. However, most states adopt *Weingarten* for all public employees.

GARRITY RIGHTS

The Fifth Amendment to the United States Constitution provides that no person shall be compelled to give testimony against himself in a criminal case. Occasionally a firefighter is accused of some wrongdoing that may constitute a crime, and as part of an investigation is ordered by a superior officer to answer questions about what happened. In these circumstances, firefighters are in a no-win situation. If they answer the questions, they may be convicted of a crime, and yet if they refuse to answer the questions they may be disciplined for insubordination.

Firefighters do not surrender their constitutional rights by virtue of being public employees. The Garrity Rule is a rule applicable to public employees—most commonly firefighters and police officers—that resulted from *Garrity v. New Jersey*, 385 U.S. 493, decided in 1967 by the United States Supreme Court. *Garrity* involved police officers who were accused of fixing traffic tickets. During the investigation, the officers were informed of their right to remain silent, but informed that if they remained silent they would lose their jobs.

The Supreme Court ruled that it was unconstitutional to order police officers to answer questions under threat of losing their jobs, and then use the answers to incriminate them. As a result, the investigators have a choice: compel the employee to answer questions, in which case those statements may not be used in the criminal prosecution of the individual officer, or allow the employee the right to remain silent without penalty or threat of penalty **(Figure 11-9)**.

In order to invoke one's Garrity rights, the employee must announce that he or she wants the protections under *Garrity*. By invoking the Garrity rule, the employee is invoking his or her right against self-incrimination. Any statements made after invoking Garrity may be used only for department investigation purposes, but not for criminal prosecution purposes.

Fire service managers should address two important considerations with regard to Garrity rights before questioning begins. First, if a firefighter is

Figure 11-9 *A firefighter accused of criminal wrongdoing has a right to remain silent.*

compelled to answer questions as a condition of employment, the firefighter's answers and any evidence generated from those answers may not be used against the firefighter in any subsequent criminal prosecution. Second, the department managers become limited as to what subjects they may inquire about. When a firefighter is ordered to answer questions involving possible wrongdoing, the questions must be specifically, narrowly, and directly tailored to the job-related matter being investigated. The questioning cannot extend into areas that involve matters unrelated to the original investigation. Fire service managers confronted with a Garrity situation should obtain competent legal counsel to assist them in making a determination about the proper course to follow.

SUMMARY

Collective bargaining is the process whereby an employer and the duly appointed representative of the employees negotiate an agreement pertaining to wages, hours, and other terms and conditions of employment. Public sector employees fall under the jurisdictions of state labor relations acts, while private sector employees are subject to the National Labor Relations Act. States differ tremendously in collective bargaining laws for firefighters, in what subjects are considered bargainable, and the recourse in the event that collective bargaining does not result in a negotiated settlement.

Strikes by public employees are illegal in most states, and are particularly troubling for firefighters. Many states provide for mediation, fact-finding, or arbitration of labor disputes involving firefighters as tools to resolve collective bargaining impasses, and avoid the prospect of a strike. Underlying collective bargaining is the principle of exclusive representation, meaning that one union will represent the interests of all employees in a bargaining unit. A necessary corollary to exclusive representation is the duty of fair representation. The union must represent the interests of all bargaining unit members fairly, and without acting in an arbitrary, discriminatory, or bad-faith manner.

Employees are entitled to union representation when being disciplined, or questioned about a matter that may lead to disciplinary action. These rights are known as Weingarten rights. In addition, firefighters cannot be ordered to disclose matters that may incriminate themselves, under threat of being disciplined if they remain silent. These rights, known as Garrity rights, require the employer to make a decision when questioning such a member, and decide whether the member can remain silent without consequence in his or her employment, or be ordered to disclose information, in which case he or she cannot be prosecuted.

REVIEW QUESTIONS

1. In the absence of constitutional, legislative, or executive authorization for collective bargaining, do municipal firefighters have a right to bargain collectively with their employers?

2. What are the three basic types of impasse disputes?

3. What are the three types of subjects for collective bargaining?

4. What are the most common forms of interest dispute resolution mechanisms?

5. What type of dispute resolution mechanism is most common for grievances?

6. Explain how impact bargaining can be used to expand the scope of bargaining.

7. List the six types of union security provisions and describe each.

8. What are Weingarten rights, and when do they apply?

9. When do Garrity rights apply to a firefighter?

10. What is a strike?

DISCUSSION QUESTIONS

1. When a union has to make a choice about benefiting one member at the expense of another, how can it possibly satisfy the duty of fair representation to both?

2. Explain the differences between collective bargaining in the public sector compared with the private sector.

3. Can you identify any examples of bargaining in bad faith?

Chapter 12

EMPLOYEE RIGHTS AND DISCRIMINATION

Learning Objectives

Upon completion of this chapter, you should be able to:

- Identify the major employment discrimination laws impacting the fire service.
- Explain the difference between disparate treatment and disparate impact.
- Explain equal opportunity employer and affirmative action.
- Identify the three standards of review that courts apply to governmental actions that are challenged as being discriminatory.

INTRODUCTION

The firefighters flowed into the union hall. At the front of the room, the executive board sat stone-faced, as the union president began to speak.

"A decision has been issued in the promotion case. The 1990 consent decree that we asked to be rescinded has instead been upheld by the court. That means the promotions will go forward next week. For every white male firefighter promoted to lieutenant, one minority or woman firefighter will also be promoted. We have a commitment from all parties that after this round of promotions, the consent decree will be rescinded. According to the latest information, after next week's promotions, the city's affirmative action goals will have been met."

A mixture of emotions ran through the room. "Do you understand this?" asked one firefighter to another standing beside him.

"Not me. I don't understand any of it. How can it be a consent decree if the union didn't consent?"

"Well, at least this will be the last time. Fifteen years is long enough."

CONSTITUTIONAL RIGHTS

The United States Constitution provides us with a number of important rights, including the right to free speech, freedom of the press, freedom of religion, due process, and equal protection. These rights collectively are called our civil rights. While the Constitution identifies these rights, it is silent about how these rights are to be protected.

A variety of laws at both the state and Federal levels are aimed at protecting constitutional rights, and in particular addressing discrimination and inequities in the workplace. Workplace discrimination is a complex topic that will undoubtedly become even more complex as time goes on. Before we look at employment discrimination, we need to look historically at employment discrimination laws.

CIVIL RIGHTS LAWS

The first civil rights law was the Civil Rights Act of 1866. The primary focus of this legislation was to prohibit discrimination against the recently freed slaves by government and public officials, and provide a means of enforcing the rights that were granted. Violations of the act were punishable as misdemeanors, and jurisdiction over civil rights cases was given to the Federal courts. Over the years, the Act has been amended and updated, and is now codified in 42 USC §1981 et. seq. One important component that has been added to this act is 42 UCS §1983, which allows civil suits against those acting under "color of law" who violate the constitutional rights of others.

EXAMPLE

42 USC §1983. Civil action for deprivation of rights. Every person who, under color of any statute, ordinance, regulation, custom, or usage, of any State or Territory or the District of Columbia, subjects, or causes to be subjected, any citizen of the United States or other person within the jurisdiction thereof to the deprivation of any rights, privileges, or immunities secured by the Constitution and laws, shall be liable to the party injured in an action at law, suit in equity, or other proper proceeding for redress. . . .

While the Civil Rights Act of 1866 has been amended and updated, its scope remains limited to "state actions" or actions by governmental officials under "color of law." One of the most significant laws designed to take the theory of civil rights and transform it into tangible results was the Civil Rights Act of 1964. The scope of the Civil Rights Act of 1964 was not limited to state action, but extended protection for the first time to private acts of discrimination. The Civil Rights Act of 1964 addressed a broad range of issues, including voting rights, public accommodation, education, and employment **(Figure 12-1)**.

Title VII of the Civil Rights Act of 1964 addressed employment discrimination, and provided:

EXAMPLE

SEC. 703. (a) It shall be an unlawful employment practice for an employer

(1) to fail or refuse to hire or to discharge any individual, or otherwise to discriminate against any individual with respect to his compensation, terms, conditions, or privileges of employment, because of such individual's race, color, religion, sex, or national origin; or

(2) to limit, segregate, or classify his employees in any way which would deprive or tend to deprive any individual of employment opportunities or otherwise adversely affect his status as an employee, because of such individual's race, color, religion, sex, or national origin.

Figure 12-1 *A civil rights march in 1963. (Photo by U.S. Census Bureau.)*

Title VII of the Civil Rights Act of 1964 is codified in 42 USC 2002e-2. In addition to prohibiting employment discrimination, Title VII created the Equal Employment Opportunity Commission (EEOC), whose task it was to oversee implementation and enforcement of Title VII.

A host of other Federal statutes have since been passed to prohibit various forms of discrimination in employment. These laws include the

- Equal Pay Act of 1963 (29 U.S.C. § 206)
- Rehabilitation Act of 1973 (29 U.S.C. §§ 791, 793, 794(a))
- Americans with Disabilities Act of 1990 (42 U.S.C. Chapter 126)
- Age Discrimination in Employment Act (29 U.S.C. §§ 621–634)
- Pregnancy Discrimination Act (an amendment to Title VII of the Civil Rights Act of 1964)
- Civil Rights Act of 1991 (amended the Civil Rights Act of 1964 to strengthen and improve Federal civil rights laws, provide damages in cases of intentional employment discrimination, and clarify provisions regarding disparate impact actions)

Collectively these laws prohibit a wide range of discriminatory practices, and provide victims of discrimination with a variety of tools to remedy acts of discrimination.

WHAT IS DISCRIMINATION?

The term discrimination refers to an act that treats another person differently because of a prohibited classification. The Fourteenth Amendment's equal protection clause is the constitutional basis for the prohibition against illegal discrimination. The very word "discrimination" refers to the fact that we make distinctions. Distinctions, and thus discrimination, are a part of every-day life. We decide where we will shop, where we will eat, who cuts our hair, and even who we allow to pull out in front of us in traffic. Examinations in courses purposefully discriminate. The important question to consider is: on what grounds is the discrimination taking place? When the grounds for discrimination is a person's race, religion, national origin, sex, disability, or age, and the discrimination involves employment, housing, or other activity to which discrimination laws apply, the discrimination is illegal.

Title VII of the Civil Rights Act of 1964, the Americans with Disabilities Act (ADA), and the Age Discrimination in Employment Act (ADEA) collectively prohibit employment discrimination that is based on race, national origin, sex, religion, disability, or age. Discrimination is prohibited in

- testing, hiring, firing, and discipline of employees
- compensation, assignment, or classification of employees
- transfer, promotion, layoff, or recall
- recruiting and advertising
- training and apprenticeship programs
- fringe benefits, retirement plans, and disability leave
- other terms and conditions of employment

Illegal discrimination includes harassment of, or retaliation against, an individual who has made a complaint of discrimination, cooperated with an investigation, or opposed discriminatory practices, as well as employment decisions based on perceived stereotypes or assumptions about the abilities, traits, or performance of individuals of a certain sex, race, age, religion, or ethnic group, or individuals with disabilities.

PROOF OF DISCRIMINATION

Establishing that someone has been the victim of unlawful discrimination is usually not a simple matter. In the absence of the person responsible for the discrimination giving a blatant admission or expressing his or her actual intent in writing, circumstantial evidence is required to prove discrimination. Most authorities recognize two types of discrimination: disparate treatment and disparate impact.

Disparate Treatment

disparate treatment
a form of discrimination in which a particular victim (or group of victims) is treated differently because of a prohibited classification; disparate treatment discrimination is based upon an intentional act of discrimination

Disparate treatment refers to discrimination in which a particular victim (or group of victims) is treated differently because of a prohibited classification. Proof of disparate treatment requires proof that a decision, action, or pattern of behavior was directed at a particular person or group of people because of their race, sex, religion, or other prohibited classification. Disparate treatment discrimination is based upon an *intentional* act of discrimination.

Watts v. City of Norman
270 F.3d 1288 (10th Cir., 2001)
United States Court of Appeals for the Tenth Circuit

. . . Watts, who terms himself an Afro-American, was a captain in the Norman fire department when he became involved in a physical confrontation with one of his subordinates, whom Watts describes as Caucasian. After this incident, the department disciplined Watts by demoting him from captain to firefighter. Watts retired rather than accept the demotion. He sued the City under Title VII of the Civil Rights Act of 1964 . . . alleging that the City terminated his employment on account of his race. . . .

The City's decision to demote Watts arose out of the events of October 31, 1998. About seven o'clock that morning, firefighter Charles Wilson complained to Watts about Watts's personal use of the station laundry facilities. According to a written narrative Watts made later that day, the exchange was already acrimonious. Watts wrote, "[S]hortly after 7am [sic], Firefighter Chuck Wilson viciously and virulently, verbally blew up at me in the rear bedroom by the washing/drying machines." Watts also described Wilson's speech in the bedroom as a "vile barrage of words." Wilson said, in crude language, that Watts was annoying him and "everyone" else.

Watts left the bedroom and prepared to shave, but before shaving he went to ask two other firefighters about what Wilson said. According to Watts, "To better gauge this incident, I briefly inquired about how I was operating the Station with two other 'A' Crew Firefighters, Brian Starkey and Paul Harvey. . . ." Then Watts shaved, "while reflecting on what had transpired," as he said.

At eight o'clock, an hour after the first incident, Watts again looked up Starkey and Harvey, to talk more with them about Wilson's assertion that Watts was annoying everyone. Watts said, due to the hot temper and ill will displayed by Firefighter Wilson, at 8:00 am I asked Firefighter Starkey and then Firefighter Harvey to visit with me one at a time in the truck room. To avoid tunnel vision and to keep a broad and flexible perspective, I asked each one for their observations of how I was operating the Station.

Watts next asked Wilson to come talk to him in the truck room of the station. Wilson did not come. After waiting a while, Watts went and asked Wilson again

(continued)

(*continued*)

to come to the truck room. Still Wilson did not come. Watts sent Starkey and Harvey to ask Wilson to come. When Wilson still did not come, Watts said he went to Wilson and "told him in a clear and precise cuss language that he should comply." Wilson did not.

Watts decided to have his talk with Wilson on the spot. Watts began to pace, which according to other testimony was a habit of his. According to Watts, Wilson began to imitate his pacing:

"I used Civility, Patience and Empathy as I paced back and forth. Firefighter Wilson then began to mimic my pacing in a bizarre and erratic fashion, with a stooped-over posture and to imitate my attempts to have him communicate with me about how I was running the station. This temper tantrum display continued for a few minutes.

As I stopped pacing and stood still, Firefighter Wilson started ranting and raging in a loud contemptible voice reflected with an exaggerated facial expression, about how he felt I was paranoid. He then came in front of me, stopped mimicking my pacing, and stood up erect.

Finally, he got in my face about one foot distance, squared off, never answering the root question of how he felt I operated the station and continued to argue. Without warning Firefighter Wilson viciously head butted me. He stuck his forehead into my forehead and continued to aggressively lean into me. The head butt sounded loud and for an instant I saw stars. With no premeditation, I instinctively removed his head from my face with my opened right hand, protecting myself, and did not follow up with further re-action [sic] to being struck on my forehead by Firefighter Wilson."

Wilson immediately telephoned Assistant Fire Chief Johnny Vaughn. Vaughn came to the station to investigate. The first thing Vaughn did was talk to Watts. Vaughn testified in his deposition that Watts began talking about defending himself. Vaughn testified:

And all of a sudden and I am not positive about what it was that Greg said that triggered my thought pattern, but I asked I said, "Greg, what have you done here? Did you hit him?" And by this time, even in the conversation, Greg was very upset. He was mad. And it was something to the nature of doing his hands like this (indicating) clapping real loud. And he said, "You [sic] goddam right. Right up side his head." Vaughn testified that Watts claimed Wilson had butted his head, but Vaughn observed that "Captain Watts didn't show any redness or a knot or anything else to where I could be sure that he had been head-butted." At this point, Vaughn decided that Watts was too agitated to run the fire station that day, so he sent him home for the day. Vaughn asked Watts to provide him with a written statement about the incident, and Watts provided two different statements (which formed the basis for the foregoing statement of facts).

Next, Vaughn interviewed Wilson. Vaughan understood that Wilson had a reputation of being "about half hard to get along with" and that he was "one of the

(*continued*)

(continued)

guys that will fly off the handle real fast." According to Vaughn, Wilson said that he and Watts "got into it about the laundry." Then Wilson either said that Watts had slapped him or that he had punched him at his deposition, Vaughn wasn't sure which Wilson had said. Vaughn could see that the left side of Wilson's face was swollen and red down to his ear. He told Wilson that Watts said Wilson butted him in the head, and Wilson denied this, saying that although their heads were close, he did not touch Watts. Vaughn decided Wilson was also too agitated to stay at work, so he sent him home. Wilson, too, provided a written statement of his side of the story. Vaughn mentioned that in his written report, Wilson stated that the two men's foreheads "may have touched." (Wilson's report actually says that Watts advanced on Wilson, who stood his ground so that "our foreheads touched, eyeball to eyeball.")

Vaughn also requested written reports from Firefighters Harvey and Starkey. Their accounts were very similar. . . . Harvey wrote: "As I was walking off I saw Greg jump into Chuck's face and at the top of his lungs yell to Chuck, 'I'm f . . . in charge here. You're not f . . . in charge.'" Starkey wrote, "Before turning and leaving, I saw Greg step very close to Chuck and did what I would describe as ranting and raving, flailing his arms and screaming obscenities. The best I can recall, Greg said, "I'm running this f . . . place you. . . ." Both men said that they rounded the corner at that point and therefore could no longer see Watts and Wilson, but that they heard a loud slap. Both men also said that when they saw Wilson a short time later, Wilson asked if they had seen Watts hit him. They said that Wilson's face was "very red and puffy," "from his ear down the whole side of his cheek." Wilson was assigned to another station while the department investigated the incident. Vaughn concluded that it was impossible to say which of the two men was the aggressor since there were no third-party witnesses and the two antagonists had contradictory stories: "It was his word against his word." Vaughn conveyed to the Fire Chief, John Dutch, the results of his investigation, consisting of his memorandum and the written statements of Watts, Wilson, Harvey, and Starkey. Vaughn's personal opinion was that both Wilson and Watts should be dismissed because "the rules and regulations of our fire department stated that no firefighter would ever strike another under any circumstances. It doesn't address captain, firefighter, or anything else. It's just no firefighter shall have an altercation." However, Fire Chief Dutch's decision was that Wilson would not be disciplined because the department could not prove "through statements or visuals or anything else, about the head-butt." There was a pre-disciplinary hearing to consider Watts's case, attended by Watts and his attorney, a city attorney, a city personnel employee, the union president, Fire Chief Dutch and Vaughn. After the hearing, Fire Chief Dutch submitted a memorandum to the city personnel director, George Shirley, proposing that the City terminate Watts's employment. Dutch wrote:

> *In summary, it is my finding that Captain Watts did verbally abuse a subordinate, using loud, offensive, profane, and vulgar language; that Watts did direct physically aggressive movements toward a subordinate; that Watts'* [sic] *actions*

(continued)

(*continued*)

have, in fact, intimidated and frightened employees in the Fire Department; and that Watts did physically strike a subordinate employee.

Shirley concurred with Dutch's decision. Dutch's proposal was subject to further review by city Manager Ron Wood.

Wood determined that Watts should be disciplined, but that the discipline should be demotion from captain to firefighter, rather than termination of employment. Wood wrote in a letter to Watts that the decision to discipline him was based on two grounds: first, the evidence that Watts struck a subordinate was very strong, while Watts's claim of self-defense was not corroborated, despite investigation; and second, regardless of who made the "initial contact," Watts failed in his duty as a supervisor by allowing the conflict to escalate to the point of violence, rather than simply sending Wilson home. Despite the gravity of Watts's conduct, Wood concluded that in light of Watts's long record of service with the City, demotion to a non-supervisory position, rather than firing, was the appropriate sanction.

Watts resigned rather than accept the demotion. He brought this suit against the City, alleging discriminatory termination of his employment on account of his race, in violation of Title VII, 42 U.S.C. § 2000e-2(a) (1994). . . .

For this appeal, the City does not dispute that Watts presented a prima facie case, and Watts does not dispute that the City articulated a legitimate, nondiscriminatory reason for disciplining him. Therefore, the only issue presented on appeal is whether there is evidence of pretext—in other words, evidence that a discriminatory reason more likely motivated the City or that the reason the City gave for its treatment of Watts was unworthy of belief. . . .

One of the established methods of proving pretext is to show that the employer treated the plaintiff "differently from other similarly-situated [sic] employees who violated work rules of comparable seriousness." . . . Watts argues that he showed pretext by showing that the City did not discipline Wilson for violating the rule against fighting. Watts bears the burden of establishing that he and Wilson were similarly situated. . . .

The City responds that Watts and Wilson were not similarly situated because Watts was a supervisor and therefore had a greater responsibility not only to avoid fighting, but to actively defuse the explosive situation before it escalated into violence. Under our precedent, employees may not be "similarly situated" when one is a supervisor and the other is not. . . .

The distinction between supervisors and non-supervisors is clearly relevant to whether we would expect the employees in this case to be treated the same in the absence of discrimination. An employer who entrusts greater authority to its supervisors than to ordinary employees surely can be expected to exact greater responsibility from them. Supervisors often have to manage difficult employees. The record in this case shows that the City had heightened expectations of its supervisory employees and also gave them authority to neutralize a deteriorating situation by such measures as sending a subordinate home. . . .

(*continued*)

(*continued*)

Wood said that while Watts's long record of service and lack of prior disciplinary problems argued in his favor, "it does concern me that even in our meeting on January 8, that you failed to recognize that you mishandled the situation as a Supervisor and continue to claim that you are the 'victim.' I have no confidence in your continuing in a supervisory capacity with the City of Norman." Regardless of who hit whom, Watts's own statements gave the City reason to conclude that Watts escalated a situation rife with potential for violence and in fact used his position as supervisor to do so. . . .

Watts's managerial failure fundamentally distinguishes his situation from Wilson's. After all, the challenged discipline here was to reduce Watts to Wilson's rank. Because of the difference in their positions of employment, we cannot hold that Watts and Wilson were similarly situated or that the City's decision to discipline only Watts is proof of pretext. . . .

The City investigated the incident thoroughly before making any decision, holding a pre-disciplinary hearing at which Watts was represented by counsel. The City had the statements of Starkey and Harvey that an agitated Watts sought out Wilson and sent them away so he could be alone with Wilson; that Watts was shouting obscenities at Wilson; that they heard a slap; and that Wilson's face was red and swollen immediately afterwards. The City had Vaughn's evidence that Watts had admitted striking Wilson in the head, that Watts had no marks of a blow on his head, and that Wilson's face was red and swollen. The City had conflicting statements by Wilson and Watts, each claiming the other hit him and each denying hitting the other. . . .

On the other hand, the City lacked objective evidence that Wilson had struck Watts. Whereas Vaughn reported that Watts admitted striking Wilson, Wilson always denied striking Watts. City Manager Wood wrote to Watts, "There is no corroborating evidence to support your claim that you were head butted such as a bruise or swelling or observation by other employees. Whether you were head butted by the subordinate appears to be your word against the subordinate's word." Fire Chief Dutch decided not to discipline Wilson because the City could not prove he struck Watts "through statements or visuals or anything else," except, of course, for the word of his antagonist. . . .

This Court has no need and no authority to determine what really happened between the two men or what discipline would have been appropriate to each. . . . Our task in this case is only to say whether the men were "similarly situated" so that the City's different disposition of their two cases is evidence of pretext. The existence of corroborating evidence in Watts's case and the absence of such evidence in Wilson's case is a crucial difference from the point of view of an employer trying to decide what disciplinary measures it ought to mete out to the respective employees and, for that matter, what actions it could later defend if challenged by the disciplined employee. Wilson was therefore not similarly situated to Watts, and the Wilson case therefore does not provide the pretext evidence Watts needs. . . .

(*continued*)

> (*continued*)
>
> Because the City's announced reason for disciplining Watts was failure to live up to responsibilities which were not a part of Wilson's job, and because the evidence available to the City about the two men's misconduct differed qualitatively, different treatment of the two men does not amount to evidence of pretext. Nor does Watts adduce other evidence sufficient to carry his burden of proving that his demotion was based on intentional discrimination. . . . We must affirm the district court's entry of summary judgment against Watts.

Case Name: Watts v. City of Norman

Court: United States Circuit Court of Appeals, 10th Circuit

Summary of Main Points: A case for disparate treatment (intentional race discrimination) cannot be proven simply because two employees who engaged in an altercation received differing treatment. The differing treatment was based upon differences in their rank and factual circumstances surrounding the altercation. In other words, the two men were not similarly situated, and therefore the fact that they were treated differently cannot be presumed to have been based on race.

Employers will frequently offer seemingly legitimate reasons for an employment decision that adversely affects an employee in a protected class. The burden then shifts to the employee to establish that the stated reason was not the real reason for the employment decision, and that the stated reason was merely a pretext for an act of discrimination. Proof of a pretext can be established by showing that others who were similarly situated, but not in the protected class, were treated differently.

Disparate Impact

disparate impact
a form of discrimination that appears on its face to be nondiscriminatory, but that has the effect of discriminating based upon a prohibited classification; disparate impact can be proven only through statistical analysis

Some types of employment decisions appear to be nondiscriminatory, but have the effect of discriminating. The proof of such discrimination is evident only from looking at a statistical analysis. This type of discrimination is called **disparate impact**. In these cases, it may be difficult if not impossible to clearly identify the specific reasons for the statistical difference, and just as impossible to prove that the discrimination was intentional. For example, the entrance examination used for a particular job, or neutral-appearing prerequisites, may have a tendency to eliminate minority or protected class candidates more frequently than white males. Irrespective of the employer's actual motivations for using such an examination or prerequisites, when the statistics show that a protected class has been unlawfully impacted, the disparate impact theory will apply.

Disparate impact cases also differ from disparate treatment cases in another important regard. In disparate treatment cases, such as the *Watts* case,

class action lawsuit
a suit brought by certain named individuals on behalf of all persons similarly situated

consent decree
a court order, the terms of which have been agreed to by the parties to a lawsuit, and which is overseen and enforced by the court; once entered, consent decrees are considered to be binding decisions of the court

the identity of the injured party is usually quite clear. In disparate impact cases, it may be impossible to identify a particular victim. In fact, in many disparate impact cases, the suit is filed as a class action. A **class action lawsuit** is a suit brought by certain named individuals on behalf of all persons similarly situated.

Disparate impact cases often lead to law suits where both the employer and the individuals who are alleging discrimination agree that a statistical disparity exists, and as a remedy the employer needs to take affirmative steps to address the numerical imbalance. As a result the parties may enter into consent decrees. A **consent decree** is a court order, the terms of which have been agreed to by the parties to the suit, and which is overseen and enforced by the court. Once entered, consent decrees are considered to be binding decisions of the court.

As the *Cleveland* and the *Dallas* cases below indicate, consent decrees are frequently attacked by those who are negatively affected by the decree. Prior to the *Cleveland* case, it was unclear if courts could provide relief under Title VII that benefited individuals who were not the actual victims of discrimination.

Firefighters (IAFF Local 93) v. City of Cleveland
478 U.S. 501 (1986)
United States Supreme Court

JUSTICE BRENNAN delivered the opinion of the Court. . . .
On October 23, 1980, the Vanguards of Cleveland (Vanguards), an organization of black and Hispanic firefighters employed by the City of Cleveland, filed a complaint charging the City and various municipal officials (hereinafter referred to collectively as the City) with discrimination on the basis of race and national origin "in the hiring, assignment and promotion of firefighters within the City of Cleveland Fire Department." . . . The Vanguards sued on behalf of a class of blacks and Hispanics consisting of firefighters already employed by the City, applicants for employment, and "all blacks and Hispanics who in the future will apply for employment or will be employed as firemen by the Cleveland Fire Department." . . .

The Vanguards claimed that the City had violated the rights of the plaintiff class under the Thirteenth and Fourteenth Amendments to the United States Constitution, Title VII of the Civil Rights Act of 1964, 42 U.S.C. § 2000e et seq., and 42 U.S.C. §§ 1981 and 1983. Although the complaint alleged facts to establish discrimination in hiring and work assignments, the primary allegations charged that black and Hispanic firefighters "have . . . been discriminated against
(continued)

(*continued*)

by reason of their race and national origin in the awarding of promotions within the Fire Department." . . . The complaint averred that this discrimination was effectuated by a number of intentional practices by the City. The written examination used for making promotions was alleged to be discriminatory. The effects of this test were said to be reinforced by the use of seniority points and by the manipulation of retirement dates so that minorities would not be near the top of promotion lists when positions became available. In addition, the City assertedly limited minority advancement by deliberately refusing to administer a new promotional examination after 1975, thus cancelling out the effects of increased minority hiring that had resulted from certain litigation commenced in 1973. . . .

[T]he Vanguards' lawsuit was not the first in which the City had to defend itself against charges of race discrimination in hiring and promotion in its civil services. In 1972, an organization of black police officers filed an action alleging that the Police Department discriminated against minorities in hiring and promotions. . . . The District Court found for the plaintiffs, and issued an order enjoining certain hiring and promotion practices and establishing minority hiring goals. In 1977, these hiring goals were adjusted and promotion goals were established pursuant to a consent decree. Thereafter, litigation raising similar claims was commenced against the Fire Department and resulted in a judicial finding of unlawful discrimination and the entry of a consent decree imposing hiring quotas. . . . In 1977, after additional litigation, the . . . court approved a new plan governing hiring procedures in the Fire Department.

By the time the Vanguards filed their complaint, then, the City had already unsuccessfully contested many of the basic factual issues in other lawsuits. Naturally, this influenced the City's view of the Vanguards' case. As expressed by counsel for the City at oral argument in this Court:

> [W]hen this case was filed in 1980, the City of Cleveland had eight years at that point of litigating these types of cases, and eight years of having judges rule against the City of Cleveland. You don't have to beat us on the head. We finally learned what we had to do and what we had to try to do to comply with the law, and it was the intent of the city to comply with the law fully. . . .

Thus, rather than commence another round of futile litigation, the City entered into "serious settlement negotiations" with the Vanguards. . . .

On April 27, 1981, Local Number 93 of the International Association of Firefighters . . . which represents a majority of Cleveland's firefighters, moved . . . to intervene as a party-plaintiff. The District Court granted the motion and ordered the Union to submit its complaint in intervention within 30 days.

Local 93 subsequently submitted a three-page document entitled "Complaint of Applicant for Intervention." Despite its title, this document did not allege any causes of action or assert any claims against either the Vanguards or the City. It expressed the view that "[p]romotions based upon any criterion other than competence, such as a racial quota system, would deny those most

(*continued*)

(*continued*)

capable from their promotions, and would deny the residents of the City of Cleveland from maintaining the best possible fire fighting force, and asserted that Local #93's interest is to maintain a well trained and properly staffed fire fighting force and [Local 93] contends that promotions should be made on the basis of demonstrated competency, properly measured by competitive examinations administered in accordance with the applicable provisions of Federal, State, and Local laws." . . . The "complaint" concluded with a prayer for relief in the form of an injunction requiring the City to award promotions on the basis of such examinations. . . .

In the meantime, negotiations between the Vanguards and the City continued, and a proposed consent decree was submitted to the District Court in November, 1981. This proposal established "interim procedures" to be implemented "as a two-step temporary remedy" for past discrimination in promotions. . . . The first step required that a fixed number of already planned promotions be reserved for minorities: specifically, 16 of 40 planned promotions to Lieutenant, 3 of 20 planned promotions to Captain, 2 of 10 planned promotions to Battalion Chief, and 1 of 3 planned promotions to Assistant Chief were to be made to minority firefighters. . . . The second step involved the establishment of "appropriate minority promotion goal[s]," . . . for the ranks of Lieutenant, Captain, and Battalion Chief. The proposal also required the City to forgo using seniority points as a factor in making promotions. . . . The plan was to remain in effect for nine years, and could be extended upon mutual application of the parties for an additional 6-year period. . . .

The District Court held a 2-day hearing at the beginning of January to consider the fairness of this proposed consent decree. Local 93 objected to the use of minority promotional goals and to the 9-year life of the decree. In addition, the Union protested the fact that it had not been included in the negotiations. This latter objection particularly troubled the District Judge. Indeed, although hearing evidence presented by the Vanguards and the City in support of the decree, the Judge stated that he was "appalled that these negotiations leading to this consent decree did not include the intervenors . . . ," and refused to pass on the decree under the circumstances. . . . Instead, he concluded: "*I am going to at this time to defer this proceeding until another day, and I am mandating the City and the [Vanguards] to engage the Fire Fighters in discussions, in dialogue. Let them know what is going on, hear their particular problems.*" . . . At the same time, Judge Lambros explained that the Union would have to make its objections more specific to accomplish anything:

> "*I don't think the Fire Fighters are going to be able to win their position on the basis that, 'Well, Judge, you know, there's something inherently wrong about quotas. You know, it's not fair.' We need more than that.*" . . .

A second hearing was held on April 27. Local 93 continued to oppose any form of affirmative action. Witnesses for all parties testified concerning the proposed consent decree. The testimony revealed that, while the consent decree dealt only

(*continued*)

(*continued*)

with the 40 promotions to Lieutenant already planned by the City, the Fire Department was actually authorized to make up to 66 offers; similarly, the City was in a position to hire 32, rather than 20, Captains, and 14, rather than 10, Battalion Chiefs. After hearing this testimony, Judge Lambros proposed as an alternative to have the City make a high number of promotions over a relatively short period of time. The Judge explained that, if the City were to hire 66 Lieutenants, rather than 40, it could "plug in a substantial number of black leadership that can start having some influence in the operation of this fire department" while still promoting the same nonminority officers who would have obtained promotions under the existing system. . . . Additional testimony revealed that this approach had led to the amicable resolution of similar litigation in Atlanta, Georgia. Judge Lambros persuaded the parties to consider revamping the consent decree along the lines of the Atlanta plan. The proceedings were therefore adjourned, and the matter was referred to a United States Magistrate.

Counsel for all three parties participated in 40 hours of intensive negotiations under the Magistrate's supervision, and agreed to a revised consent decree that incorporated a modified version of the Atlanta plan. . . . However, submission of this proposal to the court was made contingent upon approval by the membership of Local 93. Despite the fact that the revised consent decree actually increased the number of supervisory positions available to nonminority firefighters, the Union members overwhelmingly rejected the proposal.

On January 11, 1983, the Vanguards and the City lodged a second amended consent decree with the court and moved for its approval. This proposal was "patterned very closely upon the revised decree negotiated under the supervision of [the] Magistrate . . . ," and thus its central feature was the creation of many more promotional opportunities for firefighters of all races. Specifically, the decree required that the City immediately make 66 promotions to Lieutenant, 32 promotions to Captain, 16 promotions to Battalion Chief, and 4 promotions to Assistant Chief. These promotions were to be based on a promotional examination that had been administered during the litigation. The 66 initial promotions to Lieutenant were to be evenly split between minority and nonminority firefighters. However, since only 10 minorities had qualified for the 52 upper-level positions, the proposed decree provided that all 10 should be promoted. The decree further required promotional examinations to be administered in June, 1984, and December, 1985. Promotions from the lists produced by these examinations were to be made in accordance with specified promotional "goals" that were expressed in terms of percentages and were different for each rank. The list from the 1985 examination would remain in effect for two years, after which time the decree would expire. The life of the decree was thus shortened from nine years to four. In addition, except where necessary to implement specific requirements of the consent decree, the use of seniority points was restored as a factor in ranking candidates for promotion. . . .

The District Court approved the consent decree on January 31, 1983. Judge Lambros found that *"[t]he documents, statistics, and testimony presented at the*
(*continued*)

(continued)

January and April, 1982, hearings reveal a historical pattern of racial discrimination in the promotions in the City of Cleveland Fire Department." . . . He then observed:

> *While the concerns articulated by Local 93 may be valid, the use of a quota system for the relatively short period of four years is not unreasonable in light of the demonstrated history of racial discrimination in promotions in the City of Cleveland Fire Department. It is neither unreasonable nor unfair to require nonminority firefighters who, although they committed no wrong, benefited from the effects of the discrimination to bear some of the burden of the remedy. Furthermore, the amended proposal is more reasonable, and less burdensome, than the nine-year plan that had been proposed originally.* . . .

The Judge therefore overruled the Union's objection and adopted the consent decree "as a fair, reasonable, and adequate resolution of the claims raised in this action." . . . The District Court retained exclusive jurisdiction for "all purposes of enforcement, modification, or amendment of th[e] Decree upon the application of any party." . . .

The Union appealed the overruling of its objections. A panel for the Court of Appeals for the Sixth Circuit affirmed, one judge dissenting. . . . The court rejected the Union's claim that the use of race-conscious relief was "unreasonable," finding such relief justified by the statistical evidence presented to the District Court and the City's express admission that it had engaged in discrimination. The court also found that the consent decree was "fair and reasonable to nonminority firefighters," emphasizing the "relatively modest goals set forth in the plan," the fact that "the plan does not require the hiring of unqualified minority firefighters or the discharge of any nonminority firefighters," the fact that the plan "does not create an absolute bar to the advancement of nonminority employees," and the short duration of the plan. . . .

Local 93 petitioned this Court for a writ of certiorari. The sole issue raised by the petition is whether the consent decree is an impermissible remedy under § 706(g) of Title VII. Local 93 argues that the consent decree disregards the express prohibition of the last sentence of § 706(g) that *"[n]o order of the court shall require the admission or reinstatement of an individual as a member of a union, or the hiring, reinstatement, or promotion of an individual as an employee, or the payment to him of any back pay, if such individual was refused admission, suspended, or expelled, or was refused employment or advancement or was suspended or discharged for any reason other than discrimination on account of race, color, religion, sex, or national origin or in violation of section 2000e-3(a) of this title."* . . . According to Local 93, this sentence precludes a court from awarding relief under Title VII that may benefit individuals who were not the actual victims of the employer's discrimination. The Union argues further that the plain language of the provision that "[n]o order of the court" shall provide such relief extends this limitation to orders entered by consent, in addition to orders issued after litigation. Consequently, the Union concludes that a consent decree

(continued)

(*continued*)

entered in Title VII litigation is invalid if—like the consent decree approved in this case—it utilizes racial preferences that may benefit individuals who are not themselves actual victims of an employer's discrimination. The Union is supported by the United States as amicus curiae.

We granted the petition in order to answer this important question of federal law. . . . [C]ourts may, in appropriate cases, provide relief under Title VII that benefits individuals who were not the actual victims of a defendant's discriminatory practices. We need not decide whether this is one of those cases, however. For we hold that, whether or not § 706(g) precludes a court from imposing certain forms of race-conscious relief after trial, that provision does not apply to relief awarded in a consent decree. We therefore affirm the judgment of the Court of Appeals.

II   We have on numerous occasions recognized that Congress intended voluntary compliance to be the preferred means of achieving the objectives of Title VII. . . . This view is shared by the Equal Employment Opportunity Commission (EEOC), which has promulgated guidelines setting forth its understanding that Congress strongly encouraged employers . . . to act on a voluntary basis to modify employment practices and systems which constituted barriers to equal employment opportunity. . . .

It is equally clear that the voluntary action available to employers and unions seeking to eradicate race discrimination may include reasonable race-conscious relief that benefits individuals who were not actual victims of discrimination. . . . [In *Weber*]. . . . we concluded that *"[i]t would be ironic indeed if a law triggered by a Nation's concern over centuries of racial injustice and intended to improve the lot of those who had "been excluded from the American dream for so long" constituted the first legislative prohibition of all voluntary, private, race-conscious efforts to abolish traditional patterns of racial segregation and hierarchy."* . . . Accordingly, we held that Title VII permits employers and unions voluntarily to make use of reasonable race-conscious affirmative action, although we left to another day the task of defin[ing] in detail the line of demarcation between permissible and impermissible affirmative action plans. . . .

[A]bsent some contrary indication, there is no reason to think that voluntary, race-conscious affirmative action such as was held permissible in *Weber* is rendered impermissible by Title VII simply because it is incorporated into a consent decree. . . .

III   Relying upon *Firefighters v. Stotts,* 467 U.S. 561 (1984), and *Railway Employees v. Wright,* 364 U.S. 642 (1961), Local 93—again joined by the United States—contends that we have recognized as a general principle that a consent decree cannot provide greater relief than a court could have decreed after a trial. They urge that, even if § 706(g) does not directly invalidate the consent decree, that decree is nonetheless void because the District Court "would have been powerless to order [such an injunction] under Title VII, had the matter actually gone to trial." . . .

(*continued*)

(continued)

We concluded above that voluntary adoption in a consent decree of race-conscious relief that may benefit nonvictims does not violate the congressional objectives of § 706(g). It is therefore hard to understand the basis for an independent judicial canon or "common law" of consent decrees that would give § 706(g) the effect of prohibiting such decrees anyway. . . . [A] federal court is not necessarily barred from entering a consent decree merely because the decree provides broader relief than the court could have awarded after a trial. . . .

IV   Local 93 and the United States also challenge the validity of the consent decree on the ground that it was entered without the consent of the Union. They take the position that, because the Union was permitted to intervene as of right, its consent was required before the court could approve a consent decree. This argument misconceives the Union's rights in the litigation.

A consent decree is primarily a means by which parties settle their disputes without having to bear the financial and other costs of litigating. It has never been supposed that one party—whether an original party, a party that was joined later, or an intervenor—could preclude other parties from settling their own disputes, and thereby withdrawing from litigation. Thus, while an intervenor is entitled to present evidence and have its objections heard at the hearings on whether to approve a consent decree, it does not have power to block the decree merely by withholding its consent. . . . Here, Local 93 took full advantage of its opportunity to participate in the District Court's hearings on the consent decree. It was permitted to air its objections to the reasonableness of the decree and to introduce relevant evidence; the District Court carefully considered these objections, and explained why it was rejecting them. Accordingly, "the District Court gave the union all the process that it was due. . . ."

The only issue before us is whether § 706(g) barred the District Court from approving this consent decree. We hold that it did not. Therefore, the judgment of the Court of Appeals is Affirmed.

Case Name: Firefighters (IAFF Local 93) v. City of Cleveland

Court: United States Supreme Court

Summary of Main Points: Employers may agree to use race-conscious affirmative action goals in an effort to correct imbalances of minority members in the workplace. Similarly, consent decrees may also embody the use of race-conscious remedies without violating the constitutional rights of nonminority members.

Employment discrimination cases tend to be very complex, with numerous parties involved. Cases may drag on for a decade or more and may involve repeated detours from trial court, up to appellate courts, and back down to the trial court, before final resolution. The *Dallas* case below is a perfect example. The controversy dates back to a 1976 consent decree, and subsequent voluntary affirmative action policies adopted by the City. Nonminority employees challenged the continued use of race- and sex-conscious promotions starting in 1990, culminating in the following decision in 1998.

Dallas Firefighters v. Dallas
150 F.3d 438 (5th Cir., 1998)
United States Court of Appeals for the Fifth Circuit

POLITZ, Chief Judge:

The Dallas Fire Department (DFD) has the following rank structure, beginning with the entry level position: (1) fire and rescue officer, (2) driver-engineer, (3) lieutenant, (4) captain, (5) battalion chief, (6) deputy chief, (7) assistant chief, and (8) chief. Positions are filled only from within the department. The city manager appoints the chief who in turn appoints the assistant and deputy chiefs. For battalion chief and below, firefighters become eligible to take a promotion examination for advancement to the next highest rank after a certain amount of time in grade. Those passing the examination are placed on an eligibility roster, listed in accordance with their scores. Vacancies occurring thereafter are filled by promoting individuals from the top of the eligibility list, unless there is a countervailing reason such as unsatisfactory performance, disciplinary problems, or non-paramedic status.

In 1988 the City Council adopted a five-year affirmative action plan for the DFD, extending same for five years in 1992 with a few modifications. In an effort to increase minority and female representation the DFD promoted black, hispanic [sic], and female firefighters ahead of male, nonminority firefighters who had scored higher on the promotion examinations. Between 1991 and 1995 these promotions occasioned four lawsuits filed by the Dallas Fire Fighters Association on behalf of white and Native American male firefighters who were passed over for promotions. These actions were consolidated by the district court.

The plaintiffs consist of four groups, three of which contend that the DFD impermissibly denied them promotions to the ranks of driver-engineer, lieutenant, and captain respectively. Additionally, a fourth group of plaintiffs challenges the fire chief's appointment of a black male to deputy chief in 1990. The plaintiffs claim that the City and the fire chief, Dodd Miller, acting in his official capacity, violated: (1) the fourteenth amendment of the United States Constitution, (2) the equal rights clause of the Texas Constitution, (3) Title VII of the Civil Rights Act of 1964, 42 U.S.C. §§ 2000e et seq., and (4) article 5221k of the Texas Civil Statutes. . . .

2. The Out-of-Rank Promotions

A. Race-Conscious Promotions

To survive an equal protection challenge under the fourteenth amendment [sic], a racial classification must be tailored narrowly to serve a compelling governmental interest. That standard applies to classifications intended to be remedial, as well as to those based upon invidious discrimination. A governmental body has a

(continued)

(*continued*)

compelling interest in remedying the present effects of past discrimination. In analyzing race conscious remedial measures we essentially are guided by four factors: (1) necessity for the relief and efficacy of alternative remedies; (2) flexibility and duration of the relief; (3) relationship of the numerical goals to the relevant labor market; and (4) impact of the relief on the rights of third parties.

We conclude that on the record before us the race-based, out-of-rank promotions at issue herein violate the equal protection clause of the fourteenth amendment. The only evidence of discrimination contained in the record is the 1976 consent decree between the City and the United States Department of Justice, precipitated by a DOJ finding that the City engaged in practices inconsistent with Title VII, and a statistical analysis showing an underrepresentation of minorities in the ranks to which the challenged promotions were made. The record is devoid of proof of a history of egregious and pervasive discrimination or resistance to affirmative action that has warranted more serious measures in other cases. We are aware that the out-of-rank promotions do not impose as great a burden on nonminorities as would a layoff or discharge. In light of the minimal record evidence of discrimination in the DFD, however, we . . . must conclude that the City is not justified in interfering with the legitimate expectations of those warranting promotion based upon their performance in the examinations.

There are other ways to remedy the effects of past discrimination. The City contends, however, that alternative measures employed by the DFD, such as validating promotion exams, recruiting minorities, eliminating the addition of seniority points to promotion exam scores, and initiating a tutoring program, have been unsuccessful, as evidenced by the continuing imbalance in the upper ranks of the DFD. That minorities continue to be underrepresented does not necessarily mean that the alternative remedies have been ineffective, but merely that they apparently do not operate as quickly as out-of-rank promotions.

B. Gender-Conscious Promotions

Applying the less exacting intermediate scrutiny analysis applicable to gender-based affirmative action, we nonetheless find the gender-based promotions unconstitutional. The record before us contains, as noted above, little evidence of racial discrimination; it contains even less evidence of gender discrimination. Without a showing of discrimination against women in the DFD, or at least in the industry in general, we cannot find that the promotions are related substantially to an important governmental interest.

C. Title VII

Having struck down the out-of-rank promotions as unconstitutional, we need not address their validity under Title VII or Texas article 5221k.

3. The Deputy Chief Appointment

The City contends . . . Chief Miller's appointment of Robert Bailey, a black male, to deputy chief violated neither Title VII nor article 5221k. To determine the

(*continued*)

(*continued*)

validity of the appointment we must examine whether it was justified by a manifest imbalance in a traditionally segregated job category and whether the appointment unnecessarily trammeled the rights of nonminorities or created an absolute bar to their advancement. The plaintiffs do not dispute that there is a manifest imbalance in the rank of deputy chief and we therefore limit our discussion to the second prong of the *Johnson* test.

The only . . . evidence specific to the Bailey appointment is the affidavit of Chief Miller in which he states:

> *In 1990, I selected Robert Bailey as Deputy Chief because I believed he was capable of performing the job responsibilities of the position of Deputy Chief, and he was recommended by my executive staff. In addition, the appointment of Chief Bailey was made pursuant to the City of Dallas Affirmative Action Plan.*

The City contends that Chief Miller's statement reflects that, in appointing Bailey, he considered race as one factor among many, making the appointment permissible under *Johnson*. The plaintiffs concede that Bailey was qualified but insist that the reference to the affirmative action plan, and the failure of Chief Miller to explain how Bailey compared to other candidates, established that Chief Miller based his final decision solely upon race. The plaintiffs also contend that the promotional goals in the affirmative action plan are out of proportion to the percentage of available candidates, demonstrating that the appointment was made to fulfill impermissible goals and, thus, unnecessarily trammeled the rights of nonminorities.

The plaintiffs' position is that any employment decision utilizing the affirmative action plan is illegal. We decline to accept that contention, particularly in light of the fact that the validity of the affirmative action plan is not in question herein. We are persuaded beyond peradventure that the mere reference to the affirmative action plan does not create a fact issue concerning whether Chief Miller had an impermissible motive in promoting Bailey. The only relevant summary judgment evidence reflects that Chief Miller chose Bailey based upon substantially more than just his race, and the opponents have failed to produce any acceptable material evidence to the contrary. We therefore conclude that the appointment did not unnecessarily trammel the rights of nonminorities or pose an absolute bar to their advancement. Accordingly, the appointment was consistent with Title VII and article 5221k. . . .

Case Name: Dallas Firefighters v. Dallas

Court: United States Circuit Court of Appeals, 5th Circuit

Summary of Main Points: Based on the facts of the case (which do not show a pervasive history of discrimination), the continued use of race-conscious remedies violates the rights of nonminority members.

strict scrutiny
a standard of review applied by courts when reviewing the constitutionality of a governmental policy, action, or law, such that the policy, action, or law must be narrowly tailored to address a compelling governmental interest in order to be upheld

intermediate level of scrutiny
a standard of review applied by courts when reviewing the constitutionality of a governmental policy, action, or law, such that the policy, action, or law must be substantially related to important governmental objectives in order to be upheld

rational basis standard
a standard of review applied by courts when reviewing the constitutionality of a governmental policy, action, or law, such that the policy, action, or law will be upheld provided it is rationally related to a legitimate governmental interest; the rational basis standard is a deferential standard that usually results in the reviewing court upholding the government's action, except where no rational basis for the governmental action exists

STANDARD OF REVIEW FOR CONSTITUTIONAL CLAIMS OF DISCRIMINATION

As mentioned in the *Dallas* case, when a constitutional challenge is brought against a governmental entity under the Fourteenth Amendment equal protection clause of the United States Constitution, the standard of review applied by the courts differs depending upon the type of discrimination under review. The most exacting scrutiny is reserved for governmental policies, actions, and laws that discriminate on the basis of race or national origin. The term often used when courts analyze race discrimination cases is **strict scrutiny**. In order for a policy, action, or law to be upheld under strict scrutiny, it must be "narrowly tailored to address a compelling governmental interest." While the language might not seem significant, when reviewing cases of race discrimination, courts are extremely demanding when looking at the justification for a policy, action, or law that results in discrimination based on racial classifications. Strict scrutiny is also applicable when courts review governmental action that impacts fundamental rights, such as freedom of speech, freedom of religion, and freedom of the press.

In cases in which a governmental policy, action, or law involves classifications based on a person's sex, an **intermediate level of scrutiny** is applied. This intermediate level of scrutiny is commonly defined as "substantially related to important governmental objectives." While not as demanding as the scrutiny applied to racial classifications, the intermediate standard nevertheless requires the governmental entity to have some "exceedingly persuasive justification" for the policy, action, or law to be upheld.

The third level of review is called the **rational basis standard**, and applies to all other types of alleged discrimination. Under the rational basis standard a governmental policy, action, or law will be upheld provided it is rationally related to a legitimate governmental interest. The rational basis standard is a deferential standard that usually results in the reviewing court upholding the government's action except where no rational basis for the governmental action exists.

The *Evanston* case below is a sex discrimination case challenging a physical abilities test, in which the court applies the intermediate level of scrutiny. Physical abilities testing will be discussed later in this chapter.

Evans v. City of Evanston
881 F. 2d 382 (7th Cir.,1989)
United States Court of Appeals for the Seventh Circuit

POSNER, Circuit Judge
This is a class action under Title VII of the Civil Rights Act of 1964 on behalf of the 39 women who failed the physical agility test given by the Evanston fire department

(continued)

(*continued*)

[sic] to applicants for firefighting jobs in 1983. Eighty-five percent of the women who took the test failed (only seven percent of the men failed) and were thereby disqualified, and there are no women among Evanston's 106 firefighters although at one time there were two. The test is conceded to have had a "disparate impact" on women. So unless the test (more specifically the method of scoring it — the focus of the plaintiff's attack) serves a legitimate interest of the employer, it violates Title VII. The district judge found a violation and gave judgment for the class. . . . The city appeals, challenging the finding of liability; the plaintiff also appeals, challenging the adequacy of the equitable relief that the judge ordered. . . .

The physical agility test that the Evanston Fire Department used in 1983 (and also in 1981 and 1985) consisted of a group of tasks which were to be performed consecutively by each applicant without a break, while wearing a firefighter's uniform. The tasks were: climbing to the top of a 70-foot ladder; climbing an extension ladder twice while carrying a hose pack; removing a ladder from a firetruck [sic], carrying the ladder to a wall, leaning it up against the wall, and then removing it and returning it to the truck; connecting a hose to a fire hydrant, turning the hydrant on and off, and disconnecting the hose; and dragging a section of hose filled with water fifty feet, dragging a tarpaulin to the top of a hill, carrying the tarp through ten tires, and again dragging a section of hose filled with water fifty feet. The test was timed. The mean time in 1983 was 628 seconds, and the Fire Department chose one standard deviation above this mean as the passing score, with the result that anyone who took more than 767 seconds to complete the test flunked.

The physical agility test is only the first hurdle an applicant must clear to become a firefighter. Next come tests of intelligence and of psychological stability, and in the end only nine of the 839 persons who applied for firefighter jobs in 1983 were hired—all men. The fire department's choice of one standard deviation above the mean as the passing score was not consistent. In 1985 the passing score was 915 seconds, which was 2.8 standard deviations above the mean for that year. In 1981 the passing score had been 890 seconds, which was 1.7 standard deviations above the mean, but had been raised in order to enable three of the four women who took the test to pass it.

The district judge found that the test itself was fine. . . . The test was designed by firefighters, consists of tasks that faithfully imitate the tasks that firefighters are called on to perform in their work, tests for speed, skill, endurance in—in a word, aptitude for—performing those tasks, and was pretested on the Evanston firefighter force before being given to applicants. It seems clearly related to the employer's legitimate need for physically strong firefighters, and the plaintiff has suggested no alternative that would serve that need as well yet be less difficult for women. . . .

The rub is in the scoring of the test. Since men are on average stronger and faster than women, the higher the passing score on a test such as Evanston's physical agility test (that is, the shorter the time in which it must be completed) the smaller the percentage of women likely to pass it. To satisfy its burden of

(*continued*)

(*continued*)

producing evidence that the test—which means all aspects of the test including the method of scoring it—served a legitimate employer purpose, the city was obliged to produce evidence that the method of determining who passed the test in 1983 was related to the city's need for a physically capable firefighting force. . . .

The city did produce evidence relating to this question but it consisted of little more than testimony that one standard deviation above the mean is a frequent cut-off point on tests and that the cut-off point for the physical agility test was generous to the candidates and quite possibly should have been lower. It is not surprising that Judge Zagel was not persuaded by this evidence. The choice of one standard deviation above the mean was a decision to pass 84 percent of the test takers, and this meant that the passing score would depend on the average performance of those who happened to take it. But the ability to perform firefighting tasks adequately depends not on relative but on absolute test performance. If one year all the applicants were superbly fit, it would be irrational to disqualify the entire bottom 16 percent. For it is not only physical abilities that the fire department is after—as is made plain by the fact that no preference is given to candidates who do exceptionally well on the physical agility test, as opposed to those who barely pass it. The department wants firefighters who are intelligent and stable, as well as strong and swift. If it cuts off from further consideration persons who are perfectly able physically—although less so than some other applicants who may, however, be their inferiors in intelligence and stability—it is shooting itself in the foot. No explanation was offered, moreover, for why different pass rates were selected in 1981 and 1985, the effect being to enlarge markedly the time allowed to complete the test compared to what it had been in 1983. There was some evidence that the weather was bad in 1985, but that would explain only why the mean would be higher—not why the department would allow a higher number of standard deviations above the mean.

One would think the rational way of scoring the physical agility test would be to determine the maximum time in which a firefighter who had no training or practice—for remember that the test is for applicants—ought to be able to complete the test, and make that the cutoff [sic] point. Applicants who passed the test would then take the other two tests (intelligence and stability), which presumably would have their own cut-offs. Among those who passed all three tests, those whose composite score, weighted by the relative importance of the tests, was the highest would be hired. This was not the procedure followed by the Evanston Fire Department (the record is unclear on what procedure was followed), and no satisfactory reasons for departing from it were presented. . . .

The judge ordered the city to submit for his consideration a new test (or rather a new method of scoring the old test), and neither side questions that relief. But he refused to order the city to hire any of the members of the plaintiff class or even to allow them to advance to the next test. The plaintiff argues that those class members whose times of completing the 1983 test were within the passing range on the 1985 test (915 seconds, compared to only 767 in 1983) should be excused from having to retake the physical agility test and be allowed to move on to the other tests.

(*continued*)

(*continued*)

Judge Zagel was within his remedial discretion in declining to take this step. When he issued his order, it was five years since the class members had taken the physical agility test, and they offered no evidence that their agility had not declined in the interim. The importance of competent firefighting to the safety of the people of Evanston, as well as of the firefighters themselves, whose safety depends in part anyway on each other's physical fitness and agility, justified the city in insisting that all applicants have taken the test in the recent rather than remote past. An equity court must always consider the possible impact of a decree on innocent third parties. . . .

While there is much talk in the cases about "make whole" relief . . . this talk has reference to cases where it is reasonably clear that, had it not been for the discriminatory behavior, the plaintiff would have got (or retained) the job or other employment benefit in issue, and where making the plaintiff whole would not unduly injure innocent third parties. . . . As only 1.2 percent of the applicants who passed the Evanston fire department's physical agility test were actually hired, what the class members lost was not a job but a long-shot chance at a job. They will be restored to the place they would have occupied if they pass a new physical agility test approved by the district court. Depending on their performance on that test and on the other tests required of applicants, they may eventually be in a position to show that but for unfair scoring of the 1983 test they would have been hired in 1983, and if so they can then claim additional backpay [sic]. . . .

The case is remanded for further consideration in light of this opinion. . . .

AUTHOR'S NOTE: On remand, the district court concluded that the passing score Evanston selected was arbitrary and had a disparate impact on women in violation of Title VII.

Case Name: Evans v. City of Evanston

Court: United States Circuit Court of Appeals, 7th Circuit

Summary of Main Points: In cases of alleged sex discrimination, the appropriate level of review is the intermediate level. While a practice may be discriminatory (in this case the physical ability test had a disparate impact upon women), the trial judge has the discretion to fashion an appropriate remedy for such discrimination. The law does not require that the women who failed the test be hired. The judge's decision to restore them to their rightful place on the list of applicants was adequate.

PROCEDURAL ISSUES IN DISCRIMINATION

Congress has established a strict enforcement procedure for all employment-based claims of race, sex, age, national origin, religious, and disability discrimination under Title VII of the Civil Rights Act of 1964, as amended, the

Civil Rights Act of 1991, the ADA, and the ADEA. Complaints must initially be filed with the Equal Employment Opportunity Commission or with designated state human rights agencies. Complaints made under Title VII must be filed within 180 days of the act of discrimination, or within 300 days provided the complaint is also covered by state or local discrimination laws. These agencies are charged with investigating the complaint and, where appropriate, prosecuting the violators.

Victims of discrimination are generally not allowed to file suit against those who committed the discrimination unless the EEOC has issued a right-to-sue letter. The right-to-sue letter authorizes the victim to file litigation against the employer. In cases of alleged discrimination involving a state or municipal employer, if the EEOC determines there is reasonable cause to believe that a violation has occurred, and efforts to resolve the dispute are unsuccessful, the EEOC will refer the complaint to the Department of Justice (DOJ). The DOJ will then either initiate litigation on the complaint or issue a right-to-sue letter.

Victims of unlawful discrimination may also choose to proceed under state anti-discrimination laws. State anti-discrimination laws frequently have longer time frames in which victims may file their claims, beyond the 300-day time frame allowed by Federal law.

AFFIRMATIVE ACTION—EQUAL OPPORTUNITY

Two terms that are commonly discussed with regard to employment discrimination are equal employment opportunity (EEO) and affirmative action. The two terms are related, but refer to different aspects of unlawful employment discrimination.

equal employment opportunity
the right of a person to compete for a job and/or to be promoted on the basis of his or her knowledge, skills, and abilities, free from unlawful discrimination

Equal employment opportunity refers to the right of a person to compete for a job and/or be promoted on the basis of his or her knowledge, skills, and abilities, free from unlawful discrimination **(Figure 12-2)**. EEO laws require the elimination of unlawful barriers to employment. Employers are required to post notices in the workplace to advise employees of their EEO rights and their right to be free from retaliation for exercising those rights **(Figure 12-3)**.

affirmative action
positive steps taken to increase the presence of minorities and women in the workforce and in education

Affirmative action refers to positive steps taken to increase the presence of minorities and women in the workforce and in education. The term was first used in 1965 in Executive Order 11246, issued by President Lyndon B. Johnson to mandate that Federal contractors "take affirmative action to ensure that applicants are employed, and that employees are treated during employment, without regard to their race, creed, color, or national origin."

Affirmative action policies go beyond equal employment opportunities and seek to increase the representation of minorities and women in schools and employment through recruiting, and the use of race, sex, or ethnicity as

Figure 12-2 *Equal employment opportunity refers to the right of a person to compete for a job and/or be promoted on the basis of his or her knowledge, skills, and abilities, free from unlawful discrimination.*

a factor, among a multitude of factors, upon which otherwise qualified candidates may be considered. In other words, race, gender, and ethnicity may be viewed as additional criteria in choosing from among the qualified candidates, just as are other factors such as grade point average, schools attended, and work experience.

Affirmative action policies could call for an employer faced with two similarly qualified applicants to choose a minority candidate over a white candidate, or for a manager to recruit and hire a qualified woman for a job instead of a man. Affirmative action decisions are not to be based upon the use of quotas, and are not supposed to give any preference to unqualified candidates. In addition, affirmative action policies must be based upon statistical analysis that indicates an underrepresentation of woman or minorities in the workforce.

AMERICANS WITH DISABILITIES ACT OF 1990 (ADA)

The Americans with Disabilities Act of 1990 prohibits discrimination against a person on account of a disability, including employment discrimination **(Figure 12-4)**. Prior to the ADA, the Rehabilitation Act of 1973 was the primary law under which a person with a disability could challenge employment decisions that were based on physical or mental abilities. However, the

Equal Employment Opportunity is

THE LAW

Employers Holding Federal Contracts or Subcontracts

Applicants to and employees of companies with a Federal government contract or subcontract are protected under the following Federal authorities:

RACE, COLOR, RELIGION, SEX, NATIONAL ORIGIN

Executive Order 11246, as amended, prohibits job discrimination on the basis of race, color, religion, sex or national origin, and requires affirmative action to ensure equality of opportunity in all aspects of employment.

INDIVIDUALS WITH DISABILITIES

Section 503 of the Rehabilitation Act of 1973, as amended, prohibits job discrimination because of disability and requires affirmative action to employ and advance in employment qualified individuals with disabilities who, with reasonable accommodation, can perform the essential functions of a job.

VIETNAM ERA, SPECIAL DISABLED, RECENTLY SEPARATED, AND OTHER PROTECTED VETERANS

38 U.S.C. 4212 of the Vietnam Era Veterans' Readjustment Assistance Act of 1974, as amended, prohibits job discrimination and requires affirmative action to employ and advance in employment qualified Vietnam era veterans, qualified special disabled veterans, recently separated veterans, and other protected veterans.

Any person who believes a contractor has violated its nondiscrimination or affirmative action obligations under the authorities above should contact immediately:

The Office of Federal Contract Compliance Programs (OFCCP), Employment Standards Administration, U.S. Department of Labor, 200 Constitution Avenue, N.W., Washington, D.C. 20210 or call (202) 693-0101, or an OFCCP regional or district office, listed in most telephone directories under U.S. Government, Department of Labor.

Private Employment, State and Local Governments, Educational Institutions

Applicants to and employees of most private employers, state and local governments, educational institutions, employment agencies and labor organizations are protected under the following Federal laws:

RACE, COLOR, RELIGION, SEX, NATIONAL ORIGIN

Title VII of the Civil Rights Act of 1964, as amended, prohibits discrimination in hiring, promotion, discharge, pay, fringe benefits, job training, classification, referral, and other aspects of employment, on the basis of race, color, religion, sex or national origin.

DISABILITY

The Americans with Disabilities Act of 1990, as amended, protects qualified applicants and employees with disabilities from discrimination in hiring, promotion, discharge, pay, job training, fringe benefits, classification, referral, and other aspects of employment on the basis of disability. The law also requires that covered entities provide qualified applicants and employees with disabilities with reasonable accommodations that do not impose undue hardship.

AGE

The Age Discrimination in Employment Act of 1967, as amended, protects applicants and employees 40 years of age or older from discrimination on the basis of age in hiring, promotion, discharge, compensation, terms, conditions or privileges of employment.

SEX (WAGES)

In addition to sex discrimination prohibited by Title VII of the Civil Rights Act of 1964, as amended (see above), the Equal Pay Act of 1963, as amended, prohibits sex discrimination in payment of wages to women and men performing substantially equal work in the same establishment.

Retaliation against a person who files a charge of discrimination, participates in an investigation, or opposes an unlawful employment practice is prohibited by all of these Federal laws.

If you believe that you have been discriminated against under any of the above laws, you should contact immediately:

The U.S. Equal Employment Opportunity Commission (EEOC), 1801 L Street, N.W., Washington, D.C. 20507 or an EEOC field office by calling toll free (800) 669-4000. For individuals with hearing impairments, EEOC's toll free TDD number is (800) 669-6820.

Programs or Activities Receiving Federal Financial Assistance

RACE, COLOR, RELIGION, NATIONAL ORIGIN, SEX

In addition to the protection of Title VII of the Civil Rights Act of 1964, as amended, Title VI of the Civil Rights Act prohibits discrimination on the basis of race, color or national origin in programs or activities receiving Federal financial assistance. Employment discrimination is covered by Title VI if the primary objective of the financial assistance is provision of employment, or where employment discrimination causes or may cause discrimination in providing services under such programs. Title IX of the Education Amendments of 1972 prohibits employment discrimination on the basis of sex in educational programs or activities which receive Federal assistance.

INDIVIDUALS WITH DISABILITIES

Sections 501, 504 and 505 of the Rehabilitation Act of 1973, as amended, prohibits employment discrimination on the basis of disability in any program or activity which receives Federal financial assistance in the federal government. Discrimination is prohibited in all aspects of employment against persons with disabilities who, with reasonable accommodation, can perform the essential functions of a job.

If you believe you have been discriminated against in a program of any institution which receives Federal assistance, you should contact immediately the Federal agency providing such assistance.

Figure 12-3   *All employers covered by EEO laws are required to display this poster in the workplace.*

Figure 12-4 *The Americans with Disabilities Act prohibits discrimination against anyone because of a disability, and requires buildings to be made accessible.*

Rehabilitation Act was limited to Federal agencies and those entities receiving Federal funds. The ADA applies to all businesses with 15 or more employees, as well as to state and local governments, and provides significantly more protection that the Rehabilitation Act.

The ADA has important ramifications for the fire service. Due to the arduous physical nature of firefighting, and the consequences to the member, other firefighters, and the public if firefighters are physically incapable of performing their duties, a person's physical abilities are a critical factor in determining whether a particular person should be hired as a firefighter, or be allowed to continue working as a firefighter. It is important to understand several key definitions under the ADA:

- *Individual with a disability* refers to a person who has a physical or mental impairment that substantially limits one or more major life activities, has a record of such an impairment, or is regarded as having such an impairment. As a general rule, persons with a temporary condition or impairment are not regarded as having a disability.

- *Major life activities* are activities that an average person can perform with little or no difficulty, such as walking, breathing, seeing, hearing, speaking, learning, and working. Recent case law has placed greater emphasis on the analysis of the degree of limitation of major life activities. The Supreme Court made it clear that a person who has a physical impairment that prevents him or her from performing one type of job, but who is not prevented from performing other types of jobs, does not have a substantial limitation of a major life activity, and thus does not have a disability under the ADA. This interpretation substantially narrows the number of people who can sue under the ADA.

- *Qualified individual with a disability* is someone who has the requisite skill, experience, education, and other job-related requirements of the position held or desired, and who, with or without reasonable accommodation, can perform the essential functions of that position.

- *Reasonable accommodation* is a modification or adjustment to a job or work environment that will enable a qualified individual with a disability to perform essential job functions. Reasonable accommodation includes changes to an application process that will allow a qualified individual with a disability to participate in the application process. Reasonable accommodation may require making existing facilities accessible to and usable by persons with disabilities, job restructuring, modifying work schedules, providing additional unpaid leave, reassignment to a vacant position, acquiring or modifying equipment or devices, adjusting or modifying examinations, training materials, or policies, and providing qualified readers or interpreters. Reasonable accommodation may be necessary to perform job functions, or to enjoy the benefits and privileges of employment that are enjoyed by people without disabilities. An employer is not required to lower production standards to make an accommodation. An employer generally is not obligated to provide personal-use items, such as eyeglasses or hearing aids. An employer is not required to reassign responsibility for performing an essential function of a job to another employee as a reasonable accommodation.

- *Essential functions* of a job are those fundamental duties of a position that a person must be able to perform, with or without the assistance of reasonable accommodations. Job duties that are not fundamental or essential are characterized as marginal functions. The determination of what duties are essential functions for a particular job must be made on a case-by-case basis. Great deference is usually given to the duties outlined in written job descriptions. In addition, other factors come into play, such as the fact that the position exists to perform that function; the amount of time actually spent performing the function; and how many other employees can or must perform the function.

- *Undue hardship* refers to an action that requires significant difficulty or expense when considered in relation to factors such as a business's size, financial resources, and the nature and structure of its operation.

An employer is required to make a reasonable accommodation to a qualified individual with a disability, unless doing so would impose an undue hardship on the operation of the employer's business.

Prohibited Inquiries and Examinations

Under the ADA, an employer may not ask job applicants whether or not they have a disability, or the nature or severity of a disability, prior to making an offer of employment. Also prohibited are questions that are calculated to cause an applicant to disclose a disability, such as asking whether the candidate has had any serious illnesses, or taken any extended time out of work. Applicants may be asked about their ability to perform the essential functions of the job.

Medical examinations may not be required of applicants prior to a job offer being extended. A job offer may be conditioned on the results of a medical examination, but only if the examination is required of all new employees in that job category. Medical examinations of employees must be job-related and consistent with business necessity.

Drug and Alcohol Use

For purposes of the ADA, employees and applicants with drug addictions, or who are currently using illegal drugs, are not included within the definition of a person with a disability. Illegal use of drugs is not protected by the ADA, and an employer is allowed to take action against employees or applicants on the basis of such use. Tests for use of illegal drugs are not considered medical examinations and, therefore, are not subject to the ADA's restrictions on pre-employment medical examinations.

Employees who abuse alcohol are treated somewhat differently than drug abusers under the ADA. An employer is not prohibited by the ADA from taking action against an employee who comes to work intoxicated, or who consumes alcohol while at work in violation of the employer's regulations, even if the employee is an alcoholic. Employers may hold individuals who are illegally using drugs and individuals with alcoholism to the same standards of performance as other employees. However, an employer cannot take action against an employee or applicant solely because of the medical condition of being an alcoholic.

In addition, employees and applicants who have a history of drug use or alcoholism, but who are no longer using such drugs, or who have successfully been rehabilitated, cannot be discriminated against based solely on their past medical history. There are many cases that have defined the line between a person who has a history of addiction to drugs or alcohol, and someone who is currently using drugs and alcohol. Most of these cases come down to a factual determination, with the understanding that the ADA does not protect

Figure 12-5 *Physical abilities testing of firefighters is important, given the arduous nature of the job.*

or immunize those who are currently using drugs or who violate an employer's rules on the use of alcohol from adverse employment actions.

Physical Abilities Testing

Physical abilities testing has become commonplace throughout the fire service for both new hires and incumbent members **(Figure 12-5)**. Any physical abilities test that a fire department uses today must be validated in order to withstand an ADA or EEO challenge. Physical abilities testing of both candidates and incumbent firefighters has come under increased scrutiny, as a result of the ADA and Title VII sex discrimination cases. Under the ADA and Title VII, physical abilities tests are limited to tests that validly measure the ability to perform essential functions. The validation process starts with a job task analysis, conducted by a credentialed expert, who identifies the essential functions for the job. A relationship must then be empirically established between the essential functions and the physical abilities test.

A physical abilities test that requires a candidate to perform 20 push-ups or run a mile in eight minutes would be more difficult to validate than one requiring a candidate to remove a roof ladder from an engine, carry a hose up several flights of stairs, or drag a 1¾" attack line a given distance, because push-ups and running involve measuring physical abilities not directly required as an essential function of firefighters. Tests that simulate

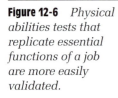
Figure 12-6 *Physical abilities tests that replicate essential functions of a job are more easily validated.*

actual fireground activities can more easily be validated **(Figure 12-6)**, provided a job task analysis is conducted, which properly establishes those tasks to be essential functions. For this reason, many fire departments have chosen to utilize a physical abilities test that simulates the execution of essential functions.

There is no generic set of essential functions applicable to every fire department in the United States. Each department must conduct its own job task analysis and develop its own set of essential functions based on local equipment and conditions, in order for its physical abilities test to be validated. In the absence of a formal job task analysis, those activities listed in a job description will be presumed to be essential functions for ADA purposes. However, use of a job description to establish essential functions can be challenged by a qualified person with a disability.

> ### SIDEBAR
>
> ### What is a Job Task Analysis?
>
> A job task analysis is essentially a process to identify and establish the various tasks and activities that make up a job. It involves observing and documenting all the actions, movements, and skills that are involved in a job. There are a variety of methodologies and techniques used to conduct a job task analysis. For some jobs, such as firefighter, the job task analysis can be a challenging and complex endeavor, while for other jobs it may be relatively simple. The results of a job task analysis can be used to identify the job functions related to a particular position. Those functions that are vital to the job are considered to be essential functions. All other functions are called marginal or non-essential functions.

Medical Requirements

Like physical abilities testing, medical requirements for firefighters implicate ADA and EEO concerns **(Figure 12-7)**, while at the same time impacting the ability of a fire department to safely fulfill its mission. Medical examinations

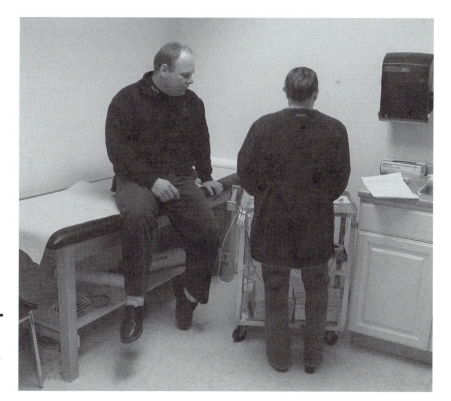

Figure 12-7 *Medical requirements for firefighters implicate ADA and EEO concerns.*

and requirements must be based on the actual demands of the job, as well as the associated environmental hazards. These issues again require consideration of essential functions and an empirical relationship between the job and the medical requirements.

In *Hegwer v. Board of Civil Service Commissioners of Los Angeles,* 5 Cal. App. 4th 1011, 7 Cal. Rptr. 2d 389 (Cal.App.Dist.2, 1992), an overweight female paramedic was suspended and disciplined for repeatedly failing to conform her weight to department standards. The paramedic sought to have her suspensions reversed based upon the city's affirmative action policy and state antidiscrimination laws. The court upheld the discipline, finding that the city had documented the relationship of weight to job performance. The court's reasoning focused on the medical evidence to support the standards, and the city's repeated efforts to assist the paramedic in complying with the requirements.

While the *Hegwer* case was going through the state court system, a second case was filed in Federal Court on Ms. Hegwer's behalf by her union, citing state and Federal constitutional issues.

United Paramedics of Los Angeles v. City of Los Angeles
936 F.2d 580 (9th Cir.,1991)
United States Court of Appeals for the Ninth Circuit

MEMORANDUM

The Los Angeles City Fire Department enforces an employee body weight limitation program. The United Paramedics of Los Angeles ("UPLA"), a labor union, represents Emergency Medical Services employees subject to the Fire Department's weight program. UPLA and four members sued to invalidate the weight program as unconstitutional. The district court granted summary judgment to the City. UPLA and the four employees appeal.

I. EMS Employees' Fundamental Right to Privacy

UPLA does not assert a privacy right to control one's own body. Instead, UPLA asserts that constitutional privacy protects the collection and dissemination of data on EMS employees' body weight. In the wake of cases permitting urinalysis of employees for drug use despite federal constitutional privacy rights, UPLA does not vigorously press this claim as a federal privacy right. The union does, however, assert the California constitution's privacy guarantee in a pendant state law claim. The parties do not dispute that California's constitutional right of privacy is broader than the federal one.

UPLA characterizes EMS employees' weight as private medical information protected by the state constitution. . . . Absent a compelling government interest,

(*continued*)

(*continued*)

UPLA argues, the California constitution protects EMS employees' private medical information from disclosure and dissemination through the Fire Department's weight control program.

. . . As in federal privacy analysis, California courts evaluate an asserted state privacy right through examination of an employee's reasonable expectation of privacy. . . . Where an employee has little or no reasonable expectation of privacy, the California courts will not vindicate the asserted right. . . . Even where employees may reasonably expect privacy, moreover, California courts balance the asserted right to privacy against the rationale for intrusion on that privacy right. . . .

In this case, we rely on several factors to hold that EMS employees have no reasonable expectation of privacy concerning their weight under these circumstances. First, as the district court reasoned, body weight is both public and generally obvious to anyone looking at an EMS employee. . . . Second, the close living quarters of EMS personnel while on duty undermines their assertion of a privacy expectation of their body weight. General assessments of EMS personnels' body weight are presumably available to all Fire Department employees with whom they closely work and live. Third, EMS employees must submit to biannual medical examinations, including weight assessment, in any event. All of these factors, undermine EMS employees' assertion of a reasonable expectation of privacy. We therefore hold that UPLA and the employees have failed to assert a privacy interest protected by the California constitution.

Even if we recognized EMS employees' privacy interest in data about their body weight, we would still have to balance any legitimate privacy interest against the Fire Department's asserted health and safety justification for the weight control program. . . . Indeed, neither party disputes the Fire Department's responsibility to assure employee health and fitness, even at the expense of privacy interests, where employee health and fitness bear on safety concerns. UPLA argues that the Fire Department failed to demonstrate a compelling relationship between its weight control program and safety. We conclude that the weight control program's intrusion upon EMS employees' asserted privacy interest in data about their body weight is, at most, minimal. Assuming arguendo that that privacy interest arises to state constitutional importance, the Fire Department has sufficiently justified the minimal intrusion.

We therefore affirm the district court's judgment on the privacy claim.

II. EMS Employees' Procedural Due Process Rights

UPLA contends that the weight control program and consequent discipline embody an unconstitutional "conclusive presumption" of EMS employee unfitness. As tenured Civil Service employees, EMS personnel hold a vested property interest in continued employment. . . .

UPLA argues that due process requires the Fire Department to determine each EMS employee's fitness individually before subjecting him or her to the punitive weight control program. Due process does not permit, UPLA argues, the

(*continued*)

(*continued*)

Fire Department to rely on a conclusive presumption that an overweight EMS employees must be unfit and therefore subject to discipline.

The parties argue extensively in their briefs about the continued viability of "conclusive presumption analysis" in due process law. Suffice it to say here that we have rejected [the UPLA's] conclusive presumption analysis [argument]. . . .

Nonetheless, UPLA urges that subjecting EMS personnel to the weight control program without an individualized determination of unfitness violates due process. We construe this claim as an allegation that the weight control program violates EMS employees' procedural due process rights. EMS employees' vested property interests do not trigger strict scrutiny of the weight control program itself, of course. Instead, assertion of the EMS employees' property interests triggers inquiry into whether sufficient process protects EMS employees prior to deprivation of those property interests.

We note that EMS personnel do not face discharge or discipline without benefit of procedural remedies. EMS employees may grieve the imposition of discipline. They may also appeal any six-day suspension or more drastic discipline to the Civil Service Commission. Moreover, EMS employees may receive individualized evaluations under the weight control program by submitting their own medical evidence. Because EMS employees have procedural remedies available and because they may rebut any "presumptions" inherent in the program with their individual medical evidence, the Union and the employees have failed to show how the program violates their procedural due process rights.

We therefore affirm the district court's judgment on the due process claim.

III. EMS Employees and Equal Protection of the Law

UPLA contends that the weight control program denies EMS employees equal protection of the law because it classifies them in a program not bearing even a rational relationship to legitimate government interests. The City failed to demonstrate, UPLA contends, that the weight control program advances the health or safety of either employees or the public. Noting that, under equal protection analysis, a classification of the type involved here need only not be irrational, the district court upheld the validity of the program.

We affirm this conclusion. Application of the rational basis test requires a two-step analysis. . . . First, we must determine whether the weight control program has a legitimate purpose. . . . The City's ostensible purpose of furthering employee and public health and safety is certainly legitimate.

Second, we must determine whether the program serves this purpose. . . . This determination does not depend on a "tight fitting relationship" between the program and its purpose. . . . We need merely discern a "plausible," "arguable," or "conceivable" relationship between the program and its purpose. . . . The general statistical evidence of increased risks of illness and injury associated with being overweight creates the necessary nexus between the weight control program and the purpose of health and safety.

UPLA argues finally that we should not pay to the weight control program the ordinary deference due to a legislative enactment. When evaluating a
(*continued*)

(*continued*)
legislative enactment for a rational basis, courts must defer to legislative wisdom and expertise. . . . No legislature, however, but rather Chief Manning alone promulgated the weight control program. UPLA urges we accord no deference to the Chief's administrative fiat.

We need not defer to a governmental entity, however, in order to perceive a rational relationship between the weight control program and a legitimate government purpose. Regardless of who proposed the program, it appears to us to be rationally related to health and safety. UPLA does not propose that administrative fiats require strict scrutiny. Reviewing for a rational relationship, and without paying anyone deference, we conclude that the weight control program passes constitutional scrutiny.

Conclusion

UPLA failed to assert a fundamental right infringed by the weight control program. Accordingly, we review the program under rational basis analysis. The Fire Department's concern for health and safety sufficiently justifies the program. Moreover, procedural due process sufficiently protects the EMS employees' property interests in their employment. Accordingly, the judgment of the district court is AFFIRMED.

Case Name: United Paramedics of Los Angeles v. City of Los Angeles

Court: United States Circuit Court of Appeals, 9th Circuit

Summary of Main Points: A fire department may institute a weight control program without infringing upon an employee's constitutional rights, nor violating any employment discrimination laws.

STATE LAW DISABILITY DISCRIMINATION

In addition to the ADA, most states have laws that prohibit discrimination based on disability in a wide variety of settings, including employment, housing, and education. Many local jurisdictions have adopted prohibitions against disability discrimination in their charters and ordinances. As a result, persons who may not be able to sue under the ADA may nonetheless have some recourse under state and local law.

AGE DISCRIMINATION

The Age Discrimination in Employment Act (ADEA) prohibits discrimination based on a person's age. While the law has been amended significantly over the years, at the present time the ADEA applies to discrimination against persons over the age of 40. Like the ADA, the ADEA has important ramifications for the fire service.

Any type of fire department policy or program that impacts personnel on the basis of age has the potential to raise a claim under the ADEA. Policies that adversely impact older workers, including physical abilities testing and medical requirements, must be developed with the ADEA in mind. As the *Smith* case below demonstrates, the business necessity of a given policy is a vitally important consideration for fire departments.

Jerry O. Smith v. City of Des Moines, Iowa
99 F.3d 1466 (8th Cir. 1996)

BOWMAN, Circuit Judge.

Appellant Jerry O. Smith brought suit against the City of Des Moines, claiming that he was fired from his position as a city firefighter in violation of the Age Discrimination in Employment Act of 1967 . . . and the Americans With Disabilities Act of 1990. . . .

At the time of his dismissal, Smith had been a firefighter with the Des Moines Fire Department for thirty-three years and had risen to the rank of fire captain. In 1988, the city began to require annual testing of all firefighters at the rank of captain or below to determine whether they could safely fight fires while wearing a self-contained breathing apparatus (SCBA). Each firefighter underwent spirometry testing, which gauges pulmonary function by measuring the capacity of the lungs to exhale. Any firefighter whose forced expiratory volume in one second (FEV1) exceeded 70% of lung capacity was approved to wear a SCBA. If a firefighter scored less than 70%, he or she was required to take a maximum exercise stress test, which measures the capacity of the body to use oxygen effectively. The city required firefighters to establish a maximum oxygen uptake (VO2 max) of at least 33.5 milliliters per minute per kilogram of body weight in order to pass the stress test.

Smith failed both tests in 1988 and was not approved to wear a SCBA that year. In 1989, 1990, and 1991, Smith passed the spirometry test and was approved for SCBA use. In August 1992, Smith narrowly failed the spirometry test and was referred to Dr. Steven K. Zorn, a consultant to the city, for further testing. In Dr. Zorn's office, Smith passed the spirometry test but registered a VO2 max of only 22.2 on the stress test. The fire department placed Smith on sick leave. In January 1993, Smith returned to Dr. Zorn but scored only 21.1 on a stress test. The fire department offered to allow Smith to remain on sick leave until April, when he would turn age fifty-five and thus be eligible for retirement.

In the interim, the fire department sent Smith to another physician, Dr. John Glazier, for a second opinion. Additionally, when Smith did not file for retirement in April, the fire chief filed an application for disability retirement on

(continued)

(*continued*)

Smith's behalf. Before ruling on this application, the state pension board required Smith to be examined by a panel of three additional physicians. Dr. Glazier did not perform a stress test, but the panel of three physicians did (Smith's VO2 max was 28.9). All four physicians concluded that Smith was physically capable of working as a firefighter. After receiving these recommendations, the pension board denied the application for disability retirement, finding that Smith was not disabled from working as a firefighter.

The fire department did not permit Smith to return to work but did offer to place him on leave of absence with benefits until July 1, 1994, when he would be eligible for maximum pension benefits. Smith did not file for retirement at that time, however, and the city discharged him on July 18, 1994 for failure to meet the fire department's physical fitness standards.

After obtaining right-to-sue letters from the Equal Employment Opportunity Commission (EEOC) and the Iowa Civil Rights Commission, Smith brought suit against the city in federal district court, raising claims under the ADEA, the ADA, and the Iowa Civil Rights Act. . . . The District Court granted summary judgment in favor of the city on all counts. The court, assuming Smith could establish that the city's testing standards have a disparate impact on older firefighters, held that the city had established a "business necessity" defense because firefighters require "a high standard of physical fitness." Similarly, Smith's ADEA disparate treatment claim failed because he was not qualified for the job, and the state law claim failed because Iowa law mirrors federal law. The District Court also concluded that Smith did not have a disability and granted summary judgment for the city on his ADA claim. Smith's appeal raises only the disparate impact and ADA claims. . . .

A. We consider first the city's argument, which the District Court rejected, that a claim of disparate impact is not cognizable under the ADEA. Disparate impact claims challenge "'employment practices that are facially neutral in their treatment of different groups but that in fact fall more harshly on one group than another and cannot be justified by business necessity.'". . . . A disparate impact plaintiff need not prove a discriminatory motive. . . .

Like Title VII of the Civil Rights Act of 1964, to which the disparate impact theory was first applied . . . the ADEA contains two prohibitions relevant here:

It shall be unlawful for an employer—

(1) to fail or refuse to hire or to discharge any individual or otherwise discriminate against any individual with respect to his compensation, terms, conditions, or privileges of employment, because of such individual's age

(2) to limit, segregate, or classify his employees in any way which would deprive or tend to deprive any individual of employment opportunities or otherwise adversely affect his status as an employee, because of such individual's age. . . .

29 U.S.C. SS 623(a) (1994).

We have on several occasions applied disparate impact analysis to age discrimination claims. . . . We conclude that disparate impact claims under the ADEA are cognizable.

(*continued*)

(*continued*)

B.   We assume, as the District Court did, that Smith has established a prima facie case of disparate impact, that is, that he has demonstrated "that a facially neutral employment practice actually operates to exclude from a job a disproportionate number of persons protected by the ADEA." . . . We therefore turn to Smith's argument that the District Court erroneously granted summary judgment to the city based on the so-called "business necessity" defense.

This defense is derived in part from the cases in which the Supreme Court developed the disparate impact doctrine under Title VII, . . . and in part from a provision of the ADEA which states that an employment practice is not unlawful "where the differentiation is based on reasonable factors other than age." . . . We recognize that in the Title VII context the business necessity defense has undergone several transformations in recent years. . . .

We conclude that the city met its burden on the business necessity defense by supporting its motion with evidence. . . . On the job-relatedness issue, the city presented undisputed evidence that a captain is frequently involved in fire suppression activities when a company arrives at a fire scene and that the captain wears a SCBA under those circumstances. . . . This evidence alone is sufficient to carry the city's burden of showing that its fitness standard has a "manifest relationship" to the position in question. . . .

The other element of the defense is whether the standard is necessary to safe and effective job performance. The city's evidence on this issue is more complicated and begins with some of the extensive regulations governing the manner in which the city operates its fire department. Federal regulations require the fire department to provide firefighters with SCBAs "when such equipment is necessary to protect the health of the employee." 29 C.F.R. SS 1910.134(a)(2) (1995). The city may not assign firefighters to tasks requiring use of a SCBA unless they are "physically able to perform the work and use the equipment." . . . The city must review the medical status of SCBA users periodically. . . . The American National Standards Institute (ANSI) standard on physical qualifications for respirator use recommends spirometry testing as a screening mechanism for SCBA users and suggests stress testing for persons who use SCBAs under strenuous conditions. . . . ANSI recommends a 70% FEV1 threshold for spirometry testing but does not specify an acceptable result for stress testing. . . .

To reach its determination that a VO2 max of 33.5 was the appropriate threshold for stress testing, the city relied on a review of the relevant medical literature by Dr. Zorn. A number of studies suggest that firefighters consume between 25 and 35 milliliters of oxygen per kilogram per minute while suppressing a fire. . . . One study in particular involved 150 firefighters performing a series of tasks in a simulated fire-suppression environment. . . . The authors of that study determined that a VO2 max of 33.5 was the minimum required to allow the firefighters to complete the simulation successfully. . . . The authors then repeated the simulation with 32 additional firefighters. Id. Of those with a VO2 max less than 33.5, only 40% (4 of 10) completed the simulation successfully.

(*continued*)

(*continued*)

Id. On the other hand, of those with a VO2 max of 33.5 or more, 86% (19 of 22) completed the simulation successfully. . . . After reviewing this study and others, Dr. Zorn concluded that 33.5 was the minimum satisfactory VO2 max requirement for the Des Moines firefighters. . . . This evidence would clearly be sufficient to entitle the city to a directed verdict on the issue of necessity if it were uncontroverted. . . .

To summarize our conclusions: fitness and the ability to perform while wearing a SCBA are undoubtedly job-related and necessary requirements for firefighters. The dispute in this case is not whether firefighters must be physically fit, but how fitness can be most appropriately measured and how the city may distinguish those firefighters who are probably capable of performing the job from those firefighters who are probably not capable. The city has not proceeded arbitrarily, but rather has carefully developed a standard based upon the available medical literature and using the best test available for measuring fitness, the stress test. . . . The literature indicates that a high proportion of firefighters with a VO2 max above 33.5 can perform fire suppression tasks successfully, but a much lower proportion of those with a VO2 max below 33.5 can do so. Smith argues, and the physicians' evaluations suggest, that some firefighters with lower VO2 max scores—Smith in particular—may be able to perform their jobs. This may well be true, but the law does not require the city to put the lives of Smith and his fellow firefighters at risk by taking the chance that he is fit for duty when solid scientific studies indicate that persons with test results similar to his are not. The lack of a precise or universally perfect fit between a job requirement and actual effective performance is not fatal to a claim of business necessity, particularly when the public health and safety are at stake. . . . We conclude that Smith has not met his burden of presenting a triable issue on the business necessity defense.

C. Smith also argues that he presented evidence of an alternative means of assessing fitness that would have less of a disparate impact on older firefighters. In particular, he suggests that the city use the spirometry and stress tests to determine which firefighters may be unfit for the job, then require those firefighters to undergo a physical examination and "a battery of tests" to determine whether they are actually fit for duty. . . .

We have not previously had the occasion to determine whether this branch of the Title VII disparate impact doctrine applies to the ADEA. For purposes of this appeal, however, we assume that the Title VII framework applies: once the defendant has met its burden of demonstrating business necessity, the plaintiff may still prevail by showing "that other selection devices without a similar discriminatory effect would also serve the employer's legitimate interest in efficient and trustworthy workmanship." . . . For several reasons, Smith's argument on this point is unavailing.

First, it does not appear from the record that Smith advanced this argument before the District Court. We will not reverse a grant of summary judgment

(*continued*)

(*continued*)

on the basis of an argument not presented below. . . . Even if the argument were proper, however, Smith has not made any showing that his proposed alternative (which is in any case rather vague) would have less of a disparate impact on older firefighters than the city's present system does. At most, Smith has asserted that he would be able to pass his proposed battery of tests, but he has not shown the effect of his system on other firefighters. Nor has he shown that his more subjective approach would serve the city's legitimate interest in the fitness of its firefighters as well as the current system. Smith has failed to raise a genuine issue of material fact on this branch of the disparate impact doctrine. . . .

The judgment of the District Court is affirmed.

Case Name: Jerry O. Smith v. City of Des Moines, Iowa

Court: United States Circuit Court of Appeals, 8th Circuit

Summary of Main Points: A fire department may impose reasonable medical requirements upon firefighters that adversely impact members on the basis of age, where such requirements are a "business necessity." There are two aspects of the business necessity defense: (1) that the requirements have a "manifest relationship" to the position in question; and (2) the requirements are necessary for safe and effective job performance.

As the *Smith* case demonstrates, whenever a fire department policy impacts older workers, the fire department must be prepared to establish valid scientific reasons to support its contention that the policy is essential due to business necessity.

Firefighters and Mandatory Retirement Age

The ADEA has been subject to a number of changes and amendments since it was first enacted. One area that has seen several changes involves mandatory retirement ages for firefighters and police officers. The current status of the ADEA is that firefighters and police officers may be required to retire at a specific age, provided it is pursuant to a bona fide retirement plan. The primary concern related to firefighters and police officers is that the retirement plan is not being used as a subterfuge to permit age-based discrimination. Provided there is a bona fide retirement plan, mandatory retirement age may be set as low as 55 years for firefighters and police officers.

The *Minch* case discusses the various changes to the ADEA, as well as the current state of the mandatory retirement exception.

Minch v. City of Chicago
Drnek v. City of Chicago
No. 02-2588 No. 02-2587 (7th Cir. 04/09/2004)

ROVNER, Circuit Judge.

In 1996, Congress restored to the Age Discrimination in Employment Act ("ADEA") an exemption permitting state and local governments to place age restrictions on the employment of police officers and firefighters. . . . Four years later, the Chicago City Council exercised its authority under this exemption to reestablish a mandatory retirement age of 63 for certain of the City's police and firefighting personnel. Police officers and firefighters who were subject to the age restriction filed two suits asserting in relevant part that the reinstated mandatory retirement program amounted to subterfuge to evade the purposes of the ADEA. . . .

I.   Historically, Chicago, like many other state and local governments, has placed age limits on the employment of its police and firefighting personnel. As early as 1939, for example, Chicago's municipal code required city firefighters to retire at the age of 63.

As it was originally enacted in 1967, the ADEA by its terms did not apply to the employees of state and local governments. Congress amended the statute to include those employees in 1974. . . . State and local rules establishing maximum hiring and retirement ages for police officers and firefighters were now vulnerable to challenge; only if it could be shown that age was a bona fide occupational qualification for these positions would the rules survive scrutiny under the ADEA. . . . The Equal Employment Opportunity Commission ("E.E.O.C.") began to challenge these age limits as discriminatory. Chicago, seeing the handwriting on the wall, raised the mandatory retirement age for its firefighters and police officers to 70, the maximum age at which employees enjoyed the protection of the ADEA at that time.

Responding to the concerns expressed by state and local governments, Congress in 1986 amended the ADEA to exempt the mandatory retirement of state and local police and firefighting personnel from the statute's coverage. . . . Congress enacted the exemption in recognition that there was, as of that time, no consensus as to the propriety of age limits on employees working in the realm of public safety. . . . The exemption thus permitted any state or local government which, as of March 3, 1983 . . . had in place age restrictions on the employment of police officers and firefighters, to restore those restrictions. In 1988, Chicago took advantage of the exemption and reinstated a mandatory retirement age of 63 for its firefighters and police officers.

Pursuant to a sunset provision in the 1986 legislation, the exemption permitting the reinstatement of these age limits expired at the end of 1993. . . . In the ensuing years, Chicago, along with other state and local governments, were again compelled to drop their age restrictions on the employment of police and firefighting personnel.

(continued)

(*continued*)

In 1996, however, Congress reinstated the exemption, this time without any sunset provision, and retroactively to the date that the prior exemption had expired in 1993. . . . The 1996 legislation also broadened the exemption, allowing cities and states which had not imposed age restrictions on their police and fire-fighters prior to the *Wyoming* decision to enact such limits. As relevant here, the exemption, codified at 29 U.S.C. § 623(j),4 permits a public employer to discharge a police officer or firefighter based on his age, subject to two principal conditions. First, section 623(j)(1) specifies that the employee must have attained either the age of retirement that the state or municipality had in place as of March 3, 1983 or, if the age limit was enacted after the date the 1996 exemption took effect, the higher of the retirement age specified in the post-1996 enactment or the age of 55. Second, section 623(j)(2) requires that the state or city discharge such an em-ployee pursuant to a bona fide retirement plan that is not a subterfuge to evade the purposes of the statute. Four years later, the Chicago City Council adopted a mandatory retirement ordinance ("MRO") reinstating a mandatory retirement age of 63 for its police officers and for its uniformed firefighting fire personnel. In the preamble to that ordinance, the City Council indicated that its purpose in restor-ing the retirement age was to protect the safety of Chicago residents.

The four plaintiffs were Chicago police officers and uniformed firefighters who were 63 or greater when the MRO took effect and thus were forced to take immediate retirement. They filed two actions against the City asserting, in rele-vant part, that the City was not actually motivated by public safety purposes in enacting the MRO. The cases were consolidated in the district court. Although the plaintiffs do not dispute at this juncture that the MRO and their involuntary retirement pursuant to the MRO satisfy the criteria set forth in section 623(j)(1), they allege that the MRO amounts to a subterfuge to evade the purposes of the ADEA and for that reason amounts to illegal age discrimination. Among other motives for enacting the MRO, the plaintiffs assert, the City wanted to get rid of what one city council member described as "old-timers" and "deadbeats" in the police and fire departments and to make room in those departments for younger, more racially and ethnically diverse individuals who would work harder and bring "fresh" ideas with them. This amounts to age discrimination in violation of the ADEA, in the plaintiffs' view. The district court denied the City's motion to dismiss the plaintiffs' ADEA claims. . . . In the court's view, the question of whether the city reinstated a mandatory retirement age of 63 as a subterfuge for age discrimination was one of fact that necessitated inquiry beyond the statement of purpose set forth in the preamble to the MRO into the true motive or motives behind the legislation. . . . "Age-based retirement is tolerated in limited circum-stances under § 623(j), but not for the wrong reasons, i.e. not for reasons that are merely a coverup for the type of ageism prohibited by the ADEA." . . . Here, the plaintiffs were able to point to the remarks of the sponsor of the MRO and of high-ranking city officials as proof that the City may have been motivated impermissibly by stereotypes and bias against older members of the police and fire departments when it enacted the MRO. The plaintiffs also represented that

(*continued*)

(*continued*)

the City had delayed reinstating the retirement age of 63 until after a close friend of the Mayor (who otherwise would have been forced to retire) voluntarily retired at age 68. The district court found these allegations, suggesting that the City did not actually enact the MRO for legitimate, safety-related reasons, sufficient to state a viable claim for subterfuge. . . .

II.   This appeal calls upon us to consider under what circumstances a mandatory retirement program for public safety personnel might constitute a subterfuge to evade the purposes of the ADEA. . . . A plaintiff can establish subterfuge if he or she can demonstrate that a state or local government took advantage of the exemption and imposed a mandatory retirement age for police and firefighting personnel in order to evade a different substantive provision of the statute. However, because the ADEA expressly permits employers like Chicago to reinstate mandatory retirement programs for police and fire personnel and thus to discharge employees based on their age, proof that local officials exercised this right for impure motives will not in and of itself suffice to establish subterfuge for purposes of section 623(j)(2). Given that the plaintiffs' theory of subterfuge in these cases relies solely on proof that Chicago City Council members and other City officials may have harbored discriminatory attitudes about older workers when they reinstated a mandatory retirement age of 63 for police officers and firefighters and that they adopted the MRO for illicit motives unrelated to public safety, the plaintiffs have failed to state an ADEA claim on which relief may be granted. . . .

Evidence that City officials had impure motives for reinstating a mandatory retirement age, however, will not by itself support an inference of subterfuge. . . . The ADEA does not forbid Chicago from making age-based retirement decisions as to its police and firefighting personnel; it expressly allows state and local governments to make such decisions so long as they act within the parameters set forth in section 623(j)(1), which Chicago did. The statute does not condition the validity of such retirement programs on proof that the public employer has adopted the program genuinely believing that it is justified in the interest of public safety. Instead, recognizing that there was not yet any national consensus as to the relationship between age and one's fitness to serve as a police officer or firefighter, Congress opted simply to restore the status quo ante, permitting states and cities to continue imposing age limits on these positions as they had been able to do prior to the ADEA's extension to state and municipal employers and *Wyoming's* 1983 holding sustaining that extension. . . .

Thus, proof that Chicago resumed mandatory retirement for police and fire personnel based in whole or in part on stereotypical thinking—that older individuals are not up to the rigors of law enforcement or firefighting and should make room for younger, "fresher" replacements—or [sic] for reasons wholly unrelated to public safety, will not establish subterfuge because it does not reveal a kind of discriminatory conduct that the ADEA by its very terms forbids. . . .

What is necessary to establish subterfuge is proof that the employer is using the exemption as a way to evade another substantive provision of the act. . . . Here

(*continued*)

(continued)

then, a viable claim of subterfuge would require the plaintiffs to allege and prove that Chicago took advantage of the statutory authorization to mandatorily retire police officers and firefighters as a means of discriminating in another aspect of the employment relationship—that is, other than in the discharge decision—in a way that the statute forbids. . . .

III.   Having answered the question certified for interlocutory review, we REMAND these cases to the district court with directions to DISMISS the plaintiffs' ADEA claims and to conduct such further proceedings as may be consistent with this opinion.

Case Name: Minch v. City of Chicago

Court: United States Circuit Court of Appeals, 7th Circuit

Summary of Main Points: A fire department may impose a mandatory retirement age, provided such is not a subterfuge for unlawful age-based discrimination. Proof of subterfuge requires more than proof that city officials had discriminatory motives in re-implementing the mandatory retirement age.

SUMMARY

Employment discrimination remains one of the most heavily litigated areas for fire departments, and will likely remain so for the foreseeable future. A variety of laws at both the state and Federal levels prohibit a broad variety of discrimination, including discrimination based on race, national origin, sex, religion, disability, and age. Collectively, these laws create a comprehensive prohibition against employment discrimination that impacts fire departments when recruiting, hiring, promoting, disciplining, terminating, and retiring personnel.

REVIEW QUESTIONS

1. What was the first law aimed at protecting and enforcing civil rights?

2. Define discrimination.

3. Explain disparate treatment and how it differs from disparate impact.

4. What standard of review is usually applied to cases that challenge a governmental action that violates a person's equal protection rights based upon racial classifications?

5. What is a reasonable accommodation under the ADA?

6. Are there any laws that prohibit an employer from giving an applicant a medical examination?

7. Whom does the Age Discrimination in Employment Act protect?

8. What did the Civil Rights Act of 1964 do that the Civil Rights Act of 1866 did not do?

9. Can the same action by an employer violate both the ADA and the ADEA?

10. Does the ADEA prohibit a fire department from implementing a mandatory retirement age?

DISCUSSION QUESTIONS

1. Why would some fire departments choose to become equal employment opportunity employers, affirmative action employers, or both? Would every fire department choosing to become an equal employment opportunity employer or affirmative action employer have to adopt the same goals, or would they vary from department to department?

2. Before people can sue their employers for a violation of Title VII, what procedural step or steps must they take? In your home state, where would these steps need to be taken?

3. The *Minch* case concerned mandatory retirement age. What if a fire department refused to provide life insurance coverage to members over the age of 60, where such coverage is provided to all members pursuant to a collective bargaining agreement? Would it matter if the union agreed to it?

Chapter 13

SEXUAL HARASSMENT AND OTHER FORMS OF EMPLOYMENT DISCRIMINATION

Learning Objectives

Upon completion of this chapter, you should be able to:

- Define sexual harassment and explain the two types of sexual harassment.
- Identify factors that contribute to a sexually hostile work environment.
- Identify the Federal laws that impact pregnancy discrimination.
- Identify the need for reasonable accommodation of religion in the workplace, and impact of undue hardship on an employer.
- Explain the constitutionality of grooming and uniform regulations.

INTRODUCTION

The captain sat at his desk, staring at the computer screen intently as he typed the incident report. There was a knock on the door.

"Captain, can I talk to you about something?" Cathy asked hesitantly, "in private."

"Sure, come on in my office. What's on your mind?" said the captain, closing the door.

"Cap, I am feeling really uncomfortable about something. I don't want to make waves, but it is really bothering me. It's causing me to lose sleep at night. To be honest, it's making me sick."

"What is it, Cathy?"

"At night, after you go in your office, the guys have started watching very explicit movies on the TV. When they started doing it, I simply left the room, but it seems like they took that as my approval about what they are doing. Last week while you were on vacation, they were watching the movies during the day. I've tried talking to them, but they say that what they choose to watch on the TV is by majority rule." Cathy's eyes started tearing. "Cap, I really like it here and I don't want to leave. I don't think I should have to leave. I don't want everyone to hate me, and I know if they find out I am complaining they will. But I shouldn't have to put up with this."

Chapter 12 provided a basic overview of civil rights legislation, and the essentials of employment discrimination. However, there are a variety of additional types of discrimination issues that impact the fire service beyond those discussed in Chapter 12. These issues include sexual harassment, pregnancy discrimination, accommodating religious practices, and uniform and grooming requirements.

SEXUAL HARRASSMENT

quid pro quo sexual harassment
one of the two types of sexual harassment; quid pro quo sexual harassment occurs when the employee's employment opportunities or benefits are granted or denied because of the employee's submission to sexual advances or requests for sexual favors

Sexual harassment is a form of gender-based sex discrimination, prohibited by Title VII of the Civil Rights Act of 1964. Sexual harassment includes unwelcome sexual advances, requests for sexual favors, and verbal or physical conduct of a sexual nature, when submission to or rejection of this conduct affects an individual's employment, unreasonably interferes with an individual's work performance, or creates an intimidating, hostile, or offensive work environment.

There are two main types of sexual harassment: quid pro quo and hostile work environment. **Quid pro quo sexual harassment** occurs when the employee's employment opportunities or benefits are granted or denied because of an individual's submission to sexual advances or requests for sexual

hostile work environment sexual harassment

one of the two types of sexual harassment; hostile work environment sexual harassment occurs when unwelcome sexual conduct unreasonably interferes with an individual's work performance or has the effect of creating an intimidating or offensive work environment

favors. The term "quid pro quo" translates literally as "this for that." Quid pro quo harassment usually involves an employer, manager, or supervisor as the harasser. A single act of quid pro quo harassment can create liability if it is linked to an employment decision. An example of quid pro quo harassment would be if a woman is denied a promotion because she refuses her supervisor's requests for sexual favors. Generally when quid pro quo harassment is committed by a supervisor, the employer is strictly liable.

Hostile work environment sexual harassment occurs when unwelcome sexual conduct unreasonably interferes with an individual's work performance or has the effect of creating an intimidating or offensive work environment. The harassment need not result in tangible or economic consequences, such as termination, loss of pay, or failure to be promoted, nor in physical or psychological injury. In *Harris v. Forklift Systems, Inc.*, 510 U.S. 17 (1993), the United States Supreme Court ruled that when the sexual harassment is sufficiently severe and pervasive to create an objectively hostile work environment, proof of psychological harm is not required. See **Table 13-1** for a list of conduct that may constitute sexual harassment.

The conduct of an employer, supervisors, coworkers, customers, or clients can create a hostile work environment. In cases of hostile work environment, liability for an employer is triggered when:

- The employer knew or should have known about the harassment.
- The employer failed to take appropriate corrective action.

Acts of supervisors, coworkers, customers, or clients may be the basis for sexual harassment where the employer knows or should have known of the conduct, and fails to take immediate and appropriate corrective action.

Table 13-1   *Conduct that may constitute sexual harassment.*

Repeated sexual innuendoes or sexually oriented comments

Obscene remarks, jokes, slurs, or language related to sex or of a sexual nature

Letters, notes, faxes, e-mail, or graffiti that is of a sexual nature or is sexually abusive

Sexual propositions, insults, and threats

Sexually oriented demeaning names

Persistent unwanted sexual or romantic requests, overtures, or attention

Leering, whistling, or other sexually suggestive sounds or gestures

Display of pornographic photographs, drawings, cartoons, videos, or other sexually oriented material in the workplace

Coerced or unwelcomed touching, patting, brushing up against, pinching, kissing, stroking, massaging, squeezing, fondling, or tickling

Subtle or overt pressure for sex or sexual favors

Coerced sexual intercourse (e.g., as a condition of employment or academic status)

AUTHOR'S COMMENTARY: The following case contains language and depictions that some may find offensive. This information is quoted directly as part of a published court case and is the original language of the court. The admitted testimony describes a pattern of inappropriate actions and behavior that met the court's standards for a hostile work environment. Its purpose here is to help firefighters recognize and prevent sexual harassment in the workplace, as well as gain a clearer understanding of the victim's perspective.

Julia M. O'Rourke v. City of Providence
235 F.3d 713 (1st Cir. 2001)
United States Court of Appeals for the First Circuit

LYNCH, Circuit Judge.

. . . Until 1990, no female firefighters had ever served in the City of Providence Fire Department. In January, 1992, O'Rourke and six other women who had passed a written examination were admitted to the City's firefighter six-month training program, along with 77 male trainees. O'Rourke was hired under the City's newly implemented affirmative action policy. . . .

In January of 1992, the Department Chief promulgated a sexual harassment policy. The policy prohibited firefighters from keeping sexually explicit books and magazines, viewing sexually explicit movies, or making sexual jokes at their respective stations. The superior officer at each station was responsible for enforcing the policy, and they had been trained to do so. New firefighters were to be instructed during two hours of sensitivity training, including a component on sexual harassment, to be incorporated into the curriculum of their six-month training program.

O'Rourke underwent this six-month training program. During this period and often in the presence of supervisors, overtly sexual behavior was directed toward O'Rourke. For example, during a class break, a male trainee, Ferro, passed around a video camera playing scenes of Ferro having sex with his girlfriend. The instructor was in the classroom and did nothing. Ferro also discussed his sexual prowess . . . and his sexual encounters during lunch breaks, just outside the training facility. These incidents occurred in the presence of officers.

O'Rourke expressed her disgust and discomfort directly to Ferro, but Ferro was undeterred. During a training exercise in a pool and in the presence of an officer of the academy, Ferro pointed to O'Rourke's breasts and commented that she was "stacked." Ferro constantly discussed sexual positions . . . O'Rourke "just blocked them out." While standing in line for roll call, which was conducted by various academy officers, Ferro, standing behind O'Rourke, would frequently expound his opinion . . . [about] women. . . .

(continued)

(continued)

Ferro's behavior was not unique. Another male trainee, McDonald, snapped O'Rourke's bra, commented on her scent, and asked O'Rourke if another female trainee, whom McDonald called a "dyke," ever looked at O'Rourke while they were changing clothes. McDonald also asked O'Rourke, in the presence of several other firefighters, if she was on birth control. . . . Some of these incidents occurred in the presence of training academy officers. The commentary made O'Rourke so uncomfortable that she began trying to camouflage her body by wearing oversized shirts. She did not complain to any of the officers because she "didn't want to cause any waves" and "just wanted to get through the academy."

After completing the training program, O'Rourke accepted a temporary assignment in the office of the Fire Department Chief, Chief Bertoncini. She worked for Chief Bertoncini from June to September, 1992, and after a brief layoff, from November, 1992, until March, 1993. She performed administrative tasks under the instruction of two women who worked directly for the Chief. In O'Rourke's presence, the Chief sat on the lap of his preferred secretary with his arm around her shoulder. . . .

During that time, McCollough, another firefighter working near O'Rourke, blew in her ear, rubbed his cheek against hers, and stood over her with their bodies squarely touching as she made copies. He also asked her out on dates at least twelve times, all of which O'Rourke declined. O'Rourke did not complain to Chief Bertoncini at the time for fear of being labeled a whiner.

Also while at Chief Bertoncini's office, O'Rourke encountered her former fellow trainee Ferro, who continued to discuss his sexual encounters in front of her. . . . O'Rourke felt distinctly uncomfortable around Ferro.

Chief Johnson, the superintendent of the carpentry shop located in the Fire Prevention Bureau, told O'Rourke that McDonald . . . talked about her a lot and was crazy about her. O'Rourke made it clear to Chief Johnson that she was not interested.

In March of 1993, O'Rourke was assigned to Engine 5, Group B. The Engine company consisted of four groups, each with an officer in charge, one of whom also served as captain of the house. O'Rourke was the only female firefighter at Engine 5. The company living quarters consisted of a common bathroom, kitchen, and sitting room; each firefighter had a private bedroom. A typical shift involved two days and two nights on, four days off. O'Rourke saw stacks of pornographic magazines in the common sitting room and bathroom, which sometimes were open to a page displaying pictures of naked men and women engaged in sexual acts. O'Rourke saw male firefighters reading the magazines, which made her "very uncomfortable." The officers knew the magazines were there and did nothing. This inaction clearly violated the harassment policy. O'Rourke, however, did not complain.

In the summer of 1993, Lieutenant Young, who was acting head of O'Rourke's group, suggested that the group take a patrol ride through their district in a fire truck. Lieutenant Young offered the other firefighters beer, which O'Rourke declined. While they were driving through a part of the district that was known as a popular hangout for men, Lieutenant Young wrote O'Rourke's name and the

(continued)

(*continued*)

station phone number on pieces of paper and threw them out the window to the men on the street below. O'Rourke asked him to stop, but Lieutenant Young laughed and continued.

In the fall of 1993, O'Rourke injured her hand and was out for four or five months, but continued to go to the station weekly to pick up her paycheck. . . . Also during this time, a chief officer told O'Rourke's brother Coley, also a City of Providence firefighter, to warn her that there was a closed circuit television hidden in her bedroom at Engine 5. O'Rourke, scheduled to return to work the next week, called her commanding officer, Lieutenant Cionfolo, and asked him to investigate the rumor. She asked him not to tell anyone or "make a big thing out of it if it was just a rumor." Lieutenant Cionfolo arranged for O'Rourke to meet him at the station the next day to investigate and he promised not to tell anyone. When she arrived, Lieutenant Cionfolo, in the presence of two other firefighters in the group, told her "we didn't find the video camera." O'Rourke accompanied Lieutenant Cionfolo to her room, which was being used by a male firefighter while O'Rourke was not there. She noticed a poster of a semi-nude woman on the wall; when she objected to Lieutenant Cionfolo, he removed it. O'Rourke was unsure whether there was a camera in her room, especially since a chief officer had warned her. She wore pants to bed, had difficulty falling asleep, and was afraid to change in her room because she felt "like I was being invaded of my privacy when I was in that room."

Shortly after returning to work, while at another station, O'Rourke was sitting with several male firefighters. The men were in regular uniform but without their outer jackets on, which were to be worn outside. O'Rourke's jacket was on the back of her chair. Station Chief Costa entered the room and asked O'Rourke if the jacket was hers, then told her "you put that jacket on and you keep it on." None of the male firefighters were told to wear their outer jackets.

Another firefighter at Engine 5, Isom, left a note on O'Rourke's bed asking her out on a date. O'Rourke discovered it upon returning from a call at 2:00 A.M. O'Rourke was concerned that Isom had been in her room late at night while she wasn't there. She took the note to the officer in charge, Lieutenant Dunne, and asked him to talk to Isom about it and tell him that she was not interested in dating him. Dunne spoke to Isom, but, a few days later, Isom verbally asked O'Rourke on a date. O'Rourke declined and again spoke to Dunne, who told her he would take care of it.

After the incident, O'Rourke avoided Isom; she was plagued by worries that she might "have to work with him somewhere else at another station in a dormitory where he might be sleeping next to me." In general, O'Rourke "felt awkward all the time" to be working with men whom she knew wanted to go out with her:

> *I tried to be nice to them as much as possible. They're my co-workers [sic], but also in the back of my mind is that I also have to go to a fire with them, and I'm inexperienced, and I want to know that these people . . . are not going to stand there and let me fall flat on my face.*

(*continued*)

(*continued*)

O'Rourke started becoming anxious when she had to go to work and had trouble sleeping when she was at the station overnight.

O'Rourke once had an accident backing the fire truck into the station; she hit the wall of the station and damaged a ladder. Department procedure required drivers to have spotters while backing into the station, but none were available at the time. When Chief Costa investigated, he was told that there had not been spotters available. He told O'Rourke that she was going to have to "take the heat for this" and say that there were spotters present, an untruth. O'Rourke complied. After a hearing before the accident review board, O'Rourke was put on probation for six months.

In April, 1994, O'Rourke fought her first major fire. That event would ultimately lead to her transfer to Engine 13, a different station. Some standard procedures were not followed during the fire: Lieutenant Cionfolo did not give O'Rourke instructions or follow protocol in engaging Engine 5's line, which he handed to a member of a different fire company instead of O'Rourke; O'Rourke left the fire floor to get a new air tank after hers ran out, and she assisted a rescue company with a fire victim before returning to her position on the fire floor. O'Rourke left her position on the line after an alarm went off indicating that she had only 5 minutes remaining in her air pack. On her way to get a new pack, she encountered Chief Costa, the district battalion chief in charge of her station, in the stairwell. Chief Costa ordered O'Rourke to "get the f*** back up those stairs." O'Rourke obeyed the order, but ultimately ran out of air and returned to Engine 5 to retrieve a new tank. After the fire was out, Chief Costa walked past O'Rourke and asked, "Why did you leave your company?" but did not give O'Rourke a chance to respond. O'Rourke had seen many mistakes by others at the fire and felt singled out by Chief Costa's comment. She told Lieutenant Cionfolo about the incident, who told her not to worry about it.

Four days after the fire, Lieutenant Cionfolo informed O'Rourke that she was to attend a meeting with him, Chief Costa, and Dave Curry, a union representative, to discuss her performance at the fire. No other firefighters in O'Rourke's group were required to attend similar meetings, and Lieutenant Cionfolo told her that she was the "biggest problem" at the fire. Chief Costa asked O'Rourke to account for her actions during the fire. When O'Rourke described the incident where Chief Costa had ordered her back upstairs after her air tank alarm went off, Costa interrupted her, saying repeatedly, "I didn't tell you to get back up those stairs. I told you to get the f*** back up those stairs." Chief Costa said to O'Rourke, "You know, I'm not doing this because you're a woman"; O'Rourke responded, "You must have [a] problem with it because you keep bringing it up that I'm a woman." Chief Costa replied, "This is how you get your reputation." O'Rourke recounted the events of the fire several times over the course of approximately two hours; Chief Costa and Lieutenant Cionfolo told her that their records did not match what she was telling them. When Chief Costa asked O'Rourke why she did not have a line, she told him that Lieutenant Cionfolo had given it to a member of another company, a violation of protocol. They also

(*continued*)

(continued)

confronted O'Rourke about a rumor that she was seeking to transfer out of the company and asked her if she wanted to go to fire prevention; O'Rourke replied that she wanted to remain a firefighter and that she "didn't come on this job to sit up in an office." Chief Costa ended the meeting by asking O'Rourke and Lieutenant Cionfolo whether they could continue working together—O'Rourke replied that she could, but Cionfolo did not reply. The union representative, Curry, opined that the problem was just a personality conflict. Chief Costa directed O'Rourke and Lieutenant Cionfolo to meet to work things out, and to report the results to him the next day.

The next day, O'Rourke met with Lieutenant Cionfolo. He accused her of having "sold him down the river" by telling Chief Costa that he gave up his line at the fire. He repeated the rumor about O'Rourke wanting to transfer; O'Rourke admitted that she was interested in going to a company that was more committed to doing drills and maintaining preparedness. Lieutenant Cionfolo told her to get out of his station and that she was off his group. He took O'Rourke's gear off of the truck and told her to report to a different station. He called Chief Costa in O'Rourke's presence and told him she was "rude, disrespectful, and didn't know what . . . she . . . was doing." O'Rourke was summoned to another meeting with Chief Costa, where she was asked to sign a transfer form on which Lieutenant Cionfolo had written that they had "reached an impasse that will affect our working together" and that it was "in the best interest of the department and Engine Company 5" that O'Rourke be transferred; O'Rourke did not agree to the transfer but understood that she had no choice.

O'Rourke submitted a form to Department Chief Bertoncini describing her version of the events during the fire and requesting a meeting with him, but was told by Chief Bertoncini's secretary that the chief would not respond. When O'Rourke met with the union vice-president, George Farrell, to file a grievance with the union about her transfer, he discouraged O'Rourke from filing a grievance, telling her "you don't want to do that, do you," so she did not. After a temporary detail, she was transferred permanently to Engine 13.

O'Rourke arrived at Engine 13 on May 8, 1994, and met with the officer in charge that day, Lieutenant Gonsalves, as well as the other firefighters in her group, all male. At the meeting, they asked O'Rourke whether she minded that they slept in their underwear; the sleeping quarters at Engine 13 consisted of one room of beds, without partitions. O'Rourke did not object. The showers and bathrooms were not private. O'Rourke slept in her full uniform and did not use the showers. Lieutenant Gonsalves and other firefighters asked her about the fire, stating that they heard she had "bailed out." The captain of the station, Captain Hiter, who also headed O'Rourke's group, informed O'Rourke that she came to Engine 13 with a "black cloud," a bad reputation because she had bailed out of the fire. Before O'Rourke arrived, Captain Hiter told her brother Vincent, also a Providence firefighter, that the group was not happy about her coming to work there, and that Captain Hiter himself had a problem working with a woman.

(continued)

(*continued*)

O'Rourke asked Captain Hiter for a locker to store personal belongings, including personal hygiene items. He told her none were available and that lockers were issued by seniority; O'Rourke was the least senior in her group. Others in her group had lockers issued by the City. O'Rourke brought in her own.

On her first day, O'Rourke found pornographic magazines in the drawers of the kitchen and sitting area. She did not say anything because "it was accepted . . . it was everywhere" and because she "just got shacked out of Engine 5 to Engine 13" and did not want to begin her stay at Engine 13 on a bad note. On about three occasions, O'Rourke witnessed the male firefighters in her group watching pornographic movies in the common sitting area. O'Rourke had to pass through the area in order to access the kitchen. Captain Hiter knew that these materials were in the station and that the City's sexual harassment policy prohibited them, but did nothing.

Shortly after O'Rourke arrived, Lieutenant Gonsalves told O'Rourke that he and the other firefighters in her group were trying to find a way to have O'Rourke put on detail out of Engine 13 on July 4 because they wanted to light off fireworks and did not want O'Rourke there. Lieutenant Gonsalves also told O'Rourke, "We didn't take you over here lying down. We don't want you here. You were just transferred over here." Captain Hiter confirmed that O'Rourke was to take a detail on July 4, even though it was not her turn, because the male firefighters wanted to celebrate together as a group. O'Rourke was upset because the group was supposed to make decisions collectively about who took details, without the Captain's influence, and because she was "supposed to be a part of that company now" but was obviously unwanted. Nonetheless, she did as she was instructed.

On another occasion, returning from a call, O'Rourke and her group passed Cheaters, a topless bar. Lieutenant Gonsalves, who was the officer in charge at the time, commented that "our sister has a VIP pass to get us into Cheaters." O'Rourke was embarrassed but did not say anything because she wanted to be "as nice as possible to these guys" and "want[ed] them to accept me."

Within a month of her arrival at Engine 13, O'Rourke and Captain Hiter began meeting to discuss various issues, including complaints about O'Rourke by the other firefighters in her group that O'Rourke was "trying to get away with things" when Captain Hiter was not there. Captain Hiter asked O'Rourke about whether her locker was city property because other, more senior firefighters in the company complained that she was not entitled to a locker; it was later determined that the locker was not city property, and O'Rourke was allowed to keep it.

By September, O'Rourke noticed that the male firefighters frequently met behind closed doors in Captain Hiter's office without inviting her in; these meetings occurred daily, with or without Captain Hiter, but always excluding O'Rourke. O'Rourke asked Captain Hiter what was motivating the exclusionary meetings and why they never talked as a group, but Captain Hiter did not respond.

O'Rourke's ostracism intensified. On one call, when O'Rourke was driving, she asked Lieutenant Gonsalves for directions, as she was new to the district and did not know her way around; other firefighters routinely helped each other with

(*continued*)

(*continued*)

directions while driving. Lieutenant Gonsalves responded, "I'm not the . . . chauffeur, you are." When O'Rourke approached her colleagues after a fire, they all walked away from her. One of the chiefs from another station referred to O'Rourke as the "Maytag man" because she was always alone. If O'Rourke rode in the back of the fire truck, the firefighters in the cab closed the window so that she could not hear what was being said. When the group met after a fire to debrief, they ignored O'Rourke's questions about her performance and rolled their eyes.

Other incidents contributed to the hostile environment. One firefighter kept pictures of nude women in his locker, which O'Rourke saw; she described the effect of seeing those pictures:

> I see these pictures of these women there, and is this what they think of women? Is this how they're viewing me? There was no respect. . . .

O'Rourke's car was damaged, and she suspected it was vandalized by one of her co-workers [sic]. In addition, her locker was glued shut. Lieutenant Gonsalves and other firefighters in her group frequently referred to food O'Rourke was eating as "lesbian food."

The ostracism took its toll on O'Rourke. In addition to gaining a significant amount of weight, she had difficulty sleeping because she was "up all night agonizing about going to work." She became exhausted and was a "nervous wreck."

The atmosphere at Engine 13 continued to deteriorate. In September 1994, O'Rourke once again was assigned to a detail when it was not her turn, as she had been on July 4. When she attempted to check the station's records to confirm whose turn it was, Lieutenant Gonsalves simply repeated that she had to take the detail. . . . Someone hung a poster in the dormitory of a semi-nude woman in a provocative pose, entitled "Miss Julie Stratton," but with the last name crossed out so that it read "Miss Julie." O'Rourke, whose first name is Julia, understood this to be a reference to her. O'Rourke's brother Vincent removed the poster. Vincent also found mail taped to O'Rourke's locker that was addressed, "firefighter Julia AWOL."

Also at that time, O'Rourke's brothers Vincent and Coley came to Engine 13 to confront the male firefighters about their treatment of O'Rourke. Coley had discussed his concerns with Chief Johnson on several occasions, and Chief Johnson finally suggested that Coley should go talk to the members of Engine 13 and "try to straighten the situation out." The brothers had a heated exchange with Lieutenant Gonsalves and Captain Hiter; as a result of that incident, O'Rourke's brothers were disciplined.

After that, O'Rourke's brother-in-law DiSilva met with Chief Cotter to discuss his concerns about the male firefighters' treatment of O'Rourke in general. . . . At Chief Cotter's request, DiSilva submitted a written complaint to the Department Chief. No one from the Chief's office contacted DiSilva about his complaint.

On September 18, 1994, shortly after the incident involving her brothers, O'Rourke decided to seek help from the City's EEO officer, Gwen Andrade. She did not follow the chain of command by complaining to Captain Hiter because he

(*continued*)

(*continued*)

knew what was happening at Engine 13 but had done nothing about it. O'Rourke met with Andrade, along with a union representative, Stephen Day, a union attorney (whose presence O'Rourke did not request), and O'Rourke's brother-in-law DiSilva. O'Rourke showed them the altered poster but got no response. They laughed when O'Rourke told them about the . . . comments. Union representative Day told O'Rourke the meeting was limited to discussing Engine 13 conduct only, although O'Rourke wanted to discuss the earlier incidents. EEO Officer Andrade did not ask any questions during the meeting. Andrade concluded that O'Rourke's complaints were related to "social issues," not "work issues," and ended the meeting by telling O'Rourke to come up with solutions. O'Rourke then retained her own attorney.

After O'Rourke's meeting with Andrade, O'Rourke's group continued to exclude her, spending time in Captain Hiter's office with the door closed. Because O'Rourke was excluded from group discussions about upcoming drills, she would be the only member of her group who was unprepared when it was time to perform the drills. During one such drill, one of the firefighters screamed at O'Rourke in front of others for not being prepared.

O'Rourke still attempted to discuss her concerns with Captain Hiter, but he did nothing. She began receiving crank phone calls at home and at the station, with the caller whining or making crude noises. She was a "nervous wreck" while at work and sometimes felt her body shake uncontrollably. While responding to a call, O'Rourke took a wrong turn and had an accident while trying to turn around. That incident was the breaking point. O'Rourke felt she was no longer able to function and left work on injured-on-duty status in December, 1994. She began seeing a psychiatrist, with whom she continues treatment. When O'Rourke went to the station to retrieve her belongings, she discovered pornographic mail belonging to a fellow firefighter had been placed in her locker.

In January of 1995, O'Rourke, with her attorney, met with Chiefs Bennett and Cotter, as well as the city attorney, to discuss O'Rourke's complaint. The two chiefs were assigned to investigate O'Rourke's allegations. O'Rourke had prepared an outline and discussed many of the incidents that had occurred at both Engine 5 and Engine 13, and answered the chiefs' questions. She emphasized that she was eager to have the matter resolved and return to work, and asked the chiefs to tell her if she was doing something wrong. At the end of the meeting, O'Rourke was told that she would be contacted when their investigation was completed. The two chiefs scheduled no further meetings with her.

Approximately one month after the meeting, O'Rourke called Chief Bennett to discuss the status of the investigation. Chief Bennett told her that he was unable to proceed with the investigation because "they're refusing to speak . . . there's no use." Chief Bennett also told O'Rourke that there was a gag order put out prohibiting him from speaking to firefighters, lieutenants, and captains—anyone lower than a chief. The new chief of the Department, Chief DiMascolo, had told Bennett and Cotter that the union did not want its members involved in the investigation, and Chief DiMascolo declined to take any action to encourage their participation.

(*continued*)

(*continued*)

Frustrated, Chief Bennett and Chief Cotter withdrew from the investigation in February, 1995. Shortly thereafter O'Rourke filed her administrative charge of discrimination.

O'Rourke remained out of work for over two years. As a result of the stress she experienced, O'Rourke gained a total of 80 pounds. She was anxious and afraid to leave the house, and was particularly anxious about encountering Providence firefighters.

O'Rourke returned to work at the fire department in 1997. Following her psychiatrist's advice she no longer works as a line firefighter but instead joined fire prevention, where she remains today.

II. On July 10, 1995, O'Rourke filed a discrimination charge with the Rhode Island Commission of Human Rights and with the U.S. Equal Employment Opportunity Commission and received notice of right to sue. O'Rourke also filed a federal court complaint against the City on June 30, 1995, and an amended complaint on July 17, 1995, also naming four firefighters as individual defendants. The complaint included claims of disparate impact and sex discrimination under Title VII and Rhode Island law, as well as §1983 claims.

A jury trial began July 14, 1997. Defendants filed a motion . . . to exclude all evidence of harassment occurring before O'Rourke's tenure at Engine 13 on the grounds that those acts were outside Title VII's 300-day limitations period and that such evidence was unduly prejudicial. . . . O'Rourke invoked the continuing violation doctrine, alleging that there were acts occurring before and during the 300-day period, and arguing that the evidence was not unduly prejudicial . . . but relevant to proving that theory. The court denied defendants' motion.

O'Rourke introduced evidence of harassment spanning the entire duration of her employment at the fire department from 1992 to 1994—including her time spent in training in 1992, the period in which she worked in the Chief's office later that year, her year-long stay at Engine 5 in 1993, and finally, her seven months at Engine 13 in 1994. At the close of O'Rourke's case, defendants moved for judgment as a matter of law. The court granted the motion as to the §1983 counts against the City and individual defendants on the ground that there was no evidence that the City of Providence or the individual defendants had intentionally discriminated against O'Rourke. O'Rourke does not appeal this decision. The court also dismissed O'Rourke's Title VII disparate impact claims as duplicative of her hostile work environment sexual harassment claim. . . .

The court also ruled that "[t]he statute of limitation clearly applies in this case, and limits the plaintiff's claims to September 13, 1994, forward for sexual harassment." . . . Counting back 300 days from when O'Rourke filed her complaint with the EEOC on July 10, 1995, the court found O'Rourke's claim was limited to the incidents occurring at Engine 13 since September 13, 1994. Because the state limitations period was 360 days, and for the sake of simplicity, the court decided that evidence from May, 1994 (the time O'Rourke transferred to Engine 13) could be considered. The district court instructed the jury only to consider

(*continued*)

(*continued*)

Engine 13 evidence and not to consider any of the evidence it had heard about the two years prior to O'Rourke's Engine 13 tenure. Having been successful in its motion to exclude evidence for the period before O'Rourke worked at Engine 13, the City was not permitted to put on evidence in its case relating to pre-Engine 13 incidents. The City had, of course, cross-examined plaintiff's witnesses on this point. The City made no offer of proof.

The jury awarded O'Rourke $275,000 against the City on her hostile work environment sexual harassment claim. The City moved for a new trial. The district court granted the motion on the ground that it had committed a prejudicial error of law by allowing pre-Engine 13 evidence of sexual harassment, and because it believed its curative instruction to the jury was ineffective. The $275,000 verdict was excessive, the court thought, because that sum made it clear that the jury had considered pre-Engine 13 evidence in awarding compensatory damages, despite the court's contrary instruction.

O'Rourke's second trial began in April 1998. The district court limited the evidence to O'Rourke's tenure at Engine 13, from May to December of 1994. Once again, the jury found in favor of O'Rourke, awarding her $200,000 against the City on her hostile work environment claim.

The City again moved for a new trial, which the district court denied. O'Rourke sought an award of attorneys' fees and costs for both trials. The district court awarded O'Rourke $99,685 in attorneys' fees and $10,214.50 in costs for the second trial only; no costs were awarded for the first trial.

The City appeals the district court's denial of its motion for a new trial after the second trial, and O'Rourke seeks on cross-appeal reinstatement of the first jury award, as well as attorneys' fees and costs for the first trial.

III. . . . Defendants had filed a motion . . . to exclude all evidence of events before the 300-day charge filing period, arguing that such evidence was barred by the statute of limitations. . . . Granting this motion would have meant no evidence could be introduced of the events taking place before September 13, 1994, four months after O'Rourke was assigned to Engine 13.

O'Rourke objected, arguing that the case came within the continuing violation doctrine, an equitable exception to the 300-day filing period, because there was an ongoing pattern of discrimination. . . .

[T]he district court held that the pre-charge period evidence was admitted in error because the statute of limitations barred the evidence. Like the City, the district court never explicitly addressed O'Rourke's invocation of the continuing violation doctrine or why it would not apply. That omission was itself error. Indeed, the only reason expressed by the district court in support of its decision was its feeling that O'Rourke had brought these events on herself . . . and that therefore the events were irrelevant to her discrimination claim. Insofar as the court thought that to be a reason for applying the statute of limitations, it erred. Indeed, the district court expressed its feeling that plaintiff's case as to pre-Engine 13 events failed on its merits, calling it "all eyewash." Any determination of the merits was for the jury, not the court.

(*continued*)

(*continued*)

These errors could prove to be harmless if O'Rourke was in fact not entitled to use the continuing violation doctrine. . . . We conclude that a reasonable jury could have found that O'Rourke was a victim of a continuing violation. . . .

A. Hostile Work Environment

. . . Courts have long recognized that sexual harassment is "a form of gender discrimination prohibited by Title VII." . . .

Title VII sexual harassment law has evolved considerably from its early focus on quid pro quo sexual harassment, where an employee or supervisor uses his or her superior position to extract sexual favors from a subordinate employee, and if denied those favors, retaliates by taking action adversely affecting the subordinate's employment. . . . Title VII also allows a plaintiff to prove unlawful discrimination by showing that "the workplace is permeated with 'discriminatory intimidation, ridicule, and insult' that is 'sufficiently severe or pervasive to alter the conditions of the victim's employment and create an abusive working environment.'". . . Further, Title VII protection is not limited to "economic" or "tangible" discrimination.

The Supreme Court has outlined the tests a plaintiff must meet to succeed in a hostile work environment claim: (1) that she (or he) is a member of a protected class; (2) that she was subjected to unwelcome sexual harassment; (3) that the harassment was based upon sex; (4) that the harassment was sufficiently severe or pervasive so as to alter the conditions of plaintiff's employment and create an abusive work environment; (5) that sexually objectionable conduct was both objectively and subjectively offensive, such that a reasonable person would find it hostile or abusive and the victim in fact did perceive it to be so; and (6) that some basis for employer liability has been established. . . . It is undisputed that O'Rourke is a member of a protected class and that she considered defendants' conduct unwelcome; thus, the first two elements of her claim are met. The evidence is compelling that she suffered harassment based on sex. . . .

In hostile environment cases, the fourth and fifth elements are typically the most important. They must be determined by the fact-finder "in light of the record as a whole and the totality of the circumstances.". . . Several factors typically should be considered in making this determination: "the frequency of the discriminatory conduct; its severity; whether it is physically threatening or humiliating, or a mere offensive utterance; and whether it unreasonably interferes with an employee's work performance." . . .

As part of its evaluation, a jury may consider a broad range of conduct that can contribute to the creation of a hostile work environment. Indeed, "harassing conduct need not be motivated by sexual desire to support an inference of discrimination on the basis of sex." . . . Evidence of sexual remarks, innuendoes, ridicule, and intimidation may be sufficient to support a jury verdict for a hostile work environment. . . . The accumulated effect of incidents of humiliating, offensive comments directed at women and work-sabotaging pranks, taken together, can constitute a hostile work environment. . . .

(*continued*)

(*continued*)

Still, conduct that results from "genuine but innocuous differences in the ways men and women routinely interact with members of the same sex and of the opposite sex" does not violate Title VII. . . . Thus, "offhand comments, and isolated incidents" are not sufficient to create actionable harassment; the hostile work environment standard must be kept "sufficiently demanding to ensure that Title VII does not become a 'general civility code.'" . . .

B. The Continuing Violation Doctrine

A plaintiff who brings a hostile work environment claim under Title VII must file her claim within 300 days of an act of discrimination, and in general cannot litigate claims based on conduct falling outside of that period. . . . The limitations period serves to "protect employers from the burden of defending claims arising from employment decisions that are long past." . . . But where a Title VII violation is "of a continuing nature, the charge of discrimination filed . . . may be timely as to all discriminatory acts encompassed by the violation so long as the charge is filed during the life of the violation or within the statutory period." . . . The continuing violation doctrine is an equitable exception that allows an employee to seek damages for otherwise time-barred allegations if they are deemed part of an ongoing series of discriminatory acts and there is "some violation within the statute of limitations period that anchors the earlier claims." . . . This "ensures that these plaintiffs' claims are not foreclosed merely because the plaintiffs needed to see a pattern of repeated acts before they realized that the individual acts were discriminatory." . . .

[T]here was no error in the district court's original decision to admit the pre-Engine 13 evidence and there was error in its later decision to instruct the jury not to consider the evidence. Because the grant of the new trial was based on the later erroneous decision, the order granting the new trial on that basis was in error. What to do about that error requires further analysis. . . .

Reinstatement of First Verdict

Nonetheless, the first jury verdict may not be automatically reinstated if the first trial was otherwise fatally flawed as to the City. The City's main argument is that it was precluded from putting on its version of the pre-Engine 13 events. That is only partly true. The City disputed those events on cross-examination of plaintiff's witnesses, and it introduced evidence of its 1992 sexual harassment policy. What it did not do was put on its own witnesses. That might well be enough to carry the City's argument against reinstatement of the verdict save for one thing: the City did not preserve the argument.

The City failed to make an appropriate offer of proof and so it has waived the argument. . . . Similarly, the City failed to properly object to the introduction of most of the pre-Engine 13 evidence, as it was required to do after an unsuccessful motion [to exclude], and thus failed to preserve its objection. . . .

We touch briefly on two of the City's other arguments. First, there is no merit to the City's argument at the first trial that it was entitled to a jury instruction that

(*continued*)

(*continued*)

the firefighters' conduct should be evaluated in the context of a blue collar environment, as one court has held. . . . We decline to adopt such a rule for the same reasons the Sixth Circuit rejected it:

> *We do not believe that a woman who chooses to work in the male-dominated trades relinquishes her right to be free from sexual harassment; indeed, we find this reasoning to be illogical, because it means that the more hostile the environment, and the more prevalent the sexism, the more difficult it is for a Title VII plaintiff to prove that sex-based conduct is sufficiently severe or pervasive to constitute a hostile work environment. Surely women working in the trades do not deserve less protection from the law than women working in a courthouse.*

Williams, . . . As always, regardless of the setting, "[t]he critical issue, Title VII's text indicates, is whether members of one sex are exposed to disadvantageous terms or conditions of employment to which members of the other sex are not exposed." . . .

We also reject the City's contention that the firefighters' reading of pornography in public spaces of the fire station is protected by the First Amendment, placing the burden on O'Rourke to avoid it if it offended her. The City relies on *Johnson v. County of Los Angeles Fire Dep't*, 865 F. Supp. 1430 (C.D. Cal. 1994), where a male firefighter successfully challenged a policy categorically banning the possession and reading of sexually explicit magazines in the fire station. The court held the policy an impermissible content-based regulation because the county failed to offer credible testimony that "mere exposure to the cover of Playboy directly contributes to a sexually harassing atmosphere." . . . But Johnson cannot bear the weight of the City's argument. The Johnson court emphasized that the plaintiff was "merely seeking to read and possess Playboy quietly and in private . . . [and] not seeking to expose the contents of the magazine to unwitting viewers"; it allowed to stand that portion of the County's policy prohibiting the public display of nude pictures. . . . In contrast, at Engines 5 and 13, O'Rourke was surrounded by pornographic magazines, sexually explicit movies, and nude pictures displayed, with no way to avoid them. That evidence was probative of the City's knowledge that the proscribed materials existed and relevant to O'Rourke's hostile work environment claim; *Johnson* dealt with a constitutional challenge to the policy itself. Moreover, the fact that the City's sexual harassment policy prohibited the keeping of pornographic materials at stations undercuts the City's argument that it was up to O'Rourke to avoid it.

Employer's Liability

The City also argues that the district court erred in instructing the jury that there were two ways the City could be made liable:

> *If the harasser is a superior, a supervisory employee, then that alone makes the city liable. If it is a superior officer who harassed her, the city is responsible for*
> (*continued*)

(*continued*)

that individual's conduct. If it is her coworker, or coworkers, who are guilty of this harassing conduct, then the city is only liable if a superior officer knew, or should have known, of the harassment and failed to take prompt remedial action. . . .

We find no error in the court's instruction, let alone plain error that "affects substantial rights and which has resulted in a miscarriage of justice or has undermined the integrity of the judicial process." . . . The "knew or should have known" instruction to hold the city liable for actions of coworkers was correct. . . . Further, where there is an actionable hostile environment attributable to a supervisor, an employer is subject to vicarious liability to a victimized employee where, as here, it fails to exercise reasonable care to prevent it. . . . The City cannot show a miscarriage of justice resulted from the jury's verdict for O'Rourke because there was ample evidence to support the City's vicarious liability for the hostile work environment created by both coworkers and supervisors. . . .

Accordingly, O'Rourke established at the first trial the City's liability for the hostile work environment, the final element of her sexual harassment claim, and so we reinstate the verdict. . . .

We reverse the district court's judgment as a matter of law and its grant of a new trial after the first trial, direct reinstatement of the first jury award of $275,000, affirm the attorneys' fees award for the second trial and the award of prejudgment interest, and remand for calculation of an appropriate award of attorneys' fees and costs for the first trial.

Case Name: O'Rourke v. City of Providence

Court: United States Court of Appeals for the First Circuit

Summary of Main Points: Sexual harassment in the workplace that is sufficiently pervasive and severe can be the basis for a violation of Title VII. Where the harassment is ongoing over a long period of time, the continuing violation doctrine will apply to allow consideration of events that occurred prior to the 300 day limitation period.

AUTHOR'S COMMENTARY: The O'Rourke case stands for three important propositions: (1) the organizational culture of a fire department must come in line with the expectations of society; (2) firefighters need to be aware that seemingly random acts or comments made years apart in different stations and on different shifts can be strung together to create a damaging indictment of an entire organization; and (3) fire service leaders need to act quickly and decisively when confronted with complaints about harassment.

Officers who witness conduct that could constitute sexual harassment do not stand in the same shoes as a coworker, and should not wait for the victim to complain before intervening. Observing such conduct and doing nothing can be construed as condoning the conduct. All personnel need training in sexual harassment, but officers in particular need additional training, as well as access to human relations specialists and legal counsel to provide guidance in difficult cases.

Sexual harassment is not limited to men who harass women. Harassment of men by other men can occur and is actionable, as is harassment of women by other women and harassment of men by women. See *Bianchi v. City of Philadelphia*, No. 02-2687 (3d Cir., 2003). In addition, a person's sexual orientation is not always relevant in such cases, as heterosexual men have been found guilty of sexually harassing other heterosexual men. See *Doe v. City of Belleville*, Illinois, No. 94-3699 (7th Cir., 1997). Furthermore, the victim of sexual harassment need not have been the target of offensive conduct. Any person who is exposed to a sexually hostile work environment, even where the hostility is directed at another, may have an action for sexual harassment. Appendix B, Chapter 13 contains the Federal regulations related to sexual harassment.

In response to problems with sexual harassment in the workplace, most employers have adopted sexual harassment policies and provide ongoing sexual harassment training. Occasionally these policies may be challenged on various grounds as going "too far." In a well-known California case, a sexual harassment policy that banned sexually oriented magazines in fire stations was struck down as violating a fire captain's right to free speech. The case, *Johnson v. County of Los Angeles Fire Department*, 865 F.Supp. 1430, (C.D., CA, 1994), held that in balancing the captain's First Amendment rights against the department's responsibility to maintain a workplace free of unlawful discrimination, the department must establish a causal link between the reading of *Playboy* magazines and the poor treatment of female firefighters to warrant such a policy. When the reading is done in private, without exposing the contents to unwitting viewers, an employee's First Amendment rights prevail.

PREGNANCY DISCRIMINATION

The Pregnancy Discrimination Act was an amendment to the Civil Rights Act of 1964, and prohibits employers subject to Title VII from discriminating on the basis of pregnancy, childbirth, or related conditions. The act requires employers to treat pregnant workers the same as any other employee with a temporary medical condition.

Under the Pregnancy Discrimination Act, so long as a woman can perform the essential functions of a job, she cannot be refused employment due to pregnancy, concern about future pregnancy, perceived or assumed inabilities to perform the job, or the prejudices of coworkers, customers, or clients. Pregnant employees must be permitted to work as long as they are capable of performing their jobs. Concerns about the safety of the employee or the baby are not adequate grounds for prohibiting the employee from working.

Because firefighting poses serious risks to a developing fetus in a variety of ways, fire departments should provide pregnant firefighters with the information they and their doctors need to make an informed decision about when to come off the line. However, the decision is one that must be made by the firefighter herself.

In addition to the Federal law on pregnancy discrimination, many states also have laws prohibiting pregnancy-based discrimination. Discrimination on the basis of pregnancy can also constitute sex discrimination.

RELIGIOUS ACCOMMODATION

Title VII prohibits discrimination on the basis of a person's religion **(Figure 13-1)**. Employers may not treat employees or applicants more or less favorably because of their religious beliefs or practices. For example, an employer may not refuse to hire individuals of a certain religion, may not impose stricter requirements on persons of a certain religion for promotion, and may not impose more or different work requirements on an employee

Figure 13-1 *Title VII prohibits discrimination on the basis of a person's religion.*

because of that employee's religious beliefs or practices. Employees cannot be forced to participate, or forced not to participate, in a religious activity as a condition of employment.

Problems commonly arise involving the accommodation of religious practices, such as scheduling changes around sabbath days, refusal to work assignments that conflict with religious beliefs, grooming issues, and the wearing of clothing required by religion **(Figure 13-2)**. As originally interpreted, the Constitution required governmental neutrality toward religion and religious beliefs. Laws that attempted to affirmatively authorize or recognize religious practices, or required accommodation of certain religious practices, were considered to be a violation of the First Amendment's prohibition of any law pertaining to the establishment of a religion. The more modern trend is toward a mandated tolerance of religious diversity.

In recent years, the scope of congressional authority to dictate and govern religious issues has come to the forefront. In 1993 the Religious Freedom Restoration Act (RFRA) was passed. The act was intended to be a religion-neutral act, which required the government to respect all religious practices unless they contravened a compelling governmental interest. The RFRA also required employers to accommodate religious beliefs and practices in employment, unless this would impose an undue hardship on an employer. In 1997 the RFRA was found to be unconstitutional; as a result the law is in a state of flux. Since 1997, Congress has been considering similar legislation, and many states have adopted their own laws requiring religious accommodation. In addition, Title VII still prohibits employment discrimination based on religion.

Figure 13-2 *Religious beliefs may conflict with a fire department's uniform requirements.*

The following information is provided as the state of the law as of press time, as interpreted by the EEOC. However, due to the complexity of the issues involved, changes from both the courts and Congress are anticipated.

- Employers must reasonably accommodate employees' sincerely held religious beliefs or practices, unless doing so would impose an undue hardship on the employer. A reasonable religious accommodation is any adjustment to the work environment that will allow the employee to practice his religion. Flexible scheduling, voluntary substitutions or swaps, job reassignments and lateral transfers, and modifying workplace practices, policies, and/or procedures are examples of how an employer might accommodate an employee's religious beliefs.

- An employer is not required to accommodate an employee's religious beliefs and practices if doing so would impose an undue hardship on the employer's legitimate business interests. An employer can show undue hardship if accommodating an employee's religious practices requires more than ordinary administrative costs, diminishes efficiency in other jobs, infringes on other employees' job rights or benefits, impairs workplace safety, causes coworkers to carry the accommodated employee's share of potentially hazardous or burdensome work, or if the proposed accommodation conflicts with another law or regulation.

- Employers must permit employees to engage in religious expression if employees are permitted to engage in other personal expression at work, unless the religious expression would impose an undue hardship on the employer. Therefore, an employer may not place more restrictions on religious expression than on other forms of expression that have a comparable effect on workplace efficiency.

- Employers must take steps to prevent religious harassment of their employees. An employer can reduce the chance that employees will engage in unlawful religious harassment by implementing an anti-harassment policy and having an effective procedure for reporting, investigating, and correcting harassing conduct.

  It is also unlawful to retaliate against an individual for opposing employment practices that discriminate based on religion or for filing a discrimination charge, testifying, or participating in any way in an investigation, proceeding, or litigation under Title VII.

In *Endres v. Indiana State Police,* No. 02-1247 (7th Cir. 06/27/2003), a police officer refused to accept an assignment to work at a casino, citing a conflict with his religious beliefs. The court upheld his termination, finding that the accommodation sought was unreasonable, and would work an undue hardship on the employer. The court cited a recent case involving a Catholic police officer who refused to protect an abortion clinic, who was similarly disciplined. The court stated:

> *Law enforcement agencies need the cooperation of all members. Even if it proves possible to swap assignments on one occasion, another may arise when personnel are not available to cover for selective objectors, or when . . . seniority systems or limits on overtime curtail the options for shuffling personnel. Beyond all of this is the need to hold police officers to their promise to enforce the law without favoritism—as judges take an oath to enforce all laws, without regard to their (or the litigants') social, political, or religious beliefs. Firefighters must extinguish all fires, even those in places of worship that the firefighter regards as heretical. Just so with police. The public knows that its protectors have a private agenda; everyone does. But it would like to think that they leave that agenda at home when they are on duty—that Jewish policemen protect neo-Nazi demonstrators, that Roman Catholic policemen protect abortion clinics, that Black Muslim policemen protect Christians and Jews, that fundamentalist Christian policemen protect noisy atheists and white-hating Rastafarians, that Mormon policemen protect Scientologists, and that Greek-Orthodox policemen of Serbian ethnicity protect Roman Catholic Croats. We judges certainly want to think that U.S. Marshals protect us from assaults and threats without regard to whether, for example, we vote for or against the pro-life position in abortion cases. Rodriguez v. Chicago, 156 F.3d 771 (7th Cir. 1998).*

United States v. City of Albuquerque
545 F. 2d 110 (10th Cir., 1976)
United States Court of Appeals for the Tenth Circuit

McWILLIAMS, Circuit Judge.

This is an action brought by the United States against the City of Albuquerque and its fire chief for alleged religious discrimination in its employment practices within the city fire department, in violation of 42 U.S.C. § 2000e-2. One Salomon Zamora, a fireman first class in the Albuquerque fire department, was discharged after he (Zamora) failed to report for work on the day shift for Saturday, October 28, 1972. Zamora, a Seventh Day Adventist, had refused to appear for work on October 28, 1972, because such, in his view, would have violated one of the practices of his particular religion which forbade working on the Sabbath, except for emergencies. The Sabbath as observed by the Seventh Day Adventists is from sundown Friday until sundown Saturday. . . .

(continued)

(*continued*)

Zamora joined the Seventh Day Adventist Christian Church in 1961. In 1968 Zamora and his wife were divorced, and he was disfellowshipped from his church. In March 1969 Zamora became a member of the Albuquerque Fire Department. At the time of his employment Zamora was not a member of the Seventh Day Adventists, and he indicated in his application that he could work any day of the week, and apparently did work whichever shift he was called on until around September 1971. At this time Zamora remarried his former wife and he thereafter rejoined his church.

As indicated above, one tenet of the Seventh Day Adventist is that he observe the Sabbath Day, which commences at sundown on Friday and ends at sundown on Saturday. "Observe" means to refrain from unnecessary work on the Sabbath, although a Seventh Day Adventist may engage in so-called "emergency" work on that day. Just what constitutes "emergency" work is apparently a matter between the member and his God. However, the present case does not turn on this distinction between unnecessary work and emergency work.

Zamora, a fireman first class, was assigned to Division 1 of the Fire Suppression Department, which works on a 56-hour work week. The day shift is from 8:00 a.m. to 6:00 p.m., and the night shift is from 6:00 p.m. to 8:00 a.m. The work force at Division 1 is divided into three platoons: A, B, and C, which rotate on the basis of working three consecutive day shifts, next working three night shifts, and then having three days off. Consequently, no fireman has the same days off each week, since a nine-day work cycle is imposed on a seven-day week.

On the work schedule outlined above, Zamora would be called on to work either the Friday night shift or the Saturday day shift some 35 times in a year. A minor problem did arise in connection with the Friday day shift, since in winter the sun would set before the end of the Friday day shift, i.e. 6:00 p.m. Similar problems arose in summer when sundown did not occur until sometime after the Saturday night shift commenced. However, there was no particular problem in this connection as Zamora, with the apparent approval of his superior, was in each instance, though on duty, not required to perform so-called menial work after sundown on Friday, nor before sundown on Saturday. The present controversy, then, stems from those occasions when Zamora was called on to work either the Friday night shift or the Saturday day shift.

From October 1971 until October 1972, Zamora used sick leave some 13 times in order to avoid working Friday nights or Saturday days. During this period he also took annual leave several times and he traded shifts once in order to avoid work on his Sabbath. However, in early September 1972, the matter of Zamora's not working on Friday nights or Saturday days became a subject of dialogue between Zamora and his supervisors in the fire department. Whether Zamora himself brought up the question, or whether the matter surfaced when Zamora, after taking sick leave, was not found at his home, but in church one Saturday morning, is not really material. In any event both Zamora and his supervisor, after discussion, agreed that it was a misuse of sick leave to take sick leave, when he was not in fact ill, in order to avoid working Friday nights and Saturday days.

(*continued*)

(*continued*)

On October 9, 1972, Zamora submitted a request for unscheduled vacation leave for the day shift on Saturday, October 28, 1972. On October 23, 1972, this request was denied, the assigned reason therefor being that if the request were granted, District 1 would be undermanned on that particular date. Two other firemen and a lieutenant, who, prior to Zamora, had similarly applied for unscheduled vacation, were also denied leave for the same reason. However, these men obtained time off on October 28, 1972, by trading shifts. Zamora's request for unscheduled leave on October 28, 1972, was kept open, even though denied on the 23rd, in order to allow the request to still be granted if a vacancy occurred. However, none developed and on October 27, 1972, at approximately 5:30 p.m. he was formally notified that his request for unscheduled leave for October 28, 1972, had been finally denied. Zamora at that time informed his superior that he would not report for work on October 28, 1972, and he did not. Thereafter Zamora was suspended and eventually he was discharged for such absence. Prior to the actual discharge there was discussion back and forth between Zamora and the fire chief, with the latter urging Zamora to reconsider and try to work it out within the rules and regulations of the department. Zamora remained adamant and his formal discharge followed.

As indicated above, the trial judge found that the City of Albuquerque had attempted to make reasonable accommodations to Zamora's religious practices and that such reasonable accommodations were in fact embodied in the rules and regulations of the fire department. In this regard there was a rules committee in the City's fire department, which consisted of one member from each rank in the department, which committee worked in conjunction with the fire chief. If the committee and the chief agreed on a rule or regulation, it was adopted. In the case of a disagreement between the committee and the chief, the matter was referred to a personnel director, who was the final authority.

Under the rules and regulations of the Albuquerque Fire Department there were three ways in getting time off, i.e., not having to work an assigned shift. These were using a day of one's vacation leave, taking leave without pay, and trading shifts with another fireman of the same rank. There were some minor restrictions on getting time off. For example, there was a minimum manning level and if a request for time off would mean going below that level, the request would be denied. But on the whole, the trial court found, and we agree, that the fire department had a fairly liberal time off policy, both in actual practice and as concerns the rules and regulations themselves. It is true that Zamora testified that he had never had much luck in trading shifts with other firemen. His testimony, however, was countered by several firemen who testified as defense witnesses that they also worked in Division 1 and that they frequently traded shifts, and would have traded with Zamora, but that they were never asked. In this latter regard the record indicates that for some reason Zamora was indeed reluctant to ask others to trade shifts and his ultimate position, as well as that of counsel in this appeal, is that it was up to his supervisor to take the initiative and find another fireman for him who would trade shifts. For understandable reasons, the policy

(*continued*)

(*continued*)

within the department was for the firemen to arrange their own trade-offs, as the supervisors did not want to be in the position of coercing trade-offs.

The record also indicates, as above referred to, that Zamora's supervisors were quite reluctant to be forced into the position where Zamora would inevitably be discharged for insubordination, i.e., refusing to report for work when ordered. Zamora was constantly urged by his supervisors to "work it out" within the existing rules of the department, even after he failed to report for work on October 28, 1972. And it was in this general setting that Zamora became "intransigent," as the trial court characterized it, and demanded that he never be required to work on a Friday night shift or a Saturday day shift. His supervisors stated that under the existing rules and regulations they could not give in to this demand and when he then refused to report for work on October 28, 1972, discharge followed.

Did the City of Albuquerque and its fire department demonstrate that they had, through their rules and regulations, made reasonable efforts to accommodate Zamora's religious practices and that further accommodation would result in undue hardship? The trial court found and concluded that such had been "convincingly demonstrated." We think such finds support in the record.

In *Williams v. Southern Union Gas Co.* . . . we stated that the phrases "reasonably accommodate" and "undue hardship" as used in 42 U.S.C. § 2000e(j) are relative terms and cannot be given any hard and fast meaning. Each case necessarily depends upon its own facts and circumstances, and in a sense every case boils down to a determination as to whether the employer has acted reasonably. Here the employer did not stubbornly insist that Zamora work on his Sabbath, come what may. On the contrary, as a result of interaction between the fire chief and the rules committee, made up of representatives of the firemen, rules and regulations were promulgated which granted a fireman considerable latitude in getting excused from reporting for work for a particular shift assignment. A fireman could use up a portion of his leave with pay, if he be so inclined. Or he could simply take leave without pay. When an employee for personal reasons simply prefers not to work on a given day, it is not too much to suggest that he receive no pay for the shift he does not work. And, perhaps most importantly, firemen were permitted to trade shifts with other firemen of the same grade. Trading shifts was a prevalent practice, even though Zamora was himself not particularly interested in trading shifts with others. By trading shifts, a fireman would of course suffer no loss in pay. But Zamora would have none of this. He wanted Friday nights and Saturday days off as a matter of right. We conclude that the trial court's finding that the City of Albuquerque made reasonable accommodation efforts is not clearly erroneous and should not be overturned by us.

The trial court also concluded that if the City made further efforts to accommodate Zamora there would be "undue hardship" on the "business." The "business" was that of fire suppression, and such is obviously a matter of great public interest. In our view when the "business" of an employer is protecting the lives and property of a dependent citizenry, courts should go slow in restructuring his employment practices. . . .

(*continued*)

(*continued*)

But beyond its rules and regulations, just what could the City have done to accommodate Zamora's religious practices? The City of course could have given in and set up a schedule that would never have called on Zamora to work any Friday night or Saturday day shift. Counsel says such might have "inconvenienced" the City, but would not have really amounted to "undue hardship." However, the trial court concluded that such scheduling to fit Zamora's demands would have amounted to undue hardship. The trial court's pertinent conclusions on this particular matter are worthy of note. . . .

We regard the trial court's several conclusions as to "undue hardship," as well as its earlier determination of "reasonable accommodation," to be essentially findings of fact. On review such should be accepted unless they be deemed clearly erroneous. In our review such are not clearly erroneous and hence should be upheld. . . .

We recognize that the problems arising from the fact that Seventh Day Adventists are forbidden to work on Saturdays are troublesome ones and that the courts have not been in accord in their thinking on the subject. . . .

A reading of all these cases leads us to conclude that to a very great degree each case turns on its own particular facts and circumstances. Under the facts and circumstances here present, if we reversed the trial court we would simply be substituting our best judgment of the facts for that of the trial court. This we should not do. . . .

Judgment affirmed.

Case Name: United States v. City of Albuquerque

Court: United States Circuit Court of Appeals for the Tenth Circuit

Summary of Main Points: The requirement that an employer reasonably accommodate the religious beliefs of employees does not require a fire department to create a special schedule for the benefit of a particular member.

CLOTHING, UNIFORMS, AND GROOMING ISSUES

Fire departments are paramilitary organizations, and as such commonly have grooming and uniform policies. Reasonable and nondiscriminatory uniform, clothing, and grooming requirements are usually upheld, even though the requirements are different for men and women **(Figure 13-3)**. For example, men may be required to wear ties and women allowed to wear dresses, without violating Title VII. The key factor in most cases is that the regulations are in accordance with commonly accepted social norms and reasonably related to the employer's business needs. See *Carroll v. Talman Federal Savings & Loan Assn.*, 604 F2d 1028 (7th Cir., 1979).

Figure 13-3 *The courts usually uphold reasonable uniform and grooming requirements. (Courtesy of Edwina Davis.)*

Sex-based differences in dress and grooming requirements will be struck down when the requirements are not based on legitimate business necessity, are overly burdensome, or draw a distinction that reflects stereotypical attitudes toward the sexes. For example, requiring male employees to wear suits, while requiring females in the same position to wear smocks was found to be discriminatory because it created the appearance that the men were at a higher degree of professional status. See *Carroll,* 604 F2d 1028 (7th Cir., 1979).

Over the years, clothing, uniform, and grooming policies have been challenged in a variety of ways, including sex discrimination, religious discrimination, race discrimination, equal protection, and due process. In *Goldman v. Weinberger,* 475 U.S. 503 (1986), the United States Supreme Court upheld a military order prohibiting the wearing of headgear indoors, which served to prohibit an orthodox Jewish rabbi from wearing a yarmulke. The court ruled that the military had no duty to accommodate the religious beliefs of military personnel.

In *Kelley v. Johnson,* 425 U.S. 238 (1976), a police department grooming policy restricted hair length, sideburns, and moustaches, and prohibited beards. The policy was challenged as being a violation of due process, equal protection, and the First Amendment right of free expression. The United States Supreme Court ruled that as long as there is a rational basis for the police department's grooming policy, there is no constitutionally protected right that is violated. The police department's desire to maintain esprit de corps was considered to constitute a rational basis.

Consider the *Atlanta* case, which challenges a grooming requirement based on race and disability discrimination.

Fitzpatrick v. City of Atlanta
2 F. 3d 1112 (11th Cir.,1993)
U.S. Court of Appeals, Eleventh Circuit

ANDERSON, Circuit Judge:

. . . In order to breathe in smoke-filled environments, firefighters must wear respirators, otherwise known as positive pressure self-contained breathing apparatuses ("SCBA's"). For the SCBA mask to operate properly and safely, its edges must be able to seal securely to the wearer's face. The parties do not dispute that a wearer's long facial hair can interfere with the forming of a proper seal. In an attempt to address the hazard posed by such hair, the City Fire Department until 1982 enforced a policy requiring all male firefighters to be completely clean-shaven. . . .

The twelve plaintiff-appellant firefighters in this case are all African-American men who suffer from pseudofolliculitis barbae ("PFB"), a bacterial disorder which causes men's faces to become infected if they shave them. It is generally recognized that PFB disproportionately afflicts African-American men. At least one of the appellants, firefighter Darryl Levette, has been fighting with the City over its no-beard policy for more than ten years. Levette first challenged the requirement in 1982. In response to his complaints, the City modified its policy in order to accommodate firefighters with PFB. . . .

Under the modified policy, firefighters with PFB were permitted to participate in a program known as the "shaving clinic." Shaving clinic participants were allowed to wear very short "shadow" beards, which were not to exceed length limits specified by a dermatologist employed by the City. To enforce these limits, the Fire Department subjected the participating firefighters to a series of periodic beard inspections. It was believed that so long as the shadow beards were kept very short, the SCBA masks would still be able to seal sufficiently well to enable the firefighters to use them safely.

In 1988, after one of the appellant firefighters, William Hutchinson, complained that he had been wrongly refused permission to participate in the shaving clinic, the City decided to reconsider the shadow beard policy. On the recommendation of Del Corbin, the City's then-Assistant Commissioner of Public Safety, the Fire Department decided that shadow beards would no longer be permitted, on the grounds that even shadow beards may interfere with the safe use of SCBA's.

On November 4, 1988, the Department of Public Safety issued Special Order 3.9, directing the Fire Department to resume enforcement of Bureau of Fire Services Standard Operating Procedure 88.9, the no-beard rule. Under the new policy, firefighters who cannot be clean-shaven must be removed from firefighting duty. Such persons may be transferred to non-firefighting positions within the Department, if suitable openings are available. They may also apply for other available positions with the City but are accorded no special priority and must compete on an equal basis with other eligible candidates. Under the new policy such persons are granted the right to be temporarily reassigned from firefighting duties for a one-time period of ninety days. . . . Male firefighters who cannot shave and for whom

(continued)

(*continued*)

non-firefighting positions are not available within the Department are terminated, once they have exhausted their ninety days of temporary reassignment.

Firefighter Hutchinson challenged the new policy by filing a charge with the U.S. Equal Employment Opportunity Commission ("EEOC") on December 14, 1988. In March 1989, the EEOC certified the charge as a "class" charge on behalf of all city firefighters adversely affected by the policy change. . . . The appellant firefighters initiated this suit on December 29, 1989. The district court issued and then extended a restraining order prohibiting the City from changing the terms or conditions of the plaintiff firefighters' employment during the pendency of the litigation before the district court. . . . The City has kept the appellant firefighters on the payroll and has permitted them to continue reporting for work at their regular fire stations, but it has required them to perform various janitorial duties instead of their regular jobs. . . .

A. Title VII Disparate Impact Claim

Title VII of the Civil Rights Act of 1964 prohibits employers covered by the statute from taking actions or engaging in practices that discriminate against workers or job applicants on the basis of their race, color, religion, sex, or national origin. 42 U.S.C. § 2000e-2. This ban on employment discrimination extends, not just to actions taken or practices instituted for discriminatory reasons, but also to otherwise nondiscriminatory actions or practices that have discriminatory effects. . . . In order to establish Title VII liability under this effects-based definition of discrimination, a plaintiff must first demonstrate that a challenged employment action or practice has a disproportionate adverse impact on a category of persons protected by the statute. . . . Once such a prima facie case has been made out, the defendant must show that the challenged action is demonstrably necessary to meeting a goal of a sort that, as a matter of law, qualifies as an important business goal for Title VII purposes.

Title VII . . . provide[s] that, once a plaintiff makes out a prima facie case, the full burden of proof shifts to the defendant who must demonstrate business necessity in order to avoid liability. . . .

Upon a showing of "business necessity," the challenged action or practice is deemed justifiable, its regrettable discriminatory effects notwithstanding. However, even after such a showing, the plaintiff may still overcome a proffered business necessity defense by demonstrating that there exist alternative policies with lesser discriminatory effects that would be comparably as effective at serving the employer's identified business needs. Upon such a showing, Title VII liability is established. . . .

The City defends its decision to ban shadow beards on the ground that the prohibition is required to protect the firefighters from health and safety risks. If true, these safety claims would afford the City an affirmative defense, for protecting employees from workplace hazards is a goal that, as a matter of law, has been found to qualify as an important business goal for Title VII purposes. . . . Measures demonstrably necessary to meeting the goal of ensuring worker safety are therefore deemed to be "required by business necessity" under Title VII.

(*continued*)

(*continued*)

Whether the no-beard rule is demonstrably necessary to meeting the acknowledged business goal of worker safety is a factual issue on which, for the purposes of this case, we have assumed that the City, the movant, would bear the burden of proof at trial. . . . The City has supported its safety allegations with evidence in the form of an affidavit from an expert in the field of occupational safety and health and with a citation to a U.S. Occupational Safety and Health Administration ("OSHA") regulation concerning use of respirators by persons with facial hair. . . .

The City's expert, Kevin Downes, swore that, "Based upon my research and experience in training on the proper use of SCBA's, it is my opinion that the SCBA should not be worn with any amount of facial hair that contacts the sealing surface of the face piece." . . . In the affidavit Downes detailed particular safety risks that he maintained were posed by use of SCBA's by men with facial hair. Such use would be dangerous, asserted Downes, because facial hair is likely to interfere with the forming of a proper seal between the SCBA mask and the wearer's face. An imperfect seal may permit air from the outside environment to leak into the mask—when this occurs the wearer is said to have "overbreathed"—thereby risking exposing the wearer to contaminants. . . .

As support for his opinion, Downes noted that three national organizations that set occupational safety and health standards—the American National Standards Institute ("ANSI"), the National Institute for Occupational Safety and Health ("NIOSH"), and OSHA—all recommend that SCBA's should not be worn with facial hair which contacts the sealing surface of the face piece. . . . In addition to submitting the Downes affidavit, the City in its summary judgment memorandum also referred to the OSHA, NIOSH, and ANSI recommendations, and cited directly to the OSHA respirator standard. The OSHA regulation provides: "Respirators shall not be worn when conditions prevent a good face seal. Such conditions may be a growth of beard. . . ." OSHA Occupational Safety and Health Standards, Respiratory Protection, 29 C.F.R. § 1910.134(e)(5)(i).

We hold that this evidence that safety concerns necessitate the ban on shadow beards is "credible evidence . . . that would entitle [the City] to a directed verdict if not controverted at trial." . . . The City has thus carried its movant's initial summary judgment burden on the business necessity issue.

At this point, responsibility devolves upon the firefighters to come forward with evidence that, when considered together with the City's evidence, is sufficient to create a genuine issue as to the reality of the City's safety claims. The only real evidence invoked by the firefighters to counter the City's claims is the fact that for the six years between 1982 and 1988 the City permitted firefighters with PFB to wear their SCBA's over shadow beards. The firefighters argue that the fact that the shadow beard program was tested over this period, apparently without mishap or reported problems obtaining adequate seals, creates at least a genuine issue that shadow beards may in fact be safe. . . . We disagree. The firefighters have not adduced evidence showing how carefully the firefighters' seals were monitored over this period, or whether examinations were made that would have uncovered any resulting safety or health problems. The mere absence of unfortunate

(*continued*)

(*continued*)

incidents is not sufficient to establish the safety of shadow beards; otherwise, safety measures could be instituted only once accidents had occurred rather than in order to avert accidents. Although the six-year history is not irrelevant to the question of whether it is unsafe to wear SCBA's over shadow beards, we hold that when considered in the context of the totality of the evidence, it would not be sufficient to prevent the City from obtaining a directed verdict at trial.

In reaching this conclusion we are swayed particularly by the recommendations of the occupational safety and health standards organizations. Although public employers such as the City are not required by law to comply with OSHA standards, see 29 U.S.C. § 652(5) (excluding states and their political subdivisions from definition of OSHA "employer"), such standards certainly provide a trustworthy bench mark for assessing safety-based business necessity claims. It is true that the OSHA and ANSI standards speak in somewhat general terms about "facial hair" and "growths of beard" and do not specifically address the case of very short shadow beards; however, the NIOSH standard provides that "even a few days growth of stubble should not be permitted." . . . At least in the absence of any evidence showing that safety experts view shadow beards as a special case, we hold that the only reasonable inference supported by the OSHA, ANSI, and NIOSH standards is that shadow beards are encompassed by the prohibitions.

This is not to say that allegations that a challenged practice is required for safety are by any means unassailable. Expert testimony or results from adequately conducted field tests tending to show that shadow beards do not prevent SCBA's from sealing to the face would be sufficient to create a genuine issue as to the reality of the City's safety claims. However, the firefighters have come forward with no such evidence. We thus hold that the firefighters have failed to carry their non-movant's summary judgment rebuttal burden and that, therefore, there was in the record before the district court at the time of the summary judgment motion no evidence creating a genuine issue as to whether safety requires the ban on shadow beards. . . .

The firefighters have proposed two possible alternatives to the City's rule requiring firefighters to be clean-shaven. The first is simply reinstitution of the shadow beard shaving clinic. However, in order for the shadow beard program to constitute a legitimate less discriminatory alternative, shadow beards must adequately serve the Fire Department's acknowledged business need, namely, safety. As we have explained above in addressing the City's business necessity defense, the firefighters have failed to create a genuine issue that shadow beards are safe. Thus, for the same reason, they have also failed to create a genuine issue that the shaving clinic would be a comparably effective alternative to the shadow beard ban.

The second possible alternative suggested by the firefighters is shaving only the portion of the face where the SCBA seal would come into contact with the skin. However, in the two sentences of their summary judgment papers in which they propose this alternative, the firefighters cite no evidence to show that

(*continued*)

(*continued*)

partial shaving would be a viable and safe alternative. Moreover, as a matter of common knowledge, it is apparent that partial shaving would pose the same PFB problems as full-face shaving, and thus it is doubtful that the firefighters could have adduced evidence that partial shaving constitutes a viable less discriminatory alternative. Thus, the firefighters have failed to carry their summary judgment rebuttal burden of creating a genuine issue as to the viability of either of the two less discriminatory alternatives they propose. Having concluded (1) that the City has carried its initial summary judgment burdens on the business necessity and less discriminatory alternative issues, and (2) that the firefighters have failed to carry their summary judgment rebuttal burdens on either of these two points, we affirm the grant of summary judgment on the Title VII disparate impact claim.

B. Title VII Disparate Treatment Claim

1. Elements of Claim.

The firefighters also challenge the no-beard rule on the ground that it was allegedly adopted for racially discriminatory reasons in violation of Title VII. . . .

Because the firefighters allege disparate treatment but do not possess any direct evidence showing that the no-beard rule was instituted for discriminatory reasons, they are permitted to proceed . . . using circumstantial evidence. However, the particular employment practice targeted by the firefighters is not of the sort typically challenged under disparate treatment theory. In most disparate treatment cases, a plaintiff alleges that an otherwise legitimate rule or policy is being invoked pretextually as an excuse for what is really discrimination. In this case, however, it is the employment policy itself—the no-beard rule—with which plaintiffs take issue. They charge that the rule was adopted for the purpose of removing from the Fire Department a group of African-American men: the plaintiff firefighters suffering from PFB. Such challenges are more typically brought as disparate impact claims and, of course, appellants have also chosen to proceed under that theory. . . .

2. Propriety of Summary Judgment

. . . Assuming arguendo that the firefighters have succeeded in making out a prima facie case of disparate treatment, the City has adduced evidence showing that at trial it would readily be able to carry its light burden of coming forward with a legitimate, nondiscriminatory reason—namely, safety—for its reinstitution of the shadow beard ban. . . . The City is therefore entitled to summary judgment on the disparate treatment claim if there is no genuine issue that its proffered safety justification is not a pretext for discrimination. . . .

As discussed above, the City has adduced evidence that safety considerations require firefighters to be clean-shaven, and plaintiffs have not adduced evidence sufficient to call that contention into question. . . .

In light of the substantial evidence adduced by the City in support of its safety justification, we hold that no reasonable finder of fact could find the justification pretextual solely on the basis of the underinclusiveness of the City's

(*continued*)

(*continued*)

SCBA safety rule. Accordingly, the City is entitled to summary judgment on the Title VII disparate treatment claim. . . .

CONCLUSION

For the foregoing reasons, we affirm the ruling of the district court granting summary judgment for the City on each of the firefighters' claims.

Case Name: Fitzpatrick v. City of Atlanta

Court: United States Court of Appeals, Eleventh Circuit

Summary of Main Points: Fire departments may impose reasonable restrictions on the wearing of facial hair consistent with business necessity, even where such may result in a disparate impact on minority firefighters.

Another challenge to beards that frequently comes up is based on religious discrimination. A case that is often cited by advocates of allowing beards in the fire service is the *Newark* case. The *Newark* case involved a challenge to a police department's rule prohibiting beards, by two Muslim officers who wanted to grow beards for religious purposes. Below is the case summary:

Fraternal Order of Police Newark Lodge No. 12 v. City of Newark
170 F.3d 359 (3d Cir. 1999)
United States Court of Appeals, Third Circuit

This appeal presents the question whether the policy of the Newark (N.J.) Police Department regarding the wearing of beards by officers violates the Free Exercise Clause of the First Amendment. Under that policy, which the District Court held to be unconstitutional, exemptions are made for medical reasons (typically because of a skin condition called pseudo folliculitis [sic] barbae), but the Department refuses to make exemptions for officers whose religious beliefs prohibit them from shaving their beards. Because the Department makes exemptions from its policy for secular reasons and has not offered any substantial justification for refusing to provide similar treatment for officers who are required to wear beards for religious reasons, we conclude that the Department's policy violates the First Amendment. Accordingly, we affirm the District Court's order permanently enjoining the Department from disciplining two Islamic officers who have refused to shave their beards for religious reasons. . . .

Case Name: Fraternal Order of Police Newark Lodge No. 12 v. City of Newark

Court: United States Court of Appeals, Third Circuit

Summary of Main Points: When an employer makes exceptions to grooming regulations for some employees for medical reasons, it cannot refuse to afford the same accommodations to members on religious grounds.

The key factor in the *Newark* decision was the fact that, unlike the fire department in *Atlanta,* the police department already allowed an exception for officers with medical conditions to have beards. The department could not cite a safety condition or other business necessity sufficient to justify not accommodating the religious needs of its officers to the same extent that they accommodated their medical needs. The court concluded that it was discriminatory to refuse to allow an exception to the no-beard rule for medical reasons, but not for religious reasons.

SUMMARY

Sexual harassment claims have posed problems for the fire service as women have integrated into the mainstream fire service over the past 30 years. Sexual harassment is a form of sexual discrimination prohibited by Title VII of the Civil Rights Act of 1964. There are two types of sexual harassment: quid pro quo, and hostile work environment. The accommodation of various religious practices has also been a contentious issue for fire departments, impacting issues such as scheduling, uniforms, and grooming.

Grooming and uniform standards are important considerations in the fire service, but may be challenged on a variety of fronts, including sexual discrimination, religious discrimination, and even the ADA. As a general rule, reasonable nondiscriminatory regulations based upon commonly accepted social norms and reasonably related to the employer's business needs will be upheld.

REVIEW QUESTIONS

1. What are the two types of sexual harassment?

2. Can a heterosexual man commit sexual harassment of another heterosexual man?

3. Can a fire department have different grooming and uniform requirements for men and women without violating discrimination laws?

4. What is the statute of limitations on filing a sexual harassment allegation?

5. Can a person who alleges sexual harassment immediately file suit in Federal District Court?

6. Are there any reasons why the statute of limitations for sexual harassment may be extended?

7. Identify at least two laws that could apply to a woman who is terminated or otherwise discriminated against because she becomes pregnant.

8. What factors contribute to a sexually hostile work environment?

9. Can a firefighter refuse to work at a fire station that provides fire protection to a casino, on the grounds that it is against his or her religious beliefs to go to a casino?

10. What was the key factor in the *Newark* case that distinguished it from the *Atlanta* case?

DISCUSSION QUESTIONS

1. Where is the line that is drawn between the *Johnson* case and the *O'Rourke* case with regard to magazines such as *Playboy*? Should an employer be able to prohibit certain types of material in the workplace?

2. Can a fire department prohibit a firefighter from wearing a religious headdress while on duty? What other issues come into play?

Chapter

14

FAIR LABOR, FAMILY MEDICAL LEAVE, RESIDENCY, AND DRUG TESTING

Learning Objectives

Upon completion of this chapter, you should be able to:

- Identify the maximum hour limits of the Federal Fair Labor Standards Act (FLSA), and when overtime compensation is required.
- Define compensatory time and explain the appropriate guidelines for comp time under the FLSA.
- Explain the firefighter exemption, the emergency medical (ambulance) exception, and the executive exemption, including the effect of recent changes in the laws.
- Identify what hours are compensable under the FLSA.
- Explain how volunteers are treated under the FLSA.
- Explain what the Family and Medical Leave Act (FMLA) is, and what benefits it provides.
- Explain the constitutionality of residency requirements.
- Explain when firefighters can be asked to submit to drug testing.

INTRODUCTION

"Hey Chief, do you have a minute?" the firefighter said to the chief. "I have a question."

"Sure Charlie, what's up?" replied the chief.

"I've been a volunteer firefighter here for twelve years, and now all of a sudden because I took the job as a full-time driver during the day, I can't volunteer on nights and weekends anymore? I don't think that's right. How can they keep someone from volunteering if they want to?" said the firefighter, the frustration building in his voice.

The chief replied, "The law says if you are a paid firefighter for us, you can't volunteer for us. I asked our attorney, and she says that I cannot allow you to volunteer, that it's my responsibility."

"I don't understand, why not?" replied the firefighter.

"I'm not sure I do, either," said the chief, "but I know you can't. There's a concern that if employers could get employees to 'volunteer,' then places like supermarkets and factories would start coercing their employees to 'volunteer' extra hours."

"Yeah, but this is different."

This chapter addresses four employment-related subjects that impact firefighters: the Federal Fair Labor Standards Act, state and Federal Family and Medical Leave Acts, residency requirements, and drug testing.

FAIR LABOR STANDARDS ACT

The Federal Fair Labor Standards Act (FLSA) was originally passed by Congress in 1938 to address concerns over minimum wage, maximum hours, overtime, and child labor. Enforcement of the FLSA is the responsibility of the United States Department of Labor. The FLSA authorizes the Department of Labor to interpret the act through issuance of regulations. These interpretive regulations are given considerable deference by the courts, which may be called upon to apply them in the context of a wage and hour case.

Until 1974, the FLSA applied only to private sector employers. In 1974, Congress extended the minimum wage and maximum hours protections of the FLSA to state and local employers. However, a 1976 Supreme Court decision, *National League of Cities v. Usery*, 426 U.S. 833, struck down the FLSA as it applied to public sector employers, based on the lack of Federal authority under the Constitution's interstate commerce clause to regulate employment relationships between states and their employees. In an unusual reversal, the Supreme Court revisited the validity of the FLSA for public employers in 1986, and upheld the authority of Congress in the case of *Garcia v. San Antonio Metropolitan Transit Authority*, 469 U.S. 528.

In the years since, fire service cases have played a leading role in shaping the FLSA landscape. There are many reasons why the FLSA so heavily impacts the fire service, including the long hours that firefighters commonly work, their unconventional work shifts, and the fact that most departments allow firefighters to sleep and eat while on duty. In addition, unlike the private sector, in which supervisors and managers are frequently exempt from the FLSA overtime requirements, many fire service officer ranks meet the definition and test for hourly employees, and thus remain eligible for overtime compensation.

Who Is an Employee for FLSA Purposes?

The FLSA definition of employee includes anyone who is employed by an employer, without distinction between full-time, part-time, or temporary personnel. The phrase frequently used in FLSA cases to define an employee is anyone who is "suffered or permitted to work."

Volunteers

The definition of employee does not include persons who volunteer their services for a public agency, whether it be a state, a political subdivision of a state, or an interstate governmental agency. Individuals are considered to be volunteers if they receive no compensation or are paid only for expenses, reasonable benefits, or a nominal fee **(Figure 14-1)**.

Figure 14-1 *The FLSA definition of employee does not include people who volunteer their services.*

A person who qualifies as an employee is not allowed to "volunteer" to work for his or her employer if the volunteering involves providing the same type of services for which the person is employed. Thus a career firefighter cannot "volunteer" to work as a firefighter for his or her employer while off duty without triggering the FLSA. If the employer allows the firefighter to volunteer, the firefighter is entitled to overtime compensation for all hours worked over the statutory maximum. While quite controversial in some parts of the country, the rule is intended to prevent unscrupulous employers from coercing employees into "volunteering" on their time off.

Employees are entitled to volunteer for their employers in a capacity other than their regular employment without violating FLSA. For example, a career municipal firefighter may agree to coach a town-sponsored youth baseball team on a volunteer basis without running afoul of the FLSA.

In addition, a career firefighter in one fire department may volunteer with another fire department, even when that fire department provides mutual aid to the department in which the person is employed, without violating the FLSA.

Maximum Hours

One of the cornerstone requirements of the FLSA is that employees who work more than 40 hours per week must receive overtime compensation at a rate of one and one-half times the employees' regular hourly rate of pay. The FLSA does not require overtime pay for work on Saturdays, Sundays, holidays, or regular days of rest, nor for hours worked in excess of eight hours a day. The focus of the FLSA is solely on maximum hours per week.

The maximum hour issue impacts the fire service to an unprecedented degree. Consider that there are 168 hours in a week, and that most fire departments provide 24-hours-a-day, seven-days-a-week coverage with either three or four working shifts or platoons. Each shift in a department with three platoons or shifts averages 56 hours a week, while each shift in a department with four platoons averages 42 hours a week.

In addition, due to the rotating nature of firefighters' work shifts, the normal hours for each shift fluctuate significantly from week to week. One week a firefighter in a three-platoon system might work 72 hours, the next week 48 hours. In a four-platoon system, a firefighter might work 34 hours one week, and 48 the next. Thus, some accommodation in terms of maximum hours is necessary in order for most fire departments to comply with the FLSA.

When are Firefighters Entitled to Overtime?

Under the FLSA, not all employees are entitled to overtime after 40 hours per week. There are numerous exceptions within the FLSA for various occupations, such as firefighters and police officers, as well as for certain types of positions, such as executives, administrators, and professionals.

firefighters exemption
an exception to the overtime requirements of the Federal Fair Labor Standards Act (FLSA), which applies to any personnel engaged in fire protection activities for a public employer; public sector firefighters are entitled to overtime only if their average weekly hours exceed 212 hours in a 28-day period, or a corresponding lesser number of hours for a shorter work period; the firefighters exemption is often called the "7(k) exemption" because it is under Section 207(k) of the FLSA

Firefighters Exemption The **firefighters exemption** applies to any personnel engaged in fire protection activities for a public employer. The firefighters exemption is often referred to as the "7(k) exemption," because it is Section 207(k) of the FLSA. Public sector firefighters are entitled to overtime only if their average weekly hours exceed 212 hours in a 28-day period, or a corresponding lesser number of hours for a shorter work period **(see Table 14-1)**. For example, if a fire department determines that its work period is one week, overtime is required after a firefighter works 53 hours. If the work period is two weeks, overtime is required after 106 hours.

A fire department is required to formally adopt a work period for FLSA purposes. A work period may—but does not have to—coincide with pay periods. In the absence of a stated FLSA work period, the work period that applies to a particular fire department will be a question of fact.

The definition of "persons engaged in fire protection activities" includes firefighters, engineers, hose or ladder operators, fire specialists, fire inspectors, lieutenants, captains, inspectors, fire marshals, battalion chiefs, deputy chiefs, and fire chiefs, and applies regardless of their assignment to support

Table 14-1 *Maximum hour standards for firefighters and law enforcement.*

Work Period (days)	Maximum Hours	
	Fire Protection	Law Enforcement
28	212	171
27	204	165
26	197	159
25	189	153
24	182	147
23	174	141
22	167	134
21	159	128
20	151	122
19	144	116
18	136	110
17	129	104
16	121	98
15	114	92
14	106	86
13	98	79
12	91	73
11	83	67
10	76	61
9	68	55
8	61	49
7	53	43

activities or line functions. It also includes trainees and probationary members, as well as wildland firefighters, wildland pilots, bulldozer operators, equipment operators, fire lookouts, and others engaged in wildland firefighting. However, it generally does not include civilian support personnel such as dispatchers, mechanics, camp cooks, or clerks.

Emergency Medical Exception to Firefighters Exemption The most highly controversial and frequently litigated issue regarding the definition of a "public agency engaged in fire protection activities" involves fire personnel assigned to ambulances or rescue squads. In the 1990s, numerous cases were filed on behalf of fire-based emergency medical personnel seeking compensation for all hours worked in excess of 40 hours, who claimed that they were not engaged in fire protection activities. Cases interpreting the ambiguity applied the *80-20 rule*. The 80-20 rule is used when an employee engages in both exempt and non-exempt activities, to determine if a FLSA exemption should apply. If the employee engages in a non-exempt activity (such as emergency medical transport) more than 20 percent of the time, the exemption would not be applicable. Thus, a fire department ambulance staffed with trained firefighters who spent at least 80 percent of their time responding to fires, vehicle accidents, hazmat incidents, or standing by for such emergencies qualified under the "fire protection activity" exemption, while those who spent more than 20 percent of their time handling non-fire-related emergency medical matters were subject to the 40-hour rule.

The ambulance and rescue service issue generated so much controversy that Congress amended the FLSA on December 9, 1999, to make it clear that firefighters who are emergency medical technicians, paramedics, and ambulance personnel, as well as hazardous materials workers, are to be included under the firefighters exception. The FLSA now reads as follows:

EXAMPLE

29 USC 203 (y). "Employee in fire protection activities" means an employee, including a firefighter, paramedic, emergency medical technician, rescue worker, ambulance personnel, or hazardous materials worker, who—

(1) is trained in fire suppression, has the legal authority and responsibility to engage in fire suppression, and is employed by a fire department of a municipality, county, fire district, or State; and

(2) is engaged in the prevention, control, and extinguishment of fires or response to emergency situations where life, property, or the environment is at risk.

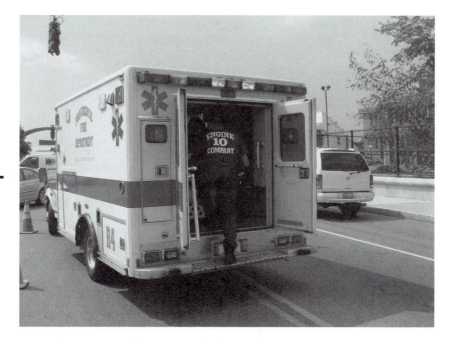

Figure 14-2 *Fire personnel assigned to perform emergency medical duties remain eligible for the firefighters exemption under DOL regulations issued in December 1999.*

As interpreted by Department of Labor regulations, the term "employee in fire protection activities" now includes all emergency medical and ambulance service personnel, if such personnel form an integral part of the public agency's fire protection activities **(Figure 14-2)**. Fire department ambulances that perform only routine patient transfers and all non-fire-department-based ambulances are still subject to the 40-hour requirement.

Public Agency The firefighters exemption applies only to firefighters who work for a public agency. A public agency is defined as a state, a political subdivision of a state, or an interstate governmental agency. Thus state, county, municipal and fire district employees would qualify under the firefighters exemption, while employees who work for a volunteer fire company or other private sector entity would not be subject to the 7k exemption. The following regulation is quite explicit:

⚖ EXAMPLE

29 CFR Sec. 553.202 Limitations. The application of sections 13(b)(20) and 7(k), by their terms, is limited to public agencies, and does not apply to any private organization engaged in furnishing fire protection or law enforcement services. This is so even if the services are provided under contract with a public agency.

Executive, Administrative, and Professional Exemption The FLSA provides an exemption from the overtime requirements for employees who are engaged in an executive, administrative, or professional capacity **(Figure 14-3)**. Unlike the firefighters exemption, which exempts firefighters from the 40-hour requirement—but imposes a higher maximum hour requirement of 53 hours—the executive, administrative, and professional employee exemptions completely exempt such employees from any maximum hour requirement.

Prior to August, 2004, most of the fire service cases focused on whether fire officers should fall under the executive exemption. However, new Department of Labor (DOL) regulations have significantly altered the FLSA landscape for fire officers. As we will see in the following sections, the full impact of these regulations will have to await full analysis by the courts.

Executive employees are those involved in the overall management of the business. Administrative personnel are those working in nonmanual jobs that involve the exercise of discretion and independent judgment. Professionals are highly skilled employees, such as attorneys, doctors, and engineers, whose work is primarily intellectual in nature. To determine whether a particular employee qualifies as an executive, administrator, or professional, the Department of Labor has developed a series of tests. These tests are included in Appendix B, Chapter 14, along with a discussion of the pre-2004 cases.

Figure 14-3 *The FLSA provides an exemption from the overtime requirements for employees who are engaged in an executive, administrative, or professional capacity.*

Recent Changes to Executive, Administrative, and Professional Employee Exemption for Firefighters

On August, 23, 2004, new Department of Labor regulations went into effect to try to clarify some of the thorny FLSA issues that have developed. In particular, these new regulations sought to address and correct the inconsistencies that have arisen with regard to the executive exception for firefighters. The new regulation related to the executive exemption reads as follows.

 EXAMPLE

29 CFR Sec.541.3 (b)(1). The section 13(a)(1) exemptions and the regulations [that govern executive, administrative, and professional employees] in this part . . . do not apply to police officers, detectives, deputy sheriffs, state troopers, highway patrol officers, investigators, inspectors, correctional officers, parole or probation officers, park rangers, fire fighters, paramedics, emergency medical technicians, ambulance personnel, rescue workers, hazardous materials workers and similar employees, regardless of rank or pay level, who perform work such as preventing, controlling or extinguishing fires of any type; rescuing fire, crime or accident victims; preventing or detecting crimes; conducting investigations or inspections for violations of law; performing surveillance; pursuing, restraining and apprehending suspects; detaining or supervising suspected and convicted criminals, including those on probation or parole; interviewing witnesses; interrogating and fingerprinting suspects; preparing investigative reports; or other similar work.

(2) Such employees do not qualify as exempt executive employees because their primary duty is not management of the enterprise in which the employee is employed or a customarily recognized department or subdivision thereof as required under Sec. 541.100. Thus, for example, a police officer or fire fighter whose primary duty is to investigate crimes or fight fires is not exempt under section 13(a)(1) of the Act merely because the police officer or fire fighter also directs the work of other employees in the conduct of an investigation or fighting a fire.

(3) Such employees do not qualify as exempt administrative employees because their primary duty is not the performance of work directly related to the management or general business operations of the employer or the employer's customers as required under Sec. 541.200.

(4) Such employees do not qualify as exempt professionals because their primary duty is not the performance of work requiring knowledge of an advanced type in a field of science or learning customarily acquired by a prolonged course of specialized intellectual instruction or the performance of work requiring invention, imagination, originality or talent in a recognized field of artistic or creative endeavor as required under Sec. 541.300. Although some police officers, fire fighters [sic], paramedics, emergency medical technicians and similar employees have college degrees, a specialized academic degree is not a standard prerequisite for employment in such occupations.

These new regulations will have to await full analysis by the courts before we can say with certainty what their impact will be. However, it would appear that the executive, administrative, and professional exemptions will no longer apply to line fire personnel, regardless of rank. It is unclear whether staff officers and fire chiefs who are not first responders may still be subject to the executive or administrative exemption. The application of these regulations to fire chiefs and fire commissioners will no doubt require additional clarification. Under the FLSA definition of employee, elected officials, as well as appointed officials who are in a policymaking capacity, are not subject to the FLSA. Thus a fire chief or fire commissioner may be considered to be an appointed official of policymaking capacity, and completely exempt from the FLSA.

Small Fire Department Exception

An additional exception from the FLSA is provided under 29 USC Section 213 (b)(20) for employees of public fire protection agencies with fewer than five employees during a given workweek **(Figure 14-4)**. Such fire departments are completely exempt from the FLSA.

What Hours are Included in the Calculation of Maximum Hours?

In calculating what hours must be included in maximum hours calculations, a variety of fire service issues arise, including waiting time, sleep time, break time, mealtime, training time, travel time, and on-call time. As a general rule, all hours during which an employee is required to be on the employer's premises or at a prescribed workplace are included in the calculation of maximum hours. Compensable time includes pre-shift and post-shift activities

Figure 14-4 *Public fire agencies with fewer than five employees are exempt from the FLSA.*

that are integral to work, including attending roll calls, checking equipment, preparing reports, washing hose, rolling hose, and readying equipment for the next run. If the employer permits an employee to continue working after the scheduled hours are over, the time is considered compensable, even though such additional time is not employer-mandated.

Waiting time: Time that an employee spends at an employer's workplace waiting to work is compensable. It does not matter that an employee is permitted to read a book, watch television, or participate in other activities, as long as the employee is engaged by the employer to wait.

Sleep time: An employee who is permitted to sleep or engage in other personal activities when not busy is still considered to be engaged to wait, and thus time spent sleeping is compensable. However, where employees are on duty for more than 24 hours, they may enter into an agreement with the employer to exclude sleep time for periods of up to 8 hours per night, provided the employees can usually enjoy an uninterrupted night's sleep. If the employees must be awakened to resume working, such as for a fire or emergency, they must be compensated for the time they are engaged.

Break time: The FLSA does not require that rest breaks, coffee breaks, and/or meal breaks be provided or taken. However, where short breaks lasting from 5 to 20 minutes are provided, such break time is considered to be compensable.

Mealtime: Mealtimes may be excluded from the calculation of hours worked, provided they are at least 30 minutes long, and no work activity is required or performed **(Figure 14-5)**. A person who takes a lunch break but

Figure 14-5
Mealtimes may be excluded from the calculations of hours worked, provided they are at least 30 minutes long and no work activity is required or performed.

remains responsible for answering phones, or who is otherwise engaged to wait while eating, is considered to be working. In the case of firefighters, mealtimes are usually considered compensable because the firefighters are required to remain at the station, available to respond to emergency calls. Only if the employee is completely relieved of duty and not expected to respond to emergencies would the time be considered deductible.

Training time: Attendance at job-related training, lectures, and meetings is generally considered to be compensable. Time spent at such activities may be excluded only if four criteria are met: the activity is outside normal hours, it is voluntary, it is not job-related, and no other work is concurrently performed. See 29 CFR 785.27.

Travel time: Time spent traveling to and from work is not considered to be compensable. However, time spent traveling as part of one's principal employment is considered to be work time. Time spent traveling from one's principal place of employment to another site is also compensable as travel time. During travel that includes an overnight stay for a work-related purpose, travel time is compensable to the extent that it involves time that would otherwise have been scheduled work time. Time spent traveling outside of normal work hours will not ordinarily be compensable, except under special circumstances. See 29 CFR 785.36.

What if, at the end of a work shift, an employee is unable to go home due to the distance from the workplace to the employee's residence? This situation arises frequently when fire personnel respond to major wildland fires out of their jurisdictions, or as part of an urban search and rescue (USAR) team **(Figure 14-6)**. It also arises when personnel attend training classes away from home. Ordinarily, when an employee is at a work location, all hours are considered compensable, including sleep, meal, and waiting time. However, when an employee is at a distant work location for an extended period of time, and while off-duty, the employee may "engage in normal private pursuits and thus have enough time for eating, sleeping, entertaining, and other periods of complete freedom from all duties when he may leave the premises for purposes of his own," hours not spent actually working are not compensable. See 29 CFR 785.23. The primary focus is on whether the employee is given enough time and freedom to be "completely relieved from duty and which are long enough to enable him to use the time effectively for his own purposes." See 29 CFR 785.16; 29 CFR 553.221.

On call: On-call time, when a firefighter may be at home or engaging in his or her own personal leisure activities subject to being recalled, is usually not compensable unless the restrictions placed upon the employee's freedom while on call are so significant that the member is considered to be the equivalent of on duty. The DOL regulations provide a good explanation of the on-call requirements.

Figure 14-6
Personnel traveling long distances to respond to major incidents cannot simply go home when their shifts are over.

Time spent at home on call may or may not be compensable depending on whether the restrictions placed on the employee preclude using the time for personal pursuits. Where, for example, a firefighter has returned home after the shift, with the understanding that he or she is expected to return to work in the event of an emergency in the night, such time spent at home is normally not compensable. On the other hand, where the conditions placed on the employee's activities are so restrictive that the employee cannot use the time effectively for personal pursuits, such time spent on call is compensable.

The *Renfro* case provides a good example of a fire department that pushed the on-call requirements a bit too far.

Renfro v. City of Emporia
948 F. 2d 1529 (10th Cir., 1991)
United States Court of Appeals for the Tenth Circuit

BARRETT, Circuit Judge. . . .
The Fair Labor Standards Act . . . requires that employers pay their employees overtime for additional hours worked over forty hours per week. Section 207 addresses the maximum work hours allowed under FLSA, and § 207(k) specifically applies to law enforcement and fire protection agencies. . . .

Firefighters employed with City were regularly scheduled to work six shifts of twenty-four hours each in a 19-day cycle, for a total of 144 hours. Each firefighter also appeared on a mandatory callback list for each 24-hour period following a regularly scheduled tour of duty. During this on-call period, the firefighters were not required to remain on the stationhouse [sic] premises. However, they were required to carry pagers and return to work within twenty minutes if called or be subject to discipline. Firefighters who were late or missed a callback received a "white slip." . . . Firefighters were paid overtime for on-call time only when actually called back to work. Firefighters called back to work were paid a minimum of an hour of overtime. . . .

City has two fire stations and each station maintains a separate on-call list. Five firefighters were normally on the list at one station and three at the other. The callbacks were made in the order the firefighters appeared on the list, and the order was rotated for each 24-hour on-call period. Firefighters were called in when the on duty staff at the stations fell below the required minimum. . . .

In January, 1987, firefighters filed this action for declaratory judgment . . . , and for compensation and other relief under FLSA. . . . In their Complaint, firefighters alleged that: the on-call policy was so restrictive they were unable to effectively use the time for personal activities; the on-call duty was time spent

(*continued*)

(*continued*)

working for City in excess of the hourly levels set forth in FLSA and therefore compensable under FLSA; and, City's "actions and omissions were done in a knowing, willful, purposeful, intentional and bad faith [sic] manner." . . .

City answered, denying firefighters' allegations of bad faith and contending it had acted in good faith when implementing the on-call policy in that firefighters had been paid any overtime to which they were entitled. . . . City stated the on-call policy was designed to be "non-restrictive so as to allow the firefighters to effectively engage in their own personal pursuits." . . .

City contended that at the time of this lawsuit: eleven of the firefighters had secondary employment; firefighters were not required to remain on the station-house premises during their on-call duty but were to report within twenty minutes from the time they were called back; firefighters had traded their on-call duties with other firefighters; and that while on on-call duty, firefighters had "participated in sports activities, socialized with friends and relatives, attended business meetings, gone shopping, gone out to eat, babysitted, and performed maintenance or other activities around their home." . . .

In their motion for partial summary judgment, firefighters set forth a "statement of material facts to which there existed no genuine issue," . . . , including statements that firefighters were required to return to the stationhouse within twenty minutes after receiving a callback; white slips were a form of discipline used when a firefighter missed or was late to a call; and firefighters were paid overtime only when they were actually called back to duty and not for the time spent on-call. . . . Firefighters further stated that on occasion a firefighter must answer 12–13 calls in one day; and that on the average, on-call firefighters receive 3–5 calls per day. . . .

Firefighters argued . . . that the on-call policy greatly restricted their personal activities: that due to the twenty minute time constraint and the large number of callbacks, they could not go out of town; they could not do simple things such as change their oil or work on their cars; they could not go to a movie or go out to dinner for fear of being called back; they could not be alone with their children unless they had a babysitter "on-call"; they could not drive anywhere with anyone when on-call (i.e., they must take separate cars in case of a callback); and, they were reluctant to participate in group activities for fear of being called away. . . .

Firefighters further contended the disciplinary "white slips" were logged by the fire chief and that a firefighter who received four or more within a four-month period could be terminated. . . . Firefighters asserted that trading the mandatory on-call shifts was very difficult because there was no economic incentive to trade and the on-call duty was so restrictive that the other firefighters would rather not trade. . . .

[T]he district court found that the relevant facts were largely undisputed and concluded that the on-call time was compensable time under FLSA. . . . The district court found that while firefighters were not required to remain on the stationhouse premises, the conditions of the on-call duty were such that firefighters'

(*continued*)

(*continued*)

personal pursuits were restricted. . . . The court set forth the conditions it believed restricted the firefighters, including: "the firefighter must be able to hear the pager at all times; that the firefighter must be able to report to the stationhouse within twenty minutes of being paged or be subject to discipline; that the on-call periods [were] 24-hours in length; and primarily, that the calls [were] frequent—a firefighter may receive as many as 13 calls during an on-call period, with a stated average frequency of 3–5 calls per on-call period." . . .

The district court determined that, based on these conditions, the firefighters were "engaged to wait," and therefore entitled to compensation under FLSA . . . and that the frequency with which firefighters were subject to callbacks distinguished this case from other cases which have held on-call time as noncompensable. . . .

In assessing actual damages, the district court found sleep and meal times were compensable . . . since the on-call shifts were 24-hours in length. The district court also determined the proper calculation of actual damages was the number of hours each firefighter was subject to on-call duty, multiplied by 1.5 times each firefighters' hourly rate of pay at that time. The amount of overtime pay actually received by each firefighter was to be subtracted from that figure. . . .

The district court determined it was required to award liquidated damages together with the actual damages under 29 U.S.C. § 216(b), unless City demonstrated its actions were in good faith and City had reasonable grounds for believing its on-call policy was not in violation of FLSA. The court found that City had not met its burden, and found that City's alleged attempt to consult the Department of Labor regarding the legality of its on-call policy was insufficient. . . .

I. City contends the district court erred by entering summary judgment against them on the issue of City's liability for overtime pay. . . . .

City contends on appeal the district court erred in finding firefighters were called back to work from on-call status on average 3–5 times per on-call period; it was difficult, if not impossible for firefighters to work in other employment while on-call; and that firefighters could not effectively use their on-call time for personal pursuits. City further contends the district court erred in failing to consider that firefighters were permitted to trade, and did trade, on-call shifts; and in failing to draw all inferences from the underlying facts in favor of City. . . .

a. . . . .We hold the district court did not err in its determination that the 3–5 callbacks per 24-hour period was an undisputed fact.

b. . . . Firefighters contended in their memorandum in opposition to City's motion for summary judgment that maintaining secondary employment while on-call was difficult or impossible. . . . City's only statement regarding this issue was that City was waiting for the further documentation concerning secondary employment it had requested from firefighters. . . .

While City stated that eleven of the firefighters had secondary employment, it did not address the specific issue firefighters set forth in their argument, i.e., the difficulty firefighters experienced in obtaining or maintaining secondary employment.

(*continued*)

(*continued*)
The record is void of any rebuttal from City regarding the secondary employment issue. The mere statement by City that some of the firefighters have secondary employment does not refute the contention that obtaining and maintaining secondary employment is difficult. Absent any specific rebuttal from City, we hold the district court did not err in determining it was difficult, if not impossible, for firefighters to obtain secondary employment.

c. City contends the district court erred by finding firefighters could not effectively use their time on-call for personal pursuits.

The Department of Labor (DOL) promulgated regulations applicable to on-call time for public employees to determine whether such hours were compensable under FLSA. These regulations state in part,

§ 785.17 On-call Time: An employee who is required to remain on call on the employer's premises or so close thereto that he cannot use the time effectively for his own purposes is working while 'on call.' An employee who is not required to remain on the employer's premises but is merely required to leave word at his home or with company officials where he may be reached is not working while on call.

§ 553.221 Compensable Hours of Work:

(c) Time spent away from the employer's premises under conditions that are so circumscribed that they restrict the employee from effectively using the time for personal pursuits also constitutes compensable hours of work.

(d) An employee who is not required to remain on the employer's premises but is merely required to leave word . . . where he or she may be reached is not working while on call. Time spent at home on call may or may not be compensable depending on whether the restrictions placed on the employee preclude using the time for personal pursuits.

The district court found the on-call time was compensable under the above regulations based upon specific undisputed facts set forth by both parties. The undisputed facts included "the firefighter must be able to report to the stationhouse within twenty minutes of being paged or be subject to discipline; that the on-call periods are 24-hours in length; and primarily that the calls are frequent—a firefighter may receive as many as 13 calls during an on-call period, with a stated average frequency of 3–5 calls per on-call period." . . . Based on the undisputed facts and the applicable law, we hold the district court did not err in determining the on-call time was compensable under FLSA.

d. City contends the district court erred in denying its motion for summary judgment by failing to take into account that firefighters were permitted to trade, and did trade, on-call shifts.

Firefighters set forth a specific issue regarding the difficulty experienced with trading on-call shifts . . . to which City responded that it was the personal decision of a firefighter whether or not to trade shifts. City also stated firefighters had in fact traded shifts. . . . The statement that the firefighters were able to trade their shifts did not rebut the contention by firefighters that it was difficult to trade on-call
(*continued*)

(*continued*)

shifts. We hold that without specific rebuttal of the contentions set forth by firefighters, the district court did not err in finding the trading of on-call shifts was difficult if not impossible. . . .

II. City contends the district court erred by misapplying the applicable law to the facts under consideration.

While FLSA does not provide a definition for "work," the courts have interpreted the meaning of the word to be "physical or mental exertion (whether burdensome or not) controlled or required by the employer and pursued necessarily and primarily for the benefit of the employer and his business." . . . The courts have further developed this definition into a method for testing whether on-call time is compensable under FLSA.

The United States Supreme Court has stated "an employer . . . may hire a man to do nothing, or to do nothing but wait for something to happen. . . . Readiness to serve may be hired, quite as much as service itself, and time spent lying in wait for threats to the safety of the employer's property may be treated by the parties as a benefit to the employer." . . . The Court further observed that whether time is spent predominantly for the employer's benefit or for the employee's is a question dependent on the circumstances of the case. . . .

To aid in the determination of whether on-call time is compensable, the DOL promulgated the regulations set forth, supra, which indicate on-call time is compensable if the employee is required to remain on the employer's premises . . . or if the employee, although not required to remain on the employer's premises, finds his time on-call away from the employer's premises is so restricted that it interferes with personal pursuits. . . . An agency's construction of its own regulations is entitled to substantial deference. . . .

We have stated the test to determine whether on-call time is compensable "requires consideration of the agreement between the parties, the nature and extent of the restrictions, the relationship between the services rendered and the on-call time and all surrounding circumstances." . . . We have further stated that the facts and circumstances of each case should determine whether periods of waiting for work should be compensable under FLSA. . . . Accordingly, we stated the "resolution of the matter involve[s] determining the degree to which the employee could engage in personal activity while subject to being called." . . . The Supreme Court has stated the "facts may show that the employee was engaged to wait, or they may show that he waited to be engaged." . . .

Firefighters in this case were required to report to a callback within twenty minutes, and to answer each callback or be subject to discipline (they were subject to discipline also if they were late to a callback); and they were not compensated for any of the waiting time. The on-call shifts were 24 hours in length and the average number of callbacks was 3–5 times per 24-hour period.

The district court found the frequency of callbacks to be an important distinction. . . . This determination by the district court is supported by two recent "Letter Rulings" issued by the DOL. In a Letter Ruling . . . the DOL determined

(*continued*)

(*continued*)

that firefighters who were on call between 5:00 a.m. and 8:00 a.m. during each scheduled workday must be compensated. The DOL determined the employer established such restrictive conditions on the firefighters that the employer effectively controlled the firefighters' time. The conditions established by the employer included, "the firefighters must be dressed and waiting for a call, they must report to the nearest fire station within ten minutes from the receipt of the call and are subject to discipline if they are late or fail to respond." . . .

A second Letter Ruling from the DOL . . . concluded that firefighters who were required to wear electronic paging devices with a reception distance of 30 miles, and who were required to respond to at least one-half of the calls, were not eligible for compensation. The letter further stated, however, if the callbacks were "so frequent that the [firefighter] is not really free to use the off-duty time effectively for [his] own benefit, the intervening periods as well as the time spent in responding to calls would be counted as compensable hours of work."

The district court also found the nature of the firefighters' employment a relevant factor in determining that the on-call time was compensable. . . . Firefighters must be alert and ready to protect the community, and the time firefighters spend lying in wait for emergencies could be considered a benefit to the employer and thus compensable under FLSA.

Therefore . . . we hold the district court did not err in its application of the law to the facts. . . .

IV. City contends the district court erred in its assessment of damages.

Specifically, City contends (a) firefighters should not receive damages for periods when they were at the bottom of the call list and not likely to be called in for duty; (b) firefighters should not receive pay for meal and sleeping time while on the "on-call" list but not on a regular "tour of duty"; (c) firefighters should not receive pay for periods when they did not restrict their normal personal activities or when they worked for other employers while on-call. . . .

a. City contends the firefighters on the bottom of the list would not be called in as frequently as a firefighter on the top of the list. While the district court acknowledged this, it found there was an undisputed average of 3–5 callbacks per 24-hours. . . .We agree. Accordingly, we hold that the district court correctly applied this 3–5 callback figure to all firefighters.

b. City contends firefighters should not receive pay for meal and sleeping time while on the "on-call" list but not on a regular "tour of duty." The district court found the meal and sleep time compensable under federal regulations 29 C.F.R. § 553.222-23. The regulations state in part,

§ 553.222 Sleep time. *(b) Where the employer has elected to use the section 7(k) exemption, sleep time cannot be excluded from the compensable hours of work. . . . (2) Where the employee is on a tour of duty of exactly 24 hours. . . .*

§ 553.223 Meal time. *(c) . . . Where the public agency elects to use the section 7 (k) exemption for firefighters, meal time cannot be excluded from the compensable hours of work . . . (2) where the firefighter is on a tour of duty of exactly 24 hours. . . .*

(*continued*)

(*continued*)

City argues that the firefighters were not on a "tour of duty" as required by the regulations, and therefore they are not entitled to compensation for sleep and meal times. The federal regulations define "tour of duty" as "the period of time during which an employee is considered to be on duty for purposes of determining compensable hours." 29 C.F.R. § 553.220 (b). The district court found the 24-hour on-call shifts were compensable hours under FLSA. . . . Further, the court found that since the on-call shifts were 24-hours in length, the sleep and meal times were compensable under 29 C.F.R. §§ 553.222-223.

Having held that the district court properly found that the 24-hour on-call shifts were compensable under FLSA, we also hold that the district court properly determined that sleep and meal time were compensable.

c. City further contends the firefighters should not receive pay for periods when they did not restrict their normal personal activities or when they worked for other employers while on-call. The district court found the on-call time so restrictive of the firefighters' personal time that the on-call time was compensable under FLSA. . . . This issue was properly addressed by the district court. Thus we hold that the district court did not err in finding that all of the on-call time was compensable under FLSA. . . .

V. City contends the district court erred by awarding firefighters liquidated damages. . . .

We hold the district court properly determined City had not met its burden in showing it had reasonable grounds to believe the on-call policy was not in violation of the FLSA, and, therefore, we further hold the district court did not err in awarding liquidated damages.

We Affirm.

Case Name: Renfro v. City of Emporia

Court: United States Court of Appeals, Tenth Circuit

Summary of Main Points: When an off-duty firefighter is required to be on call, and the limitations placed upon the firefighter restrict their normal personal activities to the extent that they are not free to use the time for their own personal benefit, the FLSA requires that the person be compensated for all hours while on call.

Substitutions

Most career fire departments allow firefighters to voluntarily substitute for one another with little or no restriction **(Figure 14-7)**. This widespread practice has the potential to create havoc from a wage and hour standpoint. Congress recognized the problem and passed a specific provision, Section 7(p)(3) of the FLSA, which allows the fire department to ignore the substitution

Figure 14-7 *Under the FLSA, substitutions and early reliefs are ignored when calculating hours worked by firefighters.*

when tracking the hours of the firefighters involved. For FLSA purposes, all firefighters are treated as having worked the hours that they were scheduled to work on their regular shifts. For example, Firefighter A's hours worked while substituting for Firefighter B do not count, for overtime purposes, toward maximum hours worked for Firefighter A, nor can they be deducted from the hours that Firefighter B must work before being eligible for overtime in that pay period. The fire department simply credits the firefighter who was assigned to work with having worked the hours, rather than counting the hours of the firefighter who actually worked.

For Section 7(p)(3) of the FLSA to be applicable, the substitution must be a completely voluntary transaction between the two employees. There is no requirement that the fire department track the substitution hours. See 29 CFR 553.31. Early reliefs are treated the same as substitutions under the FLSA, to the extent that the relief is voluntary. However, if the fire department mandates the early relief, the hours involved must be tracked, and the personnel compensated accordingly. See 29 CFR 553.225.

Medical Attention

Firefighting is a dangerous profession: approximately 100 firefighters are killed—and more than 100,000 are injured—every year. An issue that frequently arises concerns compensation for time that firefighters spend seeking

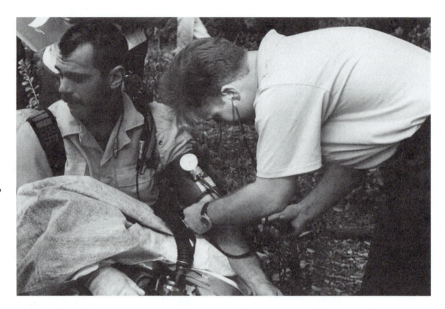

Figure 14-8   *Under the FLSA, compensable hours end when an employee is no longer performing work for the employer.*

medical treatment for job-related injuries or illnesses. For FLSA purposes, compensable hours end when an employee is no longer performing work for the employer **(Figure 14-8)**. An employee who is seeking medical attention, even for a job-related injury or illness, is not considered to be performing work for the employer.

A DOL regulation on this matter reads as follows:

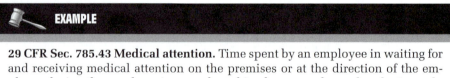

EXAMPLE

29 CFR Sec. 785.43 Medical attention. Time spent by an employee in waiting for and receiving medical attention on the premises or at the direction of the employer during the employee's normal working hours on days when he is working constitutes hours worked.

Thus, Sec. 785.43 requires that a firefighter who is injured or becomes ill at work be compensated while receiving medical attention at the station or at an emergency scene, but once the firefighter is transported to the hospital or leaves work, the time is no longer compensable.

Time spent by employees who are ordered to seek medical attention by their superiors during regularly assigned hours, or to see a doctor while off duty, is ordinarily compensable. However, when a member is off duty because of a job-related illness or injury, the matter must be considered under workers' compensation laws—not the FLSA—because while off duty, the injured member cannot be considered to be performing work for the employer. Workers' compensation laws universally permit employers to require employees to appear for reasonable periodic medical visits with the employer's doctor without additional compensation, as a condition of receiving workers' compensation benefits.

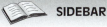 **SIDEBAR**

A firefighter is working an overtime shift from 0800 to 1800 hours. The firefighter is injured at a fire at 1000 hours, is transported to a hospital at 1030 hours, and remains there until 1600 hours, at which time he returns to the station and goes home. According to DOL regulations, the firefighter is only entitled to compensation until 1030 hours. In the event the member is placed in an off-injured status and receives workers' compensation benefits, the department has a right to require the firefighter to attend periodic medical visits without the need for additional compensation. The reasonableness and frequency of the medical visits are usually governed by the state's workers' compensation laws, not the FLSA.

The subject of line-of-duty injuries for firefighters raises issues related to workers' compensation coverage, collective bargaining agreements, and municipal ordinances that may dictate compensation practices that differ from what the FLSA requires. For example, workers' compensation laws may require that injured employees working their regular shifts be compensated by the employer for the remainder of the day in which they are injured, even though the FLSA does not require such compensation. The impact of other laws must always be considered whenever a fire department needs specific answers to complex legal issues.

Compensatory Time

Public employees may receive, in lieu of overtime compensation, a corresponding credit of compensatory time off, or comp time. Comp time must accrue at a rate of one and one-half hours of comp time for every hour of overtime worked. Public safety workers, including firefighters, may accrue a maximum of 480 hours of comp time, while non-public safety personnel are limited to 240 hours. For comp time to be an option for a fire department, there must be an agreement in place between the employer and the employees before the comp time may accrue.

Public safety workers must be permitted to use any accrued comp time within a reasonable period after making the request, provided it does not "unduly disrupt" the operations of the agency. Employees must not be coerced into accepting more comp time than an employer can realistically and in good faith expect to be able to grant within a reasonable period of time. Whether a request to use comp time has been granted within a "reasonable period" will be determined based on the facts and circumstances in each case.

If it becomes necessary to pay an employee for accrued comp time, the employee must be paid at the rate earned at the time the employee receives the payment, not the rate the employee received when the time was accrued. In addition, upon termination of employment, an employee with accrued comp time shall be paid for the unused comp time at a rate not less than either (A) the average wage received by such employee during the preceding three years; or (B) the final regular wage received by such employee.

FLSA Anti-Discrimination and Retaliation

The FLSA prohibits an employer from discriminating against employees because they have filed a complaint or otherwise availed themselves of FLSA coverage. Section 15 of the FLSA states: ". . . it shall be unlawful for any person to . . . discharge or in any other manner discriminate against any employee because such employee has filed any complaint or instituted or caused to be instituted any proceedings under or related to this Act, or has testified or is about to testify in any proceedings." The wording of this section is consistent with other Federal and state "whistleblower" provisions, which are designed to encourage and protect those who come forward to report violations.

FLSA Record Keeping

Employers must keep FLSA-required records for three years for all employees who work under the wage and overtime requirements of the FLSA.

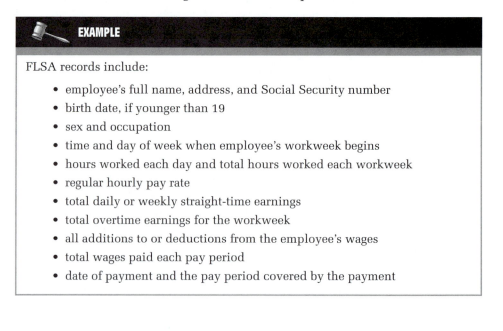

EXAMPLE

FLSA records include:

- employee's full name, address, and Social Security number
- birth date, if younger than 19
- sex and occupation
- time and day of week when employee's workweek begins
- hours worked each day and total hours worked each workweek
- regular hourly pay rate
- total daily or weekly straight-time earnings
- total overtime earnings for the workweek
- all additions to or deductions from the employee's wages
- total wages paid each pay period
- date of payment and the pay period covered by the payment

Enforcement

The Wage and Hour Division of the Department of Labor enforces the FLSA. The Wage and Hour Division will investigate complaints about FLSA violations and commence actions for enforcement. Actions may be commenced against an employer, either administratively or through court action, to recover back wages for employees who have been underpaid. The Department of Labor may also seek civil or criminal penalties against employers for willful violations. In addition, employees may pursue private actions against employers who violate the FLSA.

Normally, claims for violations of the FLSA are subject to a two-year statute of limitations, unless the DOL or the claimant employee can prove a willful or reckless violation, in which case the statute of limitations is extended to three years. See 29 U.S.C. §255. From the date the suit is filed in court, the back pay claim recovery period runs backward in time to include workweeks within the prior two or three years.

FLSA and Collective Bargaining Agreements

The FLSA provides the minimum requirements for wages and maximum hours. Collective bargaining agreements may provide for greater benefits that the FLSA requires, such as requiring overtime after 40 hours per week, or double time for hours worked in excess of the maximum hours. Employers may choose to provide greater benefits than required by the FLSA. However, to the extent that employers or collective bargaining agreements provide fewer benefits than the FLSA allows, the FLSA provisions will supersede. Collective bargaining agreements may provide for FLSA disputes to be submitted to binding arbitration.

FAMILY MEDICAL LEAVE

In 1993, Congress passed the Family and Medical Leave Act (FMLA). The FMLA was intended to protect workers who need to take time off for certain family or personal medical conditions, without risking the loss of their jobs. The FMLA provides employees with up to 12 weeks of unpaid leave per year. It requires an employer to maintain the position and any group health benefits of an employee who takes an FMLA authorized leave.

The FMLA applies to all Federal, state, and local governments, public and private elementary and secondary schools, and employers with 50 or more employees. Employers are required to provide an eligible employee with up to 12 weeks of unpaid leave each year for any of the following reasons:

- the birth and care of the newborn child of an employee
- the placement with the employee of a child for adoption or foster care

- to care for an immediate family member (spouse, child, or parent) with a serious health condition
- to take medical leave when the employee is unable to work because of a serious health condition

In order to be eligible for FMLA coverage, an employee must have worked for the employer for at least 12 months, and worked at least 1,250 hours over the preceding 12 months. In determining whether or not an employee has worked the minimum 1,250 hours, FLSA principles for determining compensable hours or work are used.

Under some circumstances, employees may take FMLA leave intermittently. In other words, where intermittent leave is allowed, employees may take leave periodically as needed, or by reducing their normal weekly or daily work schedule. Intermittent FMLA leave may be taken whenever medically necessary to care for a seriously ill family member, or because the employee is seriously ill. If FMLA leave is taken for the birth and care of a child, or placement for adoption or foster care, the use of intermittent leave is subject to the employer's approval.

Any leave that is for birth and care of a child, or placement for adoption or foster care, cannot be taken after 12 months of the birth or placement.

The FMLA allows employees to use accrued paid leave (such as sick or vacation leave) to cover some or all of the FMLA leave. However, such leave is not mandated and may only be used if provided by the employer.

Upon return from FMLA leave, an employee must be restored to his or her original position, or to an equivalent position with equivalent pay, benefits, and other terms and conditions of employment. In addition, the employer is required to maintain group health insurance benefits to the same extent as if the employer had remained employed for the 12-week period. The employee remains liable for any copayment that is otherwise required, and may be required to reimburse the employer for the health insurance in the event the employee does not return at the conclusion of the 12-week period.

FMLA Enforcement

The Department of Labor's Employment Standards Administration, Wage and Hour Division, is responsible for enforcing the FMLA for all private, state, and local government employees, and some Federal employees. The FMLA prohibits an employer from interfering with, restraining, or denying the exercise of any right provided by FMLA. An employer may not discharge or discriminate against any individuals who assert their rights under the FMLA, or based on their involvement in any FMLA-related proceeding.

As with FLSA overtime pay violations, individual employees can also file their own lawsuits to attempt to enforce FMLA compliance and seek recovery of back pay and attorneys' fees. Usually, a two-year statute of limitations applies, unless the employee can use a three-year statute of limitations by proving that the employer knowingly and intentionally committed the FMLA violation.

State Medical Leave Laws

Many states now offer medical leave protections to employees that are comparable to the FMLA. Under these state law counterparts, the definitions of employers subject to the law may differ, as may the length of time allowed for leave, and the definition of a serious illness. As a general rule, the more generous of the laws will apply in a given situation, but an employee cannot benefit from both laws to qualify for additional extended unpaid leave. For example, if a state offers 13 weeks of unpaid leave, while the FMLA allows 12 weeks, an employee in that state may get 13 weeks of unpaid leave, but would not be eligible to combine the two for a total of 25 weeks.

RESIDENCY REQUIREMENTS

Many fire departments have imposed residency requirements on their personnel as a condition of employment. Some residency requirements limit the employees to living within a certain city, town, county, region, or state, while other residency requirements limit the distance that employees can live from their employer's community. A range of legal challenges against residency requirements have been made under both state and Federal law **(Figure 14-9)**. The most significant challenges are those that test the constitutionality of residency requirements. The United States Supreme Court decided the principal residency case, which involved firefighters, in 1976.

McCarthy v. Philadelphia Civil Service Comm'n
424 U.S. 645 (1976)
United States Supreme Court

After 16 years of service, appellant's employment in the Philadelphia Fire Department was terminated because he moved his permanent residence from Philadelphia to New Jersey in contravention of a municipal regulation requiring employees of the city of Philadelphia to be residents of the city. He challenges the constitutionality of the regulation and the authorizing ordinances as violative of his federally protected right of interstate travel. The regulation was sustained
(continued)

Figure 14-9
Residency requirements for firefighters are common, yet frequently challenged on a variety of grounds. (Photo by Rick Blais.)

(*continued*)
by the Commonwealth Court of Pennsylvania, and review was denied by the Pennsylvania Supreme Court. . . .

The Michigan Supreme Court held that Detroit's similar requirement for police officers was not irrational, and did not violate the Due Process Clause or the Equal Protection Clause of the Fourteenth Amendment. We dismissed the appeal from that judgment because no substantial federal question was presented. . . . We have therefore held that this kind of ordinance is not irrational. . . .

We have not, however, specifically addressed the contention made by appellant in this case that his constitutionally recognized right to travel interstate . . . is impaired. Each of [our prior] cases involved a statutory requirement of residence in the State for at least one year before becoming eligible either to vote . . . or to receive welfare benefits. . . . Neither in those cases nor in any others have we

(*continued*)

(*continued*)

questioned the validity of a condition placed upon municipal employment that a person be a resident at the time of his application. In this case, appellant claims a constitutional right to be employed by the city of Philadelphia while he is living elsewhere. There is no support in our cases for such a claim.

We have previously differentiated between a requirement of continuing residency and a requirement of prior residency of a given duration. . . .

This case involves that kind of bona fide continuing residence requirement. The judgment of the Commonwealth Court of Pennsylvania is therefore affirmed.

Case Name: McCarthy v. Philadelphia Civil Service Comm'n

Court: United States Supreme Court

Summary of Main Points: There is no constitutional prohibition to residency requirements. The standard of review for challenges of this nature is rational basis.

Residency requirements have been implemented through state law, local charter, municipal ordinance, and collective bargaining. As the *McCarthy* case demonstrates, residency requirements have generally been upheld as constitutional. In some cases, residency requirements have been successfully challenged when they were used unlawfully to exclude minority applicants. See *U.S. v. City of Warren,* 759 F. 2d 355 (E.D. Mich. 1991).

Challenges to residency rules have also been successful on state law grounds, typically where a municipal ordinance or charter conflicts with a state law on the subject. See *Smith v. City of Newark,* 344 A.2d 782, 136 N.J. Super.; 107 (N.J. Super. App. Div., 1975); and 128 N.J. Supp. 417, 320 A 2d. 212 (N.J. Super. App. Div., 1974).

domicile

the location where a person has a residence with intent to reside permanently; a person may have several residences, but can have only one domicile

One of the primary issues that arise in residency cases is the amount of evidence necessary to establish residency. Legal residency is usually associated with the concept of **domicile**. A person may have several residences, but can have only one domicile. Domicile is where a person has a residence with intent to reside permanently. Because the definition of domicile involves an examination of the person's intent or state of mind, the area is fertile ground for debate. In general, domicile is a question of fact, which in a challenged case must be decided by a jury. Among the factors that courts will commonly look at are:

- ownership, leasing or occupancy of residential property
- where the person sleeps
- where the person is registered to vote
- where the person's cars are registered and the address on the person's drivers license

- the receipt of mail, bills, and other items consistent with someone living in a certain location
- the payment of taxes and insurance

Courts have found a firefighter's domicile to be a city apartment, despite the fact that his family resides in a house the firefighter owns outside the city. Other cases have found a firefighter's domicile to be with his family outside the city, and not in an apartment in the city. In each case, the determining factors have been the details of the case that tend to show the firefighter's connection to one location or the other.

DRUG TESTING

The problem of drugs and alcohol in the workplace has prompted many employers to undertake drug testing of employees. Drug testing of public sector employees implicates a variety of concerns, including constitutional issues involving privacy and search and seizure; contractual issues involving labor relations; the Americans with Disabilities Act; and state law drug testing issues.

In some fire departments, drug screenings have become part of an annual medical exam. Some fire departments conduct random drug tests. Other fire departments allow drug testing only upon reasonable suspicion that a member is impaired. Prior to 1989, there was a great deal of uncertainty about the constitutionality of random drug testing of public employees. Many legal experts claimed that mandatory drug testing of public employees in the absence of probable cause—or at least some suspicion of drug usage—was a violation of Fourth Amendment search and seizure requirements, and of the Fourteenth Amendment's due process clause. Courts that addressed firefighter cases routinely struck down random drug testing on Fourth and Fourteenth Amendment grounds. See *Lovvorn v. City of Chattanooga,* 846 F.2d 1539 (6th Cir., 1988).

In 1989, the Supreme Court decided two important drug testing cases, *Skinner v. Railway Labor Executives' Assn.,* 489 U.S. 602, (1989) and *Treasury Employees v. Von Raab,* 489 U.S. 656, (1989). While neither case involved firefighters, the cases made it clear that drug testing of public employees, under some situations, is constitutionally permissible without a warrant or probable cause. The Court's focus was on balancing the employees' Fourth and Fourteenth Amendment rights with valid governmental interests.

The Court concluded that any time Federal, state, or local government compels a person to submit to drug testing, Fourth Amendment search and seizure issues are implicated. In other words, drug testing is considered to be a Fourth Amendment search, and in order to be upheld, drug testing must be reasonable.

The Supreme Court identified three factors that must be considered when determining the reasonableness of drug tests: (1) the nature of the privacy interest that the search impacts; (2) the extent to which the search intrudes on the employee's privacy; and (3) the importance of the governmental interest, along with the ability of the procedure used to address that concern. In *Skinner* the Court stated: "In limited circumstances, where the privacy interests implicated by the search are minimal, and where an important governmental interest furthered by the intrusion would be placed in jeopardy by a requirement of individualized suspicion, a search may be reasonable despite the absence of such suspicion."

The *Wilmington* case below is an example of the approach that courts now take toward challenges to drug testing.

AUTHOR'S COMMENTARY: The following case contains language and depictions that some may find offensive. This information is quoted directly as part of a published court case and is the original language of the court. Its purpose here is to help firefighters understand the constitutional issues associated with drug-testing in the workplace.

Wilcher v. City of Wilmington
139 F.3d 366 (3d Cir. 03/17/1998)
U.S. Court of Appeals, Third Circuit

In July 1990, the City and the Wilmington Fire Fighters Association (the firefighters' union) agreed in a Collective Bargaining Agreement [sic] that firefighters would be subject to random drug testing through urinalysis in order to ensure that members of the Fire Department were drug free. Prior to January 1994, the City had employed a procedure whereby a randomly selected firefighter was notified he would be tested when he arrived at the station to begin his shift. A battalion chief would then stay with the firefighter and take him to Occupational Health Services at the Medical Center of Delaware ("Occupational Health") where the test was performed. There, the battalion leader would conduct the firefighter to a "dry room" to produce the urine specimen. The sink in the dry room did not contain water and the toilet bowl contained blue dye to prevent cheating by dilution. The firefighters provided their urine specimens in private; no observer was present in the dry room. Occupational Health's method of collecting urine in this manner followed the guidelines of the National Institute of Drug Abuse.

In November 1993, in an attempt to reduce the cost of random drug testing, the City solicited bids from drug testing facilities. . . . The City accepted SODAT's bid.

In January 1994, SODAT began drug testing the City's firefighters. The parties have given substantially different descriptions of how the SODAT employees
(*continued*)

(*continued*)

carried out this procedure. The male firefighters, for example, claim that the SODAT monitor looked over the firefighter's shoulder at his genitals while he urinated. SODAT, on the other hand, claims that the monitors stood to the back or the right of the firefighters but did not directly observe their genitalia.

Although SODAT employees are directed to observe the urine collection process by looking in the firefighter's general direction as he or she commences urination, the monitors are neither directed nor expected to focus on the firefighter's genitals. At trial, the SODAT monitors maintained that they had acted within the company's guidelines. . . .

[T]he district court accepted SODAT's portrayal of the monitoring process as accurate. . . . The district court further stated, "Although [the collection process] may have involved some observation of the genitalia area generally, this observation was only a by-product of the general observation of the donor." . . .

The district court further concluded, "The Court is convinced that the testimony concerning the position of the SODAT employee during the specimen collection is corroborated and demonstrates that genital observation was not the purpose nor the practice of the SODAT policy." . . .

The plaintiffs filed suit on March 18, 1994, against the City and the individual defendants. . . . In an Order and Stipulation filed on April 15, 1994, the parties agreed that the City should direct SODAT to refrain from using direct observation of urination while this case was pending. . . .

III. THE CONSTITUTIONALITY OF DIRECT OBSERVATION

The gravamen of the plaintiffs' complaint is that the direct observation method of urine collection violates the firefighters' right under the Fourth Amendment, as incorporated by the Fourteenth Amendment, to be free from unreasonable searches and seizures. The district court held that the direct observation method, as executed by SODAT, did not constitute an "unreasonable" search. . . .

The Fourth Amendment guarantees the "right of the people to be secure in their persons . . . against unreasonable searches and seizures." U.S. Const. Amend. IV. It is well established that the government's collection and testing of an employee's urine constitutes a "search" under the Fourth Amendment. . . . Ordinarily, the Constitution requires the government to obtain a warrant supported by probable cause to search a person or his property. There are, however, several well-established exceptions to the warrant and probable cause requirements. . . .

Supreme Court's jurisprudence directs us to consider three factors when judging the constitutionality of employee drug tests: (1) the nature of the privacy interest upon which the search intrudes; (2) the extent to which the search intrudes on the employee's privacy; and (3) the nature and immediacy of the governmental concern at issue, and the efficacy of the means employed by the government for meeting that concern. . . .

The firefighters do not dispute the reasonableness of compulsory drug testing per se. To the contrary, the firefighters have agreed to drug testing in their Collective Bargaining Agreement with the City. Rather, the plaintiffs challenge the

(*continued*)

(*continued*)

City's method of testing, which entails visual observation of the firefighters as they provide their urine samples. This issue has been described as "distinct and clearly severable from those that govern reasonable suspicion testing generally." . . . For this reason, we apply the Fourth Amendment's reasonableness test solely to the direct observation method utilized by SODAT and not to the broader issue of compulsory drug testing. . . .

A. The Nature of the Firefighters' Privacy Interest

"Reasonableness" entails a three pronged inquiry. First, a court examines the individual's privacy interest upon which the search at issue allegedly intrudes. . . . This expectation of privacy must be legitimate as measured by objective standards. "The Fourth Amendment does not protect all subjective expectations of privacy, but only those that society recognizes as 'legitimate.'" . . .

The district court properly concluded that firefighters enjoy only a diminished expectation of privacy. "Because they are in a highly regulated industry, and because they had consented to random testing in their collective bargaining agreement, the firefighters had a reduced privacy interest." . . . Plaintiffs now argue on appeal that the firefighting industry is not "highly regulated" and that the firefighters therefore did not have a diminished expectation of privacy.

Plaintiffs' argument lacks merit. Even though extensive regulation of an industry may diminish an employee's expectation of privacy . . . we have never held that regulation alone is the sole factor that determines the scope of an employee's expectation of privacy. It is also the safety concerns associated with a particular type of employment—especially those concerns that are well-known to prospective employees—which diminish an employee's expectation of privacy. Supreme Court precedent demonstrates this principle. In *National Treasury Employees v. Von Raab,* the Court held that a government employee's expectation of privacy depended in part on the nature of his employment and whether it posed an attendant threat to public safety. . . . Upholding the drug testing of customs officials, the Court explained:

> We think Customs employees who are directly involved in the interdiction of illegal drugs or who are required to carry firearms in the line of duty likewise have a diminished expectation of privacy in respect to the intrusions occasioned by a urine test. Unlike most private citizens or government employees in general, employees involved in drug interdiction reasonably should expect effective inquiry into their fitness and probity. . . . Because successful performance of their duties depends uniquely on their judgment and dexterity, these employees cannot reasonably expect to keep . . . personal information that bears directly on their fitness.
>
> . . . Customs officials enjoyed a reduced expectation of privacy because of the sensitive nature of their duties and of the information they received. We have held that railway employees also enjoy a diminished expectation of privacy because of the safety concerns associated with those who operate trains.

(*continued*)

(*continued*)

Certainly, a firefighter with a drug problem poses as great a threat to public safety as does a customs official or a rail operator. A firefighter whose drug use is undetected is a source of danger both to his colleagues and to the community at large. In addition, the firefighter puts himself at great risk of harm. Since the perils associated with firefighting are well known, we have no trouble concluding that firefighters enjoy a diminished expectation of privacy. . . .

B. The Character of the Search

The second factor we must consider is the character of the government's search and the extent to which it intrudes on the employee's privacy. The Supreme Court has held that the degree of intrusion "depends upon the manner in which production of the urine sample is monitored." . . .

[W]e must concede that the direct observation method represents a significant intrusion on the privacy of any government employee. Urination has been regarded traditionally by our society as a matter "shielded by great privacy." . . . Few cases have dealt with the issue of the specific method used by the government to test its employees for drugs. In *Vernonia School District 47J v. Acton*, the Supreme Court upheld the constitutionality of a mandatory random drug testing program that a school district employed to reduce drug use among its student athletes . . . [as follows]:

> *The student to be tested completes a specimen control form which bears an assigned number. . . . The student then enters an empty locker room accompanied by an adult monitor of the same sex. Each boy selected produces a sample at a urinal, remaining fully clothed with his back to the monitor, who stands approximately 12 to 15 feet behind the student. Monitors may (though do not always) watch the student while he produces the sample, and they listen for normal sounds of urination. Girls produce samples in an enclosed bathroom stall, so that they can be heard but not observed.*

. . . The Supreme Court concluded that this method of testing was not unreasonable under the Fourth Amendment. "Under such conditions, the privacy interests compromised by the process of obtaining the urine sample are in our view negligible." . . .

Relying on *Vernonia*, the district court stated, "The Court finds the SODAT collection method no more intrusive on the firefighters' privacy than was the high school's drug testing program found to be constitutional in" [*Veronia*]. . . . We agree with the district court insofar as its analogy to *Vernonia* applies to male firefighters. In a world where men frequently urinate at exposed urinals in public restrooms, it is difficult to characterize SODAT's procedure as a significant intrusion on the male firefighters' privacy.

We must admit that we are more cautious about the reasonableness of the direct observation method as it applies to female firefighters. We simply cannot characterize the presence of a monitor in a bathroom while a female urinates as an ordinary aspect of daily life. Indeed, *Vernonia* noted with approval the fact

(*continued*)

(*continued*)

that female student athletes provided urine behind a stall as monitors stood out-side listening. . . . Nevertheless, nothing in *Vernonia* suggests that the presence of a female monitor in a bathroom when an adult female firefighter provides a urine specimen is per se unconstitutional under the Fourth Amendment. Moreover, the facts of this case suggest that SODAT took substantial measures to minimize the in-trusion of privacy to female firefighters caused by the direct observation procedure. The district court found that the female monitors stood to the side of the female firefighters and that the monitors did not look at the firefighters' genitalia as they urinated, but rather in their general direction. . . . Finally, SODAT provided a nurse-practitioner as a monitor for plaintiff Wilcher when she expressed discom-fort with her first female monitor. Thus, although we find SODAT's intrusion of the female firefighters' privacy to be significant, we nevertheless agree with the defendants that SODAT has carried out its testing procedure in an appropriate and professional manner.

C. The Governmental Concern

The third and final component of the "reasonableness" test under the Fourth Amendment is the government's interest, which must be compelling. . . .

In this case, we do not review the constitutionality of drug-testing per se, but rather, the procedure by which firefighters are tested. According to the City and to SODAT, visual observation is necessary to prevent cheating. At trial, the defen-dants' expert, Dr. Closson, testified that visual monitoring is necessary to catch employees who attempt to fool the test by substituting someone else's urine or adding a chemical adulterant to their own urine.

On appeal, the plaintiffs argue that cheating can be detected by testing the urine's temperature since substitutes make the specimen colder than it should be. According to Dr. Closson, a forensic toxicologist, cheaters still can avoid detec-tion by warming substitute urine through a heating pack hidden on their body, or by keeping the urine close to their body so that it takes on the body's temperature. . . . Closson further maintained that direct observation was the most accurate col-lection method for ensuring the integrity of a urine sample. Finally, Closson tes-tified that direct observation procedures are used by the New York City Police Department, the New York City Department of Corrections, and several other New York agencies.

Like the district court, we find the defendants' expert testimony persuasive. Cheating is a significant concern. The City understandably wishes to take as many steps as possible to eliminate potential violations of the drug testing program. The plaintiffs argue that the cheating described by Dr. Closson is unlikely, as Wilmington firefighters do not receive notice that they are to be tested until the day of the test, and they remain in the company of a superior officer from the moment they are notified of the test until the time that they actually provide their urine specimen. Although this argument is strong, it does not prove that the incidences of cheating, described by Dr. Closson, are impossible or even implausible. . . . Although such cheating calls for fairly sophisticated equipment, it is possible for

(*continued*)

(*continued*)
a firefighter with a drug problem to carry a catheter or an artificial bladder taped to his body on the days following drug use, just in case he is tested on that day. Indeed, Dr. Closson stated that cheating has been known to take place within the New York agencies, which use the direct observation method.

Under Supreme Court jurisprudence, the City of Wilmington need not wait for a cheating problem to develop in order to justify its use of direct observation. . . . Moreover, the fact that there exists a less intrusive method of achieving the government's goal is not relevant to the Court's Fourth Amendment analysis. . . .

Because we find that SODAT's direct observation method . . . meets the three elements of the Fourth Amendment reasonableness test, we hold that the plaintiffs' Fourth Amendment rights have not been violated. The City's significant interest in preserving the integrity of its firefighters' drug tests outweighs their expectations of privacy. . . . Accordingly, the district court did not err when it ruled in the defendants' favor on the issue of constitutionality under the Fourth Amendment. . . .

Case Name: Wilcher v. City of Wilmington

Court: United States Court of Appeals, Third Circuit

Summary of Main Points: Drug testing of firefighters is constitutionally permissible provided the three-prong test set forth by the Supreme Court is satisfied.

Criticism of Drug Testing

Critics of drug testing frequently point to the inability of drug tests to accurately determine the level of a person's impairment. In this regard, all drug tests are not the same. Consider the difference between a drug screening that can determine if the member is under the influence of drugs or alcohol, versus one that can only determine if the member has smoked marijuana within the past 30 days. Most urine tests sample for metabolites that indicate past drug usage, but are not capable of measuring impairment. In other words, someone could have smoked marijuana a week ago while off duty, and because traces of it are still in his or her system, the employee will test positive. Meanwhile, a person who just snorted cocaine while on duty—and is currently impaired—may test negative because the drug has not been metabolized and excreted into the urine. The Supreme Court in *Skinner* dismissed this argument to invalidate the testing process.

Several states have adopted laws that prohibit random drug testing of both public and private employees. These states include Montana, Iowa, Vermont, and Rhode Island, which allow drug testing only upon reasonable suspicion. Three other states—Minnesota, Maine, and Connecticut—allow random testing only of employees in "safety-sensitive" positions. See Appendix B, Chapter 14 for copies of some state laws that prohibit random drug testing.

SUMMARY

The Fair Labor Standards Act provides the minimum hourly wage and maximum hours that employees can work without being paid overtime, and specifies a variety of other work-related matters, including child labor standards and rates for overtime compensation. Due to the varied hours that firefighters have traditionally worked, the FLSA carries major implications for firefighters and fire departments. While most employees must receive overtime compensation after 40 hours per week, the FLSA includes a firefighters exemption, which does not require overtime to paid until after 53 hours per week, or 212 hours in a 28-day period. In addition, recent updates to DOL regulations provide that all line fire personnel, as first responders, cannot be classified as FLSA-exempt executive or administrative personnel, as had been the case for many years.

The Family and Medical Leave Act (FMLA) guarantees an employee up to 12 weeks of unpaid leave per year to deal with a personal or family medical problem without risk of losing his or her job. Many states have adopted comparable protections. Residency rules are commonplace in the fire service, and have generally been upheld as constitutional. The constitutional restrictions on unreasonable searches and seizures apply to drug testing of public employees. Nevertheless, drug testing is permissible provided it complies with the three-part test for reasonableness established by the U.S. Supreme Court.

REVIEW QUESTIONS

1. What was the significance of the *Garcia v. San Antonio Metropolitan Transit Authority* case to municipal firefighters?

2. What is the maximum number of hours per week that the typical employee can be required to work without triggering FLSA overtime?

3. How is overtime compensation calculated?

4. Under the FLSA, how many hours can a firefighter be required to work without triggering the overtime requirements in a:

   a. 7-day period?

   b. 14-day period?

   c. 28-day period?

5. Can a career firefighter serve as a volunteer firefighter in his or her own department?

6. Does the FLSA require overtime compensation for a career firefighter who works for a volunteer fire company after 40 hours per week? Are there any exceptions?

7. What is the maximum number of hours of compensatory time that a firefighter may accrue? How does this differ from non-public safety employees? How about a private sector employee?

8. Is it possible for an employee to take FMLA time off one day at a time, or must the employee take the entire time off as a block?

9. Are residency requirements unconstitutional restrictions on the freedom of firefighters?

10. What are the three factors that the Supreme Court has identified as part of the balancing test for reasonableness in determining if public employees can be required to submit to drug testing?

DISCUSSION QUESTIONS

1. Can a full-time police officer serve as a volunteer firefighter in his or her own town? Would it matter if the town ran the fire department or if it was a volunteer fire company? What if a police department gives officers who volunteer as firefighters in the same community additional credit toward their next promotion?

2. At what point does a member who is required to be on call become eligible to be compensated for his or her on-call time?

   Should the frequency of being recalled be the determining factor? Should the length of time that the member has to respond be the determining factor?

3. Assume you are sitting on a jury deciding a case about whether a firefighter has complied with a residency requirement. What factors would you consider to be the most important in determining the firefighter's residency? Which ones are least important?

Chapter 15

PUBLIC ACCOUNTABILITY LAWS

Learning Objectives

Upon completion of this chapter, you should be able to:

- Identify the most common types of public accountability laws.
- Explain the difference between conflicts of interest laws and ethical codes.
- Explain the purposes and functions of open meetings and open records laws, and the types of penalties for violations of each.
- Identify the two common types of financial disclosure requirements.
- Define whistleblower acts and whistleblower provisions.

INTRODUCTION

The monthly fire company meeting started with the president banging his gavel on the table to bring the meeting to order. In the room, two dozen fire-fighters sat in small groups. Some were grouped by age, some by family, and some by coincidence.

As the secretary began to read the minutes from the previous month's meeting, the door opened, and in walked Tom Smith. Tom was a former member and political activist, known by all in the room for his frequent antics before the town council.

The president stopped the meeting as Tom started to sit down.

"Tom, this is a fire company meeting, and you are not a member of the fire company, so I have to ask you to leave."

Tom stood up, and replied, "Mr. President, with all due respect, I am a tax-payer, and this fire company is subsidized by taxpayer funds, so I have every right in the world to be here to discuss how my tax money will be spent."

"Tom, we are a private fire company and you have no right to be here."

"Mr. President, the state open meetings law applies to fire companies. You are in violation of that law for not advertising this meeting, and you will be in further violation of that law if you refuse to allow me to attend."

"Tom, our lawyer disagrees and you are welcome to take the issue up with him, but this meeting will not go on while you are in the room."

Public accountability laws are a diverse group of statutes, regulations, ordinances, executive orders, and occasionally constitutional and charter provisions, whose fundamental purpose is to foster integrity in government and promote public confidence. Much of the work of government has historically been conducted behind closed doors. At common law, the public had no right to attend meetings of governmental bodies, nor to access public records.

EXAMPLE

The Watergate scandal in the 1970s prompted a number of organizations that had been advocating open government throughout much of the 1950s and 1960s to unite under the common banner of ethics in government. A variety of subjects that fall under this category include laws pertaining to

- open meetings and open records
- the disclosure and regulation of political campaign contributions
- the mandatory disclosure of financial information by public officials, public employees, political candidates, and nominees for governmental positions

- the registration of lobbyists and the regulation and disclosure of lobbying activities
- standards of ethical conduct for public officials and employees
- conduct of campaigns and elections
- public financing of political campaigns
- conflicts of interest for public officials and employees

All these laws impact the fire service. Election and campaign laws impact fire districts where members of the district board are elected by the public. Ethics and conflict-of-interest laws impact all public officials and public employees, including firefighters. Open records laws pertain to all public agencies, including fire departments. Open meetings laws pertain to fire districts, as well as to many state and municipal boards. In many states, financial disclosure requirements pertain to appointed department heads such as fire chiefs and fire commissioners, as well as elected officials.

While there are Federal public accountability laws such as the Freedom of Information Act and the Ethics in Government Act of 1978, most public accountability laws that impact fire departments are matters of state law. Each state has its own laws relating to public accountability, and its own mechanisms for enforcing these laws.

Many states have created specialized commissions to help manage ethical issues. Typically, the commission is called the state ethics commission. In addition, other public agencies, such as the board of elections and the attorney general's office, may manage various aspects of public accountability laws.

In most states, either the state attorney general or the ethics commission issues advisory opinions interpreting ethics laws. Responsibility for overseeing the submission of financial disclosure reports may fall either upon the ethics commission or a board of elections, or in some cases both. The same agencies may also be authorized to receive disclosure forms or conflict of interest statements and disclosures.

conflict of interest
a generalized term referring to a situation in which someone has a potential or actual conflict between the responsibilities imposed by law—by virtue of the person's office or position—and some other duty or interest he or she may have, whether it be financial, business, family, or otherwise

CONFLICTS OF INTEREST AND ETHICAL CODES

The term **conflict of interest** is a generalized term referring to a situation in which someone has a potential or actual conflict between the responsibilities imposed by law—by virtue of the person's office or position—and some other duty or interest he or she may have, whether it be financial, business, family, or otherwise. Most governmental conflicts of interest occur when a public official stands to profit personally by taking a certain official action, or by promoting a certain cause. However, a conflict can also exist when the interests of relatives, business associates, or employers stand to be affected. Furthermore, from the perspective of public perception, the *appearance* of a conflict of interest can be as harmful as an actual conflict.

Conflict-of-interest laws are enacted to prohibit conflicts of interest from arising for public officials and governmental employees. In some states, conflict-of-interest laws have been supplemented by comprehensive ethical codes, while other states have merged conflict-of-interest and ethics statutes into a single law. **Ethics codes** go beyond addressing conflicts of interest, and involve matters such as the acceptance of gifts, purchasing procedures, employment practices, and governmental decision making. Ethical codes go so far as to regulate the conduct of public officials and public employees on the job, after hours, in their off-duty employment, and after they leave public service.

ethics codes

laws that establish comprehensive guidelines for the conduct of public officials and employees, often addressing conflicts of interest as well

Conflict-of-Interest Laws

Conflict-of-interest laws focus primarily on financial conflicts, that is, conflicts arising from particular kinds of economic interests. Financial conflicts of interest may involve interests that are personal, family-related, or of business associates.

EXAMPLE

While each state differs on definitions and what specific conduct is prohibited, the following are some common conflict-of-interest principles:

- A public official may not vote or participate in a decision on any matter involving a business transaction, business entity, or real estate in which the official, a family member, or a business partner has a financial interest.

- A public official or employee may be prohibited from receiving compensation from a party in connection with any particular matter in which the officials' agency is a party or has a direct and substantial interest. Conflict-of-interest rules would apply when a public official

  - makes a governmental decision (for example, by voting or making an appointment)

  - participates in making a governmental decision (for example, by giving advice or making recommendations to the decision maker)

  - influences a governmental decision by communicating with the decision maker

- Former public officials and employees may be prohibited from accepting employment from a person or firm connected to any matter in which the person's agency is or was a party, or has or had a direct and substantial interest, and in which the person participated as an employee

- A public employee may be prohibited from involvement with any matter in which the employee; his or her immediate family; a business partner; a business organization in which the employee serves as an officer, director, trustee, partner, or employee; or any person or organization with whom the employee is negotiating or has any arrangement concerning prospective employment has a financial interest.

Ethical Codes

Closely related to conflict-of-interest laws, ethics codes establish comprehensive guidelines for the conduct of public officials and employees, often addressing conflicts of interest as well.

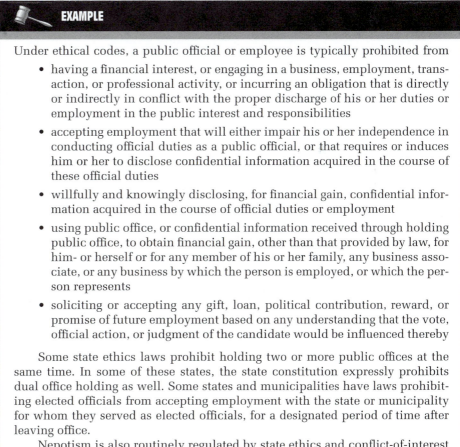

EXAMPLE

Under ethical codes, a public official or employee is typically prohibited from

- having a financial interest, or engaging in a business, employment, transaction, or professional activity, or incurring an obligation that is directly or indirectly in conflict with the proper discharge of his or her duties or employment in the public interest and responsibilities
- accepting employment that will either impair his or her independence in conducting official duties as a public official, or that requires or induces him or her to disclose confidential information acquired in the course of these official duties
- willfully and knowingly disclosing, for financial gain, confidential information acquired in the course of official duties or employment
- using public office, or confidential information received through holding public office, to obtain financial gain, other than that provided by law, for him- or herself or for any member of his or her family, any business associate, or any business by which the person is employed, or which the person represents
- soliciting or accepting any gift, loan, political contribution, reward, or promise of future employment based on any understanding that the vote, official action, or judgment of the candidate would be influenced thereby

Some state ethics laws prohibit holding two or more public offices at the same time. In some of these states, the state constitution expressly prohibits dual office holding as well. Some states and municipalities have laws prohibiting elected officials from accepting employment with the state or municipality for whom they served as elected officials, for a designated period of time after leaving office.

Nepotism is also routinely regulated by state ethics and conflict-of-interest laws. A public official is prohibited from hiring, appointing, confirming the appointment of, or voting for the appointment or confirmation of appointment of someone who is a relative by blood or marriage. States differ about the degree of relation that would trigger an ethical violation.

Ethics codes also prohibit accepting gifts or gratuities from a person or business with a direct financial interest in a decision that the public official or employee is authorized to make—or to participate in—in his or her official capacity. In addition, specific dollar amounts may be placed upon gifts of any kind that public officials may accept, from anyone. The law may also prohibit gifts to spouses and relatives of public officials and employees.

Codes of ethics may be adopted at the state level, but are more commonly adopted at the local level. Many communities adopt ethical codes for their public officials and all public employees. See Appendix B, Chapter 15 for a sample code of ethics. The result is a fairly wide variety of ethics laws that exist from community to community, that interrelate and overlap to varying degrees with state ethics codes and conflict-of-interest laws.

Ethics Commissions

Many states have established ethics commissions to oversee and enforce ethics and conflict-of-interest laws. In some states, the ethics commission is established by the state constitution, leading critics to complain about the creation of a fourth branch of government. Others have claimed that ethics commissions alter the delicate balance of power, in effect serving as a "superbranch" of government. Ethics commissions are typically staffed by unelected officials, yet have constitutional or statutory authority to police the other three branches.

However, an ethics commission's jurisdiction is actually quite limited. In addition, there must be adequate procedural safeguards, checks and balances placed on ethics commissions to ensure that the constitutional balance of power is maintained. For example, the governor commonly has the authority to appoint ethics commissioners, often with input from the legislature. Funding and supportive legislation for the ethics commission comes from the legislature, which in turn has the power to hold hearings on matters related to the ethics commission. Any decision or action taken by the ethics commission is subject to judicial review by the courts.

In enforcing state ethics laws, ethics commissions commonly have the authority to

- issue regulations relating to ethical issues
- investigate violations of ethics laws
- hold hearings on alleged violations, and cite ethics violators
- issue advisory opinions
- collect and manage financial disclosure information mandated by law

In addition, in some states the ethics commission assumes certain responsibilities that other states assign to the state board of elections. These election-related responsibilities include enforcing campaign and campaign finance laws, as well as regulating lobbyists.

Local Ethics Commissions

In some jurisdictions, ethics commissions are established at the county or municipal level. Typically these local ethics commissions serve to enforce county or local ethical codes. Local ethics commissions function in a manner analogous to state ethics commissions, investigating complaints, enforcing the applicable laws, and issuing advisory opinions for county or local officials.

Conflict-of-Interest Disclosures

Conflicts of interest are considered to be worse when undisclosed or concealed. Some states allow public officials to avoid conflict-of-interest problems by disclosing the potential conflict, without withdrawing from participation in a matter. Officials disclose potential conflicts of interest by filing a public disclosure form with the state ethics commission or other appropriate state agency. These public disclosures must be submitted prior to any official participation or action in regard to the matter at hand. Once a public disclosure has been made, the official may be allowed to participate in the matter despite the appearance of a possible conflict of interest. Not all states allow officials to act on a matter in which there is a possible conflict. Even where allowed to participate, officials must nonetheless be careful to act objectively and not use their official positions to secure any unwarranted privilege or benefit.

Conflict-of-Interest and Ethics Violations

Violations of state conflict-of-interest laws as well as violations of state codes of ethics may be considered to be civil violations, for which the guilty party can be fined by the ethics commission, or they may be prosecuted as criminal offenses. Typically the state ethics commission is assigned responsibility to investigate the civil violation of state ethics laws, while the attorney general or district attorney would oversee any criminal investigation. States differ tremendously in regard to the strictness of their interpretation and enforcement of ethics codes. Generally, more serious and intentional violations are treated as criminal offenses.

Financial Disclosure Laws

As part of their public accountability laws, many states have imposed financial disclosure requirements on certain elected and public officials. Financial disclosure requirements actually arise out of two separate but related issues. One type of financial disclosure requirement arises out of campaign and election laws, while the other involves ethics-based disclosures.

Financial disclosure requirements are commonly placed upon candidates for public office, elected officials, and certain high-ranking appointed public officials. In addition, many states require financial disclosure reports from lobbyists. Commonly, elected officials such as fire district board members, as well as high-ranking government officials such as municipal fire chiefs and fire commissioners, are subject to financial disclosure requirements.

Mandatory disclosure reports may be submitted to a state elections board, a state ethics commission, or in many cases to both agencies. Some states have created specific state agencies to monitor and manage public disclosures. For example, the State of Washington maintains a Public Disclosure Commission.

The frequency of public disclosures is based upon a number of factors, and varies greatly from state to state. Campaign financial disclosures may need to be filed weekly in some states for certain elective offices, during the weeks leading up to an election. More typically, an annual financial disclosure filing is required of most elected officials, high-ranking public officials, and others subject to disclosure laws.

The contents of financial disclosures include sources of income, business interests, investments, financial interests, bank accounts, stock ownership, real estate ownership, liabilities, and creditors. Failure to comply with financial disclosure requirements can result in administrative charges and fines, based upon the severity and willfulness of the violation. In some states, it can also lead to removal from office and the institution of criminal charges. It is common for the agency to whom the disclosure report is submitted to have investigative and enforcement power over violations. In other jurisdictions, it is the attorney general or district attorney who has the responsibility to investigate and prosecute violations.

OPEN GOVERNMENT LAWS

Open meetings and open records laws are often referred to as "sunshine laws." The name refers to the fact that these laws are intended to open up the inner operations of government to public observation and scrutiny. A few jurisdictions refer to both open meetings and open records laws as "freedom of information" laws, although other jurisdictions reserve that term for open records laws. Open meetings and open records laws are generally enforced by the state's attorney general or district attorney. Private citizens may also be authorized to file suit to force compliance with sunshine laws.

Open Meetings Laws

Open meetings laws mandate that public and governmental bodies hold their meetings and deliberations in public **(Figure 15-1)**. The purpose of open meetings laws is to enhance public confidence in government by allowing the public to observe, participate in, and critique governmental decision making.

The organization of state open meetings laws is relatively simple and surprisingly universal from state to state.

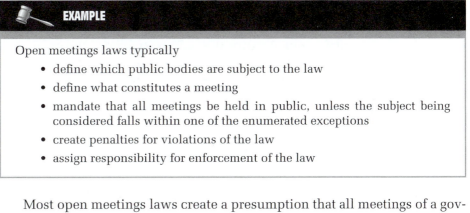

EXAMPLE

Open meetings laws typically
- define which public bodies are subject to the law
- define what constitutes a meeting
- mandate that all meetings be held in public, unless the subject being considered falls within one of the enumerated exceptions
- create penalties for violations of the law
- assign responsibility for enforcement of the law

Most open meetings laws create a presumption that all meetings of a governmental body should be open to the public, unless specifically exempted.

Figure 15-1 *Open meetings laws require that public bodies meet and deliberate in an open and public forum.*

The specific governmental and public bodies that are subject to the law are defined in the statute, and differ slightly from state to state. The typical definition of a governmental or public body includes all elected bodies, governmental units, quasi-governmental units, governmental sub-units, committees, subcommittees, and any other body created by the constitution, statute, charter, ordinance, rule, or executive order.

The question of which sub-units and subcommittees of government are subject to the open meetings law often poses a particularly difficult problem. If a fire district board appoints a subcommittee of board members to serve as a personnel board, that personnel board would be subject to the open meetings laws in many states. If a city charter establishes a purchasing committee for a municipality, it would be subject to the open meetings law. However, a fire department health and safety committee—created pursuant to a collective bargaining agreement or at the fire chief's direction—would not be subject to the open meetings law, because it was not created pursuant to a constitution, statute, charter, ordinance, rule, or executive order.

Volunteer Fire Companies The applicability of open meetings laws to private volunteer fire companies is unsettled, and varies considerably from state to state. In Wisconsin, the attorney general has rendered an opinion that a volunteer fire department that is created by municipal ordinance is subject to the open meetings law, while a volunteer fire company that is organized by private citizens is not considered to be a governmental body, and thus not subject to open meetings requirements. See 66 Op. Att'y. Gen. 113 (Wisconsin, 1977).

In *In re Anderson v. Weathersfield*, FIOC, CT, Docket #FIC 88-489 (1989), the Connecticut Freedom of Information Commission ruled that a volunteer fire company was sufficiently intertwined with local government, in terms of funding and control, to be acting in a governmental capacity, and thereby subject to the state's open meetings requirement. The Commission also noted that the fire company could not conclusively establish that it was privately chartered because the original charter, dating back more than 186 years, could not be located.

In another case from Connecticut, the Connecticut Supreme Court upheld a Freedom of Information Commission decision that a volunteer fire company was subject to the open meetings law.

Cos Cob Volunteer Fire Co. No. 1, Inc. v. FOIC
212 Conn. 100, 561 A.2d 429 (Conn., 1989)
Supreme Court of Connecticut

This is an appeal from a decision of the Superior Court overruling an order of the Freedom of Information Commission (FOIC) that required the named plaintiff,

(continued)

(*continued*)

the Cos Cob Volunteer Fire Company No. 1 (CCVFC), to open to the public certain portions of its meetings. . . .

The CCVFC is a volunteer fire department in the Cos Cob section of the town of Greenwich. On April 2, 1987, the complainant, Lawrence Orrico, a nonmember, attended a meeting of the CCVFC. Orrico was asked to leave the meeting.

Orrico thereafter filed a complaint with the FOIC alleging that the CCVFC violated his right to attend the meeting as provided by General Statutes 1-21 (a). The CCVFC responded that it was not subject to the provisions of 1-21 (a) because it was not a public agency within the meaning of General Statutes 1-18a (a). It further claimed that even if it were a public agency its meetings would be exempt from the Freedom of Information Act (FOIA) by reason of General Statutes 7-314 (b) which provides that: "Any operational meeting of active members of a volunteer fire department shall not be subject to the provisions of sections 1-15, 1-18a, 1-19 to 1-19b, inclusive, and 1-21 to 1-21k, inclusive."

The FOIC held a hearing on the matter and concluded that with respect to the bulk of its activities the CCVFC was a public agency within the meaning of 1-18a (a). The FOIC also made the following factual findings: "[T]his meeting included, among others, the following activities: a. a report by the chief on the fires responded to, including the number of fires and the manpower used[;] b. a report by the treasurer on the budget[;] c. a report by the house committee on the condition of the firehouse[;] d. reports by other committees. . . ."

The FOIC concluded that these activities were not operational in nature "and to the extent they [did] not concern fraternal activities, [were] subject fully to the open meetings provisions of the Freedom of Information Act." The FOIC found, however, that certain other activities were operational and therefore not subject to the open meeting [sic] requirement. The Commission found that the meeting included "e. [a] discussion of the current radio procedures and returning to a town-wide paging system[;] f. [and a] discussion of social activities." The FOIC concluded that the discussion of the radio procedures was an operational activity and, therefore, that portion of the meeting was not subject to the open meeting [sic] requirement. Finally, the FOIC found that "respondents are not [acting as] public agencies within the meaning of 1-18a (a) when they are performing purely fraternal or social functions," and thus concluded that the discussion of social activities was not subject to the open meeting [sic] provision.

On July 8, 1987, the FOIC accepted and adopted the hearing officer's recommendation and ordered the CCVFC to "allow the public to attend those portions of the meetings in which they undertake those "non-fraternal" activities listed [above]," i.e., the reports on the budget, the condition of the firehouse, and the fires to which the department responded. The plaintiffs appealed this decision to the Superior Court.

The trial court agreed with the FOIC's conclusion that the CCVFC was a public agency but found that the FOIC had failed to adopt any definitive meaning for the term "operational meeting." The trial court concluded that "in failing to reveal the criteria it utilized in making its determination, [with respect to which

(*continued*)

(*continued*)

activities on the agenda were "operational,"] the FOIC acted arbitrarily." The trial court sustained the appeal in accordance with General Statutes 4-183 (g). The FOIC appealed. . . .

On appeal the FOIC claims that the trial court erred in interpreting the term "operational" and in sustaining the plaintiffs' appeal and reversing the FOIC's decision. We agree. . . .

Words of a statute "must be construed according to their commonly approved usage. . . . The FOIC found that various reports on the fires responded to, the budget, and the condition of the firehouse directly related to a governmental function and, thus, were not "operational" matters. Therefore, the FOIC concluded, those portions of the meetings were subject to the open meeting [sic] requirement of the FOIA. In disputing the accuracy of the FOIC's interpretation, the trial court supplied its own definition of "operational meeting." The trial court concluded that "operational meetings are those that deal with the operating or functioning of the fire company, i.e. all meetings at which fire department operations are discussed."

It is axiomatic that the trial court may not substitute its judgment for that of the agency. . . . In excluding "operational meeting[s]" of volunteer fire departments from the open meeting [sic] requirement, the legislature could have provided criteria for determining what activities constituted "operational" activities or it could have otherwise defined the term. It chose [to] do neither. Thus the FOIC was left to decide whether the facts and circumstances surrounding the plaintiff's meeting gave rise to a conclusion that the meeting, or certain portions of the meeting, were "operational" within the broad meaning of that term. When the legislature uses a broad term, such as "operational meeting[s]," in an administrative context, without attempting to define that term, it evinces a legislative judgment that the agency should define the parameters of that term on a case-by-case basis. . . . Thus we cannot agree with the trial court that the FOIC acted arbitrarily in failing to reveal the criteria it used in making its determination.

"'The agency's practical construction of the statute, if reasonable, is 'high evidence of what the law is.' . . . In view of the record as a whole, we conclude that the FOIC acted well within the discretion delegated to it in interpreting the term "operational meeting" and applying that term to the facts before it.

There is error, the judgment is set aside and the case is remanded with direction to render judgment dismissing the plaintiffs appeal.

Case Name: Cos Cob Volunteer Fire Co. No. 1, Inc. v. FOIC

Court: Connecticut Supreme Court

Summary of Main Points: A volunteer fire company may be subject to a state open meetings requirement. As we discussed in Chapter 4, the Connecticut Supreme Court recognized in Cos Cob that when courts are asked to review agency decisions, they are to give agency decisions a great deal of deference.

While these cases may be of limited significance outside of Connecticut, the ultimate determination of whether the open meetings law will apply to a volunteer fire company in another state will rest on an analysis of the open meetings law, and the factual details of the relationship between the fire company and the community. Among the legal issues to be considered is the description of agencies subject to the open meetings law, and whether the fire company was publicly or privately chartered. Factual issues such as the degree of control exercised by the town over the fire company, as well as the extent to which public funds are used to subsidize the fire company's operation, will also be relevant to the ultimate determination.

Secret and Informal Meetings Open meetings laws prohibit not only secret meetings of public bodies, but informal meetings as well. If a fire district board is subject to the open meetings requirement, it would be a violation for the board members to meet informally to discuss any matter that may come before them. Deliberations and verbal exchanges of views between board members must be conducted in public, in the context of a formal meeting **(Figure 15-2)**. Courts have concluded that the use of a formal meeting simply to ratify decisions made in private is a violation of the open meetings law. It is usually permissible for two members of a board to discuss an issue in private, but when additional members join such a meeting, or when a quorum of the board is present, an open meetings law violation may be deemed to occur.

Figure 15-2
Deliberations and verbal exchanges of views between board members must be conducted in public in the context of a formal meeting.

Figure 15-3
*Conference calls
or teleconferencing
with a quorum of
board members
constitutes a
meeting under open
meetings laws.*

Modern Technology and Open Meetings Laws Modern technology has added to the complexity of open meetings laws, raising questions over whether conference calls, teleconferencing, and the exchange of e-mails constitutes a "meeting" **(Figure 15-3)**. Conference calls with a quorum of board members, as well as teleconferencing, is universally recognized as constituting a meeting, and therefore is subject to the open meetings requirements.

What if the chairman of a fire board engages in a series of preplanned, consecutive, one-on-one encounters with board members intended to circumvent the open meetings laws? What if the chairman does this by telephone? Admittedly, a one-on-one meeting or telephone call is nothing more than the normal lobbying and advocacy that are integral parts of our democracy. However, some states maintain that holding sequential meetings, in which one member contacts each other member individually to discuss an issue under consideration, solicit input, or take a vote intended to circumvent the open meetings laws is illegal. Other states are more lenient in this regard.

The exchange of e-mails as a tool to share information and ideas has recently become the focus of cases and ethics opinions **(Figure 15-4)**. The concern is that matters of public policy can be actively deliberated and discussed through the back-and-forth of e-mails among board members, without public knowledge. Some states have ruled that the use of e-mail to exchange information, deliberate over issues, and take polls of board members is a violation of the open meetings law. These states include Massachusetts, California, Colorado, and Montana.

Figure 15-4 *Some states consider the exchange of e-mail messages between board members to constitute a meeting.*

The state of Virginia has taken the opposite approach. In *Beck v. Shelton* 593 S.E. 2d 195 (Va., 2004), a Virginia Circuit Court upheld an attorney general's opinion that the exchange of e-mails did not constitute a meeting within the meaning of the state's open meeting law. The court determined that the concept of a meeting denotes an assemblage of individuals, which an e-mail does not entail. Both the court and the attorney general concluded that an e-mail is more in the nature of a written correspondence like a letter, and thus does not constitute a meeting. Still open in states such as Virginia are issues such as the use of forum bulletin boards and chat rooms as a means of sharing information.

Open Meetings Requirements Where the open meetings law applies, there are two basic requirements. First, there must be advance notice of any meeting. Second, all business must be conducted in the open meeting, unless a specific exemption exists.

Notice of any meeting must be made to the public, to members of the media who have requested to be notified, and to the officially designated newspaper or news medium for dissemination to the public. Generally, a notice of the meeting is posted in a public place at the city or town hall, and a notice is placed in the newspaper. The open meetings law usually designates how many days in advance the notice must be issued. There is always a provision for emergency meetings to be held upon shorter notice for specific, designated types of situations.

The notice must include the time, date, place, and subject matter of the meeting. The subject matter description must be specific enough to inform the average person who may be interested in the subject that it is being discussed. If it is anticipated that part of the meeting will be held in closed session, that fact must also be disclosed in the notice.

Public Participation Open meetings laws allow the public to be present, but do not require that the public be allowed to participate **(Figure 15-5)**. It is not a violation of the open meetings law to prohibit members of the public from speaking. However, some states have statutes that specifically require public input at certain types of meetings and hearings. Any requirement that the public be allowed to participate in a public meeting, such as when public comment or a formal public hearing is required, exists independently of the open meetings laws.

Figure 15-5 *Open meetings laws do not mandate public participation.*

Open meetings laws generally require that minutes of all meetings be kept and maintained. The amount of detail required in maintaining minutes varies from state to state. For example, Wisconsin requires only that attendance and a record of motions and votes be maintained, while other states require more comprehensive documentation of minutes.

Closed or Executive Session Open meetings laws allow a board to meet in closed session, often referred to as executive session. All meetings must begin in open session, and a vote must be taken to close the meeting based upon a specific grounds allowed by the open meetings law to enter executive session.

EXAMPLE

The grounds for closed sessions **(Figure 15-6)** commonly include:

- discussions about job performance, character, or physical or mental health of a person or persons
- any investigative proceedings regarding allegations of misconduct, either civil or criminal
- discussions of personnel matters, employee grievances, or discipline
- discussions of matters related to security, including but not limited to the deployment of security personnel or security devices
- discussions related to the acquisition or sale of public properties, or the investment of public funds, when premature disclosure may be detrimental to the public interest
- discussions related to the relocation of prospective business or industry, when premature disclosure may have a detrimental effect on the public interest
- judicial or quasi-judicial hearings
- public business with competitive or collective bargaining implications
- matters relating to pending litigation involving the town or agency

---

Once in closed session, boards are prohibited from discussing any matter that should be discussed in open session.

Penalties for Open Meetings Violations Open meetings laws typically specify how complaints about open meetings violations are to be reported and investigated. In most states, the attorney general receives the complaint and conducts an investigation. The attorney general may then file suit against the public body. In addition, persons affected by the violation may file suit

Figure 15-6　*The reasons for a public body to meet in executive session are limited and listed in the open meetings law.*

against the public body. Among the remedies and penalties for violations of open meetings law violations are:

- The court may issue an injunction against the public body and declare null and void any actions or decisions made by the board in violation of the open meetings law.

- The court may impose a civil fine against a public body or any of its members found to have committed a willful or knowing violation of the open meetings law.

Some states consider violations of the open meetings laws to be criminal in nature, and authorize charges to be filed against the individual board members who knowingly participated in the violation. See Tex. Gov't Code §§ 551.143 and 551.144. As a practical matter, however, the criminal sanction option is rarely used.

Open Records Laws

Open records laws are intended to allow the public to have access to all records normally created, maintained, received, collected, or compiled by any agency, branch, or part of government, except when specifically

Figure 15-7 *Open records laws give the public access to many fire department records and documents.*

protected **(Figure 15-7)**. The best-known open records law is the Federal Freedom of Information Act, applicable to Federal agencies. Some states have also used the term "freedom of information act" to name their open records laws, while other states simply call the law the open records act.

The scope of agencies subject to the open records laws is considerably broader than for open meetings laws. Generally, open records laws apply to all state and local governmental offices and agencies, including executive, legislative, judicial, or administrative bodies. In some states, open records laws also extend to any private agency, person, partnership, corporation, or business entity acting on behalf of and/or in place of any public agency. This may include entities such as private volunteer fire companies, who are basically acting in the place of a local public fire department.

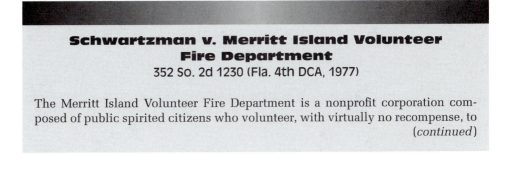

Schwartzman v. Merritt Island Volunteer Fire Department
352 So. 2d 1230 (Fla. 4th DCA, 1977)

The Merritt Island Volunteer Fire Department is a nonprofit corporation composed of public spirited citizens who volunteer, with virtually no recompense, to
(continued)

(continued)

man the county owned fire fighting equipment on Merritt Island. In addition to performing this most valuable and essential function, thereby saving the county and taxpayers vast sums of money, these citizens also sell Christmas trees, hold chicken fries and cater banquets, donating much of the net income per annum generated therefrom to a large number of charitable projects. Included in such projects we find seven or eight leagues and associations such as lassie leagues, midget football, crippled children and a host of others.

The appellant is an investigative reporter for the Today newspaper who wishes to engage in a fishing expedition into all of the volunteer fire department's records because he claims that the department is subject to the Florida Public Records Act, § 119 Florida Statutes (1975), and we are compelled to agree.

Any fire department cannot help but be classified as an agency under the following definition in § 119.011(2):

(2) "Agency" shall mean any state, county, district, authority, or municipal officer, department, division, board, bureau, commission, or other separate unit of government created or established by law and any other public or private agency, person, partnership, corporation, or business entity acting on behalf of any public agency.

Likewise, "public records" in § 119.011(1) include all the following:

(1) "Public records" means all documents, papers, letters, maps, books, tapes, photographs, films, sound recordings or other material, regardless of physical form or characteristics, made or received, pursuant to law or ordinance or in connection with the transaction of official business by any agency.

The volunteer department argues that its membership files, minutes of its meetings and its charitable activities, for example, are not "official business" and that it is not an "agency" under the Act insofar as those matters are concerned. It also points out that other civic organizations use the fire station premises owned by the county for similar fund raising activities and functions.

We are most sympathetic to these arguments, but we cannot ignore the language of the enacted statute and have no doubt that this volunteer fire department is at the very least "acting on behalf of (a) public agency." Inevitably, the present Act covers an organization entrusted with the sole stewardship over firefighting and funded in part by public moneys. The whys and wherefores of how membership is obtained with a view to firefighting expertise, the fact that the county pays $850.00 per month toward its support and supplies all the equipment, the placing of county funds in a common bank account along with the fish fry money, and the conduct of all its activities on county owned property (even if the department did raise $65,000 to help construct the building) all establish the public agency nature of the organization.

(continued)

(continued)

Perhaps not any one of the above factors set forth in the preceding paragraph would bring the volunteer department within the purview of the public records law, but the totality of them leads irresistibly to the conclusion that this department is subject to the Public Records Act.

Lest we be misunderstood, or misquoted, we would comment that this opinion is limited to the organization and the facts of this case and should not be interpreted to subject all civic and charitable organizations to the Public Records Act.

Reversed with instructions to permit the inspection and copying of all "public records" of the Merritt Island Volunteer Fire Department.

Case Name: Schwartzman v. Merritt Island Volunteer Fire Department

Court: Florida Court of Appeals, 4th District

Summary of Main Points: Volunteer fire companies are subject to open records laws.

For another open records case against a volunteer fire company with the same outcome, see *Westchester Rockland Newspapers v. Kimball,* 50 NY 2d 575 (1980).

Records Subject to Open Records Laws Open records laws apply to all documents, forms, applications, permits, papers, letters, maps, books, tapes, photographs, films, sound recordings, magnetic or other tapes, electronic data processing records, computer-stored data, electronic mail messages, or other materials that are created by a governmental agency or that are received by a governmental agency, pursuant to a law or ordinance, or in connection with the transaction of official business **(Figure 15-8)**.

Each agency subject to the law is required to establish a procedure for providing access to public records, and complying with the open records law. The public is allowed to examine, and in most cases copy, any public record. Limitations are commonly placed on how much an agency can charge for copies of public records, and how much time the agency has to provide access to the records.

Exceptions to Open Records Laws Most open records laws have a number of exceptions, for which governmental agencies do not have to release records.

Figure 15-8 *Records include documents, forms, applications, permits, papers, letters, photographs, and a host of other items.*

 EXAMPLE

These exceptions commonly include records pertaining to:

- personal or medical information relating to an individual's medical or psychological condition, personal finances, welfare, employment security, student performance, or information in personnel files maintained to hire, evaluate, promote, or discipline any employee
- client/attorney relationship
- doctor/patient relationship
- trade secrets, and commercial or financial information received or gathered about an individual or business
- adoption and child custody
- criminal investigations
- scientific and technology secrets of military or public safety and security importance
- strategy or negotiation involving labor negotiations or collective bargaining
- strategy or negotiation involving investments or borrowing, or the purchase of real estate or other property
- preliminary drafts, work product

- test questions, scoring keys, and other examination data used to administer a licensing examination, examination for employment or promotion, or academic examinations
- all investigatory records of public bodies (besides law enforcement agencies)

Violations of Open Records Laws Like violations of open meetings laws, violations of open records laws may be enforced in one of two ways. First, the attorney general may investigate alleged violations and take action against a noncompliant agency. Second, an aggrieved third party may file suit against the public agency that refuses to comply, in order to force the production of the required records. In addition to issuing a court order compelling production of the requested documents, courts may fine the parties responsible for noncompliance.

WHISTLEBLOWER ACTS AND PROVISIONS

The term "whistleblower" refers to individuals who expose some governmental or industry wrongdoing. Whistleblowers are typically insiders of the organization they are exposing, frequently employees who have observed the wrongdoing firsthand. Whistleblowers have historically paid a heavy price for their actions, including being terminated, demoted, reassigned, threatened, or discriminated and retaliated against by their employers.

Since the 1970s, laws have been passed at both the state and Federal levels to protect, safeguard, and encourage whistleblowing. Some of these laws have been formally named "whistleblower acts," and are intended to protect individuals who have exposed some wrongdoing, without specifying the law that is being violated. The following is an example from the Rhode Island Whistleblower Protection Act.

EXAMPLE

§ 28-50-3 **Protection.** An employer shall not discharge, threaten, or otherwise discriminate against an employee regarding the employee's compensation, terms, conditions, location, or privileges of employment:

(1) Because the employee, or a person acting on behalf of the employee, reports or is about to report to a public body, verbally or in writing, a violation which the employee knows or reasonably believes has occurred or is about to occur, of a law or regulation or rule promulgated under the law of this state, a political

subdivision of this state, or the United States, unless the employee knows or has reason to know that the report is false, or

(2) Because an employee is requested by a public body to participate in an investigation, hearing, or inquiry held by that public body, or a court action, or

(3) Because an employee refuses to violate or assist in violating federal, state or local law, rule or regulation, or

(4) Because the employee reports verbally or in writing to the employer or to the employee's supervisor a violation, which the employee knows or reasonably believes has occurred or is about to occur, of a law or regulation or rule promulgated under the laws of this state, a political subdivision of this state, or the United States, unless the employee knows or has reason to know that the report is false. Provided, that if the report is verbally made, the employee must establish by clear and convincing evidence that the report was made.

In addition to generalized whistleblower acts, there are literally hundreds of laws that include a whistleblower provision. A whistleblower provision is a part of a law that prohibits discrimination or retaliation against anyone who filed a complaint, supported a complaint, testified on behalf of someone who filed a complaint, or who otherwise took an action under the specific law. For example, the following passage is part of the Federal Occupational Safety and Health Act.

EXAMPLE

29 USC 660 (c) Discharge or discrimination against employee for exercise of rights under this chapter; prohibition; procedure for relief.

(1) No person shall discharge or in any manner discriminate against any employee because such employee has filed any complaint or instituted or caused to be instituted any proceeding under or related to this chapter or has testified or is about to testify in any such proceeding or because of the exercise by such employee on behalf of himself or others of any right afforded by this chapter.

The degree of whistleblower protection varies considerably from state to state, and from law to law. In addition, the availability of protection will vary based upon the nature and seriousness of the violation that is being exposed. It is not uncommon in whistleblower cases for the individual to be a disgruntled employee. However, in the interest of encouraging the uncovering of wrongdoing, whistleblower protections are afforded to all persons who avail themselves of the opportunity to report the matter, without regard to motive.

Whistleblower protections are usually not available to those who falsely report a problem, nor to those who are aware of a problem but fail to report it.

However, in the interests of encouraging people to come forward to report matters of public concern, whistleblower protections may be afforded to those who are mistaken or incorrect about the reported violation.

SUMMARY

Public accountability laws are intended to improve and enhance public confidence in government. States differ in the organization and structure of public accountability laws. Laws related to ethics, conflicts of interest, campaign finance, lobbying, open meetings, and open records all seek to improve the openness and transparency of government. Civil and criminal penalties may attach to violations of public accountability laws. All these public accountability laws impact the fire service.

In addition, whistleblower acts are laws that are intended to protect persons who expose governmental or industry wrongdoing.

REVIEW QUESTIONS

1. Name four types of laws that could be considered public accountability laws.
2. Define conflict of interest.
3. What is the difference between a conflict-of-interest law and a code of ethics?
4. What constitutional concerns are raised by ethics commissions?
5. How are the constitutional concerns about ethics commissions addressed?
6. Which law is usually broader, an open meetings law or an open records law?
7. Identify four reasons why a public board that is subject to an open meetings law may hold part of a meeting in closed session.
8. Identify the five exceptions to the open records law.
9. What are the two types of financial disclosure laws?
10. What is the difference between a whistleblower act and a whistleblower provision?

DISCUSSION QUESTIONS

1. Three members of a five-member board of wardens for a fire district attend a fire district barbeque. During the course of the cookout, they are overheard discussing the need for a new ladder truck, and the impact such a purchase will have on the tax rate. What are the consequences of their discussion?
2. Should members of a public body be able to discuss issues via e-mail?
3. What causes some people to complain that ethics commissions threaten the balance of power between the three branches of government?

Appendix A

CONSTITUTION OF THE UNITED STATES OF AMERICA

We the People of the United States, in Order to form a more perfect Union, establish Justice, insure domestic Tranquility, provide for the common defence, promote the general Welfare, and secure the Blessings of Liberty to ourselves and our Posterity, do ordain and establish this Constitution for the United States of America.

Article I.

Section 1

All legislative Powers herein granted shall be vested in a Congress of the United States, which shall consist of a Senate and House of Representatives.

Section 2

The House of Representatives shall be composed of Members chosen every second Year by the People of the several States, and the Electors in each State shall have the Qualifications requisite for Electors of the most numerous Branch of the State Legislature.

No Person shall be a Representative who shall not have attained to the Age of twenty five Years, and been seven Years a Citizen of the United States, and who shall not, when elected, be an Inhabitant of that State in which he shall be chosen.

Representatives and direct Taxes shall be apportioned among the several States which may be included within this Union, according to their respective Numbers, which shall be determined by adding to the whole Number of free Persons, including those bound to Service for a Term of Years, and excluding Indians not taxed, three fifths of all other Persons.

The actual Enumeration shall be made within three Years after the first Meeting of the Congress of the United States, and within every subsequent Term of ten Years, in such Manner as they shall by Law direct. The Number of Representatives shall not exceed one for every thirty Thousand, but each State shall have at Least one Representative; and until such enumeration shall be made, the State of New Hampshire shall be entitled to chuse three, Massachusetts eight, Rhode Island and Providence Plantations one, Connecticut five, New York six, New Jersey four, Pennsylvania eight, Delaware one, Maryland six, Virginia ten, North Carolina five, South Carolina five and Georgia three.

When vacancies happen in the Representation from any State, the Executive Authority thereof shall issue Writs of Election to fill such Vacancies.

The House of Representatives shall chuse their Speaker and other Officers; and shall have the sole Power of Impeachment.

Section 3

The Senate of the United States shall be composed of two Senators from each State, chosen by the Legislature thereof, for six Years; and each Senator shall have one Vote.

Immediately after they shall be assembled in Consequence of the first Election, they shall be divided as equally as may be into three Classes. The Seats of the Senators of the first Class shall be vacated at the Expiration of the second Year, of the second Class at the Expiration of the fourth Year, and of the third Class at the Expiration of the sixth Year, so that one third may be chosen every second Year; and if Vacancies happen by Resignation, or otherwise, during the Recess of the Legislature of any State, the Executive thereof may make temporary Appointments until the next Meeting of the Legislature, which shall then fill such Vacancies.

No person shall be a Senator who shall not have attained to the Age of thirty Years, and been nine Years a Citizen of the United States, and who shall not, when elected, be an Inhabitant of that State for which he shall be chosen.

The Vice President of the United States shall be President of the Senate, but shall have no Vote, unless they be equally divided.

The Senate shall chuse their other Officers, and also a President pro tempore, in the absence of the Vice President, or when he shall exercise the Office of President of the United States.

The Senate shall have the sole Power to try all Impeachments. When sitting for that Purpose, they shall be on Oath or Affirmation. When the President of the United States is tried, the Chief Justice shall preside: And no Person shall be convicted without the Concurrence of two thirds of the Members present.

Judgment in Cases of Impeachment shall not extend further than to removal from Office, and disqualification to hold and enjoy any Office of honor, Trust or Profit under the United States: but the Party convicted shall nevertheless be liable and subject to Indictment, Trial, Judgment and Punishment, according to Law.

Section 4

The Times, Places and Manner of holding Elections for Senators and Representatives, shall be prescribed in each State by the Legislature thereof; but the Congress may at any time by Law make or alter such Regulations, except as to the Place of Chusing Senators.

The Congress shall assemble at least once in every Year, and such Meeting shall be on the first Monday in December, unless they shall by Law appoint a different Day.

Section 5

Each House shall be the Judge of the Elections, Returns and Qualifications of its own Members, and a Majority of each shall constitute a Quorum to do Business; but a smaller number may adjourn from day to day, and may be authorized to compel the Attendance of absent Members, in such Manner, and under such Penalties as each House may provide.

Each House may determine the Rules of its Proceedings, punish its Members for disorderly Behavior, and, with the Concurrence of two-thirds, expel a Member.

Each House shall keep a Journal of its Proceedings, and from time to time publish the same, excepting such Parts as may in their Judgment require Secrecy; and the Yeas and Nays of the Members of either House on any question shall, at the Desire of one fifth of those Present, be entered on the Journal.

Neither House, during the Session of Congress, shall, without the Consent of the other, adjourn for more than three days, nor to any other Place than that in which the two Houses shall be sitting.

Section 6

The Senators and Representatives shall receive a Compensation for their Services, to be ascertained by Law, and paid out of the Treasury of the United States. They shall in all Cases, except Treason, Felony and Breach of the Peace, be privileged from Arrest during their Attendance at the Session of their respective Houses, and in going to and returning from the same; and for any Speech or Debate in either House, they shall not be questioned in any other Place.

No Senator or Representative shall, during the Time for which he was elected, be appointed to any civil Office under the Authority of the United States which shall have been created, or the Emoluments whereof shall have been increased during such time; and no Person holding any Office under the United States, shall be a Member of either House during his Continuance in Office.

Section 7

All bills for raising Revenue shall originate in the House of Representatives; but the Senate may propose or concur with Amendments as on other Bills.

Every Bill which shall have passed the House of Representatives and the Senate, shall, before it become a Law, be presented to the President of the United States; If he approve he shall sign it, but if not he shall return it, with his Objections to that House in which it shall have originated, who shall enter the Objections at large on their Journal, and proceed to reconsider it. If after such Reconsideration two thirds of that House shall agree to pass the Bill, it shall be sent, together with the Objections, to the other House, by which it shall likewise be reconsidered, and if approved by two thirds of that House, it shall become a Law. But in all such Cases the Votes of both Houses shall be determined by Yeas and Nays, and the Names of the Persons voting for and against the Bill shall be entered on the Journal of each House respectively. If any Bill shall not be returned by the President within ten Days (Sundays excepted) after it shall have been presented to him, the Same shall be a Law, in like Manner as if he had signed it, unless the Congress by their Adjournment prevent its Return, in which Case it shall not be a Law.

Every Order, Resolution, or Vote to which the Concurrence of the Senate and House of Representatives may be necessary (except on a question of Adjournment) shall be presented to the President of the United States; and before the Same shall take Effect, shall be approved by him, or being disapproved by him, shall be repassed by two thirds of the Senate and House of Representatives, according to the Rules and Limitations prescribed in the Case of a Bill.

Section 8

The Congress shall have Power To lay and collect Taxes, Duties, Imposts and Excises, to pay the Debts and provide for the common Defence and general Welfare of the United States; but all Duties, Imposts and Excises shall be uniform throughout the United States;

To borrow money on the credit of the United States;

To regulate Commerce with foreign Nations, and among the several States, and with the Indian Tribes;

To establish an uniform Rule of Naturalization, and uniform Laws on the subject of Bankruptcies throughout the United States;

To coin Money, regulate the Value thereof, and of foreign Coin, and fix the Standard of Weights and Measures;

To provide for the Punishment of counterfeiting the Securities and current Coin of the United States;

To establish Post Offices and Post Roads;

To promote the Progress of Science and useful Arts, by securing for limited Times to Authors and Inventors the exclusive Right to their respective Writings and Discoveries;

To constitute Tribunals inferior to the supreme Court;

To define and punish Piracies and Felonies committed on the high Seas, and Offenses against the Law of Nations;

To declare War, grant Letters of Marque and Reprisal, and make Rules concerning Captures on Land and Water;

To raise and support Armies, but no Appropriation of Money to that Use shall be for a longer Term than two Years;

To provide and maintain a Navy;

To make Rules for the Government and Regulation of the land and naval Forces;

To provide for calling forth the Militia to execute the Laws of the Union, suppress Insurrections and repel Invasions;

To provide for organizing, arming, and disciplining the Militia, and for governing such Part of them as may be employed in the Service of the United States, reserving to the States respectively, the Appointment of the Officers, and

the Authority of training the Militia according to the discipline prescribed by Congress;

To exercise exclusive Legislation in all Cases whatsoever, over such District (not exceeding ten Miles square) as may, by Cession of particular States, and the acceptance of Congress, become the Seat of the Government of the United States, and to exercise like Authority over all Places purchased by the Consent of the Legislature of the State in which the Same shall be, for the Erection of Forts, Magazines, Arsenals, dock-Yards, and other needful Buildings; And

To make all Laws which shall be necessary and proper for carrying into Execution the foregoing Powers, and all other Powers vested by this Constitution in the Government of the United States, or in any Department or Officer thereof.

Section 9

The Migration or Importation of such Persons as any of the States now existing shall think proper to admit, shall not be prohibited by the Congress prior to the Year one thousand eight hundred and eight, but a tax or duty may be imposed on such Importation, not exceeding ten dollars for each Person.

The privilege of the Writ of Habeas Corpus shall not be suspended, unless when in Cases of Rebellion or Invasion the public Safety may require it.

No Bill of Attainder or ex post facto Law shall be passed.

No capitation, or other direct, Tax shall be laid, unless in Proportion to the Census or Enumeration herein before directed to be taken.

No Tax or Duty shall be laid on Articles exported from any State.

No Preference shall be given by any Regulation of Commerce or Revenue to the Ports of one State over those of another: nor shall Vessels bound to, or from, one State, be obliged to enter, clear, or pay Duties in another.

No Money shall be drawn from the Treasury, but in Consequence of Appropriations made by Law; and a regular Statement and Account of the Receipts and Expenditures of all public Money shall be published from time to time.

No Title of Nobility shall be granted by the United States: And no Person holding any Office of Profit or Trust under them, shall, without the Consent of the Congress, accept of any present, Emolument, Office, or Title, of any kind whatever, from any King, Prince or foreign State.

Section 10

No State shall enter into any Treaty, Alliance, or Confederation; grant Letters of Marque and Reprisal; coin Money; emit Bills of Credit; make any Thing but gold and silver Coin a Tender in Payment of Debts; pass any Bill of Attainder, ex post facto Law, or Law impairing the Obligation of Contracts, or grant any Title of Nobility.

No State shall, without the Consent of the Congress, lay any Imposts or Duties on Imports or Exports, except what may be absolutely necessary for executing it's inspection Laws: and the net Produce of all Duties and Imposts, laid by any State on Imports or Exports, shall be for the Use of the Treasury of the United States; and all such Laws shall be subject to the Revision and Controul of the Congress.

No State shall, without the Consent of Congress, lay any duty of Tonnage, keep Troops, or Ships of War in time of Peace, enter into any Agreement or Compact with another State, or with a foreign Power, or engage in War, unless actually invaded, or in such imminent Danger as will not admit of delay.

Article II.

Section 1

The executive Power shall be vested in a President of the United States of America. He shall hold his Office during the Term of four Years, and, together with the Vice-President chosen for the same Term, be elected, as follows:

Each State shall appoint, in such Manner as the Legislature thereof may direct, a Number of Electors, equal to the whole Number of Senators and Representatives to which the State may be entitled in the Congress: but no Senator or Representative, or Person holding an Office of Trust or Profit under the United States, shall be appointed an Elector.

The Electors shall meet in their respective States, and vote by Ballot for two persons, of whom one at least shall not lie an Inhabitant of the same State with themselves. And they shall make a List of all the Persons voted for, and of the Number of Votes for each; which List they shall sign and certify, and transmit sealed to the Seat of the Government of the United States, directed to the President of the Senate. The President of the Senate shall, in the Presence of the Senate and House of Representatives, open all the Certificates, and the Votes shall then be counted. The Person having the greatest Number of Votes shall be the President, if such Number be a Majority of the whole Number of Electors appointed; and if there be more than one who have such Majority, and have an equal Number of Votes, then the House of Representatives shall immediately chuse by Ballot one of them for President; and if no Person have a Majority, then from the five highest on the List the said House shall in like Manner chuse the President. But in chusing the President, the Votes shall be taken by States, the Representation from each State having one Vote; a quorum for this Purpose shall consist of a Member or Members from two-thirds of the States, and a Majority of all the States shall be necessary to a Choice. In every Case, after the Choice of the President, the Person having the greatest Number of Votes of the Electors shall be the Vice President. But if there should remain two or more who have equal Votes, the Senate shall chuse from them by Ballot the Vice-President.

The Congress may determine the Time of chusing the Electors, and the Day on which they shall give their Votes; which Day shall be the same throughout the United States.

No person except a natural born Citizen, or a Citizen of the United States, at the time of the Adoption of this Constitution, shall be eligible to the Office of President; neither shall any Person be eligible to that Office who shall not have attained to the Age of thirty-five Years, and been fourteen Years a Resident within the United States.

In Case of the Removal of the President from Office, or of his Death, Resignation, or Inability to discharge the Powers and Duties of the said Office, the same shall devolve on the Vice President, and the Congress may by Law provide for the Case of Removal, Death, Resignation or Inability, both of the President and Vice President, declaring what Officer shall then act as President, and such Officer shall act accordingly, until the Disability be removed, or a President shall be elected.

The President shall, at stated Times, receive for his Services, a Compensation, which shall neither be increased nor diminished during the Period for which he shall have been elected, and he shall not receive within that Period any other Emolument from the United States, or any of them.

Before he enter on the Execution of his Office, he shall take the following Oath or Affirmation:

"I do solemnly swear (or affirm) that I will faithfully execute the Office of President of the United States, and will to the best of my Ability, preserve, protect and defend the Constitution of the United States."

Section 2

The President shall be Commander in Chief of the Army and Navy of the United States, and of the Militia of the several States, when called into the actual Service of the United States; he may require the Opinion, in writing, of the principal Officer in each of the executive Departments, upon any subject relating to the Duties of their respective Offices, and he shall have Power to Grant Reprieves and Pardons for Offenses against the United States, except in Cases of Impeachment.

He shall have Power, by and with the Advice and Consent of the Senate, to make Treaties, provided two thirds of the Senators present concur; and he shall nominate, and by and with the Advice and Consent of the Senate, shall appoint Ambassadors, other public Ministers and Consuls, Judges of the supreme Court, and all other Officers of the United States, whose Appointments are not herein otherwise provided for, and which shall be established by Law: but the Congress may by Law vest the Appointment of such inferior Officers, as they think proper, in the President alone, in the Courts of Law, or in the Heads of Departments.

The President shall have Power to fill up all Vacancies that may happen during the Recess of the Senate, by granting Commissions which shall expire at the End of their next Session.

Section 3

He shall from time to time give to the Congress Information of the State of the Union, and recommend to their Consideration such Measures as he shall judge necessary and expedient; he may, on extraordinary Occasions, convene both Houses, or either of them, and in Case of Disagreement between them, with Respect to the Time of Adjournment, he may adjourn them to such Time as he shall think proper; he shall receive Ambassadors and other public Ministers; he shall take Care that the Laws be faithfully executed, and shall Commission all the Officers of the United States.

Section 4

The President, Vice President and all civil Officers of the United States, shall be removed from

Office on Impeachment for, and Conviction of, Treason, Bribery, or other high Crimes and Misdemeanors.

Article III.

Section 1
The judicial Power of the United States, shall be vested in one supreme Court, and in such inferior Courts as the Congress may from time to time ordain and establish. The Judges, both of the supreme and inferior Courts, shall hold their Offices during good Behavior, and shall, at stated Times, receive for their Services a Compensation which shall not be diminished during their Continuance in Office.

Section 2
The judicial Power shall extend to all Cases, in Law and Equity, arising under this Constitution, the Laws of the United States, and Treaties made, or which shall be made, under their Authority; to all Cases affecting Ambassadors, other public Ministers and Consuls; to all Cases of admiralty and maritime Jurisdiction; to Controversies to which the United States shall be a Party; to Controversies between two or more States; between a State and Citizens of another State; between Citizens of different States; between Citizens of the same State claiming Lands under Grants of different States, and between a State, or the Citizens thereof, and foreign States, Citizens or Subjects.

In all Cases affecting Ambassadors, other public Ministers and Consuls, and those in which a State shall be Party, the supreme Court shall have original Jurisdiction. In all the other Cases before mentioned, the supreme Court shall have appellate Jurisdiction, both as to Law and Fact, with such Exceptions, and under such Regulations as the Congress shall make.

The Trial of all Crimes, except in Cases of Impeachment, shall be by Jury; and such Trial shall be held in the State where the said Crimes shall have been committed; but when not committed within any State, the Trial shall be at such Place or Places as the Congress may by Law have directed.

Section 3
Treason against the United States, shall consist only in levying War against them, or in adhering to their Enemies, giving them Aid and Comfort. No Person shall be convicted of Treason unless on the Testimony of two Witnesses to the same overt Act, or on Confession in open Court.

The Congress shall have power to declare the Punishment of Treason, but no Attainder of Treason shall work Corruption of Blood, or Forfeiture except during the Life of the Person attainted.

Article IV.

Section 1
Full Faith and Credit shall be given in each State to the public Acts, Records, and judicial Proceedings of every other State. And the Congress may by general Laws prescribe the Manner in which such Acts, Records and Proceedings shall be proved, and the Effect thereof.

Section 2
The Citizens of each State shall be entitled to all Privileges and Immunities of Citizens in the several States.

A Person charged in any State with Treason, Felony, or other Crime, who shall flee from Justice, and be found in another State, shall on demand of the executive Authority of the State from which he fled, be delivered up, to be removed to the State having Jurisdiction of the Crime.

No Person held to Service or Labour in one State, under the Laws thereof, escaping into another, shall, in Consequence of any Law or Regulation therein, be discharged from such

Service or Labour, But shall be delivered up on Claim of the Party to whom such Service or Labour may be due.

Section 3

New States may be admitted by the Congress into this Union; but no new States shall be formed or erected within the Jurisdiction of any other State; nor any State be formed by the Junction of two or more States, or parts of States, without the Consent of the Legislatures of the States concerned as well as of the Congress.

The Congress shall have Power to dispose of and make all needful Rules and Regulations respecting the Territory or other Property belonging to the United States; and nothing in this Constitution shall be so construed as to Prejudice any Claims of the United States, or of any particular State.

Section 4

The United States shall guarantee to every State in this Union a Republican Form of Government, and shall protect each of them against Invasion; and on Application of the Legislature, or of the Executive (when the Legislature cannot be convened) against domestic Violence.

Article V.

The Congress, whenever two thirds of both Houses shall deem it necessary, shall propose Amendments to this Constitution, or, on the Application of the Legislatures of two thirds of the several States, shall call a Convention for proposing Amendments, which, in either Case, shall be valid to all Intents and Purposes, as part of this Constitution, when ratified by the Legislatures of three fourths of the several States, or by Conventions in three fourths thereof, as the one or the other Mode of Ratification may be proposed by the Congress;

Provided that no Amendment which may be made prior to the Year One thousand eight hundred and eight shall in any Manner affect the first and fourth Clauses in the Ninth Section of the first Article; and that no State, without its Consent, shall be deprived of its equal Suffrage in the Senate.

Article VI.

All Debts contracted and Engagements entered into, before the Adoption of this Constitution, shall be as valid against the United States under this Constitution, as under the Confederation.

This Constitution, and the Laws of the United States which shall be made in Pursuance thereof; and all Treaties made, or which shall be made, under the Authority of United States, shall be the supreme Law of the Land; and the Judges in every State shall be bound thereby, any Thing in the Constitution or Laws of any State to the Contrary notwithstanding.

The Senators and Representatives before mentioned, and the Members of the several State Legislatures, and all executive and judicial Officers, both of the United States and of the several States, shall be bound by Oath or Affirmation, to support this Constitution; but no religious Test shall ever be required as a Qualification to any Office or public Trust under the United States.

Article VII.

The Ratification of the Conventions of nine States, shall be sufficient for the Establishment of this Constitution between the States so ratifying the Same.

Done in Convention by the Unanimous Consent of the States present the Seventeenth Day of September in the Year of our Lord one thousand seven hundred and Eighty seven and of the Independence of the United States of America the Twelfth. In Witness whereof We have hereunto subscribed our Names.

Go Washington - President and deputy from Virginia

New Hampshire - John Langdon, Nicholas Gilman

Massachusetts - Nathaniel Gorham, Rufus King

Connecticut - Wm Saml Johnson, Roger Sherman

New York - Alexander Hamilton

New Jersey - Wil Livingston, David Brearley, Wm Paterson, Jona. Dayton

Pensylvania - B Franklin, Thomas Mifflin, Robt Morris, Geo. Clymer, Thos FitzSimons, Jared Ingersoll, James Wilson, Gouv Morris

Delaware - Geo. Read, Gunning Bedford jun, John Dickinson, Richard Bassett, Jaco. Broom

Maryland - James McHenry, Dan of St Tho Jenifer, Danl Carroll

Virginia - John Blair, James Madison Jr.

North Carolina - Wm Blount, Richd Dobbs Spaight, Hu Williamson

South Carolina - J. Rutledge, Charles Cotesworth Pinckney, Charles Pinckney, Pierce Butler

Georgia - William Few, Abr Baldwin

Attest: William Jackson, Secretary

Amendment I
Congress shall make no law respecting an establishment of religion, or prohibiting the free exercise thereof; or abridging the freedom of speech, or of the press; or the right of the people peaceably to assemble, and to petition the Government for a redress of grievances.

Amendment II
A well regulated Militia, being necessary to the security of a free State, the right of the people to keep and bear Arms, shall not be infringed.

Amendment III
No Soldier shall, in time of peace be quartered in any house, without the consent of the Owner, nor in time of war, but in a manner to be prescribed by law.

Amendment IV
The right of the people to be secure in their persons, houses, papers, and effects, against unreasonable searches and seizures, shall not be violated, and no Warrants shall issue, but upon probable cause, supported by Oath or affirmation, and particularly describing the place to be searched, and the persons or things to be seized.

Amendment V
No person shall be held to answer for a capital, or otherwise infamous crime, unless on a presentment or indictment of a Grand Jury, except in cases arising in the land or naval forces, or in the Militia, when in actual service in time of War or public danger; nor shall any person be subject for the same offense to be twice put in jeopardy of life or limb; nor shall be compelled in any criminal case to be a witness against himself, nor be deprived of life, liberty, or property, without due process of law; nor shall private property be taken for public use, without just compensation.

Amendment VI
In all criminal prosecutions, the accused shall enjoy the right to a speedy and public trial, by an impartial jury of the State and district wherein the crime shall have been committed, which district shall have been previously ascertained by law, and to be informed of the

nature and cause of the accusation; to be confronted with the witnesses against him; to have compulsory process for obtaining witnesses in his favor, and to have the Assistance of Counsel for his defence.

Amendment VII

In Suits at common law, where the value in controversy shall exceed twenty dollars, the right of trial by jury shall be preserved, and no fact tried by a jury, shall be otherwise re-examined in any Court of the United States, than according to the rules of the common law.

Amendment VIII

Excessive bail shall not be required, nor excessive fines imposed, nor cruel and unusual punishments inflicted.

Amendment IX

The enumeration in the Constitution, of certain rights, shall not be construed to deny or disparage others retained by the people.

Amendment X

The powers not delegated to the United States by the Constitution, nor prohibited by it to the States, are reserved to the States respectively, or to the people.

Amendment XI

The Judicial power of the United States shall not be construed to extend to any suit in law or equity, commenced or prosecuted against one of the United States by Citizens of another State, or by Citizens or Subjects of any Foreign State.

Amendment XII

The Electors shall meet in their respective states, and vote by ballot for President and Vice-President, one of whom, at least, shall not be an inhabitant of the same state with themselves; they shall name in their ballots the person voted for as President, and in distinct ballots the person voted for as Vice-President, and they shall make distinct lists of all persons voted for as President, and of all persons voted for as Vice-President and of the number of votes for each, which lists they shall sign and certify, and transmit sealed to the seat of the government of the United States, directed to the President of the Senate;

The President of the Senate shall, in the presence of the Senate and House of Representatives, open all the certificates and the votes shall then be counted;

The person having the greatest Number of votes for President, shall be the President, if such number be a majority of the whole number of Electors appointed; and if no person have such majority, then from the persons having the highest numbers not exceeding three on the list of those voted for as President, the House of Representatives shall choose immediately, by ballot, the President. But in choosing the President, the votes shall be taken by states, the representation from each state having one vote; a quorum for this purpose shall consist of a member or members from two-thirds of the states, and a majority of all the states shall be necessary to a choice. And if the House of Representatives shall not choose a President whenever the right of choice shall devolve upon them, before the fourth day of March next following, then the Vice-President shall act as President, as in the case of the death or other constitutional disability of the President.

The person having the greatest number of votes as Vice-President, shall be the Vice-President, if such number be a majority of the whole number of Electors appointed, and if no person have a majority, then from the two highest numbers on the list, the Senate shall choose the Vice-President; a quorum for the purpose shall consist of two-thirds of the whole number of Senators, and a majority of the whole number shall be necessary to a choice. But no person constitutionally ineligible to the office of President shall be eligible to that of Vice-President of the United States.

Amendment XIII

1. Neither slavery nor involuntary servitude, except as a punishment for crime whereof the party shall have been duly convicted, shall exist within the United States, or any place subject to their jurisdiction.

2. Congress shall have power to enforce this article by appropriate legislation.

Amendment XIV

1. All persons born or naturalized in the United States, and subject to the jurisdiction thereof, are citizens of the United States and of the State wherein they reside. No State shall make or enforce any law which shall abridge the privileges or immunities of citizens of the United States; nor shall any State deprive any person of life, liberty, or property, without due process of law; nor deny to any person within its jurisdiction the equal protection of the laws.

2. Representatives shall be apportioned among the several States according to their respective numbers, counting the whole number of persons in each State, excluding Indians not taxed. But when the right to vote at any election for the choice of electors for President and Vice-President of the United States, Representatives in Congress, the Executive and Judicial officers of a State, or the members of the Legislature thereof, is denied to any of the male inhabitants of such State, being twenty-one years of age, and citizens of the United States, or in any way abridged, except for participation in rebellion, or other crime, the basis of representation therein shall be reduced in the proportion which the number of such male citizens shall bear to the whole number of male citizens twenty-one years of age in such State.

3. No person shall be a Senator or Representative in Congress, or elector of President and Vice-President, or hold any office, civil or military, under the United States, or under any State, who, having previously taken an oath, as a member of Congress, or as an officer of the United States, or as a member of any State legislature, or as an executive or judicial officer of any State, to support the Constitution of the United States, shall have engaged in insurrection or rebellion against the same, or given aid or comfort to the enemies thereof. But Congress may by a vote of two-thirds of each House, remove such disability.

4. The validity of the public debt of the United States, authorized by law, including debts incurred for payment of pensions and bounties for services in suppressing insurrection or rebellion, shall not be questioned. But neither the United States nor any State shall assume or pay any debt or obligation incurred in aid of insurrection or rebellion against the United States, or any claim for the loss or emancipation of any slave; but all such debts, obligations and claims shall be held illegal and void.

5. The Congress shall have power to enforce, by appropriate legislation, the provisions of this article.

Amendment XV

1. The right of citizens of the United States to vote shall not be denied or abridged by the United States or by any State on account of race, color, or previous condition of servitude.

2. The Congress shall have power to enforce this article by appropriate legislation.

Amendment XVI

The Congress shall have power to lay and collect taxes on incomes, from whatever source derived, without apportionment among the several States, and without regard to any census or enumeration.

Amendment XVII

The Senate of the United States shall be composed of two Senators from each State, elected by the people thereof, for six years; and each Senator shall have one vote. The electors in each State shall have the qualifications requisite for electors of the most numerous branch of the State legislatures.

When vacancies happen in the representation of any State in the Senate, the executive authority of such State shall issue writs of election to fill such vacancies: Provided, That the legislature of any State may empower the executive thereof to make temporary appointments until the people fill the vacancies by election as the legislature may direct.

This amendment shall not be so construed as to affect the election or term of any Senator chosen before it becomes valid as part of the Constitution.

Amendment XVIII

1. After one year from the ratification of this article the manufacture, sale, or transportation of intoxicating liquors within, the importation thereof into, or the exportation thereof from the United States and all territory subject to the jurisdiction thereof for beverage purposes is hereby prohibited.

2. The Congress and the several States shall have concurrent power to enforce this article by appropriate legislation.

3. This article shall be inoperative unless it shall have been ratified as an amendment to the Constitution by the legislatures of the several States, as provided in the Constitution, within seven years from the date of the submission hereof to the States by the Congress.

Amendment XIX

The right of citizens of the United States to vote shall not be denied or abridged by the United States or by any State on account of sex.

Congress shall have power to enforce this article by appropriate legislation.

Amendment XX

1. The terms of the President and Vice President shall end at noon on the 20th day of January, and the terms of Senators and Representatives at noon on the 3d day of January, of the years in which such terms would have ended if this article had not been ratified; and the terms of their successors shall then begin.

2. The Congress shall assemble at least once in every year, and such meeting shall begin at noon on the 3d day of January, unless they shall by law appoint a different day.

3. If, at the time fixed for the beginning of the term of the President, the President elect shall have died, the Vice President elect shall become President. If a President shall not have been chosen before the time fixed for the beginning of his term, or if the President elect shall have failed to qualify, then the Vice President elect shall act as President until a President shall have qualified; and the Congress may by law provide for the case wherein neither a President elect nor a Vice President elect shall have qualified, declaring who shall then act as President, or the manner in which one who is to act shall be selected, and such person shall act accordingly until a President or Vice President shall have qualified.

4. The Congress may by law provide for the case of the death of any of the persons from whom the House of Representatives may choose a President whenever the right of choice shall have devolved upon them, and for the case of the death of any of the persons from whom the Senate may choose a Vice President whenever the right of choice shall have devolved upon them.

5. Sections 1 and 2 shall take effect on the 15th day of October following the ratification of this article.

6. This article shall be inoperative unless it shall have been ratified as an amendment to the

Constitution by the legislatures of three-fourths of the several States within seven years from the date of its submission.

Amendment XXI

1. The eighteenth article of amendment to the Constitution of the United States is hereby repealed.

    2. The transportation or importation into any State, Territory, or possession of the United States for delivery or use therein of intoxicating liquors, in violation of the laws thereof, is hereby prohibited.

    3. The article shall be inoperative unless it shall have been ratified as an amendment to the Constitution by conventions in the several States, as provided in the Constitution, within seven years from the date of the submission hereof to the States by the Congress.

Amendment XXII

1. No person shall be elected to the office of the President more than twice, and no person who has held the office of President, or acted as President, for more than two years of a term to which some other person was elected President shall be elected to the office of the President more than once. But this Article shall not apply to any person holding the office of President, when this Article was proposed by the Congress, and shall not prevent any person who may be holding the office of President, or acting as President, during the term within which this Article becomes operative from holding the office of President or acting as President during the remainder of such term.

    2. This article shall be inoperative unless it shall have been ratified as an amendment to the Constitution by the legislatures of three-fourths of the several States within seven years from the date of its submission to the States by the Congress.

Amendment XXIII

1. The District constituting the seat of Government of the United States shall appoint in such manner as the Congress may direct: A number of electors of President and Vice President equal to the whole number of Senators and Representatives in Congress to which the District would be entitled if it were a State, but in no event more than the least populous State; they shall be in addition to those appointed by the States, but they shall be considered, for the purposes of the election of President and Vice President, to be electors appointed by a State; and they shall meet in the District and perform such duties as provided by the twelfth article of amendment.

    2. The Congress shall have power to enforce this article by appropriate legislation.

Amendment XXIV

1. The right of citizens of the United States to vote in any primary or other election for President or Vice President, for electors for President or Vice President, or for Senator or Representative in Congress, shall not be denied or abridged by the United States or any State by reason of failure to pay any poll tax or other tax.

    2. The Congress shall have power to enforce this article by appropriate legislation.

Amendment XXV

1. In case of the removal of the President from office or of his death or resignation, the Vice President shall become President.

    2. Whenever there is a vacancy in the office of the Vice President, the President shall nominate a Vice President who shall take office upon confirmation by a majority vote of both Houses of Congress.

    3. Whenever the President transmits to the President pro tempore of the Senate and the Speaker of the House of Representatives his written declaration that he is unable to discharge

the powers and duties of his office, and until he transmits to them a written declaration to the contrary, such powers and duties shall be discharged by the Vice President as Acting President.

4. Whenever the Vice President and a majority of either the principal officers of the executive departments or of such other body as Congress may by law provide, transmit to the President pro tempore of the Senate and the Speaker of the House of Representatives their written declaration that the President is unable to discharge the powers and duties of his office, the Vice President shall immediately assume the powers and duties of the office as Acting President.

Thereafter, when the President transmits to the President pro tempore of the Senate and the Speaker of the House of Representatives his written declaration that no inability exists, he shall resume the powers and duties of his office unless the Vice President and a majority of either the principal officers of the executive department or of such other body as Congress may by law provide, transmit within four days to the President pro tempore of the Senate and the Speaker of the House of Representatives their written declaration that the President is unable to discharge the powers and duties of

his office. Thereupon Congress shall decide the issue, assembling within forty eight hours for that purpose if not in session. If the Congress, within twenty one days after receipt of the latter written declaration, or, if Congress is not in session, within twenty one days after Congress is required to assemble, determines by two thirds vote of both Houses that the President is unable to discharge the powers and duties of his office, the Vice President shall continue to discharge the same as Acting President; otherwise, the President shall resume the powers and duties of his office.

Amendment XXVI

1. The right of citizens of the United States, who are eighteen years of age or older, to vote shall not be denied or abridged by the United States or by any State on account of age.

2. The Congress shall have power to enforce this article by appropriate legislation.

Amendment XXVII

No law, varying the compensation for the services of the Senators and Representatives, shall take effect, until an election of Representatives shall have intervened.

Appendix

B

CHAPTER 1

CASE CITATION ABBREVIATIONS

Abbreviation	Book	Court
U.S.	United States Reports	U.S. Supreme Court
S.Ct.	Supreme Court Reporter	U.S. Supreme Court
L.Ed.	Lawyer's Edition	U.S. Supreme Court
F. or F. 2d or F. 3d	Federal Reporter	U.S. Ct of Appeals
F.Supp.	Federal Supplement	U.S. District Court
A. or A.2d	Atlantic Reporter	Regional compilation of state cases
N.E. or N.E. 2d	Northeast Reporter,	Regional compilation of state cases
N.W. or N.W. 2d	Northwest Reporter	Regional compilation of state cases
P. or P. 2d	Pacific Reporter	Regional compilation of state cases
S.E. or S.E. 2d	Souththeast Reporter,	Regional compilation of state cases
So or So. 2d	Southern Reporter	Regional compilation of state cases
S.W. or S.W. 2d	Southwest Reporter	Regional compilation of state cases
N.Y. or N.Y. 2d	New York Supplement	Compilation of N.Y. state reports
Cal. Rptr.	California Reporter	Compilation of Cal. state reports
Mo.	Missouri Reports	Missouri Supreme Court
Mo. App.	Missouri Appeal Reports	Missouri Appellate Courts
R.I.	Rhode Island Reporter	Rhode Island Supreme Court
N.M.	New Mexico Reporter	New Mexico Appellate Courts
Ohio St.	Ohio State Reporter	Ohio Appellate Courts
Ill.	Illinois Reporter	Illinois Appellate Courts
Ariz.	Arizona Reporter	Arizona Appellate Courts
Wis. or Wis. 2d	Wisconsin Reporter	Wisconsin Appellate Court

EXAMPLE OF A STATE LAW ADOPTING A STANDARD INTO LAW

Rhode Island General Laws

§ 23-28.2-21 National Fire Code. Except wherever herein specifically defined or covered in this code, the provisions of the N.F.P.A. Standards included in the National Fire Code, 2003 edition, shall be used by the authority having jurisdiction as the accepted standard with regard to fire safety regarding any unforeseen condition.

CHAPTER 2

SAMPLE COMPLAINT

STATE OF _____

County of _____

Civil Action, File Number _____

Plaintiff
v.
Defendant

COMPLAINT

Now comes the Plaintiff and states his complaint as follows:

1. On July 1, 2004, at approximately 10:00 am, Defendant operated a fire truck west on Smith Street in Providence, Rhode Island.
2. Only July 1, 2004, at approximately 10:00 am Plaintiff was a passenger in a motor vehicle traveling south on River Avenue in Providence, Rhode Island.
3. The fire truck operated by Defendant and the vehicle in which Plaintiff was a passenger in, collided at the intersection of Smith Street and River Avenue.
4. This accident was occasioned by Defendant's negligence in failing to stop at the red traffic signal, as required by law, and as a direct result Plaintiff suffered serious bodily injury.

Wherefore Plaintiff demands judgment against Defendant for the sum of _____ dollars, plus interest and costs.

Signed: _____

Attorney for Plaintiff Address: _____

Date: _____

SAMPLE ANSWER

STATE OF _____

County of _____

Civil Action, File Number _____

Plaintiff
v.
Defendant

ANSWER

Now comes the Defendant and answers Plaintiff's Complaint as follows:

1. Admitted
2. Admitted
3. Admitted
4. Denied

Wherefore Defendant demands judgment against Plaintiff, and asks this honorable court to deny and dismiss Plaintiff's complaint, plus award defendant costs.

SAMPLE SUMMONS

STATE OF _____

County of _____

Civil Action, File Number _____

Plaintiff
v.
Defendant

To the Above-Named Defendant:

DISTRICT COURT

The above-named plaintiff has brought an action against you in said District Court for the _____ Division sitting at _____. You are hereby summoned and required to serve upon _____, plaintiff's attorney, whose address is _____, an answer to the complaint which is herewith served upon you, not later than 20 days after service of this summons upon you. If you fail to do so, judgment by default will be taken against you for the relief demanded in the complaint. Your answer must also be filed with the court.

Signed:

Clerk of said court

SAMPLE MOTION TO DISMISS

STATE OF _____

County of _____

Civil Action, File Number _____

Plaintiff

v.

Defendant

MOTION TO DISMISS

Defendant moves this Honorable Court to deny and dismiss Plaintiff's Complaint, and for his reason states as follows:

    1. Plaintiff's complaint fails to state a claim upon which relief may be granted.

SAMPLE INTERROGATORIES

STATE OF _____

County of _____

Civil Action, File Number _____

Plaintiff

v.

Defendant

PLAINTIFF'S INTERROGATORIES PROPOUNDED TO DEFENDANT

Now comes the Plaintiff and pursuant to Rule 33 of the Rules of Civil Procedure propounds the following interrogatories to be answered under oath by the Defendant within forty days:

    1. Please state your name and any other names, nicknames or aliases which you have ever used.

    2. Please identify all motor vehicle accidents which you have ever been involved in as a driver. For each accident, please provide the date, time, location, and describe how the accident happened.

3. Please list each and every traffic violation, ticket, or traffic offense which you have been convicted, pled guilty or nolo to, or otherwise have been fined or penalized. For each please state the date, time, location, police department involved, and the exact charge or charges.

4. Please identify the location from which you were coming, and the location to which you were traveling on July 1, 2004, at approximately 10:00 am, when the accident which is the subject of this law suit occurred.

5. Please list any and all prescription medications, drugs, pills, tablets, and medicines; non-prescription medications, drugs, pills, tablets and medicines; and alcoholic beverages, liquors, beer or wine that you had taken within twenty-four hours prior to 10:00 am on July 1, 2004. Please include the quantities/dosages taken and the time and location at which such items were taken or consumed.

6. Please state whether or not you were wearing prescription eyeglasses at the time of the accident on July 1, 2004, and if so please state:

a. The location where the eyeglasses may be examined

b. The name and address of the doctor who prescribed the eyeglasses for you

c. The date of the prescription

d. The locations of the eyeglasses at the present time

e. The date of the last eye examination which you had prior to July 1, 2004.

7. Please describe in detail the facts as to how the accident occurred giving all details or events before, at the time of, and after the accident which you consider to have any bearing on the accident.

SAMPLE MOTION FOR SUMMARY JUDGMENT

STATE OF _____

County of _____

Civil Action, File Number _____

Plaintiff
v.
Defendant

MOTION FOR SUMMARY JUDGMENT

Defendant moves this Honorable Court to order summary judgment for the Defendant, and for his reason states as follows:

1. Defendant was operating an emergency vehicle pursuant to State General Laws, responding to an alarm of fire at the time of this accident.

2. Defendant was operating said emergency vehicle with due regard for the safety of other motorists, and as such Defendant is immune from liability for Plaintiff's injuries as a matter of law. For this reason, Summary Judgement should be entered for the Defendant.

CHAPTER 3

July 20, 1999

The Honorable Delma Rios
Kleberg County Attorney
P.O. Box 1411
Kingsville, Texas 78364

OPINION NO. JC-0082

Re: Whether a commissioners court may purchase fire-fighting equipment for or pay a volunteer fire department without having contracted with the volunteer fire department to provide fire-protection services for county residents

Dear Ms. Rios:

You ask whether a county commissioners court may purchase fire-fighting equipment for or pay a volunteer fire department in the absence of a contract between the two entities under which the volunteer fire department provides fire-protection services for county residents as consideration for the equipment or payment. We conclude the county commissioners court may not.

You indicate that the Kleberg County Commissioners Court has, for several years, "supplemented [the] budget[s]" of the Ricardo and Riviera volunteer fire departments. Letter from Honorable Delma Rios, Kleberg County Attorney, to Honorable Dan Morales, Texas Attorney General (Apr. 6, 1998) (on file with Opinion Committee) [hereinafter "Request Letter"]. This year, you state that the commissioners court has budgeted $25,000 for the Ricardo organization and $20,000 for the Riviera organization. *Id.* As you describe the arrangement, the commissioners court does not simply hand the budgeted funds to the volunteer fire departments; rather, the volunteer fire departments submit purchase orders, up to the budgeted limits, to the commissioners court for its approval. *Id.* The county pays the expenditures out of its general fund, you continue, and the county auditor annually audits the volunteer fire departments' accounts. *Id.* We assume, although you do not inform us, that the county supports these volunteer fire departments because they provide fire-protection services to county residents living outside the municipal limits of Ricardo or Riviera.

Section 352.001 of the Local Government Code authorizes a county commissioners court to provide, either on its own or under a contract, fire protection to county residents who live outside a municipality (to whom we will refer simply as "county residents"):

(a) The commissioners court of a county may furnish fire protection or fire-fighting equipment to the residents of the county . . . who live outside municipalities.

(b) The commissioners court may:

(1) purchase fire trucks or other fire-fighting equipment;
 . . . and

(3) contract with the governing body of a municipality located within the county or within an adjoining county to use fire trucks or other fire-fighting equipment that belongs to the municipality.

(c) The commissioners court of a county may contract with an incorporated volunteer fire department that is located within the county to provide fire protection to an area of the county that is located outside the municipalities in the county. The court may pay for that protection from the general fund of the county.

Tex. Local Gov't Code Ann. § 352.001 (Vernon 1999).

Accordingly, a county commissioners court may provide fire protection to county residents in one of two ways. *See id.* Under section 351.001(a), a county may furnish fire protection itself. *See* Tex. Local Gov't Code Ann. § 351.001(a) (Vernon 1999). A county that furnishes fire-protection services itself may, under subsection (b)(1), do so using its own fire-fighting equipment or, under subsection (b)(3), fire-fighting equipment leased from a municipality. Alternatively, under subsection (c), a county commissioners court may furnish fire-protection services to county residents by contracting with an incorporated volunteer fire department to provide the services.

Because section 352.001(c) explicitly sets forth the means by which a county may interact with a volunteer fire department to provide fire protection to county residents, we believe the statute excludes any other means of interaction with a volunteer fire department. A county commissioners court may exercise only those powers that the state constitution and statutes confer upon it, either explicitly or implicitly. *See Canales v. Laughlin*, 214 S.W.2d 451, 453 (Tex. 1948); *Abbott v. Pollock*, 946 S.W.2d 513, 517 (Tex. App.-Austin 1997, writ denied). Moreover, where, as here, a statutory power is granted and the method of its exercise is prescribed, the prescribed method excludes all others and must be followed. *See Foster v. City of Waco*, 255 S.W. 1104, 1105 (Tex. 1923); *Cole v. Texas Army Nat'l Guard*, 909 S.W.2d 535, 539 (Tex. App.-Austin 1995, writ denied); Tex. Att'y Gen. Op. Nos. *JC-0011* (1999) at 3, *DM-424* (1996) at 4. A county that desires to furnish fire-protection services to county residents by contracting with another entity must, consistently with subsection (c), do so under a contract with an incorporated volunteer fire department. Section 352.001 does not permit a county to purchase fire-fighting equipment for a volunteer fire department or to pay a volunteer fire department absent a contract.

You suggest that Government Code section 418.109(d), which was enacted in 1995, authorizes the practice at issue in this request. That subsection authorizes a county to provide "mutual aid assistance" to a fire protection agency or organized volunteer group upon request as part of a mutual-aid arrangement. Tex. Gov't Code Ann. § 418.109(d) (Vernon 1999). Section 418.109 is designed to encourage political subdivisions not participating in interjurisdictional emergency management plans, *see id.* § 418.106, to "make suitable arrangements" to aid each other "in coping with disasters." *See id.* § 418.109(a). Thus, section 418.109 as a whole tries to facilitate both disaster mitigation prior to a disaster and a political subdivision's response in the event that a particular disaster occurs, *e.g.*, a tornado or uncontrollable wildfire:

> The problems in Florida and Louisiana in coping with the aftermath of Hurricane Andrew and extensive fires in central and west United States during 1993 and 1994 brought into focus the need for a statewide mutual aid plan.

House Comm. on Public Safety, Bill Analysis, Tex. Comm. Substitute H.B. 2872, 74th Leg., R.S. (1995); *see also* House Comm. on Public Safety, Bill Analysis, Tex. S.B. 1695, 74th Leg., R.S. (1995). Section 418.109(d) encourages political subdivisions to enter into mutual-aid agreements prior to the occurrence of any disaster and thereby permits a political subdivison to assist another, catastrophe-stricken political subdivision in its response to the disaster. But the section does not authorize a county to make a gift to a volunteer fire department in nonemergency circumstances in the absence of a mutual-aid agreement.

Nor does article III, section 52(a) of the Texas Constitution authorize the county commissioners court to make the arrangements you describe without a contract, thereby avoiding the contract requirement of section 352.001(c) of the Local Government Code. The Texas Constitution prohibits a county from making a gratuitous contribution: any payment (monetary or in-kind) must serve a public purpose, must be in exchange for a *quid pro quo*, and must be accompanied by controls sufficient to ensure that the public purpose is accomplished. *See* Tex. Const. art. III, § 52(a); *see also* Tex. Att'y Gen. LO-96-035, at 3 n.3. You suggest that, under Letter Opinion 96-035, "it might be possible [under article III, section 52(a) generally] to insure sufficient controls without a formal contract." *See* Request Letter, *supra*, at 3. But Letter Opinion 96-035 concerned a county's contribution to a nonprofit organization whose purpose was to assist industrial development; it did not involve a contribution to a volunteer fire department governed by section 352.001(c). Moreover, article III, section 52(a) establishes a minimum standard to which the legislature may, by statute, add levels of security. With respect to a county's relationship with a volunteer fire department that provides fire-protection services for county residents, the legislature has required a contract, and a county must adhere to that statutory requirement.

In sum, a county commissioners court may purchase fire-fighting equipment for or pay money to a volunteer fire department only if the volunteer fire department has contracted with the county to provide fire-protection services for county residents. This office previously has articulated other requirements, which you do not question: the payment (whether in-kind or in cash) must be reasonable; and the payment must be made in accordance with the county budgeting statutes, *see* Tex. Local Gov't Code Ann. ch. 111 (Vernon 1999) and other applicable statutes. *See* Tex. Att'y Gen. Op. No. *V-101* (1947) at 5; *see also* Tex. Att'y Gen. Op. Nos. *O-6160* (1944) at 3, *O-4300* (1942) at 3. In addition, the county must make the payment from the county's general fund. *See* Tex. Local Gov't Code Ann. § 352.001(c) (Vernon 1999); *see also* Tex. Att'y Gen. Op. Nos. *O-6160* (1944) at 3, *O-4300* (1942) at 3.

SUMMARY

A county commissioners court may purchase fire-fighting equipment for or pay a volunteer fire department only if the volunteer fire department has contracted with the county to provide fire-fighting services to county residents living outside the boundaries of a municipality.

Yours very truly,

JOHN CORNYN
Attorney General of Texas

ANDY TAYLOR
First Assistant Attorney General

CLARK KENT ERVIN
Deputy Attorney General - General Counsel

ELIZABETH ROBINSON
Chair, Opinion Committee

Prepared by Kymberly K. Oltrogge
Assistant Attorney General

RHODE ISLAND GENERAL LAWS

§ 45-18-1 Appropriations authorized. Any city or town may annually appropriate, in the manner provided by law for appropriations and expenditures by a city or town, a sum to be expended and paid to any volunteer fire company or companies or other organization or organizations created, for the purpose of and engaged in the work of extinguishing fires and suppressing fire hazards within the limits of the city or town, whether the company or companies or organization or organizations incorporated or not, to be used by the volunteer fire company or companies or organization or organizations for equipment, property, salary, or other expenses in connection with and for the work of extinguishing fires and suppressing fire hazards within the limits of the city or town; provided, that the town of Cumberland may annually appropriate funds to any and all the incorporated fire districts in the town for any purpose authorized by the charter of the district or districts.

CHAPTER 4

Region 1 BOS 2003-217
Wednesday, August 20, 2003
Contact: Ted Fitzgerald
Phone: (617) 565-2072

OSHA Cites Nightclub and Band for Workplace Safety Violations
Following Fatal Fire at 'The Station' Nightclub

PROVIDENCE, R.I.--The U.S. Occupational Safety and Health Administration has completed its inspection prompted by the Feb. 20 fire at The Station nightclub in West Warwick, R.I. The fire, which killed 100 people, including seven employees, was ignited by a pyrotechnic display staged during a performance by the band, Great White.

Derco, LLC, doing business as The Station, owns and operated the nightclub. Derco was cited for alleged willful and serious violations of the Occupational Safety and Health Act and faces $85,200 in fines, while Jack Russell Touring, Inc., Great White's corporate entity, faces a $7,000 fine for an alleged serious violation.

The willful citation to Derco is for installing a door within the exit route near the stage that did not open in the direction of travel. A fine of $70,000 is proposed. Six serious citations were issued to Derco for covering walls and an exit door with highly flammable foam; having an exit door indistinguishable from the walls due to the foam covering; no written emergency action plan; no written fire prevention plan; failing to designate and train employees to assist the evacuation of other employees; and failing to review fire hazards with employees. $15,200 in fines are proposed for these items.

Jack Russell Touring was cited for failing to safeguard employees against fire hazards from the pyrotechnic display in that unused pyrotechnic materials were not stored at least 50 feet from unprotected heat sources; no pyrotechnic plan had been developed; no walkthrough and representative demonstration was conducted prior to the display; pyrotechnic operators were not licensed and approved as required; two or more fire extinguishers were not readily accessible; personnel with a working knowledge of fire extinguishers were not present; pyrotechnic devices were not mounted so that fallout from the devices would not cause injury; each device was not separated from the audience by a minimum of 15 feet; and smoking was not prohibited within 25 feet of the pyrotechnics.

Each company has 15 business days from receipt of its citations and proposed penalties to either elect to comply with them, to request and participate in an informal conference with the OSHA area director, or to contest them before the independent Occupational Safety and Health Review Commission. The inspection was conducted by the OSHA's Providence area office.

Fact Sheet   OSHA Citations and Proposed Fines

Derco, Inc. d/b/a The Station, Inc.

One alleged willful violation, with a proposed fine of $70,000, for

- an exit door near the stage installed within the exit route did not swing in the direction of travel.

Six alleged serious violations, with $15,200 in proposed penalties, for:

- An interior exit door and surrounding walls were covered with highly flammable foam;
- An exit door was not distinguishable from the walls due to its being covered by highly flammable foam;
- no written emergency action plan was prepared
- no written fire prevention plan was prepared
- employees were not designated and trained to assist in a safe and orderly evacuation of other employees
- fire hazards were not reviewed with employees.

Total proposed fines for Derco, Inc. $85,200

Jack Russell Touring, Inc.

One alleged serious violation, with a proposed penalty of $7,000, for:

• failure to provide employment or a place of employment free from recognized hazards likely to cause death or serious physical harm in that employees were exposed to the hazards of fire and fire by-products in that pyrotechnics were used where the following precautions had not been taken: unused pyrotechnic materials were not stored at least 50 feet from unprotected heat sources, namely the ignited gerbs; no plan for the use of pyrotechnics had been developed; no walkthrough and representative pyrotechnic demonstration had been conducted; all pyrotechnic operators were not licensed and approved; two or more fire extinguishers were not readily accessible during pyrotechnic preparation and operations; personnel with working knowledge of fire extinguishers were not present; pyrotechnic devices were not mounted so that fallout would not endanger lives or cause injuries; each pyrotechnic device was not separated from the audience by at least 15 feet; and smoking was not prohibited within 25 feet of pyrotechnics.

Total proposed penalties for Jack Russell Touring $7,000

• OSHA defines a willful violation as one committed with an intentional disregard of, or plain indifference to, the requirements of the Occupational Safety and Health Act and regulations.

• A serious violation is one in which there is a substantial probability that death or serious physical harm could result, and the employer knew, or should have known, of the hazard.

• The Occupational Safety and Health Administration is dedicated to saving lives, preventing injuries and illnesses, and protecting America's workers. Safety and health add value to business, the workplace and life. For more information, visit www.osha.gov.

CHAPTER 5

DUTY TO RENDER ASSISTANCE

Rhode Island General Laws

§ 11-56-1 Duty to assist. Any person at the scene of an emergency who knows that another person is exposed to, or has suffered, grave physical harm shall, to the extent that he or she can do so without danger or peril to himself or herself or to others, give reasonable assistance to the exposed person. Any person violating the provisions of this section shall be guilty of a petty misdemeanor and shall be subject to imprisonment for a term not exceeding six (6) months, or by a fine of not more than five hundred dollars ($500), or both.

ARSON LAWS

Arizona Revised Statutes

13-1701. *Definitions.* In this chapter, unless the context otherwise requires:

1. "Damage" means any physical or visual impairment of any surface.
2. "Occupied structure" means any structure as defined in paragraph 4 in which one or more human beings either is or is likely to be present or so near as to be in equivalent danger at the time the fire or explosion occurs. The term includes any dwelling house, whether occupied, unoccupied or vacant.
3. "Property" means anything other than a structure which has value, tangible or intangible, public or private, real or personal, including documents evidencing value or ownership.
4. "Structure" means any building, object, vehicle, watercraft, aircraft or place with sides and a floor, used for lodging, business, transportation, recreation or storage.
5. "Wildland" means any brush covered land, cutover land, forest, grassland or woods.

13-1702. *Reckless burning; classification*

A. A person commits reckless burning by recklessly causing a fire or explosion which results in damage to an occupied structure, a structure, wildland or property.

B. Reckless burning is a class 1 misdemeanor.

13-1703. *Arson of a structure or property; classification*

A. A person commits arson of a structure or property by knowingly and unlawfully damaging a structure or property by knowingly causing a fire or explosion.

B. Arson of a structure is a class 4 felony. Arson of property is a class 4 felony if the property had a value of more than one thousand dollars. Arson of property is a class 5 felony if the property had a value of more than one hundred dollars but not more than one thousand dollars. Arson of property is a class 1 misdemeanor if the property had a value of one hundred dollars or less.

13-1704. *Arson of an occupied structure; classification*

A. A person commits arson of an occupied structure by knowingly and unlawfully damaging an occupied structure by knowingly causing a fire or explosion.

B. Arson of an occupied structure is a class 2 felony.

13-1705. *Arson of an occupied jail or prison facility; classification.*

A. A person commits arson of an occupied jail or prison facility by knowingly causing a fire or explosion which results in physical damage to the jail or prison facility.

B. Arson of an occupied jail or prison facility is a class 4 felony.

13-1706. *Burning of wildlands; exceptions; classification*

A. It is unlawful for any person, without lawful authority, to intentionally, knowingly, recklessly or with criminal negligence to set or cause to be set on fire any wildland other than the person's own or to permit a fire that was set or caused to be set by the person to pass from the person's own grounds to the grounds of another person.

B. This section does not apply to any of the following:

1. Open burning that is lawfully conducted in the course of agricultural operations.

2. Fire management operations that are conducted by a political subdivision.

3. Prescribed or controlled burns that are conducted with written authority from the state forester.

4. Lawful activities that are conducted pursuant to any rule, regulation or policy that is adopted by a state, tribal or federal agency.

5. In absence of a fire ban or other burn restrictions to a person on public lands, setting a fire for purposes of cooking or warming that does not spread sufficiently from its source to require action by a fire control agency.

C. A person who violates this section is guilty of an offense as follows:

1. If done with criminal negligence, the offense is a class 2 misdemeanor.

2. If done recklessly, the offense is a class 1 misdemeanor.

3. If done intentionally or knowingly and the person knows or reasonably should know that the person's conduct violates any order or rule that is issued by a governmental entity and that prohibits, bans, restricts or otherwise regulates fires during periods of extreme fire hazard, the offense is a class 6 felony.

4. If done intentionally and the person's conduct places another person in danger of death or serious bodily injury or places any building or occupied structure of another person in danger of damage, the offense is a class 3 felony.

13-1707. *Unlawful cross burning; classification*

A. It is unlawful for a person to burn or cause to be burned a cross on the property of another person without that person's permission or on a highway or any other public place with the intent to intimidate any person or group of persons. The intent to intimidate may not be inferred solely from the act of burning a cross, but shall be proven by independent evidence.

B. A person who violates this section is guilty of a class 1 misdemeanor.

13-1708. *Unlawful symbol burning; classification*

A. It is unlawful for a person to burn or cause to be burned any symbol not addressed by section 13-1707 on the property of another person without that person's permission or on a highway or any other public place with the intent to intimidate any person or group of persons. The intent to intimidate may not be inferred solely from the act of burning the symbol, but shall be proven by independent evidence.

B. A person who violates this section is guilty of a class 1 misdemeanor.

13-1709. *Emergency response and investigation costs; civil liability; definitions*

A. A person who commits an act in violation of this chapter that results in an appropriate emergency response or investigation and who is convicted of the violation may be liable for the expenses that are incurred incident to the emergency response and the investigation of the commission of the offense.

B. The court may assess and collect the expenses prescribed in subsection A. The court shall state the amount of these expenses as a separate item in any final judgment, order or decree.

C. The expenses are a debt of the person. The public agency, for profit entity or nonprofit entity that incurred the expenses may collect the debt proportionally. The liability that is imposed under this section is in addition to and not in limitation of any other liability that may be imposed. If a person is subject to liability under this section and is married, only the separate property of the person is subject to liability.

D. There shall be no duty under a policy of liability insurance to defend or indemnify any person found liable for any expenses under this section.

E. For the purposes of this section:

1. "Expenses" means reasonable costs that are directly incurred by a public agency, for profit entity or nonprofit entity that makes an appropriate emergency response to an incident or an investigation of the commission of the offense, including the costs of providing police, fire fighting, rescue and emergency medical services at the scene of the incident and the salaries of the persons who respond to the incident but excluding charges assessed by an ambulance service that is regulated pursuant to title 36, chapter 21.1, article 2.

2. "Public agency" means this state, any city, county, municipal corporation or district, any Arizona federally recognized native American tribe or any other public authority that is located in whole or in part in this state and that provides police, fire fighting, medical or other emergency services.

Connecticut General Statutes

Sec. 53a-111. Arson in the first degree: Class A felony.

(a) A person is guilty of **arson** in the first degree when, with intent to destroy or damage a building, as defined in section 53a-100, he starts a fire or causes an explosion, and (1) the building is inhabited or occupied or the person has reason to believe the building may be inhabited or occupied; or (2) any other person is injured, either directly or indirectly; or (3) such fire or explosion was caused for the purpose of collecting insurance proceeds for the resultant loss; or (4) at the scene of such fire or explosion a peace officer or firefighter is subjected to a substantial risk of bodily injury.

(b) **Arson** in the first degree is a class A felony.

Sec. 53a-112. Arson in the second degree: Class B felony.

(a) A person is guilty of arson in the second degree when, with intent to destroy or damage a building, as defined in section 53a-100, (1) he starts a fire or causes an explosion and (A) such act subjects another person to a substantial risk of bodily injury; or (B) such fire or explosion was intended to conceal some other criminal act; or (C) such fire or explosion was intended to subject another person to a deprivation of a right, privilege or immunity secured or protected by the Constitution or laws of this state or of the United States; or (2) a fire or explosion was caused by an individual hired by such person to start such fire or cause such explosion.

(b) Arson in the second degree is a class B felony.

Sec. 53a-113. Arson in the third degree: Class C felony.

(a) A person is guilty of arson in the third degree when he recklessly causes destruction or damage to a building, as defined in section 53a-100, of his own or of another by intentionally starting a fire or causing an explosion.

(b) Arson in the third degree is a class C felony.

Sec. 53a-114. Reckless burning: Class D felony.

(a) A person is guilty of reckless burning when he intentionally starts a fire or causes an explosion, whether on his own property or another's, and thereby recklessly places a building, as defined in section 53a-100, of another in danger of destruction or damage.

(b) Reckless burning is a class D felony.

Sec. 53a-54d. Arson murder.

A person is guilty of murder when, acting either alone or with one or more persons, he commits **arson** and, in the course of such **arson**, causes the death of a person. Notwithstanding any other provision of the general statutes, any person convicted of murder under this section shall be punished by life imprisonment and shall not be eligible for parole.

Florida Statutes

Section 806.01 Arson. (1) Any person who willfully and unlawfully, or while in the commission of any felony, by fire or explosion, damages or causes to be damaged:

(a) Any dwelling, whether occupied or not, or its contents;

(b) Any structure, or contents thereof, where persons are normally present, such as: jails, prisons, or detention centers; hospitals, nursing homes, or other health care facilities; department stores, office buildings, business establishments, churches, or educational institutions during normal hours of occupancy; or other similar structures; or

(c) Any other structure that he or she knew or had reasonable grounds to believe was occupied by a human being, is guilty of arson in the first degree, which constitutes a felony of the first degree, punishable as provided in s. 775.082, s. 775.083, or s. 775.084.

(2) Any person who willfully and unlawfully, or while in the commission of any felony, by fire or explosion, damages or causes to be damaged any structure, whether the property of himself or herself or another, under any circumstances not referred to in subsection (1), is guilty of arson in the second degree, which constitutes a felony of the second degree, punishable as provided in s. 775.082, s. 775.083, or s. 775.084.

(3) As used in this chapter, "structure" means any building of any kind, any enclosed area with a roof over it, any real property and appurtenances thereto, any tent or other portable building, and any vehicle, vessel, watercraft, or aircraft.

Georgia Code and Acts

16-7-60 (a) A person commits the offense of **arson** in the first degree when, by means of fire or explosive, he knowingly damages or knowingly causes, aids, abets, advises, encourages, hires, counsels, or procures another to damage:

(1) Any dwelling house of another without his consent or in which another has a security interest, including but not limited to a mortgage, a lien, or a conveyance to secure debt, without the consent of both, whether it is occupied, unoccupied, or vacant;

(2) Any building, vehicle, railroad car, watercraft, or other structure of another without his consent or in which another has a security interest, including but not limited to a mortgage, a lien, or a conveyance to secure debt, without the consent of both, if such structure is designed for use as a dwelling, whether it is occupied, unoccupied, or vacant;

(3) Any dwelling house, building, vehicle, railroad car, watercraft, aircraft, or other structure whether it is occupied, unoccupied, or vacant and when such is insured against loss or damage by fire or explosive and such loss or damage is accomplished without the consent of both the insurer and the insured;

(4) Any dwelling house, building, vehicle, railroad car, watercraft, aircraft, or other structure whether it is occupied, unoccupied, or vacant with the intent to defeat, prejudice, or defraud the rights of a spouse or co-owner; or

(5) Any building, vehicle, railroad car, watercraft, aircraft, or other structure under such circumstances that it is reasonably foreseeable that human life might be endangered.

(b) A person convicted of the offense of **arson** in the first degree shall be punished by a fine of not more than $50,000.00 or by imprisonment for not less than one nor more than 20 years, or both.

16-7-61 (a) A person commits the offense of **arson** in the second degree as to any building, vehicle, railroad car, watercraft, aircraft, or other structure not included or described in Code

Section 16-7-60 when, by means of fire or explosive, he knowingly damages or knowingly causes, aids, abets, advises, encourages, hires, counsels, or procures another to damage any building, vehicle, railroad car, watercraft, aircraft, or other structure of another without his consent or in which another has a security interest, including but not limited to a mortgage, a lien, or a conveyance to secure debt, without the consent of both.

(b) A person convicted of the offense of **arson** in the second degree shall be punished by a fine of not more than $25,000.00 or by imprisonment for not less than one nor more than ten years, or both.

16-7-62 (a) A person commits the offense of **arson** in the third degree when, by means of fire or explosive, he knowingly damages or knowingly causes, aids, abets, advises, encourages, hires, counsels, or procures another to damage:

(1) Any personal property of another without his consent or in which another has a security interest, including but not limited to a lien, without the consent of both and the value of the property is $25.00 or more;

(2) Any personal property when such is insured against loss or damage by fire or explosive and the loss or damage is accomplished without the consent of both the insurer and insured and the value of the property is $25.00 or more; or

(3) Any personal property with the intent to defeat, prejudice, or defraud the rights of a spouse or co-owner and the value of the property is $25.00 or more.

(b) A person convicted of the offense of **arson** in the third degree shall be punished by a fine not to exceed $10,000.00 or by imprisonment for not less than one nor more than five years, or both.

Indiana Code

IC 35-43-1-1 Arson   Chapter 1. Arson, Mischief, and Tampering

Sec. 1. (a) A person who, by means of fire, explosive, or destructive device, knowingly or intentionally damages:

(1) a dwelling of another person without the other person's consent;

(2) property of any person under circumstances that endanger human life;

(3) property of another person without the other person's consent if the pecuniary loss is at least five thousand dollars ($5,000); or

(4) a structure used for religious worship without the consent of the owner of the structure; commits arson, a Class B felony. However, the offense is a Class A felony if it results in either bodily injury or serious bodily injury to any person other than a defendant.

(b) A person who commits arson for hire commits a Class B felony. However, the offense is a Class A felony if it results in bodily injury to any other person.

(c) A person who, by means of fire, explosive, or destructive device, knowingly or intentionally damages property of any person with intent to defraud commits arson, a Class C felony.

(d) A person who, by means of fire, explosive, or destructive device, knowingly or intentionally damages property of another person without the other person's consent so that the resulting pecuniary loss is at least two hundred fifty dollars ($250) but less than five thousand dollars ($5,000) commits arson, a Class D felony.

Michigan Penal Code

750.71 Arson and burning; definitions. Definition of "burn". The term "burn" as used in this chapter shall mean setting fire to, or doing any act which results in the starting of a fire, or aiding, counseling, inducing, persuading or procuring another to do such act or acts.

750.72 Burning dwelling house. Any person who wilfully or maliciously burns any dwelling house, either occupied or unoccupied, or the contents thereof, whether owned by himself or another, or any building within the curtilage of such dwelling house, or the contents thereof, shall be guilty of a felony, punishable by imprisonment in the state prison not more than 20 years.

750.73 Burning of other real property. Any person who wilfully or maliciously burns any building or other real property, or the contents thereof, other than those specified in the next preceding section of this chapter, the property of himself or another, shall be guilty of a felony, punishable by imprisonment in the state prison for not more than 10 years.

750.74 Burning of personal property. (1) A person who willfully and maliciously burns any personal property, other than personal property specified in section 72 or 73, owned by himself or herself or another person is guilty of a crime as follows:

(a) If the value of the personal property burned or intended to be burned is less than $200.00, the person is guilty of a misdemeanor punishable by imprisonment for not more than 93 days or a fine of not more than $500.00 or 3 times the value of the personal property burned or intended to be burned, whichever is greater, or both imprisonment and a fine.

(b) If any of the following apply, the person is guilty of a misdemeanor punishable by imprisonment for not more than 1 year or a fine of not more than $2,000.00 or 3 times the value of the personal property burned or intended to be burned, whichever is greater, or both imprisonment and a fine:

(i) The value of the personal property burned or intended to be burned is $200.00 or more but less than $1,000.00.

(ii) The person violates subdivision (a) and has 1 or more prior convictions for committing or attempting to commit an offense under this section or a local ordinance substantially corresponding to this section.

(c) If any of the following apply, the person is guilty of a felony punishable by imprisonment for not more than 5 years or a fine of not more than $10,000.00 or 3 times the value of the personal property burned or intended to be burned, whichever is greater, or both imprisonment and a fine:

(i) The value of the personal property burned or intended to be burned is $1,000.00 or more but less than $20,000.00.

(ii) The person violates subdivision (b)(i) and has 1 or more prior convictions for violating or attempting to violate this section. For purposes of this subparagraph, however, a prior conviction does not include a conviction for a violation or attempted violation of subdivision (a) or (b)(ii).

(d) If any of the following apply, the person is guilty of a felony punishable by imprisonment for not more than 10 years or a fine of not more than $15,000.00 or 3 times the value of the personal property burned or intended to be burned, whichever is greater, or both imprisonment and a fine:

(i) The personal property burned or intended to be burned has a value of $20,000.00 or more.

(ii) The person violates subdivision (c)(i) and has 2 or more prior convictions for committing or attempting to commit an offense under this section. For purposes of this subparagraph, however, a prior conviction does not include a conviction for a violation or attempted violation of subdivision (a) or (b)(ii).

(2) The values of personal property burned or intended to be burned in separate incidents pursuant to a scheme or course of conduct within any 12-month period may be aggregated to determine the total value of personal property burned or intended to be burned.

(3) If the prosecuting attorney intends to seek an enhanced sentence based upon the defendant having 1 or more prior convictions, the prosecuting attorney shall include on the complaint and information a statement listing the prior conviction or convictions. The existence of the defendant's prior conviction or convictions shall be determined by the court, without a jury, at sentencing or at a separate hearing for that purpose before sentencing. The existence of a prior conviction may be established by any evidence relevant for that purpose, including, but not limited to, 1 or more of the following:

(a) A copy of the judgment of conviction.

(b) A transcript of a prior trial, plea-taking, or sentencing.

(c) Information contained in a presentence report.

(d) The defendant's statement.

(4) If the sentence for a conviction under this section is enhanced by 1 or more prior convictions, those prior convictions shall not be used to further enhance the sentence for the conviction pursuant to section 10, 11, or 12 of chapter IX of the code of criminal procedure, 1927 PA 175, MCL 769.10, 769.11, and 769.12.

750.75 Burning of insured property. Any person who shall wilfully burn any building or personal property which shall be at the time insured against loss or damage by fire with intent to injure and defraud the insurer, whether such person be the owner of the property or not, shall be guilty of a felony, punishable by imprisonment in the state prison not more than 10 years.

750.76 Applicability of preceding sections. The preceding sections of this chapter shall apply to a married woman who may commit any of the offenses herein described although the property burnt may belong partly or wholly to her husband; and said preceding sections shall also apply to a married man although the property burnt may belong partly or wholly to his wife; and although said property may be occupied by such married man or married woman, or by such married man and wife as a residence.

750.77 Willfully and maliciously setting fire. (1) A person who uses, arranges, places, devises, or distributes an inflammable, combustible, or explosive material, liquid, or substance or any device in or near a building or property described in section 72, 73, 74, or 75 with intent to willfully and maliciously set fire to or burn the building or property or who aids, counsels, induces, persuades, or procures another to do so is guilty of a crime as follows:

(a) If the property intended to be burned is personal or real property, or both, with a combined value less than $200.00, the person is guilty of a misdemeanor punishable by imprisonment for not more than 93 days or a fine of not more than $500.00 or 3 times the combined value of the property intended to be burned, whichever is greater, or both imprisonment and a fine.

(b) If any of the following apply, the person is guilty of a misdemeanor punishable by imprisonment for not more than 1 year or a fine of not more than $2,000.00 or 3 times the combined value of the property intended to be burned, whichever is greater, or both imprisonment and a fine:

(i) The property intended to be burned is personal or real property, or both, with a combined value of $200.00 or more but less than $1,000.00.

(ii) The person violates subdivision (a) and has 1 or more prior convictions for committing or attempting to commit an offense under this section or a local ordinance substantially corresponding to this section.

(c) If any of the following apply, the person is guilty of a felony punishable by imprisonment for not more than 5 years or a fine of not more than $10,000.00 or 3 times the combined value of the property intended to be burned, whichever is greater, or both imprisonment and a fine:

(i) The property intended to be burned is personal or real property, or both, with a combined value of $1,000.00 or more but less than $20,000.00.

(ii) The person violates subdivision (b)(i) and has 1 or more prior convictions for violating or attempting to violate this section. For purposes of this subparagraph, however, a prior conviction does not include a conviction for a violation or attempted violation of subdivision (a) or (b)(ii).

(d) If any of the following apply, the person is guilty of a felony punishable by imprisonment for not more than 10 years or a fine of not more than $15,000.00 or 3 times the combined value of the property intended to be burned, whichever is greater, or both imprisonment and a fine:

(i) The property is personal or real property, or both, with a combined value of $20,000.00 or more.

(ii) The person violates subdivision (c)(i) and has 2 or more prior convictions for committing or attempting to commit an offense under this section. For purposes of this subparagraph, however, a prior conviction does not include a conviction for committing or attempting to commit an offense for a violation or attempted violation of subdivision (a) or (b)(ii).

(2) The combined value of property intended to be burned in separate incidents pursuant to a scheme or course of conduct within any 12-month period may be aggregated to determine the total value of property intended to be burned.

(3) If the prosecuting attorney intends to seek an enhanced sentence based upon the defendant having 1 or more prior convictions, the prosecuting attorney shall include on the complaint and information a statement listing the prior conviction or convictions. The existence of the defendant's prior conviction or convictions shall be determined by the court, without a jury, at sentencing or at

a separate hearing for that purpose before sentencing. The existence of a prior conviction may be established by any evidence relevant for that purpose, including, but not limited to, 1 or more of the following:

(a) The total value of property intended to be burned.

(b) A transcript of a prior trial, plea-taking, or sentencing.

(c) Information contained in a presentence report.

(d) The defendant's statement.

(4) If the sentence for a conviction under this section is enhanced by 1 or more prior convictions, those prior convictions shall not be used to further enhance the sentence for the conviction pursuant to section 10, 11, or 12 of chapter IX of the code of criminal procedure, 1927 PA 175, MCL 769.10, 769.11, and 769.12.

750.78 Wilfully or negligently setting fire to woods, prairies or grounds. Any person who shall wilfully or negligently set fire to any woods, prairies or grounds, not his property, or shall wilfully permit any fire to pass from his own woods, prairies or grounds, to the injury or destruction of the property of any other.

750.79 Clearing of land and disposing of refuse in townships. Whenever in pursuance of the authority given by law, any township board shall, by order, rule or regulation, designate a period during which it shall be unlawful to set forest fires or fires for the purpose of clearing lands, and disposing by burning of refuse material and waste matter within its respective jurisdiction or any part thereof, any person who shall be found guilty of violating the orders, rules and regulations of such board by setting any such fire in such township contrary to the provisions thereof shall be guilty of a felony: Provided, That any person desiring to dispose of refuse material by burning the same during the time prohibited by the board of such township, may do so after first procuring permission in writing, signed by the supervisor and township clerk, or by a majority of such township board, and the said supervisor and township clerk, or a majority of the said board are hereby authorized to grant such permission in their discretion, under such conditions as they may prescribe, upon application made in writing for such purpose: Provided further, That said board is hereby authorized at any time to repeal by resolution any order, rule or regulation herein mentioned.

750.80 Setting fire to mines and mining material. Any person who shall wilfully and maliciously burn or set fire to or cause to be burned or set fire to any wood, timber, or other material in any part of a mine under ground, or shall wilfully and maliciously set fire to or burn any shaft house or other structure or materials built or placed over, or upon a shaft, adit, level or other opening into any mine, such mine being then in use or operation, shall be guilty of felony, and be punishable by imprisonment in the state prison for life or for any term of years.

Minnesota

609.561 Arson in the first degree.

Subdivision 1. First degree; dwelling. Whoever unlawfully by means of fire or explosives, intentionally destroys or damages any building that is used as a dwelling at the time the act is committed, whether the inhabitant is present therein at the time of the act or not, or any building appurtenant to or connected with a dwelling whether the property of the actor or of another, commits arson in the first degree and may be sentenced to imprisonment for not more than 20 years or to a fine of not more than $20,000, or both.

Subd. 2. First degree; other buildings. Whoever unlawfully by means of fire or explosives, intentionally destroys or damages any building not included in subdivision 1, whether the property of the actor or another commits arson in the first degree and may be sentenced to imprisonment for not more than 20 years or to a fine of not more than $35,000, or both if: (a) another person who is not a participant in the crime is present in the building at the time and the defendant knows that; or (b) the circumstances are such as to render the presence of such a person therein a reasonable possibility.

Subd. 3. First degree; flammable material. (a) Whoever unlawfully by means of fire or explosives, intentionally destroys or damages any building not included in subdivision 1, whether the property of the actor or another, commits arson in the first degree if a flammable material is used to start or accelerate the fire. A person who violates this paragraph may be sentenced to imprisonment for not more than 20 years or a fine of not more than $20,000, or both.

(b) As used in this subdivision: (1) "combustible liquid" means a liquid having a flash point at or above 100 degrees Fahrenheit;

(2) "flammable gas" means any material which is a gas at 68 degrees Fahrenheit or less and 14.7 psi of pressure and which: (i) is ignitable when in a mixture of 13 percent or less by volume with air at atmospheric pressure; or (ii) has a flammable range with air at atmospheric pressure of at least 12 percent, regardless of the lower flammable limit;

(3) "flammable liquid" means any liquid having a flash point below 100 degrees Fahrenheit and having a vapor pressure not exceeding 40 pounds per square inch (absolute) at 100 degrees Fahrenheit, but does not include intoxicating liquor as defined in section 340A.101;

(4) "flammable material" means a flammable or combustible liquid, a flammable gas, or a flammable solid; and

(5) "flammable solid" means any of the following three types of materials: (i) wetted explosives; (ii) self-reactive materials that are liable to undergo heat-producing decomposition; or (iii) readily combustible solids that may cause a fire through friction or that have a rapid burning rate as determined by specific flammability tests.

609.562 Arson in the second degree. Whoever unlawfully by means of fire or explosives, intentionally destroys or damages any building not covered by section 609.561, no matter what its value, or any other real or personal property valued at more than $1,000, whether the property of the actor or another, may be sentenced to imprisonment for not more than ten years or to payment of a fine of not more than $20,000, or both.

609.563 Arson in the third degree.

Subdivision 1. Crime. Whoever unlawfully by means of fire or explosives, intentionally destroys or damages any real or personal property may be sentenced to imprisonment for not more than five years or to payment of a fine of $10,000, or both, if: (a) the property intended by the accused to be damaged or destroyed had a value of more than $300 but less than $1,000; or (b) property of the value of $300 or more was unintentionally damaged or destroyed but such damage or destruction could reasonably have been foreseen; or (c) the property specified in clauses (a) and (b) in the aggregate had a value of $300 or more.

609.5631 Arson in the fourth degree. Subdivision 1. Definitions. (a) For purposes of this section, the following terms have the meanings given. (b) "Multiple unit residential building" means a building containing two or more apartments. (c) "Public building" means a building such as a hotel, hospital, motel, dormitory, sanitarium, nursing home, theater, stadium, gymnasium, amusement park building, school or other building used for educational purposes, museum, restaurant, bar, correctional institution, place of worship, or other building of public assembly.

Subd. 2. Crime described. Whoever intentionally by means of fire or explosives sets fire to or burns or causes to be burned any personal property in a multiple unit residential building or public building and arson in the first, second, or third degree was not committed is guilty of a gross misdemeanor and may be sentenced to imprisonment for not more than one year or to payment of a fine of not more than $3,000, or both.

609.5632 Arson in the fifth degree. Whoever intentionally by means of fire or explosives sets fire to or burns or causes to be burned any real or personal property of value is guilty of a misdemeanor and may be sentenced to imprisonment for not more than 90 days or to payment of a fine of not more than $1,000, or both.

609.5633 Use of ignition devices; petty misdemeanor. A student who uses an ignition device, including a butane or disposable lighter or matches, inside an educational building and under circumstances where there is an obvious risk of fire, and arson in the first, second, third, or fourth degree was not committed, is guilty of a petty misdemeanor. This section does not apply if the student uses the device in a manner authorized by the school. For the purposes of this section, "student" has the meaning given in section 123B.41, subdivision 11.

609.564 Exluded fires. A person does not violate section 609.561, 609.562, 609.563, or 609.5641 if the person sets a fire pursuant to a validly issued license or permit or with written permission from the fire department of the jurisdiction where the fire occurs.

609.5641 Wildfire arson. Subdivision 1. Setting wildfires. A person is guilty of a felony who intentionally sets a fire to burn out of control on land of another containing timber, underbrush, grass, or other vegetative combustible material.

Subd. 2. Possession of flammables to set wildfires. A person is guilty of a gross misdemeanor who possesses a flammable, explosive, or incendiary device, substance, or material with intent to use the device, substance, or material to violate subdivision 1.

Subd. 3. Penalty; restitution. (a) A person who violates subdivision 1 may be sentenced to imprisonment for not more than five years or to payment of a fine of not more than $10,000, or both. (b) A person who violates subdivision 2 may be sentenced to imprisonment for not more than one year or to payment of a fine of not more than $3,000, or both. (c) In addition to the sentence otherwise authorized, the court may order a person who is convicted of violating this section to pay fire suppression costs and damages to the owner of the damaged land.

609.576 Negligent fires; dangerous smoking. Subdivision 1. Negligent fire resulting in injury or property damage. Whoever is grossly negligent in causing a fire to burn or get out of control thereby causing damage or injury to another, and as a result of this: (1) a human being is injured and great bodily harm incurred, is guilty of a crime and may be sentenced to imprisonment for not more than five years or to payment of a fine of not more than $10,000, or both; (2) a human being is injured and bodily harm incurred, is guilty of a crime and may be sentenced to imprisonment for not more than one year or to payment of a fine of not more than $3,000, or both; or (3) property of another is injured, thereby, is guilty of a crime and may be sentenced as follows: (i) to imprisonment for not more than 90 days or to payment of a fine of not more than $1,000, or both, if the value of the property damage is under $300; (ii) to imprisonment for not more than one year or to payment of a fine of not more than $3,000, or both, if the value of the property damaged is at least $300 but is less than $2,500; or (iii) to imprisonment for not more than three years or to payment of a fine of not more than $5,000, or both, if the value of the property damaged is $2,500 or more.

Subd. 2. Dangerous smoking. A person is guilty of a misdemeanor if the person smokes in the presence of explosives or inflammable materials. If a person violates this subdivision and knows that doing so creates a risk of death or bodily harm or serious property damage, the person is guilty of a felony and may be sentenced to imprisonment for not more than five years or to payment of a fine of not more than $10,000, or both.

Nebraska Statutes

28-501 Building, defined. As used in this article, unless the context otherwise requires, building shall mean a structure which has the capacity to contain, and is designed for the shelter of man, animals, or property, and includes ships, trailers, sleeping cars, aircraft, or other vehicles or places adapted for overnight accommodations of persons or animals, or for carrying on of business therein, whether or not a person or animal is actually present. If a building is divided into units for separate occupancy, any unit not occupied by the defendant is a building of another.

28-502 Arson, first degree; penalty. (1) A person commits arson in the first degree if he or she intentionally damages a building by starting a fire or causing an explosion when another person is present in the building at the time and either (a) the actor knows that fact, or (b) the circumstances are such as to render the presence of a person therein a reasonable probability.

(2) A person commits arson in the first degree if a fire is started or an explosion is caused in the perpetration of any robbery, burglary, or felony criminal mischief when another person is present in the building at the time and either (a) the actor knows that fact, or (b) the circumstances are such as to render the presence of a person therein a reasonable probability.

(3) Arson in the first degree is a Class II felony.

28-503 Arson, second degree; penalty. (1) A person commits arson in the second degree if he or she intentionally damages a building by starting a fire or causing an explosion or if a fire is started or an explosion is caused in the perpetration of any robbery, burglary, or felony criminal mischief.

(2) The following affirmative defenses may be introduced into evidence upon prosecution for a violation of this section:

(a) No person other than the accused has a security or proprietary interest in the damaged building, or, if other persons have such interests, all of them consented to his or her conduct; or

(b) The accused's sole intent was to destroy or damage the building for a lawful and proper purpose.

(3) Arson in the second degree is a Class III felony.

28-504 Arson, third degree; penalty. (1) A person commits arson in the third degree if he intentionally sets fire to, burns, causes to be burned, or by the use of any explosive, damages or destroys, or causes to be damaged or destroyed, any property of another without his consent, other than a building or occupied structure.

(2) Arson in the third degree is a Class IV felony if the damages amount to one hundred dollars or more.

(3) Arson in the third degree is a Class I misdemeanor if the damages are less than one hundred dollars.

28-505 Burning to defraud insurer; penalty. Any person who, with the intent to deceive or harm an insurer, sets fire to or burns or attempts so to do, or who causes to be burned, or who aids, counsels or procures the burning of any building or personal property, of whatsoever class or character, whether the property of himself or of another, which shall at the time be insured by any person, company or corporation against loss or damage by fire, commits a Class IV felony.

28-506 Lawful burning of property; training and safety promotion purposes; permit. Property may be lawfully destroyed by burning such structures as condemned by law, structures no longer having any value for habitation or business or no longer serving any useful value in the area in which situated, and any other combustible material that will serve to be used for test fires to educate and train members of organized fire departments and promote fire safety anywhere in Nebraska. Before any structure may be destroyed by fire for training and educational purposes it must be reported to the State Fire Marshal and a permit issued for that purpose. Any expense incurred in burning a structure shall be assumed by the organized fire department requesting this type of training for members of its department.

Nevada Revised Statutes

ARSON

NRS 205.005 "Set fire to" defined. Any person shall be deemed to have "set fire to" a building, structure or any property mentioned in NRS 205.010 to 205.030, inclusive, whenever any part thereof or anything therein shall be scorched, charred or burned.

NRS 205.010 First degree. A person who willfully and maliciously sets fire to or burns or causes to be burned, or who aids, counsels or procures the burning of any:

1. Dwelling house or other structure or mobile home, whether occupied or vacant; or

2. Personal property which is occupied by one or more persons, whether the property of himself or of another, is guilty of arson in the first degree which is a category B felony and shall be punished by imprisonment for a minimum term of not less than 2 years and a maximum term of not more than 15 years, and may be further punished by a fine of not more than $15,000.

NRS 205.015 Second degree. A person who willfully and maliciously sets fire to or burns or causes to be burned, or who aids, counsels or procures the burning of any abandoned building or structure, whether the property of himself or of another, is guilty of arson in the second degree which is a category B felony and shall be punished by imprisonment in the state prison for a minimum term of not less than 1 year and a maximum term of not more than 10 years, and may be further punished by a fine of not more than $10,000.

NRS 205.020 Third degree. A person who willfully and maliciously sets fire to or burns or causes to be burned, or who aids, counsels or procures the burning of:

1. Any unoccupied personal property of another which has the value of $25 or more;

2. Any unoccupied personal property owned by him in which another person has a legal interest; or

3. Any timber, forest, shrubbery, crops, grass, vegetation or other flammable material not his own, is guilty of arson in the third degree which is a category D felony and shall be punished as provided in NRS 193.130.

NRS 205.025 Fourth degree.

1. A person who willfully and maliciously attempts to set fire to or attempts to burn or to aid, counsel or procure the burning of any of the buildings or property mentioned in NRS 205.010, 205.015 and 205.020, or who commits any act preliminary thereto or in furtherance thereof, is guilty of arson in the fourth degree which is a category D felony and shall be punished as provided in NRS 193.130, and may be further punished by a fine of not more than $5,000.

2. In any prosecution under this section the placing or distributing of any inflammable, explosive or combustible material or substance, or any device in any building or property mentioned in NRS 205.010, 205.015 and 205.020, in an arrangement or preparation eventually to set fire to or burn the building or property, or to procure the setting fire to or burning of the building or property, is prima facie evidence of a willful attempt to burn or set on fire the property.

NRS 205.030 Burning or aiding and abetting burning of property with intent to defraud insurer; penalty. A person who willfully and with the intent to injure or defraud the insurer sets fire to or burns or attempts to set fire to or burn, or who causes to be burned or who aids, counsels or procures the burning of any building, structure or personal property of whatsoever class or character, whether the property of himself or of another, which is at the time insured by any person, company or corporation against loss or damage by fire, is guilty of a category B felony and shall be punished by imprisonment in the state prison for a minimum term of not less than 1 year and a maximum term of not more than 6 years, and may be further punished by a fine of not more than $5,000. In addition to any other penalty, the court shall order the person to pay restitution.

NRS 205.034 Additional penalties. The court may, in addition to imposing the penalties set forth in NRS 205.010, 205.015, 205.020, 205.025 or 205.030, order the person to pay:

1. Court costs;

2. The costs of providing police and fire services related to the crime; or

3. The costs of the investigation and prosecution of the crime, or any combination of subsections 1, 2 and 3.

NRS 205.045 Contiguous fires. Whenever any building or structure which may be the subject of arson in either the first or second degree shall be so situated as to be manifestly endangered by any fire and shall subsequently be set on fire thereby, any person participating in setting such fire shall be deemed to have participated in setting such building or structure on fire.

NRS 205.050 Ownership of building. To constitute arson it shall not be necessary that another person than the defendant should have had ownership in the building or structure set on fire.

NRS 205.055 Preparation is attempt to commit arson. Any willful preparation made by any person with a view to setting fire to any building or structure shall be deemed to be an attempt to commit the crime of arson, and shall be punished as such.

Rhode Island General Laws

§ 11-4-2 Arson – First degree. Any person who knowingly causes, procures, aids, counsels or creates by means of fire or explosion a substantial risk of serious physical harm to any person or damage to any building the property of that person or another, whether or not used for residential purposes, which is occupied or in use for any purpose or which has been occupied or in use for any purpose during the six (6) months preceding the offense or to any other residential structure, shall, upon conviction, be sentenced to imprisonment for not less than five (5) years and may be imprisoned for life, or shall be fined not less than three thousand dollars ($3,000) nor more than twenty-five thousand dollars ($25,000), or both; provided, further, that whenever a death occurs to a person as a direct result of the fire or explosion or to a person who is directly involved in fighting the fire or explosion, imprisonment shall be for not less than twenty (20) years. In all such cases, the justice may only impose a sentence less than the minimum if he or she finds that substantial and compelling circumstances exist which justify imposition of the alternative sentence. That finding may be based upon the character and background of the defendant, the cooperation of the defendant with law enforcement authorities, the nature and circumstances of the offense, and/or the nature and quality of the evidence presented at trial. If a sentence which is less than imprisonment for the minimum term is imposed, the trial justice shall set forth on the record the circumstances which he or she found as justification for imposition of the lesser sentence.

§ 11-4-2.1 Arson – Custody. Any person who while under arrest or incarcerated knowingly causes, procures, aids, counsels or creates by means of fire or explosion the damage or destruction of any occupied or unoccupied building shall be guilty of arson in the first degree and shall be sentenced as provided in § 11-4-2.

§ 11-4-3 Arson – Second degree. Any person who knowingly causes, aids, procures or counsels by means of fire or explosion the damage or destruction of any unoccupied building, structure or facility the property of that person or another shall, upon conviction, be sentenced to imprisonment for not less than two (2) years nor more than twenty (20) years, or shall be fined not more than two thousand five hundred dollars ($2,500), or both; provided, that if death occurs to a person as a direct result of the fire or explosion or to a person who is directly involved in fighting the fire or explosion, imprisonment shall be for not less than twenty (20) years.

§ 11-4-4 Arson – Third degree. Any person who knowingly causes, procures, aids or counsels or creates by means of fire or explosion the damage or destruction of any property of that person or another with the purpose to defraud an insurer shall, upon conviction, be sentenced to imprisonment for not less than two (2) years nor more than twenty (20) years or shall be fined not more than five thousand dollars ($5,000), or both; provided, that if death occurs to a person as a direct result of the fire or explosion or to a person who is directly involved in fighting the fire or explosion, imprisonment shall be for not less than twenty (20) years.

§ 11-4-5 Arson – Fourth degree. Any person who knowingly causes, procures, aids, counsels or creates by means of fire or explosion damage to or destruction of any personal property valued in excess of one hundred dollars ($100) and the property of another person shall, upon conviction, be sentenced to imprisonment for not less than one year nor more than three (3) years, or shall be fined not more than one thousand dollars ($1,000), or both.

§ 11-4-6 Arson – Fifth degree. Any person who knowingly attempts to cause, procure, aid, or counsel by fire or explosion the damage or destruction of any property mentioned in §§ 11-4-2 – 11-4-5 shall, upon conviction, be fined not exceeding one thousand dollars ($1,000), or imprisonment for not less than one year nor more than twenty (20) years.

§ 11-4-7 Arson – Sixth degree. Any person who knowingly causes, procures, aids or counsels the destruction of woodlands by fire which shall run and spread at large shall, upon conviction, be imprisoned not exceeding two (2) years, or shall be fined not more than one thousand dollars ($1,000), or both.

§ 11-4-8 Arson – Seventh degree. Every person who shall make a bonfire in any public street, road, square, land or rotary, without special permission from the local governing body, shall be fined not exceeding one hundred dollars ($100). No complaint for a violation of any of the provisions of this section shall be sustained unless it shall be brought within thirty (30) days after the commission of the offense, and all fines for the violation shall inure one-half (1/2) of the fine to the complainant and one-half (1/2) of the fine to the state. The local governing body may appoint a designee to grant permission under the provisions of this section.

South Carolina General Statutes

§ 16-11-110. Arson.

(A) A person who wilfully and maliciously causes an explosion, sets fire to, burns, or causes to be burned or aids, counsels, or procures a burning that results in damage to a building, structure, or any property specified in subsections (B) and (C) whether the property of himself or another, which results, either directly or indirectly, in death or serious bodily injury to a person is guilty of arson in the first degree and, upon conviction, must be imprisoned not less than ten nor more than thirty years.

(B) A person who wilfully and maliciously causes an explosion, sets fire to, burns, or causes to be burned or aids, counsels, or procures the burning that results in damage to a dwelling house, church or place of worship, a public or private school facility, a manufacturing plant or warehouse, a building where business is conducted, an institutional facility, or any structure designed for human occupancy to include local and municipal buildings, whether the property of himself or another, is guilty of arson in the second degree and, upon conviction, must be imprisoned not less than five nor more than twenty-five years.

(C) A person who wilfully and maliciously:

(1) causes an explosion, sets fire to, burns, or causes a burning which results in damage to a building or structure other than those specified in subsection (A) or (B), a railway car, a ship, boat, or other watercraft, an aircraft, an automobile or other motor vehicle, or personal property; or

(2) aids, counsels, or procures a burning that results in damage to a building or structure other than those specified in subsection (A) or (B), a railway car, a ship, boat, or other watercraft, an aircraft, an automobile or other motor vehicle, or personal property with intent to destroy or damage by explosion or fire; whether the property of himself or another, is guilty of arson in the third degree and, upon conviction, must be imprisoned not less than one and not more than ten years.

(D) For purposes of this section, "damage" means an application of fire or explosive that results in burning, charring, blistering, scorching, smoking, singeing, discoloring, or changing the fiber or composition of a building, structure, or any property specified in this section.

§ 16-11-125. Making false claim or statement in support of claim to obtain insurance benefits for fire or explosion loss.

Any person who wilfully and knowingly presents or causes to be presented a false or fraudulent claim, or any proof in support of such claim, for the payment of a fire loss or loss caused by an explosion, upon any contract of insurance or certificate of insurance which includes benefits for such a loss, or prepares, makes, or subscribes to a false or fraudulent account, certificate, affidavit, or proof of loss, or other documents or writing, with intent that such documents may be presented or used in support of such claim, is guilty of a felony and, upon conviction, must be fined not more than ten thousand dollars or imprisoned for not more than five years or both in the discretion of the court.

The provisions of this section are supplemental to and not in lieu of existing law relating to falsification of documents and penalties therefor.

§ 16-11-130. Burning personal property to defraud insurer.

Any person who (a) wilfully and with intent to injure or defraud an insurer sets fire to or burns or causes to be burned or (b) aids, counsels, or procures the burning of any goods, wares, merchandise, or other chattels or personal property of any kind, whether the property of himself or of another, which is at the time insured by any person against loss or damage by fire is guilty of a felony and, upon conviction, must be imprisoned for not less than one nor more than five years.

§ 16-11-140. Burning of crops, fuel or lumber.

It is unlawful for a person to (a) wilfully and maliciously set fire to or burn or cause to be burned, or (b) aid, counsel, or procure the burning of any:

(1) barracks, cock, crib, rick or stack of hay, corn, wheat, oats, barley, or other grain or vegetable product of any kind;

(2) field of standing hay or grain of any kind;

(3) pile of coal, wood, or other fuel;

(4) pile of planks, boards, posts, rails, or other lumber.

A person who violates the provisions of this section is guilty of a misdemeanor and, upon conviction, must be imprisoned not more than three years.

§ 16-11-150. Burning lands of another without consent.

It shall be unlawful for any person without prior written consent of the landowner or his agent to intentionally set fire to lands of another, or to intentionally cause or allow fire to spread to lands of another, whereby any woods, fields, fences or marshes of any other person are burned. Any person violating the provisions of this section shall, upon conviction, be punished as follows: (a) For the first offense, by a fine of not more than one thousand dollars, or imprisonment for not more than one year, or both, (b) for a second or subsequent offense, by a fine of not more than five thousand dollars or imprisonment for not more than five years.

§ 16-11-160. Carrying fire on lands of another without permit.

It shall be unlawful for any person to carry a lighted torch, chunk or coals of fire in or under any mill or wooden building or over and across any of the enclosed or unenclosed lands of another person at any time without the special permit of the owner of such lands, mill or wooden building, whether any damage result therefrom or not. Any person violating the provisions of this section shall be guilty of a misdemeanor and, upon conviction thereof, shall be subject to imprisonment in the county jail for a term not to exceed thirty days or to a fine not to exceed one hundred dollars.

§ 16-11-170. Wilfully burning lands of another.

It is unlawful for a person to wilfully and maliciously set fire to or burn any grass, brush, or other combustible matter, causing any woods, fields, fences, or marshes of another person to be set on fire or cause the burning or fire to spread to or to be transmitted to the lands of another, or to aid or assist in such conduct.

A person who violates the provisions of this section is guilty of a felony and, upon conviction, must be fined not more than five thousand dollars or imprisoned not more than five years. A person convicted under this section is liable to any person who may have sustained damage.

§ 16-11-180. Negligently allowing fire to spread to lands or property of another.

Any person who carelessly or negligently sets fire to or burns any grass, brush, leaves, or other combustible matter on any lands so as to cause or allow fire to spread or to be transmitted to the lands or property of another, or to burn or injure the lands or property of another, or who causes the burning to be done or who aids or assists in the burning, is guilty of a misdemeanor and, upon conviction, must be imprisoned for not less than five days nor more than thirty days or be fined not less than twenty-five dollars nor more than two hundred dollars. For a second or subsequent offense the sentence must be imprisonment for not less than thirty days nor more than one year, or a fine of not less than one hundred dollars nor more than five hundred dollars, or both, in the discretion of the court.

§ 16-11-190. Attempts to burn.

It is unlawful for a person to wilfully and maliciously attempt to set fire to, burn, or aid, counsel, or procure the burning of any of the buildings or property mentioned in Sections 16-11-110 to 16-11-140 or commit an act in furtherance of burning these buildings.

A person who violates the provisions of this section is guilty of a felony and, upon conviction, must be imprisoned not more than five years or fined not more than ten thousand dollars.

§ 16-11-200. Placing or distributing combustible materials and the like in buildings and property as constituting attempt.

The placing or distributing of any inflammable, explosive or combustible materials or substance or any device in any building or property mentioned in Sections 16-11-110 to 16-11-140 in an arrangement or preparation with intent eventually wilfully and maliciously to set fire to or burn the same or to procure the setting fire to or burning of the same shall for the purposes of Section 16-11-190 constitute an attempt to burn such building or property.

CHAPTER 6

George Harvey MEEK, Appellant, v. The STATE of Texas, Appellee.
790 S.W.2d 618 (1990)
Court of Criminal Appeals of Texas, En Banc.

WHITE, Judge.

On Sunday morning, April 6, 1986, firefighters rushed to the scene of a burning house at 1304 Wedgewood in El Paso. The house's owner, Carol DeWees Meek, ("DeWees") later pleaded guilty to arson in connection with the fire. Appellant, her estranged husband at the time of the fire, was convicted as a party to the offense and sentenced to seven years' confinement in the Texas Department of Corrections. His appeal to the Court of Appeals in El Paso resulted in a reversal of his conviction. Meek v. State, 747 S.W.2d 30 (Tex.App., El Paso 1988). The State then petitioned this Court for discretionary review. After reviewing four closely related grounds for review, we reverse the judgment of the Court of Appeals and affirm appellant's original conviction.(fn1)

The contested issues on appeal revolve around the admissibility of two statements appellant gave on April 8, 1986 at the office of El Paso Fire Department Inspector Zubia. Two versions of the events surrounding the taking of the statements appear in the record.

Inspector Hector Zubia testified that the principal suspect in the arson investigation was the owner of the property, DeWees, who was also the beneficiary of the house's fire insurance policy. (The house had been acquired before DeWees' marriage to appellant and hence he possessed no proprietary interest in it. See V.T.C.A., Family Code Sec. 5.01). DeWees was the house's sole resident on the date of the fire, and the only person witnesses (other than she herself) were able to place at the scene on the morning of the fire. When Zubia questioned DeWees about the fire on Sunday evening, he found her reactions very suspicious. Because he considered her a suspect, he gave her Miranda warnings before taking her statement.

In that statement, DeWees said that appellant had come to her residence on the morning of the fire and had threatened to "torch" her house. (She later admitted on the stand that she had lied to Zubia about this). Zubia testified that DeWees' implication of appellant made little impression on him at the time; he claimed that an interview with appellant had been sought for the sole purpose of gathering more evidence against DeWees, and not that appellant was under investigation.

After leaving his card with a note at appellant's residence on Sunday night, Zubia and his assistant paid appellant a second visit on Monday. Appellant told them that he was currently unavailable because he was working for the Drug Enforcement Administration and had important business to attend to. The fire investigators proceeded to the DEA office where they inquired about appellant; the DEA personnel knew nothing about him. Appellant later admitted on the stand that his DEA story was "a little bit incorrect."

The next day, April 8, appellant appeared voluntarily at Zubia's office. From this point on, the testimony concerning events diverges.

Zubia claimed that appellant was very cooperative during the entire interview. He testified that no threats, promises, or coercion were used to obtain appellant's statements, and that appellant was never deprived of his freedom in any way during his time at the fire department. According to Zubia, appellant was free to leave at any time and could have done so on one occasion when he left the building and went to his automobile to retrieve some papers.

Appellant, on the other hand, testified that he went to Fire Marshall Jackson's office upon arrival at the fire station. Appellant maintains that Zubia handcuffed him, told him he was under arrest (without giving him Miranda warnings), generally intimidated him, and told him that he would be allowed to go if he told Zubia what Zubia wanted to hear. According to appellant's testimony, this intimidation in secret continued for a few hours, after which time appellant's handcuffs were removed and he was taken before an office secretary who recorded his statements. Appellant concedes that he was allowed to leave the station unhindered and that he suffered no formal arrest until May 16, more than a month after his statements were taken.

In the statements, appellant explicitly claims that he did not set the fire, make arrangements for it, or pay anyone to do it. He states his belief that his wife started the fire. However, he also tells of numerous instances when he goaded his wife into setting the fire, pushed her into taking out additional insurance coverage on the house, instructed her how to burn the house, inquired whether acquaintances would be interested in burning the house for her, or purchased supplies which could be used in burning the house. He even describes how he cooked a large quantity of sausage, piled all the sausage near the stove with the cooking grease and instructed his wife to use this rather unique incendiary device in her upcoming arson! These admissions implicating appellant as a party to the crime are couched within a long, disjointed, rambling statement that the fire department secretary testified was possibly the longest she had ever taken.

At a suppression hearing, the trial judge is the sole judge of the credibility of the witnesses and of the weight to be given their testimony. *Cannon v. State,* 691 S.W.2d 664, 673 (Tex.Cr.App.1985), cert. den., *Cannon v. Texas,* 474 U.S. 1110, 106 S.Ct. 897, 88 L.Ed.2d 931 (1986); *Green v. State,* 615 S.W.2d 700, 707 (Tex.Cr.App. 1981). The judge may believe or disbelieve all or any part of a witness's testimony. *Cannon,* 691 S.W.2d at 673. His findings should not be disturbed absent a clear abuse of discretion. *Dancy v. State,* 728 S.W.2d 772, 777 (Tex.Cr.App.1987), cert. den., *Dancy v. Texas,* 484 U.S. 975, 108 S.Ct. 485, 98 L.Ed.2d 484 (1987).

In this case, the trial court implicitly made a finding that appellant was neither handcuffed nor intimidated when he gave his statements to Zubia. At the end of the suppression hearing, the court stated:

All right, gentlemen. You can do a brief if you'd like. I think however, since---the Court is going to rule that this is not a custodial interrogation, that any statement the defendant may have made was probably voluntary. Had the court believed appellant's testimony concerning the handcuffs and intimidation, it could not have found that appellant was not in custody at the fire station.(fn2) Since Zubia's testimony supports the trial court's finding, the finding should not be disturbed on appeal. Therefore we, like the Court of Appeals, are bound to accept Zubia's version of events since it is definitely supported by the record.

We must now review the Court of Appeals' holding that custodial interrogation existed under the facts found by the trial court. The Court of Appeals held that appellant was in custody at least after he began making incriminating remarks because "it defies belief that, when the statement of DeWees regarding the appellant's threat and the above mentioned statement of the appellant are taken in conjunction, the focus of the investigation had not centered upon the Appellant . . . " *Meek,* 747 S.W.2d at 31. Even were we to accept this very debatable conclusion of the Court of

Appeals, we would still have to conclude that the court applied the wrong test for determining the presence of custody. Being the focus of a criminal investigation does not equate to custody. *Beckwith v. United States,* 425 U.S. 341, 96 S.Ct. 1612, 48 L.Ed.2d 1 (1976).

The correct test is much broader and more complex. Although we have used several approaches to determine whether "custody" exists, we have always considered multiple factors. One approach merges the idea of "focus" with the idea of "whether a reasonable person would believe that his freedom was being deprived in a significant way." *Shiflet v. State,* 732 S.W.2d 622, 624 (Tex.Cr.App.1985). Another approach cites four factors as relevant to the inquiry: probable cause to arrest, subjective intent of the police, focus of the investigation, and subjective belief of the defendant." *Wicker v. State,* 740 S.W.2d 779, 786 (Tex.Cr.App.1987), cert. den., *Wicker v. Texas,* 485 U.S. 938, 108 S.Ct. 1117, 99 L.Ed.2d 278 (1988); *Payne v. State,* 579 S.W.2d 932, 933 (Tex.Cr.App. 1979). By making focus solely determinative, the Court of Appeals erred since it failed to consider the other relevant factors. We must now determine whether "custody" existed considering all the relevant circumstances.

The facts of this case are extremely similar to those in *Oregon v. Mathiason,* 429 U.S. 492, 97 S.Ct. 711, 50 L.Ed.2d 714 (1977). The issue in *Mathiason* was whether the defendant was in custody when he made an incriminating confession without having first been warned about his Miranda rights. Even though the interview took place in a State policeman's office behind closed doors at the invitation of the policeman, our nation's highest court held that defendant had not been in custody because he came to the police offices voluntarily and was allowed to leave unmolested, having been told that his case would be referred to the district attorney for evaluation. *Id.* 97 S.Ct. at 713--714.

Several State cases are also on point: *Shiflet v. State,* 732 S.W.2d 622 (Tex.Cr. App. 1985); *Wicker v. State,* 740 S.W.2d 779 (Tex.Cr.App.1987); *Brown v. State,* 475 S.W.2d 938 (Tex.Cr.App.1971); *Dancy v. State,* 728 S.W.2d 772 (Tex.Cr.App.1987); see also *Livingston v. State,* 739 S.W.2d 311 (Tex.Cr.App.1987), cert. den., *Livingston v. Texas,* 487 U.S. 1210, 108 S.Ct. 2858, 101 L.Ed.2d 895 (1988); *Aredondo v. State,* 694 S.W.2d 378 (Tex.App.---Corpus Christi, 1985).

In *Shiflet,* this Court held that when a person voluntarily accompanies officers to a location, and he knows or should know that he is a suspect in the crime the officers are investigating, the person is not in custody. Shiflet accompanied officers voluntarily to Austin for a polygraph examination in connection with a homicide investigation. He was not in custody even after he began to make oral admissions at the polygraph examiner's office. *Shiflet,* 732 S.W.2d at 631.

Wicker involved a sex offender who voluntarily went to a Texas Department of Human Services social worker and confessed to having sexual intercourse with his daughter. When the confession was later used against him, Wicker objected that the lack of prior Miranda warnings rendered his confession inadmissible. We held that Mathiason controlled and that there had been no custodial interrogation because there was nothing to suggest that Wicker's freedom of action was in any way hampered before, during, or after giving his statement; he arrived and departed a free man, under no form of restraint or arrest (other than his probation for a previous offense.) *Wicker,* 740 S.W.2d at 786---787.

In Brown, the defendant and his uncle voluntarily came to the District Attorney's office to make a statement in a homicide investigation; they voluntarily left afterwards. We held that the investigation had not yet focused on defendant and that since the D.A. and police were only making a general investigation into an unsolved crime and were checking on all sources of information about the deaths involved, there was no custody for Miranda purposes when defendant's statements were made. *Brown,* 475 S.W.2d at 951. *Dancy* is very similar to the *Shiflet* case cited above. Defendant voluntarily accompanied officers to the police station after they recovered his letter jacket and comb from the scene of a murder under investigation. There was ample evidence to support a finding that Dancy was not arrested, but freely went to the station in hopes of recovering his property. *Dancy,* 728 S.W.2d at 777. We held that even after he began giving the police physical evidence for their investigation at the station, he was still not in custody. *Dancy,* 728 S.W.2d at 779. *Dancy* and the cases cited previously mandate our holding today that appellant was not in custody when he made his two statements at the fire station.

Appellant cites two inapposite cases in support of his contention that he was in custody at the fire station. In *McCrory v. State,* 643 S.W.2d 725 (Tex.Cr.App.1982), we held that defendant was in custody after he voluntarily accompanied officers to a polygraph examiner's office and confessed to committing the murder, while officers watched from behind a one way mirror. That case is distinguished from this one in that *McCrory* was never allowed to leave the officers' presence after his confession, whereas appellant remained free for over five weeks after his confession. Also, *McCrory* made a straightforward declaration "I did it" which the officers recognized immediately as a confession. Appellant, on the other hand, only made statements which could possibly make him guilty as a party, a much more subtle confession to recognize (especially for a non-lawyer), and his possibly incriminating statements were embedded in a distracting, rambling narrative. Appellant's other case, *Ruth v. State,* 645 S.W.2d 432 (Tex.Cr.App.1979), also involved a defendant who made a statement to the police and was immediately arrested, and therein lies its distinction from this case.

Examining all the relevant factors, we note that Inspector Zubia's testimony reflects the fact that DeWees was the sole focus of his investigation at the time of appellant's interviews. We also note that the trial court disbelieved appellant's story about being handcuffed and intimidated, and believed Zubia's account that no compulsion of any kind was used on appellant. The facts as found by the trial court indicate that appellant came to the fire station of his own free will at a time of his own choosing, was apparently allowed to step outside the building and go unaccompanied to his car during the interviews, was allowed to leave unhindered after the statements were taken, and was not formally arrested or detained in any way until five weeks later. These conditions and the precedent cited above mandate a holding that appellant was not in custody during his visit to Zubia's office; hence it was not error for his statements to be used against him even though Miranda warnings were not first administered before the statements were taken.

The State's first ground for review is sustained.

The judgment of the Court of Appeals is reversed and the judgment of the trial court is hereby affirmed.

TEAGUE, J., concurs in the result.

Footnotes:

1. The State's grounds for review were essentially as follows:

(1) Is an individual in custody under *Miranda v. Arizona* [under the circumstances of this case as described in Inspector Zubia's testimony]?

(2) Was appellant in custody (for purposes of Miranda v. Arizona) at the time he gave his statements where appellant was not arrested until more than a month after he gave the statements?

(3) Were appellant's statements, given under the above circumstances, admissible in evidence at his trial?

(4) Did the Court of Appeals err in giving credence to appellant's testimony [when the trial court disbelieved it]?

We find ground (2) to be a subset of ground (1), and ground (3) to be a less specific restatement of ground (1). Ground (4) presents a subissue which must be resolved in order to address ground (1). We shall therefore address only the first ground, but in so doing we shall adequately answer the State's entire petition.

2. The Supreme Court, in *Miranda,* supra, defined custodial interrogation as "questioning initiated by law enforcement officers after a person has been taken into custody or otherwise deprived of his freedom of action in any significant way." On another occasion, the Supreme Court described the deprivation of freedom required for "custody" to exist as "the restraint on freedom of movement of the degree associated with a formal arrest." *Minnesota v. Murphy,* 465 U.S. 420, 430, 104 S.Ct. 1136, 1144, 79 L.Ed.2d 409, 421 (1984). A person handcuffed by a law enforcement official in the official's office would surely meet these standards.

CHAPTER 8

NEW JERSEY LAW ABOLISHING FIREMAN'S RULE

2A:62A-21.   Additional right of action, recovery

1. In addition to any other right of action or recovery otherwise available under law, whenever any law enforcement officer, firefighter or member of a duly incorporated first aid, emergency, ambulance or rescue squad association suffers any injury, disease or death while in the lawful discharge of his official duties and that injury, disease or death is directly or indirectly the result of the neglect, willful omission, or willful or culpable conduct of any person or entity, other than that law enforcement officer, firefighter or first aid, emergency, ambulance or rescue squad member's employer or co-employee, the law enforcement officer, firefighter, or first aid, emergency, ambulance or rescue squad member suffering that injury or disease, or, in the case of death, a representative of that law enforcement officer, firefighter or first aid, emergency, ambulance or rescue squad member's estate, may seek recovery and damages from the person or entity whose neglect, willful omission, or willful or culpable conduct resulted in that injury, disease or death.

CHAPTER 9

SAMPLE TORT CLAIMS ACTS

1. Federal Tort Claims Act

28 USC 2671

Sec. 2671. Definitions

As used in this chapter and sections 1346(b) and 2401(b) of this title, the term "Federal agency" includes the executive departments, the judicial and legislative branches, the military departments, independent establishments of the United States, and corporations primarily acting as instrumentalities or agencies of the United States, but does not include any contractor with the United States.

"Employee of the government" includes (1) officers or employees of any federal agency, members of the military or naval forces of the United States, members of the National Guard while engaged in training or duty under section 115, 316, 502, 503, 504, or 505 of title 32, and persons acting on behalf of a federal agency in an official capacity, temporarily or permanently in the service of the United States, whether with or without compensation, and (2) any officer or employee of a Federal public defender organization, except when such officer or employee performs professional services in the course of providing representation under section 3006A of title 18.

"Acting within the scope of his office or employment", in the case of a member of the military or naval forces of the United States or a member of the National Guard as defined in section 101(3) of title 32, means acting in line of duty.

Sec. 2672. Administrative adjustment of claims

The head of each Federal agency or his designee, in accordance with regulations prescribed by the Attorney General, may consider, ascertain, adjust, determine, compromise, and settle any claim for money damages against the United States for injury or loss of property or personal injury or death caused by the negligent or wrongful act or omission of any employee of the agency while acting within the scope of his office or employment, under circumstances where the United States, if a private person, would be liable to the claimant in accordance with the law of the place where the act or omission occurred: Provided, That any award, compromise, or settlement in excess of $25,000 shall be effected only with the prior written approval of the Attorney General or his designee. Notwithstanding the proviso contained in the preceding sentence, any award, compromise, or settlement may be effected without the prior written approval of the Attorney General or his or her designee, to the extent that the Attorney General delegates to the head of the agency the authority to make such award, compromise, or settlement. Such delegations may not exceed the authority delegated by the Attorney General to the United States attorneys to settle claims for money damages against the United States. Each Federal agency may use arbitration, or other alternative means of dispute resolution under the provisions of subchapter IV of chapter 5 of title 5, to settle any tort claim against the United States, to the extent of the agency's authority to award, compromise, or settle such claim without the prior written approval of the Attorney General or his or her designee.

Subject to the provisions of this title relating to civil actions on tort claims against the United States, any such award, compromise, settlement, or determination shall be final and conclusive on all officers of the Government, except when procured by means of fraud.

Any award, compromise, or settlement in an amount of $2,500 or less made pursuant to this section shall be paid by the head of the Federal agency concerned out of appropriations available to that agency. Payment of any award, compromise, or settlement in an amount in excess of $2,500 made pursuant to this section or made by the Attorney General in any amount pursuant to section 2677 of this title shall be paid in a manner similar to judgments and compromises in like causes and appropriations or funds available for the payment of such judgments and compromises are hereby made available for the payment of awards, compromises, or settlements under this chapter.

The acceptance by the claimant of any such award, compromise, or settlement shall be final and conclusive on the claimant, and shall constitute a complete release of any claim against the United States and against the employee of the government whose act or omission gave rise to the claim, by reason of the same subject matter.

Sec. 2673. Reports to Congress

The head of each federal agency shall report annually to Congress all claims paid by it under section 2672 of this title, stating the name of each claimant, the amount claimed, the amount awarded, and a brief description of the claim.

Sec. 2674. Liability of United States

The United States shall be liable, respecting the provisions of this title relating to tort claims, in the same manner and to the same extent as a private individual under like circumstances, but shall not be liable for interest prior to judgment or for punitive damages.

If, however, in any case wherein death was caused, the law of the place where the act or omission complained of occurred provides, or has been construed to provide, for damages only punitive in nature, the United States shall be liable for actual or compensatory damages, measured by the pecuniary injuries resulting from such death to the persons respectively, for whose benefit the action was brought, in lieu thereof.

With respect to any claim under this chapter, the United States shall be entitled to assert any defense based upon judicial or legislative immunity which otherwise would have been available to the employee of the United States whose act or omission gave rise to the claim, as well as any other defenses to which the United States is entitled.

With respect to any claim to which this section applies, the Tennessee Valley Authority shall be entitled to assert any defense which otherwise would have been available to the employee based upon judicial or legislative immunity, which otherwise would have been available to the employee of the Tennessee Valley Authority whose act or omission gave rise to the claim as well as any other defenses to which the Tennessee Valley Authority is entitled under this chapter.

Sec. 2675. Disposition by federal agency as prerequisite; evidence

(a) An action shall not be instituted upon a claim against the United States for money damages for injury or loss of property or personal injury or death caused by the negligent or wrongful act or omission of any employee of the Government while acting within the scope of his office or employment, unless the claimant shall have first presented the claim to the appropriate Federal agency and his claim shall have been finally denied by the agency in writing and sent by certified or registered mail. The failure of an agency to make final disposition of a claim within six months after it is filed shall, at the option of the claimant any time thereafter, be deemed a final denial of the claim for purposes of this section. The provisions of this subsection shall not apply to such claims as may be asserted under the Federal Rules of Civil Procedure by third party complaint, cross-claim, or counterclaim.

(b) Action under this section shall not be instituted for any sum in excess of the amount of the claim presented to the federal agency, except where the increased amount is based upon newly discovered evidence not reasonably discoverable at the time of presenting the claim to the federal agency, or upon allegation and proof of intervening facts, relating to the amount of the claim.

(c) Disposition of any claim by the Attorney General or other head of a federal agency shall not be competent evidence of liability or amount of damages.

Sec. 2676. Judgment as bar

The judgment in an action under section 1346(b) of this title shall constitute a complete bar to any action by the claimant, by reason of the same subject matter, against the employee of the government whose act or omission gave rise to the claim.

Sec. 2677. Compromise

The Attorney General or his designee may arbitrate, compromise, or settle any claim cognizable under section 1346(b) of this title, after the commencement of an action thereon.

Sec. 2678. Attorney fees; penalty

No attorney shall charge, demand, receive, or collect for services rendered, fees in excess of 25 per centum of any judgment rendered pursuant to section 1346(b) of this title or any settlement made pursuant to section 2677 of this title, or in excess of 20 per centum of any award, compromise, or settlement made pursuant to section 2672 of this title.

Any attorney who charges, demands, receives, or collects for services rendered in connection with such claim any amount in excess of that allowed under this section, if recovery be had, shall be fined not more than $2,000 or imprisoned not more than one year, or both.

Sec. 2679. Exclusiveness of remedy

(a) The authority of any federal agency to sue and be sued in its own name shall not be construed to authorize suits against such federal agency on claims which are cognizable under section 1346(b) of this title, and the remedies provided by this title in such cases shall be exclusive.

(b)(1) The remedy against the United States provided by sections 1346(b) and 2672 of this title for injury or loss of property, or personal injury or death arising or resulting from the negligent or wrongful act or omission of any employee of the Government while acting within the scope of his office or employment is exclusive of any other civil action or proceeding for money damages by reason of the same subject matter against the employee whose act or omission gave rise to the claim or against the estate of such employee. Any other civil action or proceeding for money damages arising out of or relating to the same subject matter against the employee or the employee's estate is precluded without regard to when the act or omission occurred.

(2) Paragraph (1) does not extend or apply to a civil action against an employee of the Government—

(A) which is brought for a violation of the Constitution of the United States, or

(B) which is brought for a violation of a statute of the United States under which such action against an individual is otherwise authorized.

(c) The Attorney General shall defend any civil action or proceeding brought in any court against any employee of the Government or his estate for any such damage or injury. The employee against whom such civil action or proceeding is brought shall deliver within such time after date of service or knowledge of service as determined by the Attorney General, all process served upon him or an attested true copy thereof to his immediate superior or to whomever was designated by the head of his department to receive such papers and such person shall promptly furnish copies of the pleadings and process therein to the United States attorney for the district embracing the place wherein the proceeding is brought, to the Attorney General, and to the head of his employing Federal agency.

(d)(1) Upon certification by the Attorney General that the defendant employee was acting within the scope of his office or employment at the time of the incident out of which the claim arose, any civil action or proceeding commenced upon such claim in a United States district court shall be deemed an action against the United States under the provisions of this title and all references thereto, and the United States shall be substituted as the party defendant.

(2) Upon certification by the Attorney General that the defendant employee was acting within the scope of his office or employment at the time of the incident out of which the claim arose, any civil action or proceeding commenced upon such claim in a State court shall be removed without bond at any time before trial by the Attorney General to the district court of the United States for the district and division embracing the place in which the action or proceeding is pending. Such action or proceeding shall be deemed to be an action or proceeding brought against the United States under the provisions of this title and all references thereto, and the United States shall be substituted as the party defendant. This certification of the Attorney General shall conclusively establish scope of office or employment for purposes of removal.

(3) In the event that the Attorney General has refused to certify scope of office or employment under this section, the employee may at any time before trial petition the court to find and certify that the employee was acting within the scope of his office or employment. Upon such certification by the court, such action or proceeding shall be deemed to be an action or proceeding brought against the United States under the provisions of this title and all references thereto, and the United States shall be substituted as the party defendant. A copy of the petition shall be served upon the United States in accordance with the provisions of Rule 4(d)(4) of the Federal Rules of Civil Procedure. In the event the petition is filed in a civil action or proceeding pending in a State court, the action or proceeding may be removed without bond by the Attorney General to the district court of the United States for the district and division embracing the place in which it is pending. If, in considering the petition, the district court determines that the employee was not

acting within the scope of his office or employment, the action or proceeding shall be remanded to the State court.

(4) Upon certification, any action or proceeding subject to paragraph (1), (2), or (3) shall proceed in the same manner as any action against the United States filed pursuant to section 1346(b) of this title and shall be subject to the limitations and exceptions applicable to those actions.

(5) Whenever an action or proceeding in which the United States is substituted as the party defendant under this subsection is dismissed for failure first to present a claim pursuant to section 2675(a) of this title, such a claim shall be deemed to be timely presented under section 2401(b) of this title if—

(A) the claim would have been timely had it been filed on the date the underlying civil action was commenced, and

(B) the claim is presented to the appropriate Federal agency within 60 days after dismissal of the civil action.

(e) The Attorney General may compromise or settle any claim asserted in such civil action or proceeding in the manner provided in section 2677, and with the same effect.

Sec. 2680. Exceptions

The provisions of this chapter and section 1346(b) of this title shall not apply to—

(a) Any claim based upon an act or omission of an employee of the Government, exercising due care, in the execution of a statute or regulation, whether or not such statute or regulation be valid, or based upon the exercise or performance or the failure to exercise or perform a discretionary function or duty on the part of a federal agency or an employee of the Government, whether or not the discretion involved be abused.

(b) Any claim arising out of the loss, miscarriage, or negligent transmission of letters or postal matter.

(c) Any claim arising in respect of the assessment or collection of any tax or customs duty, or the detention of any goods, merchandise, or other property by any officer of customs or excise or any other law enforcement officer, except that the provisions of this chapter and section 1346(b) of this title apply to any claim based on injury or loss of goods, merchandise, or other property, while in the possession of any officer of customs or excise or any other law enforcement officer, if—

(1) the property was seized for the purpose of forfeiture under any provision of Federal law providing for the forfeiture of property other than as a sentence imposed upon conviction of a criminal offense;

(2) the interest of the claimant was not forfeited;

(3) the interest of the claimant was not remitted or mitigated (if the property was subject to forfeiture); and

(4) the claimant was not convicted of a crime for which the interest of the claimant in the property was subject to forfeiture under a Federal criminal forfeiture law

(d) Any claim for which a remedy is provided by sections 741-752, 781-790 of Title 46, relating to claims or suits in admiralty against the United States.

(e) Any claim arising out of an act or omission of any employee of the Government in administering the provisions of sections 1-31 of Title 50, Appendix.

(f) Any claim for damages caused by the imposition or establishment of a quarantine by the United States.

(g) Repealed

(h) Any claim arising out of assault, battery, false imprisonment, false arrest, malicious prosecution, abuse of process, libel, slander, misrepresentation, deceit, or interference with contract rights: Provided, That, with regard to acts or omissions of investigative or law enforcement officers of the United States Government, the provisions of this chapter and section 1346(b) of this title shall apply to any claim arising, on or after the date of the enactment of this proviso, out of assault, battery, false imprisonment, false arrest, abuse of process, or malicious prosecution. For the purpose of this subsection, "investigative or law enforcement officer" means any officer of the United States who is empowered by law to execute searches, to seize evidence, or to make arrests for violations of Federal law.

(i) Any claim for damages caused by the fiscal operations of the Treasury or by the regulation of the monetary system.

(j) Any claim arising out of the combatant activities of the military or naval forces, or the Coast Guard, during time of war.

(k) Any claim arising in a foreign country.

(l) Any claim arising from the activities of the Tennessee Valley Authority.

(m) Any claim arising from the activities of the Panama Canal Company.

(n) Any claim arising from the activities of a Federal land bank, a Federal intermediate credit bank, or a bank for cooperatives.

2. Nevada Tort Claims Act

Chapter 41 Nevada Revised Statutes. LIABILITY OF AND ACTIONS AGAINST THIS STATE, ITS AGENCIES AND POLITICAL SUBDIVISIONS

General Provisions

NRS 41.0305 "Political subdivision" defined. As used in NRS 41.0305 to 41.039, inclusive, the term "political subdivision" includes an organization that was officially designated as a community action agency pursuant to 42 U.S.C. § 2790 before that section was repealed and is included in the definition of an "eligible entity" pursuant to 42 U.S.C. § 9902, the Nevada Rural Housing Authority, an airport authority created by special act of the Legislature, a regional transportation commission and a fire protection district, irrigation district, school district, governing body of a charter school and other special district that performs a governmental function, even though it does not exercise general governmental powers.

NRS 41.0307 "Employee," "employment," "immune contractor," "public officer" and "officer" defined. As used in NRS 41.0305 to 41.039, inclusive:

1. "Employee" includes an employee of a:

(a) Part-time or full-time board, commission or similar body of the State or a political subdivision of the State which is created by law.

(b) Charter school.

2. "Employment" includes any services performed by an immune contractor.

3. "Immune contractor" means any natural person, professional corporation or professional association which:

(a) Is an independent contractor with the State pursuant to NRS 284.173; and

(b) Contracts to provide medical services for the Department of Corrections.

As used in this subsection, "professional corporation" and "professional association" have the meanings ascribed to them in NRS 89.020.

4. "Public officer" or "officer" includes:

(a) A member of a part-time or full-time board, commission or similar body of the State or a political subdivision of the State which is created by law.

(b) A public defender and any deputy or assistant attorney of a public defender or an attorney appointed to defend a person for a limited duration with limited jurisdiction.

(c) A district attorney and any deputy or assistant district attorney or an attorney appointed to prosecute a person for a limited duration with limited jurisdiction.

NRS 41.0308 Volunteer crossing guard for county school district deemed employee of political subdivision of State if he has completed approved training. For the purposes of NRS 41.0305 to 41.039, inclusive, a person who volunteers to a county school district or to a local law enforcement agency to serve as a crossing guard for a county school district shall be deemed an employee of a political subdivision of the State if he has successfully completed a training course in traffic safety that has been approved by a local law enforcement agency.

NRS 41.0309 Employee of or volunteer for public fire-fighting agency deemed employee of State or political subdivision of State. For the purposes of NRS 41.0305 to 41.039, inclusive, an employee of or volunteer for a public fire-fighting agency shall be deemed an employee of the State or a political subdivision of the State.

Waiver of Sovereign Immunity

NRS 41.031 Waiver applies to State and its political subdivisions; naming State as defendant; service of process; State does not waive immunity conferred by Eleventh Amendment.

1. The State of Nevada hereby waives its immunity from liability and action and hereby consents to have its liability determined in accordance with the same rules of law as are applied to civil actions against natural persons and corporations, except as otherwise provided in NRS 41.032 to 41.038, inclusive, 485.318, subsection 3 and any statute which expressly provides for governmental immunity, if the claimant complies with the limitations of NRS 41.010 or the limitations of NRS 41.032 to 41.036, inclusive. The State of Nevada further waives the immunity from liability and action of all political subdivisions of the State, and their liability must be determined in the same manner, except as otherwise provided in NRS 41.032 to 41.038, inclusive, subsection 3 and any statute which expressly provides for governmental immunity, if the claimant complies with the limitations of NRS 41.032 to 41.036, inclusive.

2. An action may be brought under this section against the State of Nevada or any political subdivision of the State. In any action against the State of Nevada, the action must be brought in the name of the State of Nevada on relation of the particular department, commission, board or other agency of the State whose actions are the basis for the suit. An action against the State of Nevada must be filed in the county where the cause or some part thereof arose or in Carson City. In an action against the State of Nevada, the summons and a copy of the complaint must be served upon:

(a) The Attorney General, or a person designated by the Attorney General, at the Office of the Attorney General in Carson City; and

(b) The person serving in the office of administrative head of the named agency.

3. The State of Nevada does not waive its immunity from suit conferred by Amendment XI of the Constitution of the United States.

Conditions and Limitations on Actions

NRS 41.032 Acts or omissions of officers, employees and immune contractors. Except as provided in NRS 278.0233 no action may be brought under NRS 41.031 or against an immune contractor or an officer or employee of the State or any of its agencies or political subdivisions which is:

1. Based upon an act or omission of an officer, employee or immune contractor, exercising due care, in the execution of a statute or regulation, whether or not such statute or regulation is valid, if the statute or regulation has not been declared invalid by a court of competent jurisdiction; or

2. Based upon the exercise or performance or the failure to exercise or perform a discretionary function or duty on the part of the State or any of its agencies or political subdivisions or of any officer, employee or immune contractor of any of these, whether or not the discretion involved is abused.

NRS 41.0321 Actions concerning incorrect date generated by computer or information system; contracts must contain immunity provision.

1. No cause of action, including, without limitation, any civil action or action for declaratory or injunctive relief, may be brought under NRS 41.031 or against an immune contractor or an officer or employee of the State or any of its agencies or political subdivisions on the basis that a computer or other information system that is owned or operated by any of those persons produced, calculated or generated an incorrect date, regardless of the cause of the error.

2. Any contract entered into by or on behalf of and in the capacity of the State of Nevada, an immune contractor or an officer or employee of the State or any of its agencies or political subdivisions must include a provision that provides immunity to those persons for any breach of contract that is caused by an incorrect date being produced, calculated or generated by a computer or other information system that is owned or operated by any of those persons, regardless of the cause of the error.

3. Any contract subject to the provisions of this section that is entered into on or after June 30, 1997, has the legal effect of including the immunity required by this section, and any provision of the contract which is in conflict with this section is void.

NRS 41.0322 Actions by persons in custody of Department of Corrections to recover compensation for loss or injury.

1. A person who is or was in the custody of the Department of Corrections may not proceed with any action against the Department or any of its agents, former officers, employees or contractors to recover compensation for the loss of his personal property, property damage, personal injuries or any other claim arising out of a tort pursuant to NRS 41.031 unless the person has exhausted his administrative remedies provided by NRS 209.243 and the regulations adopted pursuant thereto.

2. The filing of an administrative claim pursuant to NRS 209.243 is not a condition precedent to the filing of an action pursuant to NRS 41.031.

3. An action filed by a person in accordance with this section before the exhaustion of his administrative remedies must be stayed by the court in which the action is filed until the administrative remedies are exhausted. The court shall dismiss the action if the person has not timely filed his administrative claim pursuant to NRS 209.243.

4. If a person has exhausted his administrative remedies and has filed and is proceeding with a civil action to recover compensation for the loss of his personal property, property damage, personal injuries or any other claim arising out of a tort, the Office of the Attorney General must initiate and conduct all negotiations for settlement relating to that action.

NRS 41.0325 Negligence or willful misconduct of minor driver in legal custody of State. No action may be commenced pursuant to subsection 2 of NRS 483.300 against the State or an officer or employee of the State for damages caused by the negligence or willful misconduct of a minor driver whose application for a driver's license was signed by the officer or employee while the minor was in the legal custody of the State.

NRS 41.0327 Injuries arising from acts incident to certain solicitations of charitable contributions. No action may be brought under NRS 41.031 or against an officer or employee of the State or any of its agencies or political subdivisions which is based upon any injuries to any person or property arising from or incident to the act of solicitation permitted pursuant to NRS 244.3555, 268.423 or 408.601.

NRS 41.033 Failure to inspect or discover hazards, deficiencies or other matters; inspection does not create warranty or assurance concerning hazards, deficiencies or other matters.

1. No action may be brought under NRS 41.031 or against an officer or employee of the State or any of its agencies or political subdivisions which is based upon:

(a) Failure to inspect any building, structure, vehicle, street, public highway or other public work, facility or improvement to determine any hazards, deficiencies or other matters, whether or not there is a duty to inspect; or

(b) Failure to discover such a hazard, deficiency or other matter, whether or not an inspection is made.

2. An inspection conducted with regard to a private building, structure, facility or improvement constitutes a public duty and does not warrant or ensure the absence of any hazard, deficiency or other matter.

NRS 41.0331 Construction of fence or other safeguard around dangerous condition at abandoned mine. A person, the State of Nevada, any political subdivision of the State, any agency of the State or any agency of its political subdivisions is immune from civil liability for damages sustained as a result of any act or omission by him or it in constructing, or causing to be constructed, pursuant to standards prescribed by the Commission on Mineral Resources, a fence or other safeguard around an excavation, shaft, hole or other dangerous condition at an abandoned mine for which the person, State, political subdivision or agency is not otherwise responsible.

NRS 41.0332 Acts or omissions of volunteer school crossing guards. County school districts and local law enforcement agencies are not liable for the negligent acts or omissions of a person who volunteers to serve as a crossing guard, unless:

1. The volunteer made a specific promise or representation to a natural person who relied upon the promise or representation to his detriment; or

2. The conduct of the volunteer affirmatively caused the harm.

The provisions of this section are not intended to abrogate the principle of common law that the duty of governmental entities to provide services is a duty owed to the public, not to individual persons.

NRS 41.0333 Acts or omissions of members or employees of Nevada National Guard. No action may be brought under NRS 41.031 or against the State of Nevada or the Nevada National Guard or a member or employee of the Nevada National Guard which action is based upon an act or omission of the member or employee while engaged in state or federal training of the Nevada National Guard or duty as prescribed by Title 32 of U.S.C., or regulations adopted pursuant thereto, whether such training or duty is performed within or without the boundaries of this state.

NRS 41.0334 Persons engaged in certain criminal acts in or on public buildings or vehicles; exceptions.

1. Except as otherwise provided in subsection 2, no action may be brought under NRS 41.031 or against an officer or employee of the State or any of its agencies or political subdivisions for injury, wrongful death or other damage sustained in or on a public building or public vehicle by a person who was engaged in any criminal act proscribed in NRS 202.810, 205.005 to 205.080, inclusive, 205.220, 205.226, 205.228, 205.240, 205.271 to 205.2741, inclusive, 206.310, 206.330, 207.210, 331.200 or 393.410, at the time the injury, wrongful death or damage was caused.

2. Subsection 1 does not apply to any action for injury, wrongful death or other damage:

(a) Intentionally caused or contributed to by an officer or employee of the State or any of its agencies or political subdivisions; or

(b) Resulting from the deprivation of any rights, privileges or immunities secured by the United States Constitution or the Constitution of the State of Nevada.

3. As used in this section:

(a) "Public building" includes every house, shed, tent or booth, whether or not completed, suitable for affording shelter for any human being or as a place where any property is or will be kept for use, sale or deposit, and the grounds appurtenant thereto; and

(b) "Public vehicle" includes every device in, upon or by which any person or property is or may be transported or drawn upon a public highway, waterway or airway, owned, in whole or in part, possessed, used by or leased to the State or any of its agencies or political subdivisions.

NRS 41.0335 Actions against certain officers and employees of political subdivisions for acts or omissions of other persons.

1. No action may be brought against:

(a) A sheriff or county assessor which is based solely upon any act or omission of a deputy;

(b) A chief of a police department which is based solely upon any act or omission of an officer of the department;

(c) A chief of a fire department which is based solely upon any act or omission of a fireman or other person called to assist the department;

(d) A member of the board of trustees of a county school district, the superintendent of schools of that school district or the principal of a school, which is based solely upon any act or omission of a person volunteering as a crossing guard; or

(e) A chief of a local law enforcement agency which is based solely on any act or omission of a person volunteering as a crossing guard.

2. This section does not:

(a) Limit the authority of the State or a political subdivision or a public corporation of the State to bring an action on any bond or insurance policy provided pursuant to law for or on behalf of any person who may be aggrieved or wronged.

(b) Limit or abridge the jurisdiction of any court to render judgment upon any such bond or insurance policy for the benefit of any person so aggrieved or wronged.

NRS 41.0336 Acts or omissions of firemen or law enforcement officers. A fire department or law enforcement agency is not liable for the negligent acts or omissions of its firemen or officers or any other persons called to assist it, nor are the individual officers, employees or volunteers thereof, unless:

1. The fireman, officer or other person made a specific promise or representation to a natural person who relied upon the promise or representation to his detriment; or

2. The conduct of the fireman, officer or other person affirmatively caused the harm.
The provisions of this section are not intended to abrogate the principle of common law that the duty of governmental entities to provide services is a duty owed to the public, not to individual persons.

NRS 41.03365 Actions concerning equipment or personal property donated in good faith to volunteer fire department. No action may be brought under NRS 41.031 or against an immune contractor or an officer or employee of the State or any of its agencies or political subdivisions for damages caused by any equipment or other personal property that was provided by any of them, in good faith and without charge, to a volunteer fire department for use by the volunteer fire department in carrying out its duties.

NRS 41.0337 State or political subdivision to be named party defendant. No tort action arising out of an act or omission within the scope of his public duties or employment may be brought against any present or former:

1. Officer or employee of the State or of any political subdivision;

2. Immune contractor; or

3. State Legislator,

unless the State or appropriate political subdivision is named a party defendant under NRS 41.031.

Verdict, Judgment, Damages and Indemnification

NRS 41.03475 No judgment against State or political subdivision permitted for acts outside scope of public duties or employment; exception. Except as otherwise provided in NRS 41.745, no judgment may be entered against the State of Nevada or any agency of the State or against any political subdivision of the State for any act or omission of any present or former officer, employee, immune contractor, member of a board or commission, or Legislator which was outside the course and scope of his public duties or employment.

NRS 41.0348 Special verdict required. In every action or proceeding in any court of this state in which both the State or political subdivision and any present or former officer, employee, immune contractor or member of a board or commission thereof or any present or former Legislator are named defendants, the court or jury in rendering any final judgment, verdict, or other disposition shall return a special verdict in the form of written findings which determine whether:

1. The individual defendant was acting within the scope of his public duty or employment; and

2. The alleged act or omission by the individual defendant was wanton or malicious.

NRS 41.0349 Indemnification of present or former public officer, employee, immune contractor or Legislator. In any civil action brought against any present or former officer, employee, immune contractor, member of a board or commission of the State or a political subdivision or State Legislator, in which a judgment is entered against the defendant based on any act or omission relating to his public duty or employment, the State or political subdivision shall indemnify him unless:

1. The person failed to submit a timely request for defense;

2. The person failed to cooperate in good faith in the defense of the action;

3. The act or omission of the person was not within the scope of his public duty or employment; or

4. The act or omission of the person was wanton or malicious.

NRS 41.035 Limitation on award for damages in tort actions.

1. An award for damages in an action sounding in tort brought under NRS 41.031 or against a present or former officer or employee of the State or any political subdivision, immune contractor or State Legislator arising out of an act or omission within the scope of his public duties or

employment may not exceed the sum of $50,000, exclusive of interest computed from the date of judgment, to or for the benefit of any claimant. An award may not include any amount as exemplary or punitive damages.

2. The limitations of subsection 1 upon the amount and nature of damages which may be awarded apply also to any action sounding in tort and arising from any recreational activity or recreational use of land or water which is brought against:

(a) Any public or quasi-municipal corporation organized under the laws of this State.

(b) Any person with respect to any land or water leased or otherwise made available by that person to any public agency.

(c) Any Indian tribe, band or community whether or not a fee is charged for such activity or use. The provisions of this paragraph do not impair or modify any immunity from liability or action existing on February 26, 1968, or arising after February 26, 1968, in favor of any Indian tribe, band or community.

The Legislature declares that the purpose of this subsection is to effectuate the public policy of the State of Nevada by encouraging the recreational use of land, lakes, reservoirs and other water owned or controlled by any public or quasi-municipal agency or corporation of this State, wherever such land or water may be situated.

Miscellaneous Provisions

NRS 41.036 Filing tort claim against State with Attorney General; filing tort claim against political subdivision with governing body; review and investigation by Attorney General of tort claim against State; regulations by State Board of Examiners.

1. Each person who has a claim against the State or any of its agencies arising out of a tort must file his claim within 2 years after the time the cause of action accrues with the Attorney General.

2. Each person who has a claim against any political subdivision of the State arising out of a tort must file his claim within 2 years after the time the cause of action accrues with the governing body of that political subdivision.

3. The filing of a claim in tort against the State or a political subdivision as required by subsections 1 and 2 is not a condition precedent to bringing an action pursuant to NRS 41.031.

4. The Attorney General shall, if authorized by regulations adopted by the State Board of Examiners pursuant to subsection 6, approve, settle or deny each claim that is:

(a) Filed pursuant to subsection 1; and

(b) Not required to be passed upon by the Legislature.

5. If the Attorney General is not authorized to approve, settle or deny a claim filed pursuant to subsection 1, the Attorney General shall investigate the claim and submit a report of findings to the State Board of Examiners concerning that claim.

6. The State Board of Examiners shall adopt regulations that specify:

(a) The type of claim that the Attorney General is required to approve, settle or deny pursuant to subsection 4; and

(b) The procedure to be used by the Attorney General to approve, settle or deny that claim.

NRS 41.037 Administrative settlement of claims or actions.

1. Upon receiving a report of findings pursuant to subsection 5 of NRS 41.036, the State Board of Examiners may approve, settle or deny any claim or action against the State, any of its agencies or any of its present or former officers, employees, immune contractors or Legislators.

2. Upon approval of a claim by the State Board of Examiners or the Attorney General pursuant to subsection 4 of NRS 41.036:

(a) The State Controller shall draw his warrant for the payment of the claim; and

(b) The State Treasurer shall pay the claim from:

(1) The Fund for Insurance Premiums; or

(2) The Reserve for Statutory Contingency Account.

3. The governing body of any political subdivision whose authority to allow and approve claims is not otherwise fixed by statute may:

(a) Approve, settle or deny any claim or action against that subdivision or any of its present or former officers or employees; and

(b) Pay the claim or settlement from any money appropriated or lawfully available for that purpose.

NRS 41.0375 Agreement to settle: Prohibited contents; required contents; constitutes public record; void under certain circumstances.

1. Any agreement to settle a claim or action brought under NRS 41.031 or against a present or former officer or employee of the State or any political subdivision, immune contractor or state Legislator:

(a) Must not provide that any or all of the terms of the agreement are confidential.

(b) Must include the amount of any attorney's fees and costs to be paid pursuant to the agreement.

(c) Is a public record and must be open for inspection pursuant to NRS 239.010.

2. Any provision of an agreement to settle a claim or action brought under NRS 41.031 or against a present or former officer or employee of the State or any political subdivision, immune contractor or state Legislator that conflicts with this section is void.

NRS 41.039 Filing of valid claim against political subdivision condition precedent to commencement of action against immune contractor, employee or officer.

An action which is based on the conduct of any immune contractor, employee or appointed or elected officer of a political subdivision of the State of Nevada while in the course of his employment or in the performance of his official duties may not be filed against the immune contractor, employee or officer unless, before the filing of the complaint in such an action, a valid claim has been filed, pursuant to NRS 41.031 to 41.038, inclusive, against the political subdivision for which the immune contractor, employee or officer was authorized to act.

SAMPLE STATUTORY IMMUNITY STATUTES

New Jersey Permanent Statutes

TITLE 2A   ADMINISTRATION OF CIVIL AND CRIMINAL JUSTICE
2A:62A-1.2 Immunity from civil damages for firefighters at accident scenes.

1. A municipal, county or State firefighter, whether volunteer or paid, shall not be liable for any civil damages as a result of any acts or omissions undertaken in good faith in rendering care at the scene of an accident or emergency to any victim thereof, or in transporting any such victim to a hospital or other facility where treatment or care is to be rendered; provided, however, that nothing in this section shall exonerate a firefighter for gross negligence.

Ohio Revised Code §2744.01

C) (1) "**Governmental function**" means a function of a political subdivision that is specified in division (C)(2) of this section or that satisfies any of the following:

(a) A function that is imposed upon the state as an obligation of sovereignty and that is performed by a political subdivision voluntarily or pursuant to legislative requirement;

(b) A function that is for the common good of all citizens of the state;

(c) A function that promotes or preserves the public peace, health, safety, or welfare; that involves activities that are not engaged in or not customarily engaged in by nongovernmental persons; and that is not specified in division (G)(2) of this section as a proprietary function.

(2) A "**governmental function**" includes, but is not limited to, the following:

(a) The provision or nonprovision of police, fire, emergency medical, ambulance, and rescue services or protection;

(b) The power to preserve the peace; to prevent and suppress riots, disturbances, and disorderly assemblages; to prevent, mitigate, and clean up releases of oil and hazardous and extremely hazardous substances as defined in section 3750.01 of the Revised Code; and to protect persons and property;

(c) The provision of a system of public education;

(d) The provision of a free public library system;

(e) The regulation of the use of, and the maintenance and repair of, roads, highways, streets, avenues, alleys, sidewalks, bridges, aqueducts, viaducts, and public grounds;

(f) Judicial, quasi-judicial, prosecutorial, legislative, and quasi-legislative functions;

(g) The construction, reconstruction, repair, renovation, maintenance, and operation of buildings that are used in connection with the performance of a governmental function, including, but not limited to, office buildings and courthouses;

(h) The design, construction, reconstruction, renovation, repair, maintenance, and operation of jails, places of juvenile detention, workhouses, or any other detention facility, as defined in section 2921.01 of the Revised Code;

(i) The enforcement or nonperformance of any law;

(j) The regulation of traffic, and the erection or nonerection of traffic signs, signals, or control devices;

(k) The collection and disposal of solid wastes, as defined in section 3734.01 of the Revised Code, including, but not limited to, the operation of solid waste disposal facilities, as "facilities" is defined in that section, and the collection and management of hazardous waste generated by households. As used in division (C)(2)(k) of this section, "hazardous waste generated by households" means solid waste originally generated by individual households that is listed specifically as hazardous waste in or exhibits one or more characteristics of hazardous waste as defined by rules adopted under section 3734.12 of the Revised Code, but that is excluded from regulation as a hazardous waste by those rules.

(l) The provision or nonprovision, planning or design, construction, or reconstruction of a public improvement, including, but not limited to, a sewer system;

(m) The operation of a job and family services department or agency, including, but not limited to, the provision of assistance to aged and infirm persons and to persons who are indigent;

(n) The operation of a health board, department, or agency, including, but not limited to, any statutorily required or permissive program for the provision of immunizations or other inoculations to all or some members of the public, provided that a "governmental function" does not include the supply, manufacture, distribution, or development of any drug or vaccine employed in any such immunization or inoculation program by any supplier, manufacturer, distributor, or developer of the drug or vaccine;

(o) The operation of mental health facilities, mental retardation or developmental disabilities facilities, alcohol treatment and control centers, and children's homes or agencies;

(p) The provision or nonprovision of inspection services of all types, including, but not limited to, inspections in connection with building, zoning, sanitation, fire, plumbing, and electrical codes, and the taking of actions in connection with those types of codes, including, but not limited to, the approval of plans for the construction of buildings or structures and the issuance or revocation of building permits or stop work orders in connection with buildings or structures;

(q) Urban renewal projects and the elimination of slum conditions;

(r) Flood control measures;

(s) The design, construction, reconstruction, renovation, operation, care, repair, and maintenance of a township cemetery;

(t) The issuance of revenue obligations under section 140.06 of the Revised Code;

(u) The design, construction, reconstruction, renovation, repair, maintenance, and operation of any school athletic facility, school auditorium, or gymnasium or any recreational area or facility, including, but not limited to, any of the following:

(i) A park, playground, or playfield;

(ii) An indoor recreational facility;

(iii) A zoo or zoological park;

(iv) A bath, swimming pool, pond, water park, wading pool, wave pool, water slide, or other type of aquatic facility;

(v) A golf course;

(vi) A bicycle motocross facility or other type of recreational area or facility in which bicycling, skating, skate boarding, or scooter riding is engaged;

(vii) A rope course or climbing walls;

(viii) An all-purpose vehicle facility in which all-purpose vehicles, as defined in section 4519.01 of the Revised Code, are contained, maintained, or operated for recreational activities.

(v) The provision of public defender services by a county or joint county public defender's office pursuant to Chapter 120. of the Revised Code;

(w) (i) At any time before regulations prescribed pursuant to 49 U.S.C.A 20153 become effective, the designation, establishment, design, construction, implementation, operation, repair, or maintenance of a public road rail crossing in a zone within a municipal corporation in which, by ordinance, the legislative authority of the municipal corporation regulates the sounding of locomotive horns, whistles, or bells;

(ii) On and after the effective date of regulations prescribed pursuant to 49 U.S.C.A. 20153, the designation, establishment, design, construction, implementation, operation, repair, or maintenance of a public road rail crossing in such a zone or of a supplementary safety measure, as defined in 49 U.S.C.A 20153, at or for a public road rail crossing, if and to the extent that the public road rail crossing is excepted, pursuant to subsection (c) of that section, from the requirement of the regulations prescribed under subsection (b) of that section.

(x) A function that the general assembly mandates a political subdivision to perform.

(G) (1) "**Proprietary function**" means a function of a political subdivision that is specified in division (G)(2) of this section or that satisfies both of the following:

(a) The function is not one described in division (C)(1)(a) or (b) of this section and is not one specified in division (C)(2) of this section;

(b) The function is one that promotes or preserves the public peace, health, safety, or welfare and that involves activities that are customarily engaged in by nongovernmental persons.

(2) A "**proprietary function**" includes, but is not limited to, the following:

(a) The operation of a hospital by one or more political subdivisions;

(b) The design, construction, reconstruction, renovation, repair, maintenance, and operation of a public cemetery other than a township cemetery;

(c) The establishment, maintenance, and operation of a utility, including, but not limited to, a light, gas, power, or heat plant, a railroad, a busline or other transit company, an airport, and a municipal corporation water supply system;

(d) The maintenance, destruction, operation, and upkeep of a sewer system;

(e) The operation and control of a public stadium, auditorium, civic or social center, exhibition hall, arts and crafts center, band or orchestra, or off-street parking facility.

SAMPLE IMMUNITY LAW FOR AEDs

Georgia   51-1-29.3

(a) The persons described in this Code section shall be immune from civil liability for any act or omission to act related to the provision of emergency care or treatment by the use of or provision of an automated external defibrillator, as described in Code Sections 31-11-53.1 and 31-11-53.2, except that such immunity shall not apply to an act of willful or wanton misconduct and shall not apply to a person acting within the scope of a licensed profession if such person acts with gross negligence. The immunity provided for in this Code section shall extend to:

(1) Any person who gratuitously and in good faith renders emergency care or treatment by the use of or provision of an automated external defibrillator without objection of the person to whom care or treatment is rendered;

(2) The owner or operator of any premises or conveyance who installs or provides automated external defibrillator equipment in or on such premises or conveyance;

(3) Any physician or other medical professional who authorizes, directs, or supervises the installation or provision of automated external defibrillator equipment in or on any premises or conveyance other than any medical facility as defined in paragraph (2) of Code Section 31-7-1; and

(4) Any person who provides training in the use of automated external defibrillator equipment as required by subparagraph (b)(1)(A) of Code Section 31-11-53.2, whether compensated or not. This Code section is not applicable to any training or instructions provided by the manufacturer of the automated external defibrillator or to any claim for failure to warn on the part of the manufacturer.

(b) Nothing in this Code section shall be construed so as to provide immunity to the manufacturer of any automated external defibrillator or off-premises automated external defibrillator maintenance or service providers, nor shall it relieve the manufacturer from any claim for product liability or failure to warn.

Volunteer Protection Act of 1997

42 USC 139

§ 14501. Findings and purpose

(a) Findings

The Congress finds and declares that—

(1) the willingness of volunteers to offer their services is deterred by the potential for liability actions against them;

(2) as a result, many nonprofit public and private organizations and governmental entities, including voluntary associations, social service agencies, educational institutions, and other civic programs, have been adversely affected by the withdrawal of volunteers from boards of directors and service in other capacities;

(3) the contribution of these programs to their communities is thereby diminished, resulting in fewer and higher cost programs than would be obtainable if volunteers were participating;

(4) because Federal funds are expended on useful and cost-effective social service programs, many of which are national in scope, depend heavily on volunteer participation, and represent some of the most successful public-private partnerships, protection of volunteerism through clarification and limitation of the personal liability risks assumed by the volunteer in connection with such participation is an appropriate subject for Federal legislation;

(5) services and goods provided by volunteers and nonprofit organizations would often otherwise be provided by private entities that operate in interstate commerce;

(6) due to high liability costs and unwarranted litigation costs, volunteers and nonprofit organizations face higher costs in purchasing insurance, through interstate insurance markets, to cover their activities; and

(7) clarifying and limiting the liability risk assumed by volunteers is an appropriate subject for Federal legislation because—

(A) of the national scope of the problems created by the legitimate fears of volunteers about frivolous, arbitrary, or capricious lawsuits;

(B) the citizens of the United States depend on, and the Federal Government expends funds on, and provides tax exemptions and other consideration to, numerous social programs that depend on the services of volunteers;

(C) it is in the interest of the Federal Government to encourage the continued operation of volunteer service organizations and contributions of volunteers because the Federal Government lacks the capacity to carry out all of the services provided by such organizations and volunteers; and

(D) (i) liability reform for volunteers, will promote the free flow of goods and services, lessen burdens on interstate commerce and uphold constitutionally protected due process rights; and

(ii) therefore, liability reform is an appropriate use of the powers contained in article 1, section 8, clause 3 of the United States Constitution, and the fourteenth amendment to the United States Constitution.

(b) Purpose

The purpose of this chapter is to promote the interests of social service program beneficiaries and taxpayers and to sustain the availability of programs, nonprofit organizations, and governmental entities that depend on volunteer contributions by reforming the laws to provide certain protections from liability abuses related to volunteers serving nonprofit organizations and governmental entities.

§ 14502. Preemption and election of State nonapplicability

(a) Preemption

This chapter preempts the laws of any State to the extent that such laws are inconsistent with this chapter, except that this chapter shall not preempt any State law that provides additional protection from liability relating to volunteers or to any category of volunteers in the performance of services for a nonprofit organization or governmental entity.

(b) Election of State regarding nonapplicability

This chapter shall not apply to any civil action in a State court against a volunteer in which all parties are citizens of the State if such State enacts a statute in accordance with State requirements for enacting legislation—

(1) citing the authority of this subsection;

(2) declaring the election of such State that this chapter shall not apply, as of a date certain, to such civil action in the State; and

(3) containing no other provisions.

§ 14503. Limitation on liability for volunteers

(a) Liability protection for volunteers

Except as provided in subsections (b) and (d) of this section, no volunteer of a nonprofit organization or governmental entity shall be liable for harm caused by an act or omission of the volunteer on behalf of the organization or entity if—

(1) the volunteer was acting within the scope of the volunteer's responsibilities in the nonprofit organization or governmental entity at the time of the act or omission;

(2) if appropriate or required, the volunteer was properly licensed, certified, or authorized by the appropriate authorities for the activities or practice in the State in which the harm occurred, where the activities were or practice was undertaken within the scope of the volunteer's responsibilities in the nonprofit organization or governmental entity;

(3) the harm was not caused by willful or criminal misconduct, gross negligence, reckless misconduct, or a conscious, flagrant indifference to the rights or safety of the individual harmed by the volunteer; and

(4) the harm was not caused by the volunteer operating a motor vehicle, vessel, aircraft, or other vehicle for which the State requires the operator or the owner of the vehicle, craft, or vessel to—

(A) possess an operator's license; or

(B) maintain insurance.

(b) Concerning responsibility of volunteers to organizations and entities

Nothing in this section shall be construed to affect any civil action brought by any nonprofit organization or any governmental entity against any volunteer of such organization or entity.

(c) No effect on liability of organization or entity

Nothing in this section shall be construed to affect the liability of any nonprofit organization or governmental entity with respect to harm caused to any person.

(d) Exceptions to volunteer liability protection

If the laws of a State limit volunteer liability subject to one or more of the following conditions, such conditions shall not be construed as inconsistent with this section:

(1) A State law that requires a nonprofit organization or governmental entity to adhere to risk management procedures, including mandatory training of volunteers.

(2) A State law that makes the organization or entity liable for the acts or omissions of its volunteers to the same extent as an employer is liable for the acts or omissions of its employees.

(3) A State law that makes a limitation of liability inapplicable if the civil action was brought by an officer of a State or local government pursuant to State or local law.

(4) A State law that makes a limitation of liability applicable only if the nonprofit organization or governmental entity provides a financially secure source of recovery for individuals who suffer harm as a result of actions taken by a volunteer on behalf of the organization or entity. A financially secure source of recovery may be an insurance policy within specified limits, comparable coverage from a risk pooling mechanism, equivalent assets, or alternative arrangements that satisfy the State that the organization or entity will be able to pay for losses up to a specified amount. Separate standards for different types of liability exposure may be specified.

(e) Limitation on punitive damages based on actions of volunteers

(1) General rule

Punitive damages may not be awarded against a volunteer in an action brought for harm based on the action of a volunteer acting within the scope of the volunteer's responsibilities to a nonprofit organization or governmental entity unless the claimant establishes by clear and convincing evidence that the harm was proximately caused by an action of such volunteer which constitutes willful or criminal misconduct, or a conscious, flagrant indifference to the rights or safety of the individual harmed.

(2) Construction

Paragraph (1) does not create a cause of action for punitive damages and does not preempt or supersede any Federal or State law to the extent that such law would further limit the award of punitive damages.

(f) Exceptions to limitations on liability

(1) In general

The limitations on the liability of a volunteer under this chapter shall not apply to any misconduct that—

(A) constitutes a crime of violence (as that term is defined in section 16 of title 18) or act of international terrorism (as that term is defined in section 2331 of title 18) for which the defendant has been convicted in any court;

(B) constitutes a hate crime (as that term is used in the Hate Crime Statistics Act (28 U.S.C. 534 note));

(C) involves a sexual offense, as defined by applicable State law, for which the defendant has been convicted in any court;

(D) involves misconduct for which the defendant has been found to have violated a Federal or State civil rights law; or

(E) where the defendant was under the influence (as determined pursuant to applicable State law) of intoxicating alcohol or any drug at the time of the misconduct.

(2) Rule of construction

Nothing in this subsection shall be construed to effect subsection (a)(3) or (e) of this section.

§ 14504. Liability for noneconomic loss

(a) General rule

In any civil action against a volunteer, based on an action of a volunteer acting within the scope of the volunteer's responsibilities to a nonprofit organization or governmental entity, the liability of the volunteer for noneconomic loss shall be determined in accordance with subsection (b) of this section.

(b) Amount of liability

(1) In general

Each defendant who is a volunteer, shall be liable only for the amount of noneconomic loss allocated to that defendant in direct proportion to the percentage of responsibility of that defendant (determined in accordance with paragraph (2)) for the harm to the claimant with respect to which that defendant is liable. The court shall render a separate judgment against each defendant in an amount determined pursuant to the preceding sentence.

(2) Percentage of responsibility

For purposes of determining the amount of noneconomic loss allocated to a defendant who is a volunteer under this section, the trier of fact shall determine the percentage of responsibility of that defendant for the claimant's harm.

§ 14505. Definitions

For purposes of this chapter:

(1) Economic loss

The term "economic loss" means any pecuniary loss resulting from harm (including the loss of earnings or other benefits related to employment, medical expense loss, replacement services loss, loss due to death, burial costs, and loss of business or employment opportunities) to the extent recovery for such loss is allowed under applicable State law.

(2) Harm

The term "harm" includes physical, nonphysical, economic, and noneconomic losses.

(3) Noneconomic losses

The term "noneconomic losses" means losses for physical and emotional pain, suffering, inconvenience, physical impairment, mental anguish, disfigurement, loss of enjoyment of life, loss of society and companionship, loss of consortium (other than loss of domestic service), hedonic damages, injury to reputation and all other nonpecuniary losses of any kind or nature.

(4) Nonprofit organization

The term "nonprofit organization" means—

(A) any organization which is described in section 501 (c)(3) of title 26 and exempt from tax under section 501(a) of such title and which does not practice any action which constitutes a hate crime referred to in subsection (b)(1) of the first section of the Hate Crime Statistics Act (28 U.S.C. 534 note); or

(B) any not-for-profit organization which is organized and conducted for public benefit and operated primarily for charitable, civic, educational, religious, welfare, or health purposes and which does not practice any action which constitutes a hate crime referred to in subsection (b)(1) of the first section of the Hate Crime Statistics Act (28 U.S.C. 534 note).

(5) State

The term "State" means each of the several States, the District of Columbia, the Commonwealth of Puerto Rico, the Virgin Islands, Guam, American Samoa, the Northern Mariana Islands, any other territory or possession of the United States, or any political subdivision of any such State, territory, or possession.

(6) Volunteer

The term "volunteer" means an individual performing services for a nonprofit organization or a governmental entity who does not receive—

(A) compensation (other than reasonable reimbursement or allowance for expenses actually incurred); or

(B) any other thing of value in lieu of compensation, in excess of $500 per year, and such term includes a volunteer serving as a director, officer, trustee, or direct service volunteer.

CHAPTER 11

SAMPLE COLLECTIVE BARGAINING STATUTES

New York State Consolidated Laws, Article 20, Section 703. Rights of employees. Employees shall have the right of self-organization, to form, join, or assist labor organizations, to bargain collectively through representatives of their own choosing, and to engage in concerted activities, for the purpose of collective bargaining or other mutual aid or protection, free from interference, restraint, or coercion of employers, but nothing contained in this article shall be interpreted to prohibit employees from exercising the right to confer with their employer at any time, provided that during such conference there is no attempt by the employer, directly or indirectly, to interfere with, restrain or coerce employees in the exercise of the rights guaranteed by this section.

New York State Consolidated Laws, Article 20, Section 704. Unfair labor practices. It shall be an unfair labor practice for an employer:

1. To spy upon or keep under surveillance, whether directly or through agents or any other person, any activities of employees or their representatives in the exercise of the rights guaranteed by section seven hundred three.

2. To prepare, maintain, distribute or circulate any blacklist of individuals for the purpose of preventing any of such individuals from obtaining or retaining employment because of the exercise by such individuals of any of the rights guaranteed by section seven hundred three.

3. To dominate or interfere with the formation, existence, or administration of any employee organization or association, agency or plan which exists in whole or in part for the purpose of dealing with employers concerning terms or conditions of employment, labor disputes or grievances, or to contribute financial or other support to any such organization, by any means, including but not limited to the following: (a) by participating or assisting in, supervising, controlling or dominating (1) the initiation or creation of any such employee organization or association, agency, or plan, or (2) the meetings, management, operation, elections, formulation or amendment of constitution, rules or policies, of any such employee organization or association, agency or plan; (b) by urging the employees to join any such employee organization or association, agency or plan for the purpose of encouraging membership in the same; (c) by compensating any employee or individual for services performed in behalf of any such employee organization or association, agency or plan, or by donating free services, equipment, materials, office or meeting space or anything else of value for the use of any such employee organization or association, agency or plan; provided that, an employer shall not be prohibited from permitting employees to confer with him during working hours without loss of time or pay.

4. To require an employee or one seeking employment, as a condition of employment, to join any company union or to refrain from forming, or joining or assisting a labor organization of his own choosing.

5. To encourage membership in any company union or discourage membership in any labor organization, by discrimination in regard to hire or tenure or in any term or condition of employment: Provided that nothing in this article shall preclude an employer from making an agreement with a labor organization requiring as a condition of employment membership therein, if such labor organization is the representative of employees as provided in section seven hundred five.

6. To refuse to bargain collectively with the representatives of employees, subject to the provisions of section seven hundred five.

7. To refuse to discuss grievances with representatives of employees, subject to the provisions of section seven hundred five.

8. To discharge or otherwise discriminate against an employee because he has signed or filed any affidavit, petition or complaint or given any information or testimony under this article.

9. To distribute or circulate any blacklist of individuals exercising any right created or confirmed by this article or of members of a labor organization, or to inform any person of the exercise by any individual of such right, or of the membership of any individual in a labor organization for the purpose of preventing individuals so blacklisted or so named from obtaining or retaining employment.

10. To do any acts, other than those already enumerated in this section, which interfere with, restrain or coerce employees in the exercise of the rights guaranteed by section seven hundred three.

11. To utilize any state funding appropriated for any purpose to train managers, supervisors or other administrative personnel regarding methods to discourage union organization, or to discourage an employee from participating in a union organizing drive.

CALIFORNIA

Labor Code Section 1960 to 1964

1960. Neither the State nor any county, political subdivision, incorporated city, town, nor any other municipal corporation shall prohibit, deny or obstruct the right of firefighters to join any bona fide labor organization of their own choice.

1961. As used in this chapter, the term "employees" means the employees of the fire departments and fire services of the State, counties, cities, cities and counties, districts, and other political subdivisions of the State.

1962. Employees shall have the right to self-organization, to form, join, or assist labor organizations, to present grievances and recommendations regarding wages, salaries, hours, and working conditions to the governing body, and to discuss the same with such governing body, through such an organization, but shall not have the right to strike, or to recognize a picket line of a labor organization while in the course of the performance of their official duties.

1963. The enactment of this chapter shall not be construed as making the provisions of Section 923 of this code applicable to public employees.

1964. (a) The governing body of any regularly organized volunteer fire department may, but shall not be required to, adopt regulations governing the removal of volunteer firefighters from the volunteer fire department.

(b) In the event that the governing body chooses to adopt these regulations, it shall have the discretion, after soliciting comments from the membership of the volunteer fire department, to adopt any reasonable regulations which may, but need not, include some or all of the following elements, in addition to other provisions:

(1) Members of the department shall not be removed from membership, except for incompetence, misconduct, or failure to comply with the rules and regulations of the department. Removals, except for absenteeism at fires or meetings, shall be made only after a hearing with due notice, with stated charges, and with the right of the member to a review.

(2) The charges shall be in writing and may be made by the governing body. The burden of proving incompetency or misconduct shall be on the person alleging it.

(3) Hearings on the charges shall be held by the officer or body having the power to remove the person, or by a deputy or employee of the officer or body designated in writing for that purpose. In case a deputy or other employee is so designated, he or she shall for the purpose of the hearing be vested with all the powers of the officer or body, and shall make a record of the hearing which shall be referred to the officer or body for review with his or her recommendations.

(4) The notice of the hearing shall specify the time and place of the hearing and state the body or person before whom the hearing will be held. Notice and a copy of the charges shall be served personally upon the accused member at least 10 days but not more than 30 days before the date of the hearing.

(5) A stenographer may be employed for the purpose of taking testimony at the hearing.

(6) The officer or body having the power to remove the person may suspend the person after charges are filed and pending disposition of the charges, and after the hearing may remove the person or may suspend him or her for a period of time not to exceed one year.

(7) Volunteer firefighters shall serve a probationary period of a length to be specified by the governing board, not to exceed one year. A probationary volunteer firefighter may be removed from membership without specification of cause. The decision to remove a probationer shall not require notice or a hearing.

(c) The requirement of subdivision (b) to solicit comments from the membership shall not be deemed to create a duty to meet and confer with the membership.

(d) In the event that a governing body of a regularly organized volunteer fire department adopts regulations governing removal of volunteer firefighters, the regulations shall not be interpreted as creating a property right in the volunteer firefighter job or position.

(e) When regulations have been adopted, and where the regulations provide for a hearing and decision by the governing body, a volunteer firefighter may commence a proceeding in accordance with the provisions of Section 1094.5 of the Code of Civil Procedure to set aside the decision of the governing body on the ground that the decision is not supported by substantial evidence. The court shall not employ its independent judgment in reviewing the evidence. The proceeding shall be commenced within 90 days from the date that the governing body renders its decision. This remedy shall be the exclusive method for review of the governing body's decision.

PENNSYLVANIA STATUTES

§ 211.5. Rights of employes.

Employes shall have the right to self-organization, to form, join or assist labor organizations, to bargain collectively through representatives of their own choosing, and to engage in concerted activities for the purpose of collective bargaining or other mutual aid or protection.

§ 211.6. Unfair labor practices.

(1) It shall be an unfair labor practice for an employer—

(a) To interfere with, restrain or coerce employes in the exercise of the rights guaranteed in this act. [§§ 211.1 to 211.13]

(b) To dominate or interfere with the formation or administration of any labor organization or contribute financial or other material support to it: Provided, That subject to rules and regulations made and published by the board pursuant to this act, an employer shall not be prohibited from permitting employes to confer with him during working hours without loss of time or pay.

(c) By discrimination in regard to hire or tenure of employment, or any term or condition of employment to encourage or discourage membership in any labor organization: Provided, That nothing in this act, or in any agreement approved or prescribed thereunder, or in any other statute of this Commonwealth, shall preclude an employer from making an agreement with a labor organization (not established, maintained or assisted by any action defined in this act as an unfair labor practice) to require, as a condition of employment, membership therein, if such labor organization does not deny membership in its organization to a person or persons who are employes of the employer at the time of the making of such agreement, provided such employe was not employed in violation of any previously existing agreement with said labor organization.

(d) To discharge or otherwise discriminate against an employe because he has filed charges or given testimony under this act.

(e) To refuse to bargain collectively with the representatives of his employes, subject to the provisions of section seven (a) of this act.

(f) To deduct, collect, or assist in collecting from the wages of employes any dues, fees, assessments, or other contributions payable to any labor organization, unless he is authorized so to do by a majority vote of all the employes in the appropriate collective bargaining unit taken by secret ballot, and unless he thereafter receives the written authorization from each employe whose wages are affected.

(2) It shall be an unfair labor practice for a labor organization, or any officer or officers of a labor organization, or any agent or agents of a labor organization, or any one acting in the interest of a labor organization, or for an employe or for employes acting in concert—

(a) To intimidate, restrain, or coerce any employe for the purpose and with the intent of compelling such employe to join or to refrain from joining any labor organization, or for the purpose or with the intent of influencing or affecting his selection of representatives for the purposes of collective bargaining.

(b) During a labor dispute, to join or become a part of a sitdown strike, or, without the employer's authorization, to seize or hold or to damage or destroy the plant, equipment, machinery, or other property of the employer, with the intent of compelling the employer to accede to demands, conditions, and terms of employment including the demand for collective bargaining.

(c) To intimidate, restrain, or coerce any employer by threats of force or violence or harm to the person of said employer or the members of his family, with the intent of compelling the employer to accede to demands, conditions, and terms of employment including the demand for collective bargaining.

STEELWORKER TRILOGY CASES

The Steelworkers Trilogy cases are:

United Steelworkers v. Enterprise Wheel & Car Corp., 363 U.S. 593, 598-99 (1960);

United Steelworkers v. Warrior & Gulf Navigation Co., 363 U.S. 574, 577-78 (1960);

United Steelworkers v. American Mfg. Co., 363 U.S. 564, 566-67 (1960).

For a fire case, see *Wilmington v. Firefighters Local 1590,* 385 A. 2d 720 (DE, 1978).

THE SCOPE OF COLLECTIVE BARGAINING FOR FIREFIGHTERS

EXAMPLES OF MANDATORY, PROHIBITED and PERMISSIVE SUBJECTS

MANDATORY:

Retroactive pay - *Taureck v. Jersey City,* 149 N.J. Super 503, 375 A. 2d 70 (NJ, 1977)

Incentive pay for college degrees – *FOP Lodge 2 v. Scranton,* 364 A. 2d 753 (PA, 1976)

Longevity pay – *Watertown Firefighters v. Watertown,* 375 Mass. 706 (Mass. 1978); and Oliver v. City of Tulsa, 654 P. 2d 607 (Okl., 1982)

Health Insurance – *Watertown Firefighters v. Watertown,* 375 Mass. 706 (Mass. 1978)

Legal Insurance – *Town of Haverstraw v. Newman,* 75 App. Div. 2d 874 (NYAD, 1980)

Time off for union activities – *Albany v. Helsby,* 48 App. Div. 2d 998 (1975) aff'd. 38 NY 2d. 778 (1976)

Procedure for issuing notice of promotional examination – *IAFF Local 1974 v. City of Pleasanton,* 56 Cal. App. 3d 959, 129 Cal. Rptr. 68 (Cal.App.Dist.1 04/02/1976)

Right to use station facilities to wash private cars – *Local 2312, IAFF v. City of Vernon,* 107 Cal. App. 3d 802, 165 Cal. Rptr. 908 (1980)

Pension – *East Providence v. IAFF Local 850,* 117 RI 329, 366 A. 2d 1151 (1976)

Hours and shifts – *Flynn v. City of Frazier, 45 Mich.* App 346, 206 N.W. 2d 448 (1973);

Firefighters Local 1186 v. City of Vallejo, 12 Cal. 3d 608, 116 Cal. Rptr. 507, 526 P. 2d 971 (1974)

Health and safety – *N.L.R.B. v. Gulf Power,* 384 F. 2d 882 (5th Cir. 1967)

Staffing (per apparatus) – *City of Erie v. IAFF Local 293,* 74 Pa. Cmwlth. 245, 459 A. 2d 1320 (Pa. Cmwlth. Ct., 1983)

Staffing (per apparatus) - *IAFF 1052 v. PERC,* 113 Wash. 2d 197, 778 P.2d 32 (Wa., 1989)

Staffing (on duty shift size) – *Narragansett v. IAFF Local 1589,* 119 RI 506, 380 A 2d 521 (RI, 1977)

Staffing (on duty shift size) – *City of Alpina v. IAFF Local 623,* 56 Mich. App. 568

Staffing – *Firefighters Local 1186 v. City of Vallejo,* 12 Cal. 3d 608, 116 Cal. Rptr. 507, (CA, 1974)

Work rules – *City of Albany v. Helsby,* 48 App. Div. 2d 998 (1975) aff'd. 38 NY 2d. 778 (1976)

Promotions – *Cranston v. Hall,* 116 RI 183, 354 A. 2d 415 (RI, 1976)

Use of non-bargaining unit personnel to replace bargaining unit personnel – *IAFF 1885 v. Valley Community Services District,* 45 Cal. App. 3d 116, 119 Cal. Rptr. 182 (C.C.1, 1975)

Residency – *Detroit Police Officers v. Detroit,* 391 Mich. 44, 214 N. W. 2d 803 (MI, 1974)

    Contrary: *St. Bernard v. State Emp. Relations Bd.* (1991), 74 Ohio App.3d 3 (1991)

Hair length and styles – *Nashua Firefighters Local 789,* 23 GERR 1225 (NH PELRB, 1985)

Weight Standard – *City of White Plains,* 23 GERR 1784 (NY, 1985)

PROHIBITED

Parity provisions – *City of New York,* 10 PERB 3003 (NYPERB, 1977)
Compare with *City of Yonkers,* 10 PERB 3048 (NYPERB, 1978) holding a "most favored nations" clause that permitted negotiations to be reopened if another union received greater benefits

Parity provision - *Local 1219, IAFF, v. Conn. LRB,* 171 Conn. 342 (1976); *Town of Metheun,* Case No. MUP-507, 545 GERR B-17 (Mass. LRC 1974); *City of Plainfield,* 4 NJPER 255 (N.J. PERC 1978)

Total hours and shifts – *IAFF Local 330 v. City Akron,* 112 LRRM 2097 (Ohio Ct. Com. PLS., 1980)

Conflicting Legislation: Residency (via local ordinance) *St. Bernard v. State Emp. Relations Bd.* (1991), 74 Ohio App.3d 3

Conflicting Legislation: Civil Service Merit System (state Constitution) *IAFF Local 1383 v. City of Warren,* 89 Mich. App. 135, 279 N.W. 2d 556 (Mich., 1979)

Total number of personnel – *IAFF Local 669 v City of Scranton,* 59 Pa. Cmwlth 235, 429 A. 2d 779

Staffing – *IAFF v. Helsby,* 399 NYS 2d 334 (NYAD, 1977); *Portland Firefighters Local 740 v. City of Portland,* 478 A 2d 297 (ME, 1984).

Right to grieve and arbitrate a suspension – *City of Janesville v. WERC,* 535 N.W.2d 34, 193 Wis. 2d 492 (1995)

PERMISSIVE

Layoffs – *City of Brookfield v. WERC,* 87 Wisc. 2d 819, 275 NW 2d 723 (Wisc., 1979)

Establishment of paramedic unit – *Danbury v. IAFF Local 801,* 221 Conn. 244, 603 A.2d 393 (Conn., 1992)

Redefinition or constitution of bargaining unit – *Detroit IAFF 344 v. Detroit,* 96 Mich. App. 543, 294 N.W.2d 842, (Mich.App., 1980)

Staffing (per shift) – *Jackson Fire Fighters Association, IAFF Local 1306 v. City of Jackson,* 227

Staffing (per shift) – *IAFF 1088 v. City of Berlin,* 123 N.H. 404, 462 A.2d 98 (NH, 1983)

Mich.App. 520, 575 N.W.2d 823 (Mich.App., 1998)

Premium pay for ready time, hour/duties – *IAFF 610 v. Iowa PERB and Iowa City,* 554 N.W.2d 707 (IA, 1996)

EXAMPLES OF RIGHT-TO-WORK LAWS

SOUTH CAROLINA CODE ANNOTATED

41-7-10. Denial of right to work for membership or nonmembership in labor organization declared to be against public policy.

It is hereby declared to be the public policy of this State that the right of persons to work shall not be denied or abridged on account of membership or nonmembership in any labor union or labor organization.

41-7-20. Agreement between employer and labor organization denying nonmembers right to work or requiring union membership unlawful.

Any agreement or combination between any employer and any labor organization whereby persons not members of such labor organizations shall be denied the right to work for such employer or whereby such membership is made a condition of employment, or of continuance of employment by such employer, or whereby any such union or organization acquires an employment monopoly in any enterprise, is hereby declared to be against public policy, unlawful and an illegal combination or conspiracy

41-7-30. Labor organization membership as condition of employment.

(A) It is unlawful for an employer to require an employee, as a condition of employment, or of continuance of employment to:

(1) be or become or remain a member or affiliate of a labor organization or agency;

(2) abstain or refrain from membership in a labor organization; or

(3) pay any fees, dues, assessments, or other charges or sums of money to a person or organization.

(B) It is unlawful for a person or a labor organization to directly or indirectly participate in an agreement, arrangement, or practice that has the effect of requiring, as a condition of employment, that an employee be, become, or remain a member of a labor organization or pay to a labor organization any dues, fees, or any other charges; such an agreement is unenforceable.

(C) It is unlawful for a person or a labor organization to induce, cause, or encourage an employer to violate a provision of this section.

41-7-40. Deduction of labor organization membership dues from wages.

Nothing in this chapter precludes an employer from deducting from the wages of the employees and paying over to a labor organization, or its authorized representative, membership dues in a labor organization; however, the employer must have received from each employee, on whose account the deductions are made, a written assignment which must not be irrevocable for a period of more than one year or until the termination date of any applicable collective agreement or assignment, whichever occurs sooner. After one year, the employee has the absolute right to revoke the written assignment allowing for deduction of membership dues in a labor union.

ARIZONA REVISED STATUTES

23-1302. *Prohibition of agreements denying employment because of nonmembership in labor organization*

No person shall be denied the opportunity to obtain or retain employment because of nonmembership in a labor organization, nor shall the state or any subdivision thereof, or any corporation, individual, or association of any kind enter into an agreement, written or oral, which excludes a person from employment or continuation of employment because of nonmembership in a labor organization.

CHAPTER 13

29 CFR Sec. 1604.11 Sexual harassment.

(a) Harassment on the basis of sex is a violation of section 703 of title VII. Unwelcome sexual advances, requests for sexual favors, and other verbal or physical conduct of a sexual nature constitute sexual harassment when (1) submission to such conduct is made either explicitly or implicitly a term or condition of an individual's employment, (2) submission to or rejection of such conduct by an individual is used as the basis for employment decisions affecting such individual, or (3) such conduct has the purpose or effect of unreasonably interfering with an individual's work performance or creating an intimidating, hostile, or offensive working environment.

(b) In determining whether alleged conduct constitutes sexual harassment, the Commission will look at the record as a whole and at the totality of the circumstances, such as the nature of the sexual advances and the context in which the alleged incidents occurred. The determination of the legality of a particular action will be made from the facts, on a case by case basis.

(c) [Reserved]

(d) With respect to conduct between fellow employees, an employer is responsible for acts of sexual harassment in the workplace where the employer (or its agents or supervisory employees) knows or should have known of the conduct, unless it can show that it took immediate and appropriate corrective action.

(e) An employer may also be responsible for the acts of non-employees, with respect to sexual harassment of employees in the workplace, where the employer (or its agents or supervisory employees) knows or should have known of the conduct and fails to take immediate and appropriate corrective action. In reviewing these cases the Commission will consider the extent of the employer's control and any other legal responsibility which the employer may have with respect to the conduct of such non-employees.

(f) Prevention is the best tool for the elimination of sexual harassment. An employer should take all steps necessary to prevent sexual harassment from occurring, such as affirmatively raising the subject, expressing strong disapproval, developing appropriate sanctions, informing employees of their right to raise and how to raise the issue of harassment under title VII, and developing methods to sensitize all concerned.

(g) Other related practices: Where employment opportunities or benefits are granted because of an individual's submission to the employer's sexual advances or requests for sexual favors, the employer may be held liable for unlawful sex discrimination against other persons who were qualified for but denied that employment opportunity or benefit.

Appendix A to Sec. 1604.11—Background Information

The Commission has rescinded Sec. 1604.11(c) of the Guidelines on Sexual Harassment, which set forth the standard of employer liability for harassment by supervisors. That section is no longer valid, in light of the Supreme Court decisions in *Burlington Industries, Inc. v. Ellerth*, 524 U.S. 742 (1998), and *Faragher v. City of Boca Raton*, 524 U.S. 775 (1998). The Commission has issued a policy document that examines the *Faragher* and *Ellerth* decisions and provides detailed guidance on the issue of vicarious liability for harassment by supervisors.

CHAPTER 14

Fair Labor Standards

EXECUTIVE EMPLOYEE EXEMPTION TEXT

The United States Department of Labor test for an *executive employee* under the FLSA involves four parts, or subtests:

- The employee must be compensated on a salary basis (as defined in the regulations) at a rate not less than $455 per week;

- The employee's primary duty must be managing the enterprise, or managing a customarily recognized department or subdivision of the enterprise;

- The employee must customarily and regularly direct the work of at least two or more other full-time employees or their equivalent; and

- The employee must have the authority to hire or fire other employees, or the employee's suggestions and recommendations as to the hiring, firing, advancement, promotion, or any other change of status of other employees must be given particular weight.

US DOL, Fair Pay Fact Sheet #17 B, July, 2005.

In order to qualify as an executive, the employee must satisfy all four parts of the test. The salary test has often proven to be the most important of the four tests for fire officers. In order to qualify as being paid on a salary basis, the employee's pay must not be subject to reduction "because of variations in the quality or quantity of the work performed." 29 C.F.R. § 541.118(a). Implicit in this definition is the assumption that an executive employee does not "punch a clock." An executive is expected to work as many hours as are required to complete his or her job, and is paid a salary that takes any additional hours into account. Some days this means that the executive must stay late or come in early, or both. Other days it means the executive may come it late and go home early.

Evidence that an employee has to work a set number of hours each day, or that his or her salary is reduced for leaving work early or coming in late is considered evidence that an employee is actually an hourly employee. When an employee must use vacation or sick time for absences of less than a day, or else not receive full compensation, it is evidence that the employee is an hourly worker, not a salaried executive. If the employee is subject to disciplinary action for reasons other than the violation of a safety regulation, and that disciplinary action may result in a reduction in pay, it is evidence that the person is an hourly employee and not a salaried executive. The likelihood that such a reduction will actually be made must be more than a theoretical one. See *McGuire v. City of Portland*, 159 F.3d 460, (9th Cir., 1998)

There are dozens of pre–2004 fire service cases on the application of the executive exemption to fire officers. The case law establishes that rank or title is not the determinative factor in establishing whether an employee is considered an hourly or salaried employee. In some cases, deputy chiefs and battalion chiefs have been found to be hourly employees, while in other cases, captains and lieutenants have been determined to be salaried executives. *Spralding v. Tulsa*, No. 98-5204 (10th Cir. 2000) (district chiefs are executives); *Department of Labor v. City of Sapulpa, Oklahoma*, 30 F. 3d 1285 (10th Cir., 1994), (captains not executives); *Allen v. Fairfax County*, No. 93-1152, (4th cir., 1994) (lieutenants not executives); *Simmons v City of Forth Worth*, 805 F. Supp. 419 (N.D. Tex, 1992) (deputy chiefs are executives); *Keller v. City of Columbus*, 778 F. Supp. 1480 (S.D. Ind., 1991) (captains and lieutenants are executives); *McGuire v. City of Portland*, 91 F.3d 1293 (9th Cir. 08/02/1996) (battalion chiefs not executives), later reversed in *McGuire v. City of Portland*, 159 F.3d 460, (9th Cir.,1998); *Abshire v. County of Kern*, 908 F. 2d 483 (9th Cir., 1990) (battalion chiefs not executives). The key determining factor in the salary test has come down to whether or not the employee's salary can be reduced if the employee works less than a full day or for disciplinary reasons.

ADMINISTRATIVE AND PROFESSIONAL EMPLOYEE EXEMPTIONS

The Department of Labor's test for the *administrative employee* exemption, requires that all of the following be met:

• The employee must be compensated on a salary basis (as defined in the regulations) at a rate not less than $455 per week;

• The employee's primary function must be the performance of office or non-manual work that is directly related to the management or general business operations of the employer or the employer's customers; and

• The employee's primary duty includes the exercise of discretion and independent judgment with respect to matters of significance.

US DOL, Fair Pay Fact Sheet #17 C, July, 2005. Similar to the executive employee exemption, most of the cases involving the administrative employee exemption involve questions related to the salary test.

The *professional* employee exemption test consists of four parts:

1. The employee must be compensated on a salary basis (as defined in the regulations) at a rate not less than $455 per week;

2. The employee's primary duty must be the performance of work requiring advanced knowledge, defined as work which is predominantly intellectual in character and which requires the consistent exercise of discretion and judgment;

3. The advanced knowledge must be in a field of science or learning, such as law, medicine, nursing, accounting, actuarial computation, engineering, education, and various types of physical, chemical and biological sciences; and

4. The advanced knowledge must be customarily acquired by a prolonged course of specialized intellectual instruction.

US DOL, Fair Pay Fact Sheet #17 D, July, 2005. A relatively new variation on the professional employee exemption has been created for *creative professional* employees. For the creative professional employee exemption to apply, both of the following tests must be satisfied:

- The employee must be compensated on a salary basis (as defined in the regulations) at a rate not less than $455 per week;

- The employee's primary duty must be the performance of work requiring invention, imagination, originality or talent in a recognized field of artistic or creative endeavor, such as music, writing, acting, and the graphic arts.

US DOL, Fair Pay Fact Sheet #17 D, July, 2005. Neither the administrative employee exemption nor the professional employee exemption have had a major impact on the fire service.

DRUG TESTING STATUTES THAT PROHIBIT RANDOM DRUG TESTING

Rhode Island General Laws

§ 28-6.5-1 Testing permitted only in accordance with this section. (a) No employer or agent of any employer shall, either orally or in writing, request, require, or subject any employee to submit a sample of his or her urine, blood, or other bodily fluid or tissue for testing as a condition of continued employment unless that test is administered in accordance with the provisions of this section. Employers may require that an employee submit to a drug test if:

(1) The employer has reasonable grounds to believe based on specific aspects of the employee's job performance and specific contemporaneous observations, capable of being articulated, concerning the employee's appearance, behavior or speech that the employee's use of controlled substances is impairing his or her ability to perform his or her job;

(2) The employee provides the test sample in private, outside the presence of any person;

(3) Employees testing positive are not terminated on that basis, but are instead referred to a substance abuse professional (a licensed physician with knowledge and clinical experience in the diagnosis and treatment of drug related disorders, a licensed or certified psychologist, social worker, or EAP professional with like knowledge, or a substance abuse counselor certified by the National Association of Alcohol and Drug Abuse Counselors (all of whom shall be licensed in Rhode Island)) for assistance; provided, that additional testing may be required by the employer in accordance with this referral, and an employee whose testing indicates any continued use of controlled substances despite treatment may be terminated;

(4) Positive tests of urine, blood or any other bodily fluid or tissue are confirmed by a federally certified laboratory by means of gas chromatography/mass spectrometry or technology recognized as being at least as scientifically accurate;

(5) The employer provides the employee, at the employer's expense, the opportunity to have the sample tested or evaluated by an independent testing facility and so advises the employee;

(6) The employer provides the employee with a reasonable opportunity to rebut or explain the results;

(7) The employer has promulgated a drug abuse prevention policy which complies with requirements of this chapter; and

(8) The employer keeps the results of any test confidential, except for disclosing the results of a "positive" test only to other employees with a job-related need to know, and to defend against any legal action brought by the employee against the employer.

(b) Any employer who subjects any person employed by him or her to this test, or causes, directly or indirectly, any employee to take the test, except as provided for by this chapter, shall be guilty of a misdemeanor punishable by a fine of not more than one thousand dollars ($1,000) or not more than one year in jail, or both.

(c) In any civil action alleging a violation of this section, the court may:

(1) Award punitive damages to a prevailing employee in addition to any award of actual damages;

(2) Award reasonable attorneys' fees and costs to a prevailing employee; and

(3) Afford injunctive relief against any employer who commits or proposes to commit a violation of this section.

(d) Nothing in this chapter shall be construed to impair or affect the rights of individuals under chapter 5 of this title.

(e) Nothing in this chapter shall be construed to:

(1) Prohibit or apply to the testing of drivers regulated under 49 C.F.R. § 40.1 et seq and 49 C.F.R. part 382 if that testing is performed pursuant to a policy mandated by the federal government; or

(2) Prohibit an employer in the public utility or mass transportation industry from requiring testing otherwise barred by this chapter if that testing is explicitly mandated by federal regulation or statute as a condition for the continued receipt of federal funds.

§ 28-6.5-2 Testing of prospective employees. (a) Except as provided in subsections (b) and (c) of this section, an employer may require a job applicant to submit to testing of his or her blood, urine or any other bodily fluid or tissue if:

(1) The job applicant has been given an offer of employment conditioned on the applicant's receiving a negative test result;

(2) The applicant provides the test sample in private, outside the presence of any person; and

(3) Positive tests of urine, blood, or any other bodily fluid or tissue are confirmed by a federal certified laboratory by means of gas chromatography/mass spectrometry or technology recognized as being at least as scientifically accurate.

(b) The pre-employment drug testing authorized by this section shall not extend to job applicants for positions with any agency or political subdivision of the state or municipalities, except for applicants seeking employment as a law enforcement or correctional officer, firefighter, or any other position where that testing is required by federal law or required for the continued receipt of federal funds.

(c) An employer shall not be required to comply with the conditions of testing under subsection (a) of this section to the extent they are inconsistent with federal law.

§ 28-6.5-3 Severability. If any provision of this chapter or the application of it to any person or circumstances is held invalid, that invalidity shall not affect other provisions or applications of the chapter, which can be given effect without the invalid provision or application, and to this end the provisions of this chapter are declared to be severable.

Maine Revised Statutes

Title 26, Chapter 7

§683 Testing procedures. No employer may require, request or suggest that any employee or applicant submit to a substance abuse test except in compliance with this section. All actions taken under a substance abuse testing program shall comply with this subchapter, rules adopted under this subchapter and the employer's written policy approved under section 686.

1. Employee assistance program required. Before establishing any substance abuse testing program for employees, an employer with over 20 full-time employees must have a functioning employee assistance program. A. The employer may meet this requirement by participating in a cooperative employee assistance program that serves the employees of more than one employer. B. The employee assistance program must be certified by the Office of Substance Abuse under rules adopted pursuant to section 687. The rules must ensure that the employee assistance programs have the necessary personnel, facilities and procedures to meet minimum standards of professionalism and effectiveness in assisting employees

2. Written policy. Before establishing any substance abuse testing program, an employer must develop or, as required in section 684, subsection 3, paragraph C, must appoint an employee committee to develop a written policy in compliance with this subchapter providing for, at a minimum:

A. The procedure and consequences of an employee's voluntary admission of a substance abuse problem and any available assistance, including the availability and procedure of the employer's employee assistance program.

B. When substance abuse testing may occur. The written policy must describe: (1) Which positions, if any, will be subject to testing, including any positions subject to random or arbitrary testing under section 684, subsection 3. For applicant testing and probable cause testing of employees, an employer may designate that all positions are subject to testing; and (2) The procedure to be followed in selecting employees to be tested on a random or arbitrary basis under section 684, subsection 3;

C. The collection of samples.

(1) The collection of any sample for use in a substance abuse test must be conducted in a medical facility and supervised by a licensed physician or nurse. . . . [abridged]

D. The storage of samples before testing sufficient to inhibit deterioration of the sample

E. The chain of custody of samples sufficient to protect the sample from tampering and to verify the identity of each sample and test result

F. The substances of abuse to be tested for;

G. The cutoff levels for both screening and confirmation tests at which the presence of a substance of abuse in a sample is considered a positive test result.

(1) Cutoff levels for confirmation tests for marijuana may not be lower than 15 nanograms of delta-9-tetrahydrocannabinol-9-carboxylic acid per milliliter for urine samples

(2) The Department of Health and Human Services shall adopt rules under section 687 regulating screening and confirmation cutoff levels for other substances of abuse, including those substances tested for in blood samples under subsection 5, paragraph B, to ensure that levels are set within known tolerances of test methods and above mere trace amounts. An employer may request that the Department of Health and Human Services establish a cutoff level for any substance of abuse for which the department has not established a cutoff level;

H. The consequences of a confirmed positive substance abuse test result;

I. The consequences for refusal to submit to a substance abuse test;

J. Opportunities and procedures for rehabilitation following a confirmed positive result;

K. A procedure under which an employee or applicant who receives a confirmed positive result may appeal and contest the accuracy of that result. The policy must include a mechanism that provides an opportunity to appeal at no cost to the appellant; and

L. Any other matters required by rules adopted by the Department of Labor under section 687.

An employer must consult with the employer's employees in the development of any portion of a substance abuse testing policy under this subsection that relates to the employees. The employer is not required to consult with the employees on those portions of a policy that relate only to applicants. The employer shall send a copy of the final written policy to the Department of Labor for review under section 686. The employer may not implement the policy until the Department of Labor approves the policy. The employer shall send a copy of any proposed change in an approved written policy to the Department of Labor for review under section 686. The employer may not implement the change until the Department of Labor approves the change.

3. Copies to employees and applicants. The employer shall provide each employee with a copy of the written policy approved by the Department of Labor under section 686 at least 30 days before any portion of the written policy applicable to employees takes effect. The employer shall provide each employee with a copy of any change in a written policy approved by the Department of Labor under section 686 at least 60 days before any portion of the change applicable to employees takes effect. The Department of Labor may waive the 60-day notice for the implementation of an amendment covering employees if the amendment was necessary to comply with the law or if, in the judgment of the department, the amendment promotes the purpose of the law and does not lessen the protection of an individual employee. If an employer intends to test an applicant, the employer shall provide the applicant with a copy of the written policy under subsection 2 before administering a substance abuse test to the applicant. The 30-day and 60-day notice periods provided for employees under this subsection do not apply to applicants.

4. Consent forms prohibited. No employer may require, request or suggest that any employee or applicant sign or agree to any form or agreement that attempts to

A. Absolve the employer from any potential liability arising out of the imposition of the substance abuse test; or

B. Waive an employee's or applicant's rights or eliminate or diminish an employer's obligations under this subchapter.

Any form or agreement prohibited by this subsection is void

5. Right to obtain other samples. At the request of the employee or applicant at the time the test sample is taken, the employer shall, at that time

A. Segregate a portion of the sample for that person's own testing. Within 5 days after notice of the test result is given to the employee or applicant, the employee or applicant shall notify the employer of the testing laboratory selected by the employee or applicant. This laboratory must comply with the requirements of this section related to testing laboratories. When the employer receives notice of the employee or applicant's selection, the employer shall promptly send the segregated portion of the sample to the named testing laboratory, subject to the same chain of custody requirements applicable to testing of the employer's portion of the sample. The employee or applicant shall pay the costs of these tests. Payment for these tests may not be required earlier than when notice of the choice of laboratory is given to the employer; and

B. In the case of an employee, have a blood sample taken from the employee by a licensed physician, registered physician's assistant, registered nurse or a person certified by the Department of Health and Human Services to draw blood samples. The employer shall have this sample tested for the presence of alcohol or marijuana metabolites, if those substances are to be tested for under the employer's written policy. If the employee requests that a blood sample be taken as provided in this paragraph, the employer may not test any other sample from the employee for the presence of these substances. (1) The Department of Health and Human Services may identify, by rules adopted under section 687, other substances of abuse for which an employee may request a blood sample be tested instead of a urine sample if the department determines that a sufficient correlation exists between the presence of the substance in an individual's blood and its effect upon the individual's performance. (2) No employer may require, request or suggest that any employee or applicant provide a blood sample for substance abuse testing purposes nor may any employer conduct a substance abuse test upon a blood sample except as provided in this paragraph. 3) Applicants do not have the right to require the employer to test a blood sample as provided in this paragraph

5-A. Point of collection screening test. Except as provided in this subsection, all provisions of this subchapter regulating screening tests apply to noninstrumented point of collection test devices described in section 682, subsection 7, paragraph A, subparagraph (1).

A. A noninstrumented point of collection test described in section 682, subsection 7, paragraph A, subparagraph (1) may be performed at the point of collection rather than in a laboratory. Subsections 6 and 7 and subsection 8, paragraphs A to C do not apply to such screening tests. Subsection 5 applies only to a sample that results in a positive test result.

B. Any sample that results in a negative test result must be destroyed. Any sample that results in a postive test result must be sent to a qualified testing laboratory consistent with subsections 6 to 8 for confirmation testing.

6. Qualified testing laboratories required. No employer may perform any substance abuse test administered to any of that employer's employees. An employer may perform screening tests administered to applicants if the employer's testing facilities comply with the requirements for testing laboratories under this subsection. Except as provided in subsection 5-A, any substance abuse test administered under this subchapter must be performed in a qualified testing laboratory that complies with this subsection.

7. Testing procedure. A testing laboratory shall perform a screening test on each sample submitted by the employer for only those substances of abuse that the employer requests to be identified. If a screening test result is negative, no further test may be conducted on that sample. If a screening test result is positive, a confirmation test shall be performed on that sample. A testing laboratory shall retain all confirmed positive samples for one year in a manner that will inhibit deterioration of the samples and allow subsequent retesting. All other samples shall be disposed of immediately after testing.

8. Laboratory report of test results. This subsection governs the reporting of test results. A. A laboratory report of test results shall, at a minimum, state:

(1) The name of the laboratory that performed the test or tests;

(2) Any confirmed positive results on any tested sample.

(a) Unless the employee or applicant consents, test results shall not be reported in numerical or quantitative form but shall state only that the test result was positive or negative. This division does not apply if the test or the test results become the subject of any grievance procedure, administrative proceeding or civil action.

(b) A testing laboratory and the employer must ensure that an employee's unconfirmed positive screening test result cannot be determined by the employer in any manner, including, but not limited to, the method of billing the employer for the tests performed by the laboratory and the time within which results are provided to the employer. This division does not apply to test results for applicants;

(3) The sensitivity or cutoff level of the confirmation test; and

(4) Any available information concerning the margin of accuracy and precision of the test methods employed.

The report shall not disclose the presence or absence of evidence of any physical or mental condition or of any substance other than the specific substances of abuse that the employer requested to be identified. A testing laboratory shall retain records of confirmed positive results in a numerical or quantitative form for at least 2 years. B. The employer shall promptly notify the employee or applicant tested of the test result. Upon request of an employee or applicant, the employer shall promptly provide a legible copy of the laboratory report to the employee or applicant. Within 3 working days after notice of a confirmed positive test result, the employee or applicant may submit information to the employer explaining or contesting the results. C. The testing laboratory shall send test reports for samples segregated at an employee's or applicant's request under subsection 5, paragraph A, to both the employer and the employee or applicant tested. D. Every employer whose policy is approved by the Department of Labor under section 686 shall annually send to the department a compilation of the results of all substance abuse tests administered by that employer in the previous calendar year. This report shall provide separate categories for employees and applicants and shall be presented in statistical form so that no person who was tested by that employer can be identified from the report. The report shall include a separate category for any tests conducted on a random or arbitrary basis under section 684, subsection 3.

9. Costs. The employer shall pay the costs of all substance abuse tests which the employer requires, requests or suggests that an employee or applicant submit. Except as provided in paragraph A, the employee or applicant shall pay the costs of any additional substance abuse tests. Costs of a substance abuse test administered at the request of an employee under subsection 5, paragraph B, shall be paid: A. By the employer if the test results are negative for all substances of abuse tested for in the sample; and B. By the employee if the test results in a confirmed positive result for any of the substances of abuse tested for in the sample.

10. Limitation on use of tests. An employer may administer substance abuse tests to employees or applicants only for the purpose of discovering the use of any substance of abuse likely to cause impairment of the user or the use of any scheduled drug. No employer may have substance abuse tests administered to an employee or applicant for the purpose of discovering any other information.

11. Rules. The Department of Health and Human Services shall adopt any rules under section 687 regulating substance abuse testing procedures that it finds necessary or desirable to ensure accurate and reliable substance abuse testing and to protect the privacy rights of employees and applicants.

§684 Imposition of tests

1. Testing of applicants. An employer may require, request or suggest that an applicant submit to a substance abuse test only if: A. The applicant has been offered employment with the employer; or B. The applicant has been offered a position on a roster of eligibility from which applicants will be selected for employment. The number of persons on this roster of eligibility may not exceed the number of applicants hired by that employer in the preceding 6 months. The offer of employment or offer of a position on a roster of eligibility may be conditioned on the applicant receiving a negative test result.

2. Probable cause testing of employees. An employer may require, request or suggest that an employee submit to a substance abuse test if the employer has probable cause to test the employee. A. The employee's immediate supervisor, other supervisory personnel, a licensed physician or nurse, or the employer's security personnel shall make the determination of probable cause. B. The supervisor or other person must state, in writing, the facts upon which this determination is based and provide a copy of the statement to the employee

3. Random or arbitrary testing of employees. In addition to testing employees on a probable cause basis under subsection 2, an employer may require, request or suggest that an employee submit to a substance abuse test on a random or arbitrary basis if:

A. The employer and the employee have bargained for provisions in a collective bargaining agreement, either before or after the effective date of this subchapter, that provide for random or arbitrary testing of employees. A random or arbitrary testing program that would result from implementation of an employer's last best offer is not considered a provision bargained for in a collective bargaining agreement for purposes of this section;

B. The employee works in a position the nature of which would create an unreasonable threat to the health or safety of the public or the employee's coworkers if the employee were under the influence of a substance of abuse. It is the intent of the Legislature that the requirements of this paragraph be narrowly construed; or

C. The employer has established a random or arbitrary testing program under this paragraph that applies to all employees, except as provided in subparagraph (4), regardless of position.

(1) An employer may establish a testing program under this paragraph only if the employer has 50 or more employees who are not covered by a collective bargaining agreement.

(2) The written policy required by section 683, subsection 2 with respect to a testing program under this paragraph must be developed by a committee of at least 10 of the employer's employees. The employer shall appoint members to the committee from a cross-section of employees who are eligible to be tested. The committee must include a medical professional who is trained in procedures for testing for substances of abuse. If no such person is employed by the employer, the employer shall obtain the services of such a person to serve as a member of the committee created under this subparagraph.

(3) The written policy developed under subparagraph (2) must also require that selection of employees for testing be performed by a person or entity not subject to the employer's influence, such as a medical review officer. Selection must be made from a list, provided by the employer, of all employees subject to testing under this paragraph. The list may not contain information that would identify the employee to the person or entity making the selection person to serve as a member of the committee created under this subparagraph.

(4) Employees who are covered by a collective bargaining agreement are not included in testing programs pursuant to this paragraph unless they agree to be included pursuant to a collective bargaining agreement as described under paragraph A.

(5) Before initiating a testing program under this paragraph, the employer must obtain from the Department of Labor approval of the policy developed by the employee committee, as required in section 686. If the employer does not approve of the written policy developed by the employee committee, the employer may decide not to submit the policy to the department and not to establish the testing program. The employer may not change the written policy without approval of the employee committee.

(6) The employer may not discharge, suspend, demote, discipline or otherwise discriminate with regard to compensation or working conditions against an employee for participating or refusing to participate in an employee committee created pursuant to this paragraph.

4. Testing while undergoing rehabilitation or treatment. While the employee is participating in a substance abuse rehabilitation program either as a result of voluntary contact with or mandatory referral to the employer's employee assistance program or after a confirmed positive result as provided in section 685, subsection 2, paragraphs B and C, substance abuse testing may be conducted by the rehabilitation or treatment provider as required, requested or suggested by that provider.

A. Substance abuse testing conducted as part of such a rehabilitation or treatment program is not subject to the provisions of this subchapter regulating substance abuse testing.

B. An employer may not require, request or suggest that any substance abuse test be administered to any employee while the employee is undergoing such rehabilitation or treatment, except as provided in subsections 2 and 3.

C. The results of any substance abuse test administered to an employee as part of such a rehabilitation or treatment program may not be released to the employer.

5. Testing upon return to work. If an employee who has received a confirmed positive result returns to work with the same employer, whether or not the employee has participated in a rehabilitation program under section 685, subsection 2, the employer may require, request or suggest that the employee submit to a subsequent substance abuse test anytime between 90 days and one year after the date of the employee's prior test. A test may be administered under this subsection in addition to any tests conducted under subsections 2 and 3. An employer may require, request or suggest that an employee submit to a substance abuse test during the first 90 days after the date of the employee's prior test only as provided in subsections 2 and 3.

CHAPTER 15

SAMPLE CODE OF ETHICS – CITY OF RENTON, WASHINGTON

Chapter 6
Code of Ethics

SECTION:

1-6-1 DECLARATION OF PURPOSE:

It is hereby recognized and established that high moral and ethical standards among City officials are vital and essential to provide unbiased, open and honest conduct within all phases and levels of government; that a code of ethics is a helpful aid in guiding City officials and to eliminate actual conflicts of interest in public office and to improve and elevate standards of public service so as to promote and strengthen the confidence, faith and trust of the people of the City of Renton in their local government.

1-6-2 DEFINITIONS:

For the purpose of this Ordinance:

CANDIDATE: Any individual who declares himself to be a candidate for an elective office and who, if elected thereto, would meet the definition of a public official hereinabove set forth.

COMPENSATION: Anything of economic value, however designated, which is paid, loaned, advanced, granted, transferred, or to be paid, loaned, advanced, granted or transferred for or in consideration of personal services to any person.

CONTRACT: Includes any contract or agreement, sale, lease, purchase, or any combination of the foregoing.

CONTRACTING PARTY: Any person, partnership, association, cooperative, corporation, whether for profit or otherwise or other business entity which is a party to a contract with a municipality.

PUBLIC OFFICIAL: All of the elected City officials, together with all appointed officers including their deputies and assistants of such an officer who determine or are authorized to determine policy making decisions within their respective department or office, including appointive members of all municipal boards, commissions and agencies and whose appointment has been made by the Mayor and confirmed by the City Council. (Ord. 2586, 9-28-70)

1-6-3 STATEMENT OF EXPENSE OF CANDIDATE:

A. Primary Election: Every candidate for nomination at a primary election within the City shall, no later than the tenth day of the first month after the holding of such primary election at which he/she is a candidate, file an itemized statement in writing, duly sworn to as to its correctness, with the City Clerk setting forth each sum of money and item of value, or any consideration whatever, contributed, paid, advanced, promised or rendered to him/her or furnished or given to others for the benefit of such candidate and with his/her knowledge or acquiescence for the purpose of securing or influencing, or in any way affecting his/her nomination to said office. Such statement shall set forth in detail the sums or other considerations so paid to him/her, together with the name and address of the donor, and such statement shall also set forth the nature, kind and character of the expense for which such sums were expended separately, including the name and address of the payee and the purpose for each disbursement. Such statement shall likewise include any sum or other consideration as hereinabove stated, promised but not yet paid or received. In the event any such payments, services or other item of value are made to other persons on behalf of or for the benefit of such candidate, then any such information, when ascertainable, shall be furnished to such candidate and be included in any such statement or report. Cash contributions, services or anything else of value amounting to twenty five dollars ($25.00) or less are exempt from the provision of this Section in requiring the reporting of such individual contributions or services; provided, however, that every such candidate shall report the full and total amount of all contributions, services or anything else of value contributed or paid to him/her or on his/her behalf, whether such individual amount is twenty five dollars, ($25.00) or less or in excess thereof.

The filing of a duplicate copy of the completed public disclosure form with the City Clerk, as required by the State of Washington, will satisfy the requirements of this subsection.

B. General Election: The provisions of the immediately preceding paragraph governing primary elections shall likewise apply to any general election within the City and every such candidate, whether he/she is successful or not, shall file such statement or report as hereinabove set forth, and such filing to be made no later than the tenth day of the first month after the holding of such general election at which he/she is a candidate. Each such candidate is further required to file supplementary statements containing in equal detail any additional contributions of whatever type or nature in excess of twenty five dollars ($25.00), including but not limiting it to, proceeds from any fund raising dinners, meetings, parties or like arrangements, whether conducted prior or subsequent to any such election, and such supplemental statements to be filed not later than ninety (90) days after such general election.

Each such statement as hereinabove defined when so filed shall immediately be subject to the inspection and examination of any elector and shall be and become a part of the public records of the City of Renton. Any violation of this subsection shall be and constitute a misdemeanor and shall be punishable as hereinafter set forth.

The filing of a duplicate copy of the completed public disclosure form with the City Clerk, as required by the State of Washington, will satisfy the requirements of this subsection. (Ord. 4315, 6-10-91)

1-6-4 ACCEPTANCE OF GIFTS:

No public official shall receive, accept, take, seek, or solicit, directly or indirectly, anything of economic value as a gift, gratuity, or favor, from any person if such public official has reason to believe the donor would not grant or give such gift, gratuity, or favor, but for such public official's office or position within the City of Renton.

No public official shall receive, accept, take, seek, or solicit, whether directly or indirectly, anything of economic value as a gift, gratuity, or favor, from any person or from any officer or director of such person if such public official has reason to believe such person:

A. Has, or is seeking to obtain contractual or other business or financial relationship with the City of Renton; or

B. Conducts operations or activities which are regulated by the City Council, its committees or any board or commission of the City of Renton; or

C. Has interests which may be substantially affected by such public official's performance or nonperformance of his or her official duty.

1-6-5 INTEREST IN CONTRACTS PROHIBITED; EXCEPTIONS:

No public official shall be beneficially interested, directly or indirectly, in any contract which may be made by, through, or under the supervision or direction of such public official, in whole or in substantial part, or which may be made for the benefit of his office, or accept, directly or indirectly, any compensation, gratuity or reward in connection with such contract from any other person beneficially interested therein. The foregoing shall not apply to the exceptions specified in RCW 42.23.030 which are incorporated herein as if fully set forth. Remote Interest: A public official shall not be deemed to be interested in a contract as specified in the immediately preceding paragraph if he has only a remote interest in the contract and if the fact and extent of such interest is disclosed to the governing body of the City of Renton of which he is a member and noted in the official minutes or similar records of the City prior to the consummation of the contract, and thereafter the governing body authorizes, approves or ratifies the contract in good faith by a vote of its membership sufficient for the purpose without counting the vote or votes of the public official having a remote interest therein.

As used in this Section "remote interest" means:

A. That of a nonsalaried officer of a nonprofit corporation;

B. That of an employee or agent of contracting party where the compensation of such employee or agent consists entirely of fixed wages or salary;

C. That of the landlord or tenant of a contracting party;

D. That of a holder of less than one percent (1%) of the shares of a corporation or cooperative which is a contracting party.

None of the provisions of this Section shall be applicable to any public official interested in a contract, though his interest be only remote as hereinabove defined, who influences or attempts to influence, any other public official of the City of which he is an officer to enter into such contract.

Any contract made in violation of the above provisions shall be void and as otherwise provided in RCW 42.23.050 and the provisions thereof being expressly incorporated herein as if fully set forth.

1-6-6 INCOMPATIBLE SERVICE; CONFIDENTIAL INFORMATION:

No elected public official shall engage in or accept private employment or render services for any person or engage in any business or professional activity when such employment, service or activity is incompatible with the proper and faithful discharge of his official duties as such elected official, or when it would require or induce him to disclose confidential information acquired by him by reason of his official position. No such official shall disclose confidential information gained by reason of his official position, nor shall he otherwise use such information for his personal gain or benefit.

1-6-7 PERSONAL OR PRIVATE INTERESTS:

Every elected public official who has a financial or other private or personal interest in any ordinance, resolution, contract, proceeding, or other action pending before the City Council or any of its committees, shall promptly disclose such interest at the first public meeting when such matter is being considered by the City Council, on the records of the official Council minutes, the nature and extent of such personal or private interest and same shall be incorporated in the official minutes of the City Council proceedings. Such disclosure shall include, but not be limited to, the following information which shall be submitted in writing by such Councilman, sworn to under penalty of perjury, to-wit:

A. The name and address of any private business corporation, firm or enterprise affected by such councilmanic action of which the Councilman or other elected public official is or has been during the preceding twelve (12) months a shareholder, bond holder, secured creditor, partner, joint entrepreneur or sole proprietor, whenever the total value of his individual or undivided legal and equitable financial interest therein is and at any time during the preceding twelve (12) months has been in excess of one thousand five hundred dollars ($1,500.00).

B. The name of any such private business or corporation, firm or enterprise of which such elected public official or his relatives are or have been during the preceding twelve (12) months as officer, director, partner, attorney, agent, or employee, who, for services rendered during such preceding twelve (12) months or to be rendered in any such capacity, has received or has been promised compensation in excess of one thousand five hundred dollars ($1,500.00).

C. Every office or directorship held by such elected public official or his spouse in any corporation, partnership, sole proprietorship or like business enterprise, which conducts its business activities within the boundaries of the Renton School District and which is subject to any regulation or control by the City of Renton, and from which such elected public official has received compensation or has been promised compensation during the preceding twelve (12) month period in excess of one thousand five hundred dollars ($1,500.00), or services; or any other thing of value in excess of said amount.

D. A list containing a correct legal description of any and all real property located within the City limits of Renton in which any such elected public official has any interest whatsoever, as owner, purchaser, optionee, optionor, or any other proprietary interest, acquired during the preceding twelve (12) month period whenever such proprietary interest is in excess of one thousand five hundred dollars ($1,500.00). This subsection shall not apply to the residence home of such official.

The foregoing provisions shall not apply to policies of life insurance issued to such public official or his spouse or members of his family, accounts in any commercial bank, savings and loan association or credit unions, or similar financial institutions subject to regulation by the State of Washington or any other governmental agency having jurisdiction thereover.

Any such elected public official who is disqualified by reason of such personal, private or similar conflict of interest in any matter as hereinabove defined, shall, after having made the required disclosure as herein set forth, remove himself from his customary seat during such debate and, by permission of the presiding officer, leave the Council chamber until such time as the matter at hand, from which such public official has been disqualified, has been disposed of in the regular course of business.

1-6-8 FALSE CHARGE OF MISCONDUCT:

Any person who shall file with or report to the Board of Ethics a charge of misconduct on the part of any public official or other person encompassed within the definition of this Ordinance, knowing such charge to be false or to have been recklessly made without any reasonable attempt to determine relevant facts and circumstances, shall be guilty of a misdemeanor and shall be punished as hereinafter set forth.

1-6-9 PENALTY:

Any person who wilfully, knowingly and intentionally violates any provisions of this Ordinance, shall be guilty of a misdemeanor and shall, upon conviction thereof, be fined in a sum not exceeding five hundred dollars ($500.00) or be committed to jail for a period not exceeding ninety (90) days, or be penalized by both such fine and imprisonment; in addition to the foregoing, any public official found guilty of any violation of this Ordinance shall forfeit any right to his office, whether elective or appointive, as may be determined by the court at the time sentence is imposed upon such public official. (Ord. 2586, 9-28-70)

GLOSSARY

80-20 rule A rule adopted by the DOL in FLSA cases to determine if an FLSA exemption should apply to an employee who engages in both exempt and non-exempt activities; if the employee fails to engage in an exempt activity for more than 80% of his or her work time, or engages in a non-exempt activity more than 20 percent of the time, the exemption would not be applicable.

Abandonment The intentional stopping of medical care without legal excuse or justification.

Absolute privilege A privilege that arises in the law of defamation that protects certain categories of people from liability for false statements made without regard to whether they have spoken in good faith. Those with absolute privilege include judges, attorneys, witnesses, and jurors in court; members of a lawmaking body (such as Congress) for statements made on the floor; the president and his cabinet members; and paid political broadcasts, provided equal time is provided for contrary views.

Accessory Someone who helped the principal to commit a crime, but was not actually present on the scene during the commission of the crime.

Act A statute or law passed by a legislature. In addition, an act can be any affirmative action taken, or the failure to act when one is legally required to act. An act must be voluntary in nature.

Actual authority The formal legal authority of a person to bind another to a contract.

Actual malice A verbal or written statement made by a person who either knew the statement was false, or acted with reckless disregard as to whether it was true or false.

Administrative agencies Agencies that exist within the executive branch of government to assist the executive in carrying out responsibilities imposed by law. Agencies are created by the legislative branch, through laws called enabling acts.

Affirmative action Positive steps taken to increase the presence of minorities and women in the workforce and in education. The term was first used in 1965 in Executive Order 11246, issued by President Lyndon B. Johnson.

Agency shop A form of union security whereby an employer agrees to mandate that all employees pay union dues—or an amount equal to the periodic union dues—as a condition of employment.

Aider and abettor Anyone who knowingly assists someone in the commission of a crime, or who helps the perpetrator evade capture or cover it up.

Answer A legal document in which a defendant in a civil case either admits or denies the allegations that the plaintiff made in his complaint.

Apparent authority The authority of a person (called an agent) to legally bind a principal based upon the conduct of the principal in holding the agent out as having the authority. A principal who knowingly allows an agent to hold himself out as having the authority to bind the principal can be bound by the acts of the agent.

Arbitrability A labor law term used to describe challenges to whether or not an issue can be decided by an arbitrator.

Arbitration A dispute resolution mechanism whereby the parties submit the disputed issues to a neutral third party, who will render a decision.

Arraignment The initial procedure in a criminal case held before a judge or magistrate, in which the accused is formally advised of the charges, and may have the opportunity to enter a plea to the court.

Arrest The lawful seizure of one person by another, thereby depriving the person who is seized of his or her liberty.

Arrest warrant An order issued by a judge or magistrate, authorizing any peace officer to take a suspect into custody. Law enforcement officers must establish that probable cause exists to believe a crime has been committed, and that the person named in the warrant committed it, in order to obtain an arrest warrant.

Arson (at common law) The willful and malicious burning of the dwelling of another.

Articles of incorporation The organizing documents for a corporation, that when duly filed with the secretary of state's office, creates the corporation. Once accepted by the state, the articles of incorporation become the equivalent of a charter or constitution for the corporation.

Asportation The requirement that the victim of a kidnapping be transported from the site of the kidnapping to another location. At common law, the asportation had to be across some geographical boundary, such as out of a country, or across state or county lines.

Association A group of individuals who to act in furtherance of a specific purpose without incorporating. Associations may also be referred to as unincorporated associations.

Balance of power A series of checks and balances built into our Constitution that are designed to keep any one branch of government from becoming too powerful.

Beyond a reasonable doubt The burden of proof in a criminal case.

Bill A proposed law submitted before a legislative body, such as Congress or a state legislature.

Briefing a case The process of reading a case and identifying the facts, issues, holdings, and rationales that the court identifies.

Burden of proof The obligation of a party in a lawsuit to affirmatively prove the facts of the case to a required degree, in order to prevail.

Burglary A common-law crime of breaking and entering the dwelling of another in the nighttime with intent to commit a felony therein. Modern statutory burglary charges have expanded the definition to include breaking and entering, burglary during the daytime, and burglary of commercial buildings, to name a few.

By fair preponderance of the evidence The burden of proof in a civil case, also referred to as the "more likely than not" standard.

Case books Books that contain the actual text of the written decisions by judges in deciding questions of law.

Case law The written decisions of judges issued in the process of deciding actual cases.

Chain of custody A law enforcement agency's documented proof of possession of physical evidence, from the time the evidence was seized until its introduction at trial.

Charter The organizing document for a political subdivision of a state, serving the equivalent function of a constitution.

Circumstantial evidence Evidence that requires someone to infer, deduce, or presume something from the direct evidence that is presented.

Citizen's arrest An arrest made by someone other than a duly authorized peace officer. Citizens may make a lawful arrest when the person being arrested has committed a misdemeanor in their presence, or when probable cause exists to believe that the person being arrested has committed a felony.

Civil law Law that seeks to enforce private rights and remedies for some wrong done to an individual or organization. Depending upon the context of how the term is used, civil law can also be viewed as all law that is not criminal law.

Class action lawsuit A suit brought by certain named individuals on behalf of all persons similarly situated.

Clear and convincing evidence The burden of proof used in certain types of civil cases that requires the finder of fact to conclude that the party with the burden of proof has proven its case substantially more likely than not. It is an intermediate burden of proof that is greater than the normal civil burden but less than the criminal burden.

Closed shop A form of union security whereby an employer agrees to hire only workers who are already members of a particular union, and requiring that workers maintain their union membership throughout their employment.

Code of Federal Regulations The codified compilation of regulations created by Federal administrative agencies, abbreviated CFR.

Codes A compilation of standards, regulations, or statutes.

Codified Said of laws that have been passed by the legislative branch into comprehensive codes.

Collective bargaining The process whereby an employer and duly appointed representatives of the employees negotiate an agreement pertaining to wages, hours, and other terms and conditions of employment.

Common law Judge-made law that developed in England, and was in existence prior to our Declaration of Independence on July 4, 1776.

Comparative negligence Procedure whereby a jury is responsible for apportioning fault among the various parties to a lawsuit. Comparative liability is assigned on a 100 percent scale, with each party receiving a percentage of fault as determined by the jury.

Complaint A formal allegation or series of allegations by a plaintiff that the defendant did certain things that somehow injured or damaged the plaintiff, and constitute a cause of action upon which the court should take action. The complaint is what formally initiates a lawsuit.

Compulsory binding arbitration A type of arbitration in which unresolved issues are submitted to an arbitrator, or arbitration panel, whose final ruling is binding on all parties.

Conflict of interest A generalized term referring to a situation in which someone has a potential or actual conflict between the responsibilities imposed by law—by virtue of the person's office or position—and some other duty or interest he or she may have, whether it be financial, business, family, or otherwise.

Consensual search A warrantless search that occurs when the defendant, or someone who has the right to control access to a particular area, consents to a search being conducted.

Consensus standards Standards that are developed through a formal process, such as required by American National Standards Institute (ANSI), and represent generally accepted industry-wide practices and recommendations.

Consent The willingness of a person to allow certain physical contact or other conduct that might otherwise be objectionable. Consent is a defense to a charge of battery, as well as assault and false imprisonment.

Consent decree   A court order, the terms of which have been agreed to by the parties to a lawsuit, and which is overseen and enforced by the court. Once entered, consent decrees are considered to be binding decisions of the court.

Consideration   A legal term for a recognizable detriment that a party to an agreement must submit to in order for the agreement to be enforceable. The recognizable detriment can consist of a promise to do something that a party is not required to do, or actually doing something that is not required. All parties to an agreement must provide consideration or the agreement will not be enforceable.

Conspiracy   An agreement to commit a crime or break the law.

Constitution   The organizing document for the Federal government, or for a state government, that serves as the supreme law and establishes the organization of government.

Contract   An agreement between two or more people that contains mutual promises that are enforceable at law.

Contributory negligence   A defense to a negligence action in which, if the plaintiff was shown to be in any way contributorily negligent in causing his or her own injuries, the defendant could not be held liable. The contributory negligence rule was an absolute defense to a suit for negligence. It has been abolished in favor of comparative negligence.

Corporations   Legally created entities that are accepted, approved, and recognized by a state through a formalized process.

Crimes   Activities that violate a criminal law and for which society may penalize an offender.

Criminal complaint   A formal, court-approved document commonly prepared by the police that includes a statement describing what the defendant allegedly did, and the specific laws the defendant is charged with violating. It is the most common method of charging a defendant with a misdemeanor or minor offense.

Criminal law   Law designed to prevent harm to society in general, and which declares certain actions to be criminal.

Custodial interrogation   An interrogation conducted by law enforcement officers of suspects who have been taken into custody or who otherwise have been deprived of their freedom in any significant way.

Defamation   An intentional tort that involves damage to a person's reputation through the publication of false, harmful, and unprivileged statements made to others. There are two types of defamation: slander and libel.

Default judgment   A judgment in a lawsuit against a party who fails answer or defend against allegations made by the other party, resulting in the other party winning by default.

Defendant   The party who is required to answer a complaint or case in court.

Demonstrative evidence   Another name for real evidence.

Deposition   A procedure by which a party to a lawsuit may compel another party or a witness to appear to answer questions under oath as part of a process known as discovery. Most depositions are taken before a court reporter (stenographer), but in some cases videotaped depositions are also allowed.

Direct evidence   Evidence that proves a fact, without the necessity for an inference or presumption.

Discovery   The phase in a lawsuit in which each side has the opportunity to learn about the evidence and witnesses that the other side has.

Discretionary act An act for which there is no fixed requirement for a course of action, such as would eliminate the exercise of discretion. In relation to tort claims acts, discretionary acts are often considered to be decisions of a policymaking nature made by elected and appointed officials that go to the heart of our democratic form of government.

Discrimination In the employment context, the term refers to an act which treats another person differently on account of a prohibited classification. Discrimination includes harassment of, or retaliation against, an individual who has made a complaint of discrimination, cooperated with an investigation, or opposed discriminatory practices, as well as employment decisions that are based on perceived stereotypes or assumptions about the abilities, traits, or performance of individuals of a certain sex, race, age, religion, or ethnic group, or individuals with disabilities.

Disparate impact A form of discrimination that appears on its face to be non-discriminatory, but that has the effect of discriminating based upon a prohibited classification. Disparate impact can be proven only through statistical analysis.

Disparate treatment A form of discrimination in which a particular victim (or group of victims) is treated differently because a prohibited classification. Disparate treatment discrimination is based upon an intentional act of discrimination.

Domicile The location where a person has a residence with intent to reside permanently. A person may have several residences, but can have only one domicile.

Due diligence The requirement that, before entering into a contract with a person (agent) who purports to be acting on behalf of another (principal), the agent has the actual authority to bind the principal to the contract.

Dues check-off A form of union security agreement whereby the employer agrees to deduct a designated amount from an employee's paycheck, and pay the amount withheld to the union.

Element of a crime A fact that must be proven in order for a person to be convicted of a crime. Crimes are made up of elements. There are three categories of elements: acts, mental states, and attendant circumstances. All crimes have act requirements and mental state requirements. Certain crimes have attendant circumstance requirements.

Embezzlement Larceny that occurs when someone who lawfully has possession of another's property by virtue of a trust relationship fraudulently appropriates that property for his or her own use.

Enabling acts Statutes enacted by Congress or state legislatures that authorize the creation of administrative agencies, and provide the legal authority for the agencies to operate.

Equal employment opportunity The right of a person to compete for a job and/or to be promoted on the basis of his or her knowledge, skills, and abilities, free from unlawful discrimination.

Ethics codes Laws that establish comprehensive guidelines for the conduct of public officials and employees, often addressing conflicts of interests as well.

Evidence Any type of proof that can legally be presented at a trial or hearing.

Exclusionary rule A rule adopted by the United States Supreme Court, which prohibits the use of evidence in a criminal case that was obtained in violation of a defendant's constitutional rights.

Exclusivity A principle associated with workers' compensation that prohibits an employee from suing his or her employer for injuries sustained in the scope of employment; the employee's exclusive remedy is to accept workers' compensation benefits.

Executive order   A law issued by an executive, such as the president, governor, or mayor.

Exigent circumstances   An exception to the search warrant rule, holding that police officers may conduct a warrantless search when requiring them to obtain a warrant before acting would have real, immediate, and serious consequences.

Express contract   A contract formed through the actual agreement of the parties as opposed to one that is implied through their conduct.

Extortion   Obtaining of money or property, or otherwise requiring someone to do something that he or she is not legally required to do, by means of a threat.

Fact-finding   A labor impasse tool consisting of an independent party or panel appointed to investigate and report the facts related to an impasse.

Fair share   A form of union security whereby an employer agrees that employees who are not members of the union and thus do not pay union dues, will be required to pay a proportionate share of the cost of collective bargaining activities.

False arrest   A type of false imprisonment, in which a law enforcement officer or private citizen arrests an individual on criminal charges, but is mistaken about either the actual identity of the perpetrator or about whether a crime has actually been committed.

Felony murder rule   A rule adopted in many states that any killing, whether accidental or intentional, that occurs during the commission of a felony shall be considered to be first-degree murder.

Final offer arbitration (FOA)   A form of labor arbitration adopted in some states that requires each party to submit a final offer to an arbitrator or arbitration panel, who must choose one proposal or the other. A number of variations on final offer arbitration exist, such as allowing the arbitrator to choose the more reasonable solution offered by the parties on each disputed issue, or requiring the arbitrator to select one final package or the other.

Fire district   A political subdivision of the state that has the authority to impose taxes and organize fire and emergency services.

Firefighters exemption   An exception to the overtime requirements of the Federal Fair Labor Standards Act (FLSA), which applies to any personnel engaged in fire protection activities for a public employer. Public sector firefighters are entitled to overtime only if their average weekly hours exceed 212 hours in a 28-day period, or a corresponding lesser number of hours for a shorter work period. The firefighters exemption is often called the "7(k) exemption" because it is under Section 207(k) of the FLSA.

Fire protection district   A political subdivision that may be established in some states to ensure fire protection in rural or suburban areas that do not have a fire department. A fire protection district has the authority to collect and utilize taxes to contract with a nearby fire department to provide fire protection for the district. However, the fire protection district does not directly provide the service itself. The similarity between the names fire district and fire protection district has led to some confusion, and the distinction between two names is not universally recognized.

Fire scene exception   An exception to the search warrant requirement, holding that the warrantless entry by firefighters at a fire scene is lawful.

First-degree murder Murder that is committed with malice aforethought.

For-profit corporation A corporation that is owned by stockholders, whose goal is to make money for the stockholders.

Framing the issue Characterizing an issue in a case in such a way that a commonsense answer will favor a particular perspective.

Functionary act An act of carrying out established policy, as opposed to exercising discretion.

General jurisdiction Jurisdiction of a court that has the legal authority to hear and decide all cases and controversies unless specifically exempted.

General law A term commonly used for state statutes.

Governmental function An activity undertaken by a governmental agency or actor that is of a type that is generally provided by the government, such as police, fire, highways, and public health. Some jurisdictions that provide governmental immunity protection for governmental actors provide it only when the actor is involved in a governmental function.

Grand jury A type of jury used in certain types of criminal cases to determine if there is probable cause to believe that a crime has been committed and that the defendant committed it, and if so, to indict the defendant.

Grievance An allegation by one party to a collective bargaining agreement that the other side has violated the agreement.

Grievance impasse dispute A labor dispute in which the parties cannot reach an agreement over the interpretation and administration of an existing collective bargaining agreement.

Gross negligence An aggravated form of negligence that involves an extreme departure from the ordinary standard of care.

Guilty A plea entered in court serving as an admission to the commission of the crime.

Home rule charter A charter that allows the citizens of a municipality to adopt or amend their own charter, without having to go back to the state legislature.

Hostile work environment sexual harassment One of the two types of sexual harassment. Hostile work environment sexual harassment occurs when unwelcome sexual conduct unreasonably interferes with an individual's work performance or has the effect of creating an intimidating or offensive work environment. The harassment need not result in tangible or economic consequences, such as termination, loss of pay, or failure to be promoted, nor in physical or psychological injury.

Impact bargaining Mandatory bargaining over the impact of a decision that involves a management prerogative, and would otherwise be beyond the scope of bargaining.

Impasse A labor term referring to the situation in which the parties cannot reach an agreement despite bargaining in good faith.

Implied consent Consent to medical treatment that is implied by law for a patient in need of medical treatment but who lacks the capacity to consent to or decline treatment. Implied consent is limited to treatment necessary to protect life and limb.

Indictment The formal allegation of criminal wrongdoing issued by a grand jury.

Information charging A procedure in which the prosecutor can issue formal charges against a defendant. A probable cause hearing is then scheduled before a magistrate or judge, who determines the sufficiency of the evidence based upon the evidence that prosecutors have presented in the "information."

Informed consent   Consent that is informed. For consent to be valid, the person must understand what they are consenting to.

Injunction   A court order directing a party to a lawsuit to do or refrain from doing some act.

Interest bargaining   A labor term referring to the process of negotiating a collective bargaining agreement.

Interest impasse dispute   A labor dispute in which the parties cannot reach agreement over the negotiation of a collective bargaining agreement.

Intermediate level of scrutiny   A standard of review applied by courts when reviewing the constitutionality of a governmental policy, action, or law, such that the policy, action, or law must be substantially related to important governmental objectives in order to be upheld.

Interrogatories   Written questions that each party to a lawsuit is entitled to ask any other party in the suit to answer in writing under oath.

Intervening act   The act of a third person that breaks the chain of causation and eliminates liability between an original wrongdoer and an inured party.

Involuntary manslaughter   An unintentional killing that results from the reckless conduct of the defendant.

Joint enterprise   Individuals acting under an express or implied agreement with a common purpose.

Jurisdiction   The legal authority of a person, official, court, or agency to take an official action. The term may have different meaning depending upon the context of its usage. The jurisdiction of a court refers to the legal authority of a court to hear a matter. The jurisdiction of an administrative agency refers to its legal authority to take action on a certain matter.

Larceny   The taking of the property of another with intent to permanently deprive the owner.

Law   That which must be obeyed subject to sanction or consequences by the government.

Libel   Defamation through written or printed falsehoods about a person.

Maintenance of membership   A form of union security whereby an employer agrees that once employees become members of the union, they must maintain union membership as a condition of employment. However, maintenance of membership provisions do not require employees to join a union.

Malice aforethought   The mental state requirement for murder; refers to the fact that the crime was premeditated.

Malpractice   A common term for professional acts of negligence.

Mandatory subjects   Collective bargaining subjects over which both management and labor are required to bargain.

Manslaughter   Homicide crime that does not constitute murder, but may nevertheless be chargeable as a crime. There are two categories of manslaughter, voluntary and involuntary.

Mediation   A labor impasse tool whereby an independent third party helps to facilitate the parties in reaching an agreement among themselves.

Mens rea   The criminal mental state necessary to satisfy the mental state element for a crime.

Misdemeanor manslaughter rule   A rule in some states that allows prosecution of the perpetrator of a misdemeanor with involuntary manslaughter where someone dies during the commission of the crime.

Motion for summary judgment A request by a party to a lawsuit for the court to rule that as a matter of law, he or she should prevail. Motions for summary judgment may be granted when there is no genuine issue as to any material fact and the moving party is entitled to a judgment as a matter of law.

Municipal corporation A corporation that is created by the state legislature, usually at the request of the inhabitants, or at least with the approval of the inhabitants through a referendum vote. Municipal corporations are usually characterized by having inherent powers such as self-government, law making, and the ability to tax.

Murder The intentional killing of another without justification.

Name-clearing hearing A hearing required by the due process clause of the Fourth and Fourteenth Amendments whenever a public employee is accused of wrongdoing that is of a stigmatizing nature, and the accusations are made public. The interest at stake in a name-clearing hearing is a liberty interest, and the purpose of the hearing is limited to providing the employee with an opportunity to be heard relative to the stigmatizing accusations.

National Fire Protection Association (NFPA) A private, nongovernmental organization consisting of 75,000 members who are concerned about protecting people and property from fire and other hazards.

Negligence (criminal mental state) A criminal mental state in which the perpetrator should have been aware that a substantial and unjustifiable risk of harm would result from his or her conduct.

Negligence (civil tort) The failure to exercise the care that the reasonably prudent person would have exercised under the circumstances that cause damages to another.

Nolo contendere A plea entered in court, which neither admits nor denies the commission of the crime but agrees that the evidence is sufficient to establish guilt, and that the defendant does not wish to contest the matter further.

Nonprofit corporation A corporation formed for some charitable, benevolent, or other purpose not designed to make profits that will benefit its owners, directors, or officers. Also called a not-for-profit corporation.

Obtaining money under false pretenses A form of larceny that involves obtaining money, goods, wares, or other property through any false pretense or misrepresentation.

Open shop An employer that does not mandate that employees belong to a union, nor require the payment of fees or dues as a condition of employment. Opens shops are most common in right-to-work states.

Ordinance A law passed by local legislative bodies.

Past practice A labor term relating to a practice that has been followed by both labor and management consistently and over a long period of time, but which has not been incorporated in writing into the collective bargaining agreement.

Peace officer Another term for law enforcement officer in many states.

Permissive subjects Collective bargaining subjects that, while not within the scope of the mandatory subjects, are not of the type that are prohibited.

Petit jury A jury whose purpose is to serve as the fact finder at a trial.

Plaintiff The party who files a civil lawsuit.

Plain view doctrine An exception to the search warrant requirement that allows police to use evidence seized without a warrant when it is located in plain view of the officers.

Pleadings   Consists of the complaint and the answer, along with any counterclaims, third-party complaints, or motions to dismiss in a lawsuit.

Police powers   The authority of each state to govern matters related to the welfare and safety of residents. States are empowered with broad police powers to pass laws as they see fit to protect their citizens. Police powers jurisdiction includes matters pertaining to fire protection, emergency medical response, fire codes, and criminal offenses.

Precedent   An important legal principle that should be followed. Precedent is established by court decisions that have been decided in the past in similar cases in the same jurisdiction.

Preliminary injunction   Injunction issued (usually at the beginning of a lawsuit) to maintain the status quo while the case is heard.

Presumption of innocence   The principle by which a person is presumed to be innocent, until his or her guilt is established by proof beyond a reasonable doubt as to each and every element of the crime.

Principal   When more than one person commits a crime, the person who actually commits the physical act of the crime.

Private sector   Entities that are privately owned, operated, and managed.

Procedural arbitrability   A labor law term describing questions about whether the proper procedure was followed in processing and handling a grievance.

Professional standard of care   The standard of care that a person with professional skills and training is required to exercise, namely, the care that the reasonably prudent professional of like training and experience would have exercised under the circumstances.

Prohibited subjects   Collective bargaining subjects that the parties are prohibited from bargaining.

Promissory estoppel   A unilateral promise made by someone who knows that the recipient will be relying upon the promise to his or her detriment, which courts may enforce when equity requires.

Proprietary function   A function that is sometimes performed by private sector entities as well as governmental entities, including running hospitals, golf courses, swimming pools, parking garages, and utilities. In this way, a proprietary function differs from a governmental function.

Proximate cause   A legal term referring to the fact that the act in question was the legal cause of the harm that resulted.

Public accountability laws   A diverse group of laws whose fundamental purpose is to foster integrity in government and promote public confidence.

Public duty doctrine   Doctrine that when a governmental actor is engaged in a governmental function involving the exercise of discretion, the actor is not subject to liability for breaches of his or her public duty. Rather, liability only attaches if the actor breaches a duty to a particular plaintiff to whom a special duty is owed.

Public sector   Entities that are agencies and instrumentalities of the Federal, state, or local government.

Qualified privilege   A privilege that arises in the law of defamation when there are facts justifying the written or spoken statement, but that may not be available when the words are uttered with malice or without good faith. For example, a qualified privilege is granted to someone who has been defamed, in order to defend his legitimate interests.

Quasi-municipal corporation A corporation that is created by the state legislature, but not always at the request of the local inhabitants, to aid the state in its administrative functions, as opposed to serving the needs of the inhabitants for governance. Quasi-municipal corporations have some but not all the powers and authority of municipal corporations.

Quasi-public corporation A corporation that is created by the state legislature, but not always at the request of the local inhabitants, to aid the state in its administrative functions, as opposed to serving the needs of the inhabitants for governance. Quasi-public corporations have some but not all of the powers and authority of municipal corporations. They are also called quasi-municipal corporations.

Quasi-public entity An entity that has sufficient connection to the public sector to warrant it being treated as being part of the public sector for a specific purpose, such as for OSHA coverage.

Question of fact A factual determination that must be made in a case. In a jury trial, the jury decides questions of fact.

Question of law The interpretation of laws and the application of legal principles to the issues in the case. The judge decides questions of law.

Quid pro quo sexual harassment One of the two types of sexual harassment. Quid pro quo sexual harassment occurs when the employee's employment opportunities or benefits are granted or denied because of the employee's submission to sexual advances or requests for sexual favors.

Rational basis standard A standard of review applied by courts when reviewing the constitutionality of a governmental policy, action, or law, such that the policy, action, or law will be upheld provided it is rationally related to a legitimate governmental interest. The rational basis standard is a deferential standard that usually results in the reviewing court upholding the government's action, except where no rational basis for the governmental action exists.

Real evidence Tangible evidence that can be brought into court and examined. Real evidence may also referred to as demonstrative evidence.

Recklessness An aggravated form of gross negligence. Recklessness requires that the actor have knowledge that harm was likely to result from his behavior, and a conscious choice to act (or refusal to act when under a duty to act) despite the risk. It is the knowledge that harm is likely to result that separates recklessness from gross negligence. (Note: Recklessness can also be viewed as a criminal mental state.)

Recklessness (civil tort) An aggravated form of gross negligence involving willful, wanton conduct where the actor had knowledge that harm was likely to result from his behavior, and consciously chose to act despite the risk.

Referendum A process for the eligible voters in a community to vote on a given law.

Regulations Laws that are created by administrative agencies pursuant to authority delegated by the legislature.

Reporter A book that contains the actual text of the written decisions by judges in deciding questions of law. These books may also be called case books.

Representational impasse dispute A labor dispute in which the parties cannot reach agreement concerning whether employees desire union representation, who should be the bargaining representative of the employees, or which positions should rightfully be part of the bargaining unit.

Request for admission   A method of discovery that allows a party to a lawsuit to formally ask an opposing party to admit to certain statements of facts in order to narrow the issues in dispute in trial.

Request for production   A method of discovery that allows a party to obtain evidence from any other party, including documents, photographs, physical evidence, or other items relevant to the case, so that they may be examined, reviewed, and in some cases analyzed.

Respondeat superior   A legal doctrine that makes an employer liable for the torts of its employees, provided the torts are committed within the scope of the workers' employment.

Right-to-work states   States that have statutes that expressly prohibit union security agreements, as well as any agreement or requirement that an employee join or pay dues to a union as a condition of employment, whether in the public or private sector.

Robbery   Larceny through the use or threatened use of violence.

Search and seizure   The area of the law that addresses the rights and responsibilities of law enforcement authorities and governmental actors to conduct searches for evidence, and the seizure of any evidence that is found.

Search incident to arrest   A permissible warrantless search that may be conducted by a police officer upon arresting a suspect. The officer may search the suspect and the area immediately around the suspect without a search warrant.

Search warrant   A legal order issued by a judge or magistrate that authorizes law enforcement officials to search a certain location.

Second-degree murder   Any murder committed without malice aforethought. It can also be defined as any murder that is not first-degree murder.

Section number   The laws of Congress are commonly cited as 29 USC 654, the second number indicating the section number. 29 USC 654 is therefore Title 29, Section 654 of the United States Code. The symbol § is commonly used to denote a section, such as §654.

Separation of powers   A fundamental principle underlying the Constitution, designed to prevent any of the three branches of government from becoming too powerful. To accomplish this, the Constitution specifies each branch's separate powers and responsibilities.

Serve   To deliver a summons, complaint, subpoena, notice, or other official document to a person.

Service of process   Giving a person legal notice of a court or administrative action.

Slander   Defamation that is committed orally through a false spoken word or gesture. The general rule with slander is that the victim must prove an actual monetary loss in order for the case to be actionable.

Slander per se   A particular type of slander involving certain categories of falsehoods that are considered to be more serious than others. Under slander, actual damages must be proven, while under slander per se, damages will be presumed without the need to show a monetary loss.

Sovereign immunity   A common-law rule that the government is immune from liability for actions in tort. The immunity arose from the old English principle that "The king can do no wrong."

Specific jurisdiction   Jurisdiction of a court that is limited by legislation or enabling authority to hear only certain types of cases and controversies.

Standard of care A measure of what the community expects and demands from a person in a given situation, in order to avoid negligence. The ordinary standard of care is the degree of care that the reasonably prudent person would have exercised under the circumstances. See also the professional standard of care.

Standards Voluntary guidelines and recommendations that do not carry the force and effect of law.

Standing The legal right to bring a case or make a legal claim. For a person to have standing to object to a warrantless search, the person must have had a reasonable expectation of privacy in the place to be searched.

Stare decisis The principle that once a court establishes a legal principle or interpretation, other courts at the same or lower levels in the same jurisdiction must apply that decision in future cases on the same facts. Stare decisis is Latin for "let the decision stand."

Statute of frauds A state law that requires that certain types of contracts have at least some written memorandum of their existence. The statute of frauds varies somewhat from state to state, but as a general rule, a statute of frauds will apply to contracts that take over one year to complete; contracts for the sale of real estate; and contracts for the sale of goods over $500 in value.

Statutes Laws passed by Congress or state legislatures.

Strict liability The legal doctrine that makes an actor responsible for any and all harm that may occur without regard to fault. In a criminal sense, strict liability crimes are crimes for which there is no mental state requirement. In a civil sense, strict liability refers to liability for damages in tort without regard to negligence, gross negligence, recklessness, or intentional tort.

Strict scrutiny A standard of review applied by courts when reviewing the constitutionality of a governmental policy, action, or law, such that the policy, action, or law must be narrowly tailored to address a compelling governmental interest in order to be upheld.

Strike Any concerted activity by employees that serves to impact an employer in order to make demands or protest a managerial decision. This would include work stoppages, slowdowns, sick-outs, or other organized refusals to work.

Subpoena A court order commanding a person to appear at a certain time and place to give testimony or produce certain evidence.

Subrogation The right of a person to assume the legal claims of another party. The right of an insurer who pays out a claim to an insured, to file suit against who ever was responsible for causing the damage to the same extent that the insured could have sued.

Substantive arbitrability A labor law term used to describe challenges to whether or not the collective bargaining agreement allows a particular dispute to be submitted to an arbitrator.

Summons A court order that requires the defendant to answer the complaint within a designated time period, or lose the case on a default.

Temporary restraining order A form of injunction that is issued by a court on an emergency basis for a very brief period of time (7 to 14 days), until the court can hold a full hearing on the matter.

Title number The laws of Congress are commonly cited as 29 USC 654, with the first number identifying the title number. 29 USC 654 is therefore Title 29, Section 654 of the United States Code.

Tort A general term for a civil wrong. A tort is an act committed by one or more parties that causes injury to another, for which the law allows a remedy of monetary damages.

Tortfeasor The legal term for someone who commits a tort.

Union shop A form of union security whereby an employer agrees to mandate that employees become union members within a stipulated period of time after being hired.

Void contract A contract that is unenforceable even when both parties to the contract desire it to be valid. Contracts may be declared void because they are for an illegal purpose, violate a law, or are against public policy.

Voidable contract A contract that can be rescinded by one or both parties. If a party lacks the capacity to enter into a contract, enters into a contract out of duress, or if both parties are mistaken about a material factor, the agreement is voidable.

Voluntary arbitration An impasse resolution mechanism whereby the parties voluntarily agree to submit some or all of their unresolved issues to arbitration.

Voluntary manslaughter The intentional killing of another, committed in the heat of passion as a result of a severe provocation.

Volunteer fire companies Nonprofit corporations or associations whose purpose is to provide fire protection services to a given area. Volunteer fire companies exist independently of the municipality, county, or fire district in which they operate, and are part of the private sector.

Warrant A court order which commands law enforcement personnel to carry out a specific task, such as arrest a person (arrest warrant), search a certain area and seize certain evidence (search warrant), or execute a criminal defendant (death warrant).

INDEX OF CASES

INDEX